The Fairy Tale of Nuclear Fusion

L. J. Reinders

The Fairy Tale of Nuclear Fusion

 Springer

L. J. Reinders
Panningen, The Netherlands

ISBN 978-3-030-64346-1   ISBN 978-3-030-64344-7   (eBook)
https://doi.org/10.1007/978-3-030-64344-7

This Springer imprint is published by the registered company Springer Nature Switzerland AG
The registered company address is: Gewerbestrasse 11, 6330 Cham, Switzerland

When you are studying any matter or considering any philosophy, ask yourself only what are the facts and what is the truth that the facts bear out. Never let yourself be diverted either by what you think would have beneficent social effects if it were believed, but look only and solely at what are the facts.

Bertrand Russell

Preface

Science in this century has become a complex endeavour. As scientific knowledge expands, the goal of general public understanding of science becomes increasingly difficult to reach.[1]

Nuclear fusion is the process that powers the stars, including our own Sun. As soon as these stellar processes were understood (in the early 1920s), people started dreaming about harnessing their power both for the benefit and for the destruction of mankind on Earth. In the latter, we have succeeded. We now possess bombs that can destroy the Earth and all that is on it in a matter of hours or less. The other dream of an inexhaustible clean source of energy that will save mankind from the horrors of climate change and pollution has not yet become a reality. Will it remain a fantasy or is there a fair chance that the twenty-first century will see it come true? This book will tell the story, the history and content of the global efforts to realise this dream, an effort that has been going on now for close to seven decades without a solid trustworthy result being in sight. Should we despair or is it reasonable to continue to sink billions of dollars, euro, or yen into this effort?

The quest for fusion power has a long history, and its failure to live up to its early promise has apparently also diminished the interest in properly describing its science and technology for the general public. I intended my book to fill this void and describe developments in nuclear fusion from the early beginnings up to and including the latest efforts with huge international collaborations like the Joint European Torus (JET) and the International Thermonuclear Experimental Reactor (ITER) and individual small-scale enterprises with small tokamaks and other devices. The book that I had in mind at first was supposed to be a plea for fusion, an urge to put more money into this seemingly promising venture in order to speed up the process.

---

[1]From the preface to the Alfred P. Sloan Foundation Series in the delightful book by Victor Weisskopf (1991).

I was optimistic and started writing this book as a proponent of fusion. As many others who had read the same news items, I was taken in by the optimistic language and the multiple breakthroughs that were reported. Being mostly ignorant of the scientific and technical details and of its history, I saw nuclear fusion as one of the most promising ways of combating climate change. The hesitantly dawning prospect of an age of clean and affordable energy based on nuclear fusion as the world's primary energy source seemed a good time for writing a book, accessible to the general public, which describes its history and explains and summarises the progress made so far, without hiding any of the difficulties and problems.

In the course of writing, this all changed when I discovered from reading and studying some of the scientific literature that nuclear fusion was a fantasy pursued by single-minded individuals that were apparently unable to see reason and the fundamental failings of their efforts. The media only report the successes: record temperatures and confinement times which can be brought with a blazing headline like "China's 'artificial sun' sets world record with 100 second steady-state high performance Plasma", while the failures are ignored. When you only read the articles on fusion in the general press, it looks like a succession of breakthroughs without end, while the real situation kept hidden from the public mainly consists of failures. I felt fooled and decided to backtrack and write the book with a completely different message in mind: stop the way this is being done now, take some time to reflect on the facts and what they bear out. No longer continue to waste all this money and effort, which can be spent far better.

Nuclear fusion, and research into nuclear fusion, nowadays only focuses on the production of (hopefully) cheap, but in any case inexhaustible and unlimited energy without carbon dioxide emissions, to combat climate change and whatever benefits for mankind you can further think of. The only resource needed is just plain water! At least that is the story that continues to be repeated over and over again. Even purely scientific papers often start in the introduction with some waffle about nuclear fusion and the great prospects that are in store for mankind if the process could be controlled, etc. The scientists pursuing this research are all drawn into such blabbering talk, while in their hearts and even in their minds they are not one bit interested in such goals. They don't do science, at least I hope they don't, with the goal of selling cheap or any other energy to society, or to advertise their activity as something that has such a mundane result. The great power of science has always been that it could be pursued for science's sake without the promise of any 'use' in the short or long term. Had that been their goal in life, they would have been better advised to become oil merchants or power utility managers. Scientists do science to understand something about nature and that should be enough to obtain funding as long as their project proposals satisfy some clear standards. For instance, research in high energy physics, e.g. at CERN, has always been able to secure funding, even for hideously expensive accelerators, just by pointing out that it would be a great thing to find the Higgs or some other particle and get some further confirmation of the Standard Model. Nothing more was needed, and nothing more should be needed. So, also when studying nuclear fusion processes scientists hope to understand, for instance, how a plasma behaves, how the fusion processes in the stars work and whether we can copy them here on Earth, not for the sake of getting power on the grid, but in order to understand these processes.

This scientific attitude has been totally lost in today's fusion research. It has become a completely utilitarian enterprise, an engineering problem. In principle, there is nothing wrong with this, of course, but it happened even before the scientific stage had been completed. The understanding of the physics behind it, especially the physics of plasmas in the extreme conditions of a nuclear fusion reactor, is minimal. Somewhere along the line it has gone fundamentally wrong. In the USA, very sensibly, nuclear fusion research was traditionally viewed as scientific research, at least until the early 1970s when 'terrible' people (with an engineering degree) took over and wanted to build a nuclear fusion reactor, a power generating device without having even an inkling how to do this. Since then this goal has been pursued with an obstinacy verging on fanaticism, resulting in ever larger designs of reactors, culminating for the time being in the ITER monster under construction in the south of France. Since that time scientists and politicians have been fooled and are now getting company of venture capitalists, as we will see in this book. New arguments are invented along the way. Was nuclear fusion first touted as a means of delivering inexhaustible, cheap energy, now it is put forward as *the* solution to combat climate change. Something new has to be invented soon as it is crystal clear that energy production by nuclear fusion, if ever realised at all, will come far too late to contribute to this battle. It will in any case not happen in this century.

On the other hand, a lot of research has been carried out in the last 70 years or so, often beautiful research, innovative and inventive, that deserves a place among the scientific achievements of mankind. It is no mean feat to enclose an ionised gas of hundreds of millions of degrees in some complicated magnetic field configuration and keep it confined for any length of time, but to use it as a power generating device is clearly a bridge too far. The research however is worthwhile and should be continued, but with another goal.

This book is intended for a general public. At any rate, no special technical or physics training is required to understand it. A general education and a preparedness to use one's brains should be enough to be able to read it. Some words like tokamak, stellarator, divertor may be unfamiliar and sound strange at first, but will be explained and should not deter the reader. Their principles are not that difficult to understand. Jargon is often introduced in science, like in many other human endeavours, to scare off the non-initiated. In the book, physics formulas are avoided, apart from some of the obvious ones, like Einstein's mass–energy equivalence relation $E = mc^2$, and while some of the more technical issues are perhaps hard to understand, the reader should keep in mind that that is probably also true for specialists in the field, and certainly for the author. There is however no way around this. Science has become a complex endeavour, and public understanding of science is becoming increasingly difficult to achieve. I have included above a quotation from the Alfred P. Sloan Foundation that stressed this point already 30 years ago. But science, and certainly nuclear fusion, gobbles up untold amounts of public (and nowadays also private) money, and it is imperative for the public to have at least some idea how the money they have worked for is being spent. Scientists can spread all kinds of tales that nobody can or is willing to check, as they have done for close to seventy years in the case of nuclear fusion as being *the* energy source, while adding all kinds of epithets like inexhaustible, too cheap to meter and such like, that would solve most of our problems in a few decades. Nobody apparently cared to check these claims or what they are based on. And, as this book will make clear, in the case of nuclear fusion they continue to do this without having much to show for after so many years. It came as a real shock to me to realise that the emperor had no clothes and that hardly anybody has stood up, e.g. in the general press, to point this out. The general public are entitled to an insight into what has been and is being done in the field of nuclear fusion and into the prospect for the carrot held out to them being edible.

The literature on nuclear fusion is vast, going back to its very beginning more than seventy years ago, with quite a number of journals, like *Nuclear Fusion*, *Plasma Physics and Controlled Fusion*, *Journal of Fusion Energy*, *Fusion Engineering and Design*, etc., solely devoted to this kind of research. It has become even more extensive as nuclear fusion scientists, in my impression more than other physical scientists, have the tendency to publish virtually the same results multiple times. Much of the literature is very technical, indeed, and new developments make it even more so. Apart from general advertising material with catchy phrases like 'the Sun in a bottle', surprisingly little effort has been spent in trying to explain the concepts and the tortuous route towards the ultimate goal: providing energy for mankind. Few attempts have been made to make the material more accessible to a general public, and the ones that have been made have so far certainly failed to make nuclear fusion a familiar concept. If anything, the general public will hardly know what people are talking about when nuclear fusion is mentioned. The enterprise of power generation by nuclear fusion has undoubtedly failed to keep touch with the general public who are supposed to foot the bill and are eventually supposed to benefit from the efforts.

It will be clear from the above that this book is not optimistic about the prospects for power from fusion and, if ever realised, it will certainly not be "too cheap to meter" as was a much-heard claim in the early days of the fusion effort. Electricity production from nuclear fusion will most likely be so expensive and so complex that it will never become economically viable. It will not be a commercially competitive source of energy and certainly not a clean one either. In addition, the cost of a single nuclear fusion power plant will just be enormous, so huge that few countries will be able to afford even a single such plant, if such a hideously complex beast will ever be made to work reliably. It will for sure come far too late to help combat climate change, as will be made abundantly clear in this book. There is no chance whatsoever that it will contribute to the energy mix in this century, let alone before or around 2050 as required by the Paris Climate Agreement. All the activity, especially monsters like the ITER project, will have an enormous carbon footprint, and the amount of carbon dioxide produced in the course of its construction can probably not be made good.

After two introductory chapters, explaining the physics principles involved in nuclear fusion, the book mostly follows the historical chronology from the first attempts in the early fifties with the so-called pinch devices in various countries via stellarators and early tokamaks to the latest international megaproject of ITER, which originally stood for International Thermonuclear Experimental Reactor. Nowadays, the word thermonuclear has to be avoided, since to many the word 'nuclear' is as a red flag to a bull, and ITER is now just supposed to stand for the Latin word *iter* which means 'way' (in the sense of direction), i.e. the way to nuclear fusion, a way that could very well be leading us into a cul-de-sac. It will become clear that the road to the present has been bumpy, with many ups and downs, with the ups very often imagined and the downs the order of the day. This chronological overview is divided into two parts. The first of these (Part II of the book), called "Early Fusion Activity and The Rise of the Tokamak", describes the early beginnings of nuclear fusion until the momentous year 1968 when Russian results with their specific fusion device, called the tokamak, ushered in a new and optimistic age. After pausing and taking stock in an intermezzo (Chap.9), Part IV "High Noon" deals with the activity since 1968 until the present day and with prospects for the future. The final part, Part V, will delve into the criticism of the fusion efforts, economic and safety aspects, sustainability, applications and spin-offs.

If before embarking on this book, the reader wants to get an idea of the rosy view put forward by fusion proponents, I recommend them to read "A brief history of nuclear fusion", published in June 2020 in *Nature Physics* (Barbarino 2020). In the conclusion to this article, the following statement appears: "Although many challenges remain ahead on the way to a fusion-powered future, the enormous scientific and technological progress achieved through consistent high-level global partnership as well as the increased publicly and privately funded research and development demonstrate trust in fusion as a promising option to provide a sustainable, worldwide supply of energy for centuries to come." Its vacuity is typical for statements about nuclear fusion made by fusion proponents who doggedly refuse to see the glaring nakedness of the emperor. The statement's emptiness is reflected in the fact that it is almost timeless, in the sense that it could have been written at any time since the 1970s. Phrases like "many challenges remain ahead" and "enormous scientific and technological progress achieved" are completely void and have been written close to a million times about fusion since the 1970s. This book will put this progress into perspective, and although a lot has indeed been accomplished, it has not brought (commercially) viable power generation by nuclear fusion any closer. On the contrary, the lack of progress has shown that such *commercial* power generation can only be dreamed about and that the challenges that remain ahead are probably insurmountable. I would like to rephrase the statement a little into: "In view of the almost insurmountable challenges remaining ahead on the way to a fusion-powered future and the almost total lack of scientific and technological progress achieved through erratic high-level global partnership as well as the decreased publicly funded research and development, which has now attracted the vultures of venture capital, it is unwarranted and incomprehensible that there is still trust in fusion as a promising option to provide a sustainable, worldwide supply of energy for centuries to come".

I wish you happy reading.

Panningen, The Netherlands   L. J. Reinders
October 2020

Acknowledgements

I am very grateful to Prof. Van Lunteren for offering me a guest research position at Leiden University. This provided me with access to the library and other university facilities, which were indispensable for the writing of this book. I also thank Dr. Andries van Helden for carefully reading some chapters of the book. His many comments led to a number of major improvements. Special thanks go to Dr. Michael Gryaznevich, who introduced me to the world of nuclear fusion and explained many of the intricacies of tokamaks, especially spherical ones. Professor Vladimir Voitsenya, who unfortunately deceased in August 2020, provided useful information on the work with torsatrons carried out in Kharkov and on the early history of the stellarator. Finally, I want to thank the reviewers engaged by the publisher for giving their verdict on the content of the book, in particular Professor Mikhail Shifman, Dr. Leonid Zakharov and Dr. Jacques Treiner, whose comments and remarks corrected various mistakes and resulted in notable improvements to the manuscript.

Contents

Acronyms and Abbreviations

AAAPT	Asian African Association for Plasma Training
AAEC	Australian Atomic Energy Commission
ACST	Alternating Current Spherical Tokamak
AEC	Atomic Energy Commission
AERE	Atomic Energy Research Establishment
A-FNS	Advanced Fusion Neutron Source
Alcator	Alto Campo Toro
ALICE	Adiabatic Low-Energy Injection and Capture Experiment
ALPHA	Accelerating Low-cost Plasma Heating and Assembly
ANSTO	Australian Nuclear Science and Technology Organisation
ANU	Australian National University
APFRF	Australian Plasma Fusion Research Facility
ARIES	Advanced Reactor Innovation and Evaluation Study
ARPA-E	Advanced Research Projects Agency-Energy
ASDEX	Axially Symmetric Divertor Experiment
ASIPP	Institute of Plasma Physics of the Chinese Academy of Sciences
ASN	Autorité de Sûreté Nucléaire (French Nuclear Safety Authority)
ATC	Adiabatic Toroidal Compressor
ATF	Advanced Toroidal Facility
ATM	Axisymmetric Tandem Mirror
BATORM	BAby TORoidal of Masoud
BEAT	Break-Even Axisymmetric Tandem
BETHE	Breakthroughs Enabling Thermonuclear-Fusion Energy
BINP	Budker Institute of Nuclear Physics
BPX	Burning Plasma Experiment
CANDU	Canada Deuterium Uranium
CBFR	Colliding Beam Fusion Reactor
CCFE	Culham Centre for Fusion Energy
CCT	Continuous Current Tokamak
CDA	Conceptional Design Activities (ITER)

CDX	Current Drive Experiment
CDX-U	Current Drive Experiment-Upgrade
CEA	Commissariat à l'énergie atomique et aux énergies alternatives
CERN	Conseil Européen pour la Recherche Nucléaire
CFC	Carbon fibre composite
CFETR	China Fusion Engineering Test Reactor
CFS	Commonwealth Fusion Systems
CHI	Coaxial helicity injection
CIEMAT	Centro para Investigaciones Energéticas, Medioambientales y Tecnológicas
CIRCUS	CIRCUlar coil Stellarator
CIT	Compact Ignition Tokamak
CLAM	Chinese Low Activation Martensitic
CLEO	Closed Line Electron Orbit
CNEN	Comitato Nazionale per l'Energia Nucleare
CNRS	Centre National de la Recherche Scientifique
CNT	Columbia Non-neutral Torus
CPD	Compact Plasma wall interaction Device
CREST	Compact Reversed Shear Tokamak
CRP	Coordinated Research Project
CTF	Component Test Facility
CTH	Compact Toroidal Hybrid
CTR	Controlled thermonuclear research
CTX	Compact Torus Experiment
DANTE	Danish Tokamak Experiment
DARPA	Defense Advanced Research Projects Agency
DCX	Direct Current Experiment
DFD	Direct Fusion Drive
DIFFER	Dutch Institute for Fundamental Energy Research
DITE	Divertor Injection Tokamak Experiment
DND	Double Null Divertor
DOE	Department of Energy (US)
DONES	DEMO Oriented Neutron Source
DPF	Dense Plasma Focus
DREAM	Drastically Easy Maintenance Tokamak
DTT	Divertor Tokamak Test facility
EAST	Experimental Advanced Superconducting Tokamak
EBW	Electron Bernstein Wave
ECCD	Electron cyclotron current drive
ECRH	Electron Cyclotron Resonance Heating
EDA	Engineering Design Activities (ITER)
EFDA	European Fusion Development Agreement
EFRC	ENN Field-Reversed Configuration
ELiTe	EVEDA Lithium Test Loop
ELM	Edge-localised mode

ENEA	Energia Nucleare ed Energia Alternative (National Agency for New Technologies, Energy and Sustainable Economic Development, Italy)
EPR	European Pressurized Reactor/Evolutionary Power Reactor (third generation nuclear fission reactor)
ERIC	European Research Infrastructure Consortium
ESNIT	Energy Selective Neutron Irradiation Test
ESS	European Spallation Source
ETE	Experimento Tokamak Esférico
ETF	Engineering Test Facility
ETR	Engineering Test Reactor
Euratom	European Atomic Energy Community
Extrap	External Ring Trap
FDF	Fusion Development Facility
FDS	Fusion design study
FED	Fusion Engineering Device
FER	Fusion Energy Research (Facility)
FESS	Fusion Energy System Studies
FFHR	Force Free Helical Reactor
FIRE	Fusion Ignition Research Experiment
FIREX	Fast Ignition Realization EXperiment
FMIT	Fusion Materials Irradiation Test
FNSF	Fusion Nuclear Science Facility
FRC	Field-reversed configuration
FRCHX	Field-Reversed Compression and Heating Experiment
FRX-L	Field-Reversed eXperiment-Liner
FT	Frascati Tokamak
FTU	Frascati Tokamak Upgrade
FTWR	Fusion transmutation of waste reactor
GDMT	Gas Dynamic Multiple-Mirror Trap
GDT	Gas Dynamic Trap
GLAST	GLAss Spherical Tokamak
GSI	Gesellschaft für Schwerionenforschung
HBCCO	Mercury barium calcium copper oxide
HBT-EP	High Beta Tokamak
HCCB	Helium-cooled ceramic breeder
HCLL	Helium-cooled lithium lead
HED-LP	High energy density laboratory plasma
HELIAS	Helical-Axis Advanced Stellarator
HFTM	High-Flux Test Module
HHFW	High-harmonic fast wave
HHR	Horne Hybrid Reactor
HIBALL	Heavy Ion Beams and Lead Lithium
HIBLIC	Heavy Ion Beam and Lithium Curtain
HIDIF	Heavy Ion Driven Ignition Facility

HIDRA	Hybrid Illinois Device for Research and Applications
HiPER	High Power laser Energy Research
HIPGD	High-intensity plasma gun device
HIST	Helicity Injected Spherical Torus
HIT	Helicity Injected Torus
HIT-SI	Helicity Injected Torus with Steady Inductance
HSE	Heidelberg Spheromak Experiment
HSR	Helical Stellarator Reactor
HSX	Helically Symmetric Experiment
HTS	High-temperature superconductor
IAEA	International Atomic Energy Agency
IBW	Ion Bernstein Wave
ICDMP	International Centre for Dense Magnetised Plasmas
ICF	Inertial confinement fusion
ICRF	Ion Cyclotron Radio Frequency
ICRH	Ion Cyclotron Resonance Heating
IDCD	Imposed-Dynamo Current Drive
IFERC	International Fusion Energy Research Centre
IFMIF	International Fusion Materials Irradiation Facility
IFRC	International Fusion Research Council
IFSMTF	International Fusion Superconducting Magnet Test Facility
INFUSE	Innovation Network for Fusion Energy
INTOR	International Tokamak Reactor
IPA	Inductive Plasmoid Accelerator
IPCR	Institute of Physical and Chemical Research (Japan)
IPP	Max Planck Institute for Plasma Physics
IRFCM	Institut de Recherche sur la Fusion par confinement Magnétique
ISTTOK	Instituto Superior Técnico TOKamak
ISX	Impurity Study Experiment
ITER	International Thermonuclear Experimental Reactor
ITER-FEAT	ITER Fusion Energy Advanced Tokamak
IVTAN	Institute of High Temperatures of the Academy of Sciences
JAEA	Japan Atomic Energy Agency
JAERI	Japan Atomic Energy Research Institute
JET	Joint European Torus
JFT	Jaeri Fusion Torus
KAIST	Korea Advanced Institute of Science and Technology
KBM	Kinetic ballooning mode
KfA	Kernforschungsanlage
KSTAR	Korea Superconducting Tokamak Advanced Research
KTM	Kazakh Tokamak for Material studies
KTX	Keda Torus eXperiment
LAMEX	Large Axisymmetric Mirror Experiment
LANL	Los Alamos National Laboratory
LATE	Low Aspect Ratio Torus Experiment

LBCO	Lanthanum barium copper oxide
LBNL	Lawrence Berkeley National Laboratory
LCE	Lithium carbonate equivalent
LCFS	Last closed flux surface
LCOE	Levelized cost of energy or Levelized cost of electricity
LDX	Levitated Dipole Experiment
LGI	Laboratorio Gas Ionizzati
LH	Lower Hybrid
LHCD	Lower hybrid current drive
LHD	Large Helical Device
LHI	Local helicity injection
LHRF	Lower Hybrid Range of Frequencies
LIFE	Laser Inertial Fusion Energy
LIFT	Laser Inertial Fusion Test
LIPAc	Linear IFMIF Prototype Accelerator
LLCB	Lithium-lead ceramic breeder
LLE	Laboratory for Laser Energetic
LLNL	Lawrence Livermore National Laboratory
LMF	Large Mirror Facility
LOCA	Loss of coolant accidents
LOFA	Loss of coolant flow accidents
LOVA	Loss of vacuum accidents
LPP	Lawrenceville Plasma Physics
LTD	Linear Transformer Driver
LTS	Low-temperature superconductor
LTX	Lithium Tokamak Experiment
LWR	Light-water reactor
MAFIN	MAgnetic Field Intensification
MagLIF	Magnetised liner inertial fusion
MAGPIE	Magnetized Plasma Interaction Experiment
MARBLE	Multiple ambipolar recirculating beam line experiment
MAST	Mega Ampere Spherical Tokamak
MCF	Magnetic confinement fusion
MEDUSA	Madison EDUcational Small Aspect ratio
MFTF	Mirror Fusion Test Facility
MGI	Massive gas injection
MHD	Magnetohydrodynamics
MIF	Magneto-inertial fusion
MIFTI	Magneto-Inertial Fusion Technologies
MIIFED	Monaco ITER International Fusion Energy Days
MIRAPI	MInimum RAdius Pinch
MIT	Massachusetts Institute of Technology
MIX	Multipole ion-beam experiment
MPEX	Materials Plasma Exposure eXperiment
MRX	Magnetic Reconnection Experiment

MST	Madison Symmetric Torus
MST	Medium Sized Tokamak
MTF	Magnetized target fusion
MTM	Micro-tearing modes
MTR	Magnetic thermonuclear reactor
MTR	Materials test reactor
NBI	Neutral Beam Injection
NCSX	National Compact Stellarator Experiment
NET	Next European Torus
NFRI	National Fusion Research Institute (Korea)
NFTF	Nuclear Fusion Test Facility
NIFS	National Institute for Fusion Science (Japan)
NSST	Next Step Spherical Torus
NSTX	National Spherical Torus Experiment
NTFP	National Tokamak Fusion Programme (Pakistan)
NTM	Neoclassical tearing mode
NUCTE	Nihon-University Compact Torus Experiment
OH coils	Ohmic heating coils
PAEC	Pakistan Atomic Energy Commission
PALS	Prague Asterix Laser System
PBFA	Particle Beam Fusion Accelerator
PBX	Princeton Beta Experiment
PBX-M	Princeton Beta Experiment Modification
PDX	Poloidal Divertor Experiment
PF coils	Poloidal field coils
PFC	Plasma facing component
PFPP	Prototype Fusion Power Plant
PFRC	Princeton Field-Reversed Configuration
PFX	Penning fusion experiment
PHEV	Plug-in Hybrid Electric Vehicle
PHP	Pilot Hybrid Plant
PIC method	Particle in cell method
PIE	Post-Irradiation Examination
PINI	Positive Ion Neutral Injector
PKA	Primary knock-on atom
PLT	Princeton Large Torus
PLX	Plasma Liner Experiment
POPS	Periodically oscillating plasma sphere
PPCS	Power Plant Conceptual Studies
PPPL	Princeton Plasma Physics Laboratory
PSFC	Plasma Science and Fusion Center (MIT)
QHS	Quasi-Helically Symmetric
QUEST	Q-shu University Experiment with steady-state Spherical Tokamak
RAFM	Reduced activation ferritic/martensitic
ReBCO	Rare-earth barium copper oxide

RELAX	REversed field pinch of Low Aspect ratio eXperiment
RF	Radio-Frequency
RFP	Reversed Field Pinch
RFX	Reversed Field Experiment
RIKEN	Kokuritsu Kenkyū Kaihatsu Hōjin Rikagaku Kenkyūsho (National Institute of Physical and Chemical Research)
RMP	Resonant magnetic perturbation
RTP	Rijnhuizen Tokamak Project
SCR	Stellarator of Costa Rica
SEAFP	Safety and Environmental Assessment of Fusion Power
SHPD	Sustained High Power Density
SIFFER	SIno-French Fusion Energy Center
SINP	Saha Institute of Nuclear Physics
SMARTOR	Small Aspect Ratio Torus
SMBI	Supersonic molecular beam injection
SOL	Scrape-Off Layer
SPHEX	Spheromak Experiment
SPI	Shattered Pellet Injector
SSPX	Sustained Spheromak Physics Experiment
SST	Steady-state Superconducting Tokamak
SSX	Swarthmore Spheromak Experiment
ST	Spherical tokamak
ST	Symmetric Tokamak (first Princeton tokamak)
STAC	Science and Technology Advisory Committee (ITER)
STAR	Small Toroidal Atomic Reactor
START	Small Tight Aspect Ratio Tokamak
STEP	Spherical Tokamak for Energy Production
STPC-EX	Spherical tokamak with plasma centerpost experiment (Turkey)
SUNIST	Sino UNIted Spherical Tokamak
SURMAC	Surface Magnetic Confinement
SXD	Super-X divertor
TAERF	Texas Atomic Energy Research Foundation
TBCCO	Thallium barium calcium copper oxide
TBM	Test Blanket Module
TBR	Tritium-breeding ratio
TBS	Test Blanket System
TCV	Tokamak à Configuration Variable
TdeV	Tokamak de Varennes
TE	Tokamak Energy
TETR	Tokamak Engineering Test Reactor
TF coils	Toroidal field coils
TFCX	Toroidal Fusion Core Experiment
TFR	Tokamak de Fontenay-aux- Roses
TFTR	Tokamak Fusion Test Reactor
TIBER	Tokamak Ignition/Burn Experimental Reactor

TJ   Tokamak de la Junta de Energía Nuclear
TM   Tokamak Malyj (*small tokamak*)
TMP   Tor s magnitnym polem (*torus with magnetic field*)
TMX   Tandem Mirror Experiment
TORMAC   Toroidal Magnetic Cusp
TORTUR(E)   TORoidal TURbulence Experiment
TRINITI   Troitsk Institute for Innovation and Fusion Research
TST   Tokyo Spherical Tokamak
TTF   Frascati Turbulent Tokamak
TWR   Travelling wave reactor
UKAEA   United Kingdom Atomic Energy Authority
ULART   Ultra-Low Aspect Ratio Tokamak
URANIA   Unified Reduced Non-Inductive Assessment
UTST   University of Tokyo Spherical Tokamak
VASIMR   Variable Specific Impulse Magnetoplasma Rocket
VECTOR   Very Compact Tokamak Reactor
VEST   Versatile Experiment Spherical Torus (Korea)
WAM   Wisconsin Axisymmetric Mirror
WEGA   Wendelstein Experiment in Grenoble for the Application of
                    Radio-frequency Heating/Wendelstein Experiment in Greifswald
                    für die Ausbildung
WEST   Tungsten (Wolfram) Environment in Steady-state Tokamak
WHAM   Wisconsin HTS Axisymmetric Mirror
YBCO   Yttrium barium copper oxide
ZETA   Zero Energy Thermonuclear Assembly

Units and Related Quantities

appm   Atomic parts per million
dpa   Displacements per atom
dpy   Displacements per year
eV   Electronvolt (1.6×10^{-19} J)
fm   Femtometre (10^{-15} m)
gigawatt 10^9 (1000 million) W
keV   Kiloelectronvolt
MA   Megaamperes
MeV   Mega-electronvolt (1 million electronvolt)
MW   Megawatt (1 million W; 10^6 J/s)
MWe   Megawatt electric
petawatt 10^{15} W
tesla   Unit of magnetic field strength equal to 10,000 gauss

Part I
Introductory Chapters

Chapter 1
Introduction and Basic Science

1.1 Introduction

The energy that reaches us from the Sun is the product of a process called nuclear fusion. Although the particular solar process cannot be reproduced here on Earth, the idea of nuclear fusion, generating vast amounts of energy from combining light elements (especially hydrogen and its isotopes[1] deuterium and tritium) into heavier ones (in particular helium), has been a dream of mankind ever since the processes in the Sun and other stars were unravelled early in the twentieth century. In nuclear fusion the nuclei of such light elements are melted together with the simultaneous emission of some other particles, resulting in the release of a significant amount of energy. Per kilogram of consumed matter this release of energy is about ten million times larger than in a typical chemical process like the burning of fossil fuels. For two reasons fusion is sometimes called "the ultimate energy source". In the first place, since the fusion of light elements produces the harmless (but valuable) "waste" product helium and, secondly, since the basic fuel for fusion is the hydrogen isotope deuterium, which can be inexpensively obtained from water and is readily available to all nations on Earth. As will be discussed in detail in this book, for the last seventy years a lot of effort has been put into trying to control nuclear fusion on Earth, in order to harvest its energy and solve the energy problems of mankind once and for all.

Apart from nuclear fusion, there is also the more commonly known process of nuclear fission, which concerns the splitting of the nuclei of heavy elements. The best-known example is uranium, whereby the nucleus of such element is split into two smaller, more stable nuclei releasing at the same time a certain amount of energy. Per kilogram of matter this release of energy is however less than 10% of the output from fusion. Nuclear fission was discovered in Berlin in 1938, about 20 years later

---

[1] An isotope is a variant of a particular element that has the same chemical properties as that element but differs in the composition of its nucleus. This and other definitions can be found in the Glossary in the back.

than nuclear fusion, by the German physicists Otto Hahn (1879–1968) and Fritz Strassman (1902–1980).

Contrary to fusion, the discovery of nuclear fission immediately generated a lot of research, as it was soon realised that a self-sustaining nuclear chain reaction was possible, and a nuclear reactor could be built for harnessing the released energy. This was duly achieved in the first artificial nuclear reactor, the first Chicago Pile,[2] constructed in 1942 in Chicago under the direction of the Italian physicist Enrico Fermi (1901–1954) as part of the Manhattan Project, America's effort to build an atomic bomb. Since then numerous nuclear fission reactors have been built and are still being built all over the world, providing a significant portion of mankind's energy needs. The price for this is however rather high, for many too high, in the form of waste products that due to their long-lasting radioactivity are difficult to dispose of, and in the form of accidents. Power from nuclear fission reactors rose rapidly in the second half of the twentieth century, stimulated in part by the oil crisis of the early 1970s and the supposedly scary thought of being dependent on the Arab countries for oil supply. Energy supply from nuclear fission reached a peak of a 17.6% share of globally generated power in 1996 and has declined ever since. As of August 2020, there were 441 operable nuclear reactors in 31 countries, with 54 reactors still under construction and a further 109 planned, and a share of 10.2% of global electricity production. For the 28 countries of the European Union nuclear energy accounted for about 25% of total primary energy production in 2019 (down from 28.9% in 2013), its significance being particularly high in France, Belgium and Slovakia where nuclear energy accounted for respectively 71% (down from 82.5% in 2013 and to be reduced to 50% by 2035), 48% (down from 65% in 2015) and 54% (down from 62.6% in 2013) of the national production of primary energy (World Nuclear Association 2020).

One of the reasons of this decline are accidents that have greatly diminished the appetite for fission reactors, especially in Western countries. The first large accident was the partial meltdown at the Three-Mile Island plant in Pennsylvania in 1979, followed in 1986 by the horrendous Chernobyl disaster. Especially the latter event sounded the death knell for power generation from nuclear fission. The last doubters were silenced by the Fukushima disaster in 2011, which like Chernobyl has left large areas of land contaminated with radioactive materials. The Fukushima disaster led to panicky reactions of politicians all over the world, notably in Germany where plans were quickly drawn up to scrap all nuclear fission plants. Although there still is quite a lot of activity, as recalled above, its share of global electricity production will continue to fall. The public trust in nuclear fission as a power generation option has suffered a lethal blow and nobody seems to be able to turn the tide. This is unfortunate as it has been convincingly argued by many that decarbonisation of electricity generation will be a tough job, if not impossible, without nuclear fission plants (see e.g. Pinker 2018, Chap. 10). The bad reputation of fission power has also

---

[2]The word pile or atomic pile was then in use for designating such a reactor, which was not much more than a pile of blocks of uranium and graphite. Fermi described the apparatus as "a crude pile of black bricks and wooden timbers."

consequences for nuclear fusion, which likes to present itself as the safe nuclear option. That may well be the case but is a hazardous strategy as everything nuclear is viewed with suspicion by the public. Germany is a case in point where the Green Party also opposes nuclear fusion as an energy option (Schiermeier 2000).

A further reason of the decline is the recent glut of cheap (shale) oil and gas coupled to a rapid increase in wind and solar energy, forcing nuclear fission power plants out of business, a trend that can only be reversed by slapping a substantial carbon tax on fossil fuels. A deeper reason for the decline is also that the technology of commercial nuclear reactors has stagnated. Nearly every nuclear power plant built in the last half century has been a light-water reactor, a design that in rare instances can indeed allow a meltdown and was aggressively marketed by the United States, which has now all but quit the field. Meltdown-proof, cheaper and more efficient designs, like very high-temperature reactors and other Generation IV nuclear reactors, have remained on the drawing boards for years, but are now being developed mainly in China and Russia.

The reason that fission was developed in just a few years much earlier and much more extensively than fusion is in the first place that the technical realization of fusion is vastly more difficult than fission. As will become clear later, the fundamental reason for these difficulties lies in the fact that fission can be achieved by firing neutral particles (neutrons) at nuclei, while in the case of fusion positively charged nuclei must be persuaded to fuse, which is a formidable task. Secondly, as regards fission, it was realised very soon after Hahn and Strassman's discovery that an atomic bomb could be built on the basis of nuclear fission. For the latter the Manhattan Project was started up in the US in the early 1940s during World War II and fusion was put on a backburner for the duration of the war as the required temperatures, tens of millions of degrees, could only be achieved by a fission explosion. The idea of using a fission explosion to achieve fusion was first brought up by Enrico Fermi in a rather casual conversation (see Segrè and Hoerlin 2016, p. 277) and enthusiastically taken up by the Hungarian-American physicist Edward Teller (1908–2003), whose fanatic anti-Communist stance made him a great proponent of the thermonuclear or hydrogen bomb; the word thermonuclear indicating in this connection that very high temperatures are required. He went to great lengths to push the construction of such a bomb, although it was not at all clear that the technical difficulties in creating such a monster could be overcome, nor whether there was any sensible military use for a bomb that was 1000 times more powerful than a fission bomb. The technique of using a nuclear fission bomb as a 'matchstick' was eventually deployed in building the hydrogen bomb. The purpose of the fission device in its construction was not only to create the necessary temperature, but also to exert a sufficiently strong compressive force on the fusion fuel. A fission-based nuclear bomb was surrounding this fuel. Its ignition provided both the necessary temperature and pressure for the fusion bomb to go off. Thermonuclear bombs are still of this type; a pure fusion bomb (without the help of a fission bomb) has not yet seen the light, and hopefully never will. On the other hand, the fact that it hasn't, in spite of a colossal research effort, both by Western powers and by the Soviet Union, does not bode well either for fusion as an energy source. For if an uncontrolled release of fusion energy can apparently not be achieved without help from fission, how then can we have faith in controlled fusion ever being possible? Remember that in fission research it was the other way round. Fermi succeeded in keeping fission under control in his Chicago Pile, before the effort was started to construct an atomic bomb.

Fusion is conceptually a rather simple process, much simpler than fission and, more importantly, it is much cleaner. Little radioactivity is released in the fusion process, and the radioactive waste products it produces are manageable, although a future nuclear fusion reactor is not as clean in this respect as many have wanted us to believe in the past. In power generation from nuclear fusion the radioactive waste problem will certainly not be negligible, since the vast number of neutrons released in the fusion processes will make much of the material of the reactor radioactive. In view of the general public's sensitivity to radioactivity it is paramount to be clear and transparent about this from the very beginning. The mere fact however that radioactive waste is produced already implies that in most countries fusion power generation facilities, if ever realised, will require special licensing, resulting very probably in the demand for extensive and expensive safety features as are common in fission power plants. However, a comparison with nuclear fission, which indeed is much worse as regards radioactive waste, is not relevant either in this respect, for something that is better than the perceived absolute evil is of course not necessarily good. Fusion may be well advised to avoid the comparison with fission as much as possible.

In this respect, a recent Japanese report reviewing the country's commitment to the construction of a demonstration nuclear fusion power plant says: "Since the accident at Tokyo Electric Power Company's Fukushima Daiichi Nuclear Power Plant, people have been losing confidence in nuclear power. The principle of nuclear fusion differs from that of nuclear fission and can be said to be intrinsically safe. However, the accident mentioned above should be a lesson for those who are involved in nuclear fusion since it deals with neutrons and radioactive substances as is the case with nuclear fission. It should be understood while establishing the fusion DEMO reactor that the reactor is expected to have an even higher level of safety than the current level of nuclear safety technology because it would not be possible to introduce a fusion DEMO reactor in Japan without first obtaining the confidence of the public."[3] This is a typical Japanese remark and is not seen anywhere else. It must be said in this respect and will also come up later in this book that the Japanese are more honest, sensitive and transparent about any disadvantages of nuclear fusion than other participants in these endeavours.

The most important, and in the end perhaps decisive advantage of power generation from fusion over both fission and fossil fuels is however that the primary fuel, the hydrogen isotope deuterium, can be obtained cheaply from water. In the water of the Earth's oceans one atom in every 6,420 hydrogen atoms is deuterium, accounting for approximately 0.0156% (or 0.0312% on a mass basis as a deuterium atom is

[3]Executive Summary to Japan's Policy to promote R&D for a fusion DEMO reactor (December 18, 2017), https://www.mext.go.jp/component/b_menu/shingi/toushin/__icsFiles/afieldfile/2019/02/18/1400137_02.pdf, accessed 25 April 2020.

twice the mass of ordinary hydrogen) of all naturally occurring hydrogen in the oceans. No mines are needed, no miners can get trapped, no transport of fuel to be burned in power stations, and a virtually inexhaustible supply. That is indeed true for deuterium, but in the currently preferred version of nuclear fusion, as we will see, another hydrogen isotope, tritium, is also required which is in short supply and moreover dangerously radioactive.

As said, fusion is technically much more demanding than fission, the root of the problem being that unlike fission it cannot be induced by uncharged particles. The nuclei that must be brought together in a fusion process are all positively charged and, therefore, repel each other. The larger the nuclei, the larger the charge and, since the repulsive force, the Coulomb force, is proportional to the product of the charges of the nuclei,[4] even for nuclei of moderately large atoms the repulsive Coulomb force becomes prohibitive. This implies that, in order to have a chance of success, fusion fuels must be chosen from among the lightest elements–hydrogen, helium, lithium, beryllium and boron. In spite of this small number of candidate light elements, it still leaves us with more than 100 possible fusion reactions, of which those involving elements with only one charged particle in the nucleus (i.e. hydrogen and its isotopes) are the most promising.

Early progress in fusion devices for generating energy was also undoubtedly hampered by the fact that all fusion research, like fission research, not only for weapons development, but also for power generation, was kept secret until the late 1950s. The US, for instance, harboured fears that fusion reactors could be used as a neutron source to make bomb fuel. And indeed, a stimulus to most fusion research in the early days was the production of bomb-grade material for thermonuclear weapons, and the fear to be left behind in their construction.

The nuclear powers at the time, the US, Britain and the Soviet Union, all started their own fusion research programmes after World War II and jealously guarded them from the outside world. They all employed essentially the same methods and techniques, encountered the same problems and none managed to construct a working fusion reactor. It should be noted though that energy political considerations were of no importance for any of the atomic powers in their early efforts in trying to realize fusion. The argument of 'cheap and abundant' energy was put forward later, when slogans were needed to convince funding agencies and the public.

The similarity of approaches and findings arrived at independently by the three nuclear powers indicated what secrecy had cost the world for no useful purpose, as the Indian nuclear physicist Homi J. Bhabha (1909–1966) remarked in the *New York Times* in 1958. However, the fact that secrecy was lifted had undoubtedly also to do with the lack of success. If any of the parties involved in fusion research in the early 1950s had made promising progress towards a working reactor, the secrecy would undoubtedly have become tighter still. This latter point is also borne out by

---

[4]The force is also inversely proportional to the square of the distance between the nuclei. The Coulomb force is $F = k_e q_1 q_2 / r^2$, with k_e a constant called Coulomb's constant. If the charges q_1 and q_2 have the same sign, the force is repulsive. If they are of opposite sign, the force is attractive. The force also increases rather sharply when the distance decreases. All this was first formulated in 1785 by the French military engineer and physicist Charles-Augustin de Coulomb (1736–1806).

the declassification guide jointly worked out by the British and Americans in 1957, which stated that "all information except that bearing on devices exhibiting a net power gain was to be opened" (Bromberg 1982, p. 75). So, had there been any success with a working reactor, the information would have remained classified.

Since then it has gone up and down with fusion without any great success, although the proponents want us to believe otherwise. Since Homi Bhabha, who chaired the First International Conference on Peaceful Uses of Atomic Energy, predicted as early as 1955 that "a method will be found for liberating fusion energy in a controlled manner within the next two decades" (Braams and Stott 2002, p. 20), i.e. by 1975, it has become one of the clichés of nuclear fusion research that a commercial nuclear fusion reactor is ever only a few decades away. As the saying goes "nuclear fusion power is the energy of the future, and always will be". A former leader of the Tokamak Fusion Test Reactor (TFTR) at the Princeton Plasma Physics Laboratory recently stated, blaming insufficient funding, as is common practice among failing scientists and indeed people in most human endeavours, that "the goal of commercial fusion energy recedes 1 year per year" (Meade in Dean 2013, p. 222).

There is no other endeavour or project undertaken by mankind on which energy and money have been spent for close to a hundred years without any tangible results, only a dim prospect of success in another fifty years or so. The reason must be that there is a lot at stake, or perceived to be: the promise of nuclear fusion power being an abundant, inexhaustible source of energy with little or no side effects, at any rate manageable side effects, and "too cheap to meter". Although the latter argument no longer seems to hold, the rest is already too good to be true, and if true not something you would like to miss out on. No wonder that large teams of scientists in many countries are still working hard trying to solve the colossal scientific and technical problems involved in nuclear fusion. It would be a major achievement if in 15–25 years from now a working reactor for demonstration purposes would be available, meaning a reactor which demonstrates that it is possible to build reactors that consume less energy than needed to run them. This book intends to show that the chances for this to happen are very slim indeed.

In part I of this book, consisting of this chapter and the next, the basic science underlying fusion will be explained in some detail. Parts II and IV are devoted to a description of the rather chaotic history of the attempts to build a working *and* useful (i.e. energy producing) nuclear fusion device. The final part V consists of a number of chapters dealing with criticism of the fusion effort, sustainability, safety, economic effects and possible other applications.

1.2 Some Basic Nuclear Physics

All matter consists of atoms and an atom is composed of a nucleus, which is orbited by negatively charged electrons, as schematically shown in Fig. 1.1 for the isotopes of hydrogen, which play an important role in fusion. The scale of the drawings in the figure does not reflect reality. If the nuclei were as large compared to the electron

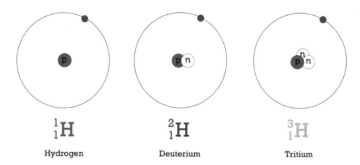

$^1_1$H $^2_1$H $^3_1$H

Hydrogen   Deuterium   Tritium

Fig. 1.1 Schematic representation of hydrogen and its isotopes (*Source* chem.libretexts.org). Ordinary hydrogen with only a proton as nucleus is also called protium

or the total atom as drawn in the figure, the electron orbit would not fit on the page. Both the nuclei and the electrons occupy only a tiny part of the total atom, which mainly consists of 'empty' space filled with the electromagnetic fields generated by the nucleus and the electrons. A more modern representation would picture the electron not as a point particle but as a cloud distribution.

All nuclei consist of just two kinds of nuclear particles: called protons and neutrons, and jointly nucleons. The nucleus of the hydrogen atom, the simplest element in the Periodic Table of Elements (Fig. 1.3), is just a single particle with positive electric charge: a proton (the first drawing in Fig. 1.1). The mass of this proton is about two thousand times as large as the electron's mass, which explains why most of the hydrogen's mass and the mass of any other element, for that matter, is concentrated in its nucleus. The nuclei of all elements are built from such positively charged protons, supplemented with a number of neutral particles, called neutrons. In the second and third drawings in Fig. 1.1 the nuclei of deuterium and tritium are shown as containing one, respectively two neutrons, in addition to a single proton. And, in general, the only difference between the isotopes of a certain element is the number of neutrons in their nuclei. The protons give the nucleus its charge. This charge determines the chemical properties of the atom and its place in the Periodic Table of Elements. Neutrons do not carry any electric charge and provide the additional mass of any given element. In the nucleus they are needed to shield the protons from each other; else the nucleus would be unstable due to the Coulomb repulsion of the charges. Protons repel each other, and for a nucleus with more than one proton to be stable, one or several neutrons are needed. So, in nature there does not exist an element which has a nucleus consisting of just two protons. The next element in the Periodic Table, helium, has at least one neutron in addition to two protons in its nucleus. Its most common form is ^{4}He, which has two neutrons. In the Earth's atmosphere there is one ^{3}He atom for every million ^{4}He atoms (both are schematically drawn in Fig. 1.2).

In this way the nuclei of all further elements in the Periodic Table, shown in Fig. 1.3, are built up, with uranium being the heaviest of the so-called primordial elements, meaning that they already existed before the Earth was formed. Uranium

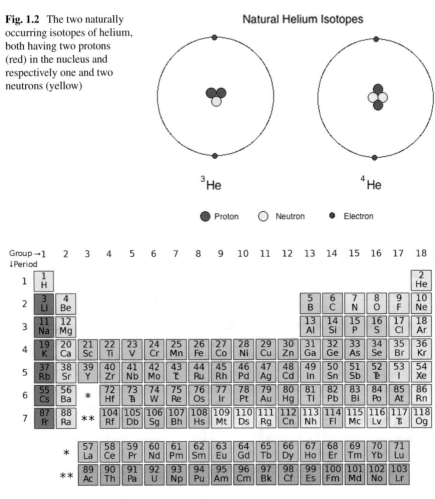

Fig. 1.2 The two naturally occurring isotopes of helium, both having two protons (red) in the nucleus and respectively one and two neutrons (yellow)

Natural Helium Isotopes

^{3}He ^{4}He

● Proton   ○ Neutron   • Electron

Fig. 1.3 Periodic table of elements in the modern standard form with 18 columns

has 92 protons and the staggering number of 146, 143 or 142 neutrons, depending on the isotope. The number of protons of an element is called the *atomic number* (indicated by the symbol Z) and is of course also equal to the number of electrons of the neutral atom. The isotopes of uranium are designated as ^{238}U or U-238, ^{235}U or U-235, and ^{234}U or U-234, where 238, 235 and 234 are the sums of the protons and neutrons in the nucleus of the particular isotope. This sum is called the *mass number* (indicated by the symbol A) of the element. More than 99% of naturally occurring uranium is U-238 and all isotopes of uranium have of course atomic number 92.

The positive charges of the protons balance the negative charges of the electrons orbiting the nucleus, making the atom of an element as a whole neutral in charge. In the early decades of the twentieth century it was thought that instead of neutrons nuclei contained only protons, a number of which were bound into a neutral structure with electrons. At the time this was quite a sensible idea since the only elementary particles known were electrons and protons. The idea was strengthened by the experimentally observed process of β-decay, in which a nucleus emits an electron and increases its electric charge by one unit. For, it was reasoned, how would a nucleus be able to emit electrons if it did not contain such particles in the first place? This process of a nucleus emitting an electron was one of the three forms of radioactive decay termed α-, β-, and γ-decay, in increasing order of their ability to penetrate matter. This terminology was introduced before it was actually known what the nature or the products of these decay processes were. It was the New Zealand born British physicist Ernest Rutherford (1871–1937) and co-workers who established that radioactivity or radioactive decay was in fact a nuclear process. In 1900 the French physicist Henry Becquerel (1852–1908) had identified the β-particle in β-decay as an electron, while one year later Rutherford and Frederick Soddy (1877–1956) had found that the α-particle emitted by the nucleus in α-decay is a ^{4}He nucleus.[5] The process of β-decay was studied extensively in the first decade of the twentieth century. The process was initially assumed to be a two-body process, since only the electron and the new nucleus were observed. One of the important features of the decay was that the kinetic energies of the emitted electrons formed a rather broad continuous spectrum. If the decay consisted in a nucleus just losing an electron, it was expected that this electron, being by far the lighter particle, would carry away most of the released energy (the nucleus would hardly move, just as, when you step out of a car, the car hardly moves[6]). In that case energy and momentum conservation would require the electron to have a certain definite energy,[7] within a fairly narrow range. So, there was a problem here and such a serious one that, when for quite some years no explanation for this continuous spectrum was forthcoming, some physicists were prepared to even give up energy conservation in these processes. The Austrian physicist Wolfgang Pauli (1900–1958) was however not among them and in 1930 he proposed what he called a "desperate remedy" to save energy conservation by proposing that in the nucleus there exist electrically neutral particles and that in β-decay such a particle is emitted together with the electron, such that the sum of the energy of the electron and this new electrically neutral particle is constant. Please note that he did not mean the neutron here, but a different, much lighter neutral particle. The neutron had not yet entered the scene. Its discovery in 1932 established the essential nature of the nucleus as consisting of neutrons and protons. Pauli still stuck to electrons and protons as elementary particles to which he now added a neutral one.

This new particle was later called *neutrino* (little neutron) by Fermi, who in 1934 wrote down a theory for the entire process by proposing that gravity and electromagnetism were not the only forces in nature. A third force, now known as the weak force, was at work here and could transform a neutron inside the nucleus into a proton, an

---

[5]The third decay process, γ-decay, involves the emission of penetrating electromagnetic radiation.
[6]Provided it isn't a Citroën 2CV.
[7]Roughly equal to $(M_A\text{-}M_B\text{-}m_e)c^2$, with M_A the mass of the original nucleus, M_B the mass of the product nucleus and m_e the mass of the electron.

electron and a neutrino. The electron and neutrino would be emitted while the proton stayed behind in the nucleus, increasing the charge of the nucleus by one unit. So, the electrons that were observed in β-decay had not been present in the nucleus after all, but had been created by the new force, simultaneously changing the nature of the nuclear particle from a neutron into a proton. It was the first time that anybody had devised a theory in which a particle could change its identity. No wonder that Fermi had some difficulty in getting his theory accepted (Segrè and Hoerlin 2016, p. 89ff).

But this is not all, as it does not yet explain how a nucleus with more than one proton can be stable. Above it was stated that neutrons are necessary to shield the protons from each other as they do not play well together because of their positive charge. Up to very small distances the resulting Coulomb repulsion will tend to force the protons apart. Only when they come even closer together, helped by the shielding effect of the neutron, a fourth force, the strong interaction or nuclear force, hits in to overcome the repulsion and bind the protons and neutrons into a stable nucleus. It was only in the 1970s that it became qualitatively clear how this works and theories of the strong force were developed. The strong interaction is powerfully attractive at distances of about 1 femtometre (1×10^{-15} m; fm), but rapidly decreases to insignificance at distances of about 2.5 fm and at distances less than 0.7 fm it becomes repulsive preventing the nucleons to come closer. This repulsion is responsible for the size of the nucleus. By comparison the size of an atom, determined by the orbits of the electrons around the nucleus, is five orders of magnitude larger (about 10^{-10} m), making, as already said above, that an atom mainly consists of 'empty' space, with a very small, but heavy nucleus, circled by electrons at rather huge distances.

How stable are such nuclei? If nuclei consist of neutrons and protons, one would naively expect that the mass of such a nucleus is equal to the sum of the masses of the protons and neutrons the nucleus consists of. That is indeed roughly the case, but not exactly, whereby this mass difference is precisely what the whole fuss is about, both in nuclear fusion and in nuclear fission. Many isotopes are lighter than would be expected from just adding up the masses of the protons and neutrons in the nucleus, so a proton or neutron bound into a nucleus has slightly less mass than a free proton or neutron. This mass difference is called the *mass defect*. The energy equivalent to this mass defect, obtained from Einstein's famous formula $E = mc^2$, was released when the nucleus was formed from its constituent protons and neutrons. It is also the binding energy of the nucleus.

In Fig. 1.4 the binding energy per nucleon, i.e. the total binding energy divided by the number of nucleons in the nucleus, has been plotted against the mass number. The figure shows that the binding energy per nucleon increases at first sharply with mass number and is largest for nuclei with mass number around 60, e.g. iron with atomic number 26 and mass number 56 (^{56}Fe), by far its most common isotope, is one of the most stable elements. Then the binding energy per nucleon slowly decreases down to mass numbers 240–250, the uranium isotopes. Nuclei of elements heavier than iron can in principle yield energy by nuclear fission (in which case they are split into two more tightly bound nuclei), while elements lighter than iron can do this in principle by nuclear fusion (in which case they are fused into one more tightly bound nucleus). So, if a uranium-235 nucleus is split by bombarding it with neutrons, as

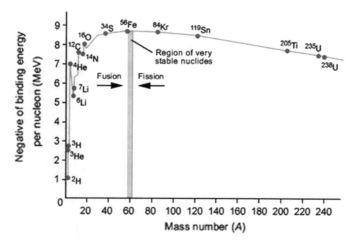

Fig. 1.4 Binding energy per nucleon plotted against the mass number

happens in a nuclear fission reactor, many fission reactions are possible, but a typical one is (with a neutron indicated by n):

$$n + {}^{235}U \rightarrow {}^{141}Ba + {}^{92}Kr + 3n$$

and shown in Fig. 1.5.

In this process the uranium nucleus first absorbs the neutron, forming the unstable uranium-236 which then splits into a barium and krypton nucleus (with respectively atomic number 56 and 36 (adding up to uranium's 92) with the simultaneous emission of three extra neutrons. These neutrons can split further uranium nuclei and set off a chain reaction. The barium and krypton nuclei are bound much more tightly than uranium resulting in the release of about 200 MeV or 200 million eV of energy, whereby the three neutrons that are released in the process carry away about 1–2 MeV each (4.8 MeV in total). An electron volt (eV) is the amount of energy gained by a single electron moving across an electric potential difference of one volt; it is equal to 1.6×10^{-19} J. This energy unit is the standard unit for the microworld of atoms and molecules, and will be used throughout this book. For comparison, 1 J is roughly equal to the kinetic energy of a tennis ball hitting the floor after falling from a height of 2 m. So, if a single tennis ball falling from a height of 2 m might be enough to dent a packet of butter, to achieve the same by splitting uranium nuclei you would need the staggering number of 30 billion of such events (whereby each event is worth about 3×10^{-11} J).

It is also important to observe, especially for the nuclear fusion discussed in this book, that ^{4}He is particularly strongly bound, certainly compared to ^{2}H (which is another way of denoting deuterium), so any fusion reaction that produces ^{4}He (e.g. from fusing two deuterium nuclei) will release a particularly large amount of energy, as will be discussed in greater detail below.

Fig. 1.5 Fission of a
uranium-235 nucleus into
barium and krypton
(*Wikipedia*)

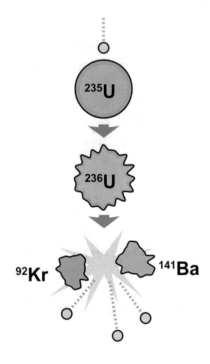

1.3 Stellar Processes and Quantum Mechanics

In 1920 in a paper in the journal *Science* the English astronomer, physicist and
mathematician Arthur Eddington (1882–1944) proposed that large amounts of energy
released by fusing small nuclei provide the energy source that powers the stars,
although he had no idea yet how this would work (Eddington 1920). This is the
more remarkable, as he was not even sure about the actual structure of atoms and
the relationship between the various elements in the Periodic Table. His proposal
predated the advent of quantum mechanics and a possible mechanism of such fusion
was unknown to him. The only forces known to Eddington were electromagnetism
and gravity. Gravitational contraction, i.e. the contraction of an astronomical object
due to the influence of its own gravity, drawing matter inwards towards the centre of
gravity, was known to be responsible for star formation, and the German physicist
Hermann von Helmholtz (1821–1894) had estimated that, if this were the source of
the Sun's radiation, it could only shine for about 20 million years. The Sun's surface
would need to drop by about 35 m per year to provide enough energy from such
gravitational contraction (see Green 2016, p. 29). So, there was a problem here as
around the same time geologists had shown that the Earth was at least two billion
years old. In actual fact both the Sun and the Earth are about 4.6 billion years old.

In his paper Eddington says: *A star is drawing on some vast reservoir of energy
by means unknown to us. This reservoir can scarcely be other than the sub-atomic
energy which, it is known, exists abundantly in all matter; we sometimes dream that*

man will one day learn how to release it and use it for his service. The store is well-nigh inexhaustible, if only it could be tapped. There is sufficient in the sun to maintain its output of heat for 15 billion years. Now a century after Eddington wrote these words, we are still dreaming of tapping this source of energy, and we are indeed getting a little closer, as the rest of this book will attempt to show, but the day of actually realizing this dream may still be far in the future or more likely remain elusive forever.

Although no real explanation of these stellar processes was possible before the advent of quantum mechanics, Eddington continues that he believes "*that some portion of this sub-atomic energy is actually being set free in the stars.*" He based his belief on experiments carried out by Francis William Aston (1877–1945), who got the Nobel Prize in chemistry in 1922 for the discovery of isotopes. In Eddington's mind these experiments had shown conclusively that "*all elements are constituted out of hydrogen atoms bound together with negative electrons*". The structure of an atomic nucleus was not yet known at the time. As already mentioned above, it was thought to consist of an assembly of protons and electrons, the only elementary particles then known.

Eddington in his paper went on to state that more importantly, Aston's precise measurements also showed that *the mass of a helium atom is less than the sum of the masses of the 4 hydrogen atoms which enter in it. (...) There is a loss of mass in the synthesis amounting to about 1 part in 120. (...) Now mass cannot be annihilated, and the deficit can only represent the mass of the electrical energy set free in the transmutation.* In the previous section we called this deficit the *mass defect*.

The mass-energy equivalence, embodied in the formula $E = mc^2$, was proposed by Einstein in 1905. Because of the factor c^2 in this formula, with c the speed of light in vacuum being equal to about 300,000 km/s, even a minuscule amount of mass is equivalent to an awesome amount of energy. Where the chemical reaction of burning 100 g of coal would release 1 million joules of energy, the mass of these 100 g would according to Einstein's formula actually be equivalent to 10 million billion joules of energy, if only we knew how to get that energy out.

Einstein's formula was of course well-known to Eddington and he used it to calculate the amount of energy released when helium is made out of hydrogen. He concludes: *If 5% of a star's mass consists initially of hydrogen atoms,[8]which are gradually being combined to form more complex elements, the total heat liberated will more than suffice for our demands, and we need look no further for the source of a star's energy. And If, indeed, the sub-atomic energy in the stars is being freely used to maintain their great furnaces, it seems to bring a little nearer to fulfilment our dream of controlling this latent power for the well-being of the human race—or for its suicide.* The final part of his paper, which as we know now contained much truth, is a rather lengthy apology on his part for having in the eyes of many gone over to speculation.

How does this energy production in stars come about? Stars start off as an interstellar cloud of gas, mainly consisting of hydrogen, which starts to collapse under

---

[8]We now know that it is actually around 75%.

the influence of gravity as soon as it is massive enough for the gravitational forces to be stronger than the internal pressure in the gas. The star becomes ever denser and hotter until at some point the temperature becomes so high that hydrogen nuclei start to fuse into helium according to the process to be described in greater detail below and energy is radiated off into space to warm planets like the Earth. In the star it increases the temperature still further and forces the gas to expand, countering the inward gravitational contraction. This results in an equilibrium situation whereby the star is held together by its own gravity and the internal gas pressure prevents it from collapsing further. This process continues until all the hydrogen has been burned away, after which a further contraction follows and other fusion processes take over, but it is in the first place the gravitational attraction that gets the process going. This is also the way it works in the Sun. The Sun being more than 300,000 times more massive than the Earth can generate sufficiently large gravitational forces. It will be clear that gravity on Earth is (fortunately) much too weak to bring such a contraction about. The gas has to be compressed in another way. That this might be extremely difficult can be surmised from the fact that there are two forces competing here, the inward compression (in stars by gravity) and the outward pressure by the gas heating up. When a star that is powered by burning hydrogen into helium, like the Sun, has exhausted all its hydrogen, its core will become denser and hotter while its outer layers expand, eventually transforming the Sun into what is called a red giant. It will become so large that it engulfs the current orbits of Mercury and Venus, rendering Earth uninhabitable. But this will not happen for another five billion years or so. After this, it will shed its outer layers and become a dense type of star known as a white dwarf. It will be very dense with a volume comparable to Earth and no longer produce energy by fusion, but still glow and give off heat from its stored thermal energy from previous fusion reactions.

A star contains a hot burning core in which the fusion processes take place, rather than that the burning occurs throughout the star. Eddington calculated that the temperature at the Sun's core would have to be about 40 million °C, which is two to three times as hot as the currently accepted number of about 15 million degrees.

But there was another rather pressing problem. How could four protons (nuclei of hydrogen), all positively charged, come together to form the nucleus of a helium atom. The protons would repel each other and there was no way in Eddington's time how this repellent Coulomb force could be overcome. Moreover, according to the classical laws of physics, the temperatures existing in the Sun were far too low for such fusion processes to take place. To find an explanation for this puzzle quantum mechanics was needed, a new theory that was developed in the 1920s, mainly in Germany, in which sheer impossible things are allowed to happen.

It was the Russian physicist George Gamow (1904–1968), who in 1928, while on leave in Göttingen from the Leningrad Physico-Technical Institute, added a vital ingredient to the solution of the puzzle by introducing the mathematical basis for what is known as *quantum tunnelling* (Gamow 1928). At Göttingen he saw that all quantum physicists were beavering away at trying to understand the quantum mechanics of atoms and molecules, and instead of joining this crowded fray he decided to have a look at what quantum theory could do for the atomic nucleus.

In the library he had come across an article by Ernest Rutherford describing an experiment on the scattering of α-particles (an alternative name, as we have seen, for nuclei of helium atoms) on uranium. From the scattering pattern it was clear that the α-particles were unable to penetrate into the nucleus of uranium. In itself not a strange result when one realizes the strong repulsive Coulomb forces between the positively charged α-particles and the positively charged uranium nucleus. But, so Gamow asked himself, if that is the case how then is it possible that uranium, being a radioactive element, does itself actually emit α-particles which have about half the energy of the α-particles Rutherford used to bombard the uranium nuclei with? Apparently the α-particles of the radioactive decay are prohibited for a rather long time from getting out of the uranium nucleus by a barrier which also stops α-particles with double the energy from getting in. He deduced that at very small distances the attractive nuclear forces become stronger than the repulsive Coulomb forces, which form a barrier at somewhat larger distances. But how then can they get out? Gamow immediately realized what the answer should be. In quantum mechanics, unlike classical Newtonian mechanics, there are no impenetrable barriers, and there is a non-zero probability for a particle to tunnel through a barrier and escape to the outside (Fig. 1.6).

The figure shows the potential barrier encountered by the particle, similar to a golf ball that has to get into a hole on the top of a little hill. The golf ball must scale the top before it can drop into the hole. Not so for a quantum mechanical particle, which has a non-zero probability to tunnel through the barrier and get into the hole, even if it does not have the energy to scale the top. So, it does not at all behave like a golf ball!

A year later Robert d'Escourt Atkinson (1898–1982) and Fritz Houtermans (1903–1966) applied Gamow's tunnelling to provide the first calculation of the rate of nuclear fusion in stars. The paper they wrote (Atkinson and Houtermans 1929) can be seen as the start of thermonuclear fusion energy research. The word thermonuclear, which was used for the first time in 1938 by Gamow (Gamow 1938), indicates the extremely high temperatures required in such nuclear processes. He defined thermonuclear reactions as nuclear transformations caused by violent thermal collisions at the extremely high temperatures existing inside a star.

Such temperatures give the particles a large enough thermal energy to overcome the Coulomb repulsion. Atkinson and Houtermans showed that, because of Gamow's

Fig. 1.6 A schematic picture of quantum tunnelling

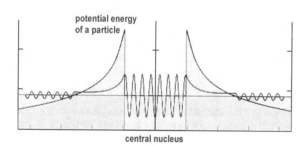

potential energy of a particle

central nucleus

tunnelling, fusion can occur at lower energies than previously believed (Eddington's 40 million degrees in the Sun's core) and that in the fusing of light nuclei energy could be created in accordance with Einstein's formula of mass-energy equivalence. The energy released in the fusion of light elements is due to the interplay of two opposing forces. As already mentioned, protons are positively charged and repel each other due to the Coulomb force, but when coming very close together, due to their high thermal energy, quantum tunnelling allows the attractive nuclear force to overcome the repulsion of the Coulomb force and attract the nuclei further towards each other. This nuclear force is short-range, i.e. is only felt when the nuclei are very close to each other (less than 10^{-15} m, a distance comparable to their size). Light nuclei are sufficiently small and have few protons. This allows them to come close enough to feel the attractive nuclear force. But to make this happen very high temperatures and pressures are needed.

1.4 Nuclear Fusion of Light Elements

At the centre of the Sun, where the fusion takes place that eventually provides us on Earth with energy and light, the temperature is around fifteen million degrees. At this temperature the electrons of the hydrogen atoms have been stripped away and the positively charged nuclei (protons) move around with very high velocities in a very dense gaseous state (ten times the density of lead) consisting of unbound negative (electrons) and positive particles (protons). This dense gaseous state is called a plasma and will be discussed in greater detail in the next chapter. The energy in the sun is created by fusing protons in the plasma into helium. The process involves three steps. It is called the "proton-proton" chain and was identified in 1939 by the German-American physicist Hans Bethe (1906–2005).[9]

The first step involves the exceedingly rare process of the fusion of two protons through the tunnelling process described by Gamow. On average it takes a billion years for a proton to fuse and the proton-proton fusion processes taking place in $1 m^3$ volume of the Sun produce just 30 W of heat, less than the heat on average given off by a human body. If the fusion rate were much larger, the Sun would burn up rather quickly and it would soon be over for us here on Earth, and we would not be able to arrogantly comment on the inefficiency of the Sun's fusion process. Fortunately, there is a huge amount of hydrogen present in the Sun, which at the Sun's temperature makes that hydrogen-hydrogen fusion can take place frequently enough to keep the Sun burning for our benefit for a very long time. In spite of the rarity of the process, the Sun fuses in its core a staggering 600 billion kilograms of hydrogen every second giving 596 billion kilograms of helium, and releasing the

---

[9]Bethe also discovered the CNO cycle as a possible fusion process that would be equally probable at solar temperatures of 16 million degrees as the p-p chain, but the p-p chain is more prominent in stars with a mass equal to the Sun's or less. In the CNO (carbon-nitrogen-oxygen) cycle four protons fuse, using carbon, nitrogen, and oxygen isotopes as catalysts, to produce one alpha particle, two positrons (positively charged electrons) and two electron neutrinos.

difference of 4 billion kilograms as energy. It has been fusing hydrogen into helium for around 4.6 billion years and has enough left to continue this process for another 4.6 billion years (Green 2016, pp. 37–38).

The precise process of the hydrogen fusion in the Sun is shown in Fig. 1.7. Every now and again two hydrogen nuclei, i.e. two protons, come close enough together for the top process in the figure to take place. Two hydrogen nuclei fuse into deuterium. In this process one hydrogen nucleus (proton) changes into a neutron by a rare weak-interaction process called inverse β-decay (the proton captures an electron and converts into a neutron and a neutrino, the latter particle, as we have seen above, was (reluctantly) postulated by Wolfgang Pauli in order to preserve energy conservation: $p + e^- \rightarrow n + \nu_e$). The reaction is so rare that it takes on average a billion years for two protons to fuse in this way in the solar core.

In the second stage of the process the deuterium nuclei formed in the first stage will in turn capture another hydrogen nucleus (proton) and form a helium-3 nucleus, which consists of two protons and a neutron. The chain then continues by two of such helium-3 nuclei fusing into a helium-4 nucleus (which compared to helium-3 has one more neutron in its nucleus) and two protons. The helium-4 atom (^{4}He) is much more

Fig. 1.7 The proton-proton chain reaction that is dominant in stars with the size of the Sun or smaller (*Wikipedia*)

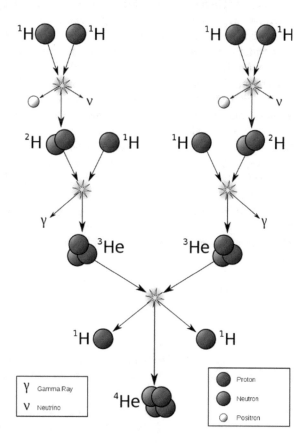

stable than helium-3 (^{3}He) and is the dominant isotope in naturally occurring helium. Hence the end result in the Sun's fusion process is that four protons (hydrogen nuclei) fuse into one helium-4 nucleus (α-particle), with the release of a certain amount of energy carried away by positrons (positively charged electrons), neutrinos and γ rays. This energy is what we are after. For a single fusion chain the release of energy amounts to the approximately 0.7% mass difference between the four protons, which formed the starting point of the process, and the helium-4 nucleus, or 26 MeV (26 million eV). This is still a very small amount of energy but compared to the most energetic chemical reactions in which only a few electron volts are released, nuclear reactions involve millions of times more energy. So, if a large number of fusion reactions takes place in a very short time or a self-sustaining chain reaction can be achieved, the total energy output will be enormous, as witnessed by the Sun and other stars. The same is true for *fission* reactions in which a heavy element (uranium or plutonium) splits into two lighter elements and about 200 million electron volts (200 MeV) is released in the splitting of a single element.

The proton-proton chain reaction of Fig. 1.7 is a rather convoluted process and not suitable for being reproduced on Earth because of the extremely low probability of the inverse β-decay mentioned above. It is simply not possible to create on Earth a sufficiently large amount of hydrogen plasma to make this process work. Gravity on Earth is much too weak and the reaction rate of the proton-proton fusion reaction into deuterium is much too small. We just have not got the time.

A simpler, much faster method is to directly start with the hydrogen isotope deuterium, which is usually, for historic reasons,[10] indicated by the symbol D (instead of ^{2}H which would be the proper designation). There is then no need to first convert a proton into a neutron. Deuterium has already a neutron in its nucleus and two deuterium nuclei already contain the necessary numbers of neutrons and protons from the start, resulting in a reaction which is 10^{24} times more probable than the Sun's p-p process (see Ongena and Ogawa 2016, p. 770). The fusion reaction of deuterium nuclei produces roughly half of the time a ^{3}He nucleus and a neutron, which respectively have an energy of 0.82 and 2.45 MeV. The other half of the time they fuse into a tritium nucleus (^{3}H) and a proton. This has been shown in the equation below, where for convenience the nuclear composition of the various particles has also been indicated. The third possibility (not indicated) of fusion into ^{4}He is very rare.

$$D(np) + D(np) \rightarrow {}^3He(npp) \ (0.82 \ MeV) + n \ (2.45 \ MeV) \ (half \ of \ the \ time)$$
$$\rightarrow {}^3H(nnp) \ (1.01 \ MeV) + p \ (3.03 \ MeV) \ (half \ of \ the \ time),$$

Even more convenient, deuterium is a stable isotope and abundantly present on Earth, where about 0.0156% of the water in the oceans is deuterated water HDO and D_2O (heavy water). The other isotope of hydrogen, tritium (^{3}H or more commonly

---

[10]It was detected in 1931, one year earlier than the neutron, and at its discovery it was not yet known that it was an isotope of hydrogen; hence its name deuterium, which derives from *deuteros*, the Greek word for two.

T), has a nucleus consisting of two neutrons and one proton. It is radioactive and decays naturally into ^{3}He with a half-life of about 12.3 years. On Earth it is present as a trace element in air and in water; about a billionth of a billionth (10^{-16}%) of natural hydrogen is tritium. The most common forms of tritium are tritium gas (HT) and tritium oxide, also called tritiated water, in analogy with deuterated water. In tritiated water, a tritium atom replaces one of the hydrogen atoms, so its chemical form is HTO, like HDO, rather than H_2O.

The interest in tritium for nuclear fusion lies in the fact that the least difficult and most profitable fusion reaction is the one between deuterium and tritium:

$$D(np) + T(nnp) \rightarrow {}^4He\,(nnpp)\,(3.5\,MeV) + n\,(14.1\,MeV),$$

The peak fusion rate of this process is 10^{25} times larger than the p-p process occurring in the Sun. Comparing it with the solar process whereby a proton fuses on average every billion years or about 10^{16} s, it implies that, if the process were available in the Sun, the latter would have been gone in a flash while releasing a staggering amount of energy, and there would have been no humans, at any rate not on this planet, to tell the tale (see Kirk 2016, p. 1). As can be seen by comparison with the D-D reaction above, both the resulting helium nucleus and neutron are much more energetic, producing in total about 17.6 MeV, which can be converted into heat, making the fusion reaction of deuterium and tritium the most energetically favourable. The neutron, which carries away about 14 MeV, about ten times the energy of the neutrons released in a uranium fission process (see above), is actually so energetic that, as we will see later, it is more a curse than a blessing. Such highly energetic neutrons activate reactor materials, i.e. make them radioactive, severely shortening their lifetime, which constitutes a serious, still largely unsolved, problem for future nuclear fusion reactors.

Tritium was produced for the first time in 1934 by Mark Oliphant (1901–2000) and Ernest Rutherford at the Cavendish laboratory in Cambridge (England). Because little tritium is naturally present, it must be produced artificially for use on a practical scale. Tritium can be made in nuclear reactors that have been specially designed to optimize the generation of tritium and other special nuclear materials. It is produced by neutron absorption of a lithium-6 atom, which was also the method used by Oliphant. This can be done by bombarding lithium-6 (7% of natural lithium is lithium-6, the more common form being lithium-7) with neutrons in a nuclear reactor, where lithium is added to the cooling water in order to regulate the pH of the water. As will be discussed in a future chapter, the hope is that in future the necessary tritium for nuclear fusion reactors can be generated in the reactor itself by embedding the reactor core in a lithium blanket which will absorb neutrons from the fusion process and produce tritium according to the reaction:

$$^6Li + n \rightarrow {}^4He(2.05\,MeV) + T\,(2.75\,MeV).$$

This is an energy producing reaction (exothermic). If ^{7}Li is used instead of ^{6}Li, tritium will also be formed, in combination with ^{4}He and an extra neutron (since the ^{7}Li nucleus has one neutron more than ^{6}Li), but in that case 2.5 MeV of energy is needed for the reaction to occur (endothermic).

The large energy release in deuterium-tritium fusion also results in minimal fuel consumption: the deuterium contained in 1 L of sea water (about 30 mg) and used in deuterium-tritium reactions will produce as much energy as burning 250 L of gasoline (see Ongena and Ogawa 2016, p. 771). The burning of gasoline, a carbohydrate, is a chemical process driven by the outer electrons orbiting the nucleus in an atom. Fossil fuels are either carbohydrates or hydrocarbons, like natural gas. Carbohydrates consist of carbon, hydrogen and oxygen. Hydrocarbons are even simpler, just carbon and hydrogen: a certain number of hydrogen atoms chemically bound with carbon atoms into a molecule. The simplest molecule of this kind is methane CH_4, the next one is called ethane and has 2 carbon atoms and 6 hydrogen atoms (C_2H_6), etc. In the burning process the chemical bonds, binding the atoms into molecules, are broken, releasing energies of the order of 10 eV. This is infinitesimal compared to fusion reactions, which fuse the nuclei of atoms and involve energies of tens of MeV, hence a *million* times more. The enormous potential of nuclear fusion as an energy source can also be seen from the following comparison. To generate 1 gigawatt of energy for one year (which is the typical need for a large industrial city) a conventional power station would have to burn 2.5 million tonnes of coal (causing an environmental pollution of 6 million tonnes of CO_2). A nuclear fission plant would need 150 tonnes of uranium (which would produce several tonnes of highly radioactive waste). A nuclear fusion power station would require just one ton of lithium[11] (to breed the quantity of tritium needed), plus 5 million litres of water and would of course not release any greenhouse gases into the atmosphere.

There is however a huge snag in all of this: the temperature of the deuterium plasma in which the fusion must take place has to be about 150–200 million degrees, more than ten times as hot as the core of the Sun, while the deuterium-tritium reaction is most probable at about 120 million degrees or 10 keV. This temperature is needed to overcome the Coulomb repulsion between the charged nuclei of tritium and deuterium. Temperature alone is however not sufficient. In addition, the density and confinement time of the plasma have to meet some stringent conditions. In the rest of this book we will see how much progress towards this goal has been made.

Other light elements are in principle also candidates for fusion and in principle there are some 100 possible fusion reactions, which are however vastly less probable. We will pay some more attention to them in Chap. 14. In the following we will almost only be concerned with deuterium-deuterium (mostly indicated as D-D) and deuterium-tritium (mostly indicated as D-T) fusion. Although the latter process is the

---

[11]Lithium is a comparatively rare element, although it is found in many rocks and some brines, but always in very low concentrations. There are a fairly large number of both lithium mineral and brine deposits but only comparatively few of them are of actual or potential commercial value. Many are very small, others are too low in grade (*Handbook of Lithium and Natural Calcium*). Although it is widely distributed on Earth, it is no foregone conclusion that there is enough lithium to meet all demands, see Chap. 19.

most favourable of the two, it has been mostly avoided because of the radioactivity of tritium. Only two experiments with tritium have to date been carried out, making the experience gained with this type of fusion fuel, which in the end is supposed to become the favoured fusion fuel, very scant indeed.

Accelerating particles like deuterium nuclei to energies needed for fusion is not difficult to achieve in the laboratory and has been done since the early thirties of the last century. Fusion reactions can be studied by bombarding a stationary tritium-containing target with accelerated deuterium nuclei. Most of these deuterium nuclei will bounce off the hill in Fig. 1.6 and only very few (one in a 100 million or so) will actually fuse with a tritium nucleus, i.e. manage to tunnel through the barrier into the hole. Most nuclei will roll off the hill and be lost. Therefore, that is not the way to achieve energy generation; it only costs energy and quite a lot. A plasma is needed, consisting of deuterium and tritium nuclei, confined in a space in which the nuclei cannot get lost, but move about randomly without losing energy, collide with other particles, bounce off the walls and for a sufficiently high fraction undergo a fusion reaction.

As said, nuclei are equally charged (positive) and hence do not play well together. The Coulomb force acting between them repels them and even at the optimum energy the nuclei are more likely to just scatter elastically than to fuse. Elastic Coulomb scattering, as it is called, is always more likely, but due to quantum mechanical tunnelling particles with sufficient energy can overcome this barrier. This implies that the nuclei have to be confined for a long time compared to the scattering time and that a significant fusion rate requires confinement of the plasma at >100 million degrees.

It is also clear that in order to prevent large heat losses a plasma must be isolated from the walls of the vessel in which it is contained, as no material would be able to withstand such temperatures. A possible solution to tackle this problem was soon clear. Magnetic fields should be used to confine the plasma, making use of the fact that the plasma consists of charged particles. Such particles, apart from creating difficulties by their own magnetic and electric fields, will also be influenced by external magnetic fields and can perhaps be shackled in a straitjacket of such fields. The further history of nuclear fusion is just that: various attempts to create such a straitjacket.

So, the stage has been set, the task is clear, create a D-T plasma of the temperature mentioned above, take care of its confinement so that a 'controlled fusion burn' can start and tap the energy released in this process. This obviously is a huge engineering challenge. Before we elaborate on the history of this challenge, we will first discuss the concept of a plasma, as it seems obvious that before trying to shackle something, you try to understand it, a maxim that is surprisingly not prevalent in the fusion community.

Chapter 2
Plasma

2.1 Introduction

2.1.1 General

At the high temperatures required for D-T and other fusion reactions, the deuterium and tritium are in a gaseous state in which all atoms have been ionised, i.e. the electrons that normally orbit the atom's nucleus have been stripped away, and a mixture of positively charged nuclei and negatively charged electrons remains. This mixture is called a plasma. A plasma is primarily an ionised gas, a gas of positive ions and electrons (although there are usually also some neutral particles present). One of the chief characteristics of a plasma is that it is electrostatically neutral. That means that to a high degree of accuracy the negative charges of the electrons balance the positive charges of the ions, ensuring that in macroscopic volumes of plasma there is only a small net electric charge.

This presence of charged particles makes a plasma very different from a solid, liquid or gas, which do not contain any or very few (partially) ionised atoms. That is why it is sometimes called the fourth state or phase of matter. The term plasma was coined in 1922 by the American physical chemist Irving Langmuir (1881–1957), who was awarded the Nobel Prize in chemistry in 1932 for his work in surface chemistry. He was the inventor of the gas-filled incandescent lamp, an electric light with a wire filament heated to such a high temperature that it glows with visible light, which explains his interest in ionised gases. He called them plasmas as they reminded him of blood plasma, a terminology that is still causing confusion. It was the beginning of the whole new branch of plasma physics, as plasmas turned out to be much more complex than originally thought.[1]

---

[1] In this book we will not discuss the properties of plasmas in any detail, only touch upon some of their properties and characteristics in so far as relevant for the fusion discussion. For a fairly comprehensive and compact survey of fusion plasma physics, see Pease in Dendy (1993), pp. 475–507.

© The Author(s), under exclusive license to Springer Nature Switzerland AG 2021
L. J. Reinders, *The Fairy Tale of Nuclear Fusion*,
https://doi.org/10.1007/978-3-030-64344-7_2

Depending on the circumstances every form of matter can exist in any of the four states mentioned (solid, liquid, gas or plasma), so any gas can be made liquid or solid or turned into a plasma. Heating a solid, it becomes liquid, heating it further it becomes a gas and upon still further heating the gas will be ionised into a plasma. The thermal energy of the particles in the plasma becomes so great that the electrostatic forces that ordinarily bind the electrons to atomic nuclei are overcome, causing the electrons to fly off. Instead of a hot gas (mainly) composed of electrically neutral atoms, we have in a plasma a swirling turbulent mix of oppositely charged particles— electrons and ionised nuclei.

This indicates that plasma physics has its roots, or at least one of its roots, in the physics of gas discharges. A gas discharge is an electric discharge in a gas (the release and transmission of electricity through a gas in an applied electric field) that occurs due to the ionisation of the gas when an electric current flows through it. In the middle of the nineteenth century physicists started to investigate the conductivity of gases. Glass tubes were partially evacuated and had metal plates sealed at the ends. The metal plates were connected to a battery causing a current to flow between the plates. With decreasing gas pressure luminous phenomena could be observed; the tube began to glow. The charged particles in the remaining gas in the tube were accelerated by the applied voltage and when colliding with neutral gas molecules ionised the latter, i.e. stripped them off their electrons. In this collision process the ions themselves gained an electron, returned to a lower energy state and released energy in the form of photons, i.e. emitting light of a characteristic frequency.

The charged products of these processes (ions and electrons) will in turn be accelerated and ionise other neutral gas molecules, causing the ionisation of the gas to grow as an avalanche. This is called the Townsend avalanche or Townsend discharge after the Irish physicist John Townsend (1868–1957). The resulting mixture of ions, electrons and neutral atoms and molecules was what Langmuir saw in his gas-filled lamp and called a plasma.

The second root of plasma physics lies in astrophysics, which is understandable as 99% of matter in the universe is in a plasma state. Early in the twentieth century astrophysicists became interested in cosmic magnetic fields, the magnetic fields of the Sun and other astronomical objects, the origin of sunspots, their magnetic fields, and suchlike.

The physics of gas discharges has been responsible for the experimental side of the study of plasmas, their production in the laboratory, the determination of density, temperature and other properties, while astrophysics has greatly contributed to the development of plasma theory and was also, until nuclear fusion research, the main field of application of plasma physics. It was astrophysics that realised the importance of magnetic fields for plasma physics and discovered a number of important phenomena in this respect, such as plasma waves. In the early days, until the 1950s, there was little contact between the two disciplines of astrophysics and the physics of gas discharges. They formed separate communities, with their own journals and conferences, and developed independently. The fusion-oriented plasma physics was mainly formulated by plasma astrophysicists with, at least in the beginning, little participation by gas discharge physicists. This has resulted in an underestimation of the experimental difficulties (Küppers in Van den Daele et al. 1979, p. 290ff). This underestimation is all too obvious in the history of nuclear fusion, among other things in the rash predictions that are commonly made.

2.1.2 Temperature

At this point it may be useful to briefly discuss the concept of temperature for a mass of particles in a gas. Temperature refers to the thermal motion (velocity) of the particles, and a gas in thermal equilibrium, i.e. at a constant temperature in space and time, has particles of all velocities. This thermal motion of the particles determines the temperature of the gas and the "temperature" of an individual particle is its velocity. The higher the temperature of the gas the more fast particles there are and the less slow particles. They obey a certain probability distribution, which in the simplest case for particle speeds in an idealised gas is a so-called Maxwell-Boltzmann velocity distribution, also called Maxwellian and by mathematicians Gaussian. It is a so-called normal distribution in the form of a bell-shaped curve, a continuous probability distribution that is symmetric around an average value, which corresponds to the temperature of the gas. This is the temperature that is meant when we speak of the temperature of a gas or a plasma. It also implies that the temperature is directly related to the energy of the particles. This energy, as we have seen in Chapter 1, is expressed in eV and an energy of 1 eV is equivalent to a temperature of about 12,000°.

An interesting phenomenon for a plasma consisting of ions and electrons is that it will have several temperatures at the same time, since the velocities of the ions and the electrons are different and obey different Maxwell-Boltzmann distributions with different average temperatures. In due course they will become equal as they collide with each other and transfer energy, but a plasma in general does not live long enough for the two temperatures to equalize. For a certain period of time each species of particle is then in its own thermal equilibrium at its own temperature (see Chen 2016, p. 6).

Another feature about temperature is that high temperature does not necessarily mean a lot of heat. The total amount of heat (the heat capacity) involved in a low-density gas, such as in a fluorescent tube, is not that much, not enough to cause burns or damage to the wall of the tube as the amount of heat that would be transferred in a collision with the wall is rather low. The same is, for instance, true for the ash from a burning cigarette dropping on a hand. Although the temperature is high enough to cause a burn, it actually doesn't as the total amount of heat is insufficient for this. For plasmas too, in spite of temperatures of millions of degrees the heating of the walls is not a serious problem.

2.1.3   Plasma on Earth and Elsewhere

Unlike the solid, liquid and gas phases, under normal conditions the plasma state does not exist freely on the Earth's surface. That does not mean though that all the other phases actually do exist under normal conditions for all matter. Helium gas, for instance, only becomes liquid at a few degrees above absolute zero (-273 °C), a temperature that does not occur under normal circumstances anywhere on the planet.

The plasma phase is different from the other three phases in that on Earth there is no known type of matter which under normal conditions is in the plasma phase, as the temperatures required for forming a plasma are extremely high. On Earth we experience plasmas in short-lived phenomena such as the flash of a lightning bolt, the soft glow of the polar aurorae (e.g. Aurora Borealis) and ionospheric lighting, but also in the light of a fluorescent tube and the pixels of a plasma TV screen, which are etched with plasmas. These are not the fully ionised plasmas needed for fusion, but are partially ionised, relatively low-temperature plasmas of about 4 eV (about 50,000°). Lamps commonly involve such low-temperature plasmas, like LED lighting, light-emitting diodes which contain solid-state plasmas.[2] Other applications are plasma etching and deposition to make semiconductors in electronic devices, and windows glazed with plasmas to transmit or reflect specific wavelengths (Chen 2016, p. 355). Figure 2.1 pictures a variety of plasmas and their temperature as a function of particle density.

Beyond the Earth's atmosphere the magnetosphere is a plasma system formed by the interaction of the Earth's magnetic field with the solar wind, the stream of charged particles released from the upper atmosphere of the Sun, the corona. Further in the universe the most common phase of matter is plasma. All matter of interstellar space is plasma, although at a very low density, while the interiors of all stars are actually made out of very dense plasma, to which we thank our existence. Without it, there would perhaps be a planet like Earth, but without the energy of the Sun, which comes to us every day, the emergence of life on this planet would not have been possible.

2.1.4   Physics of Plasmas

In physics terms a plasma is an ionised gas in which at least one of the electrons of an atom has been stripped free, leaving a positively charged ion. In any gas there is always some small degree of ionisation, but this does not mean that any ionised gas can be called a plasma. An atom in a gas can be ionised by suffering a collision of high enough energy that causes an electron to be knocked out. For nitrogen, the main

---

[2]This seemingly contradictory term is applied by convention in physics to collections of mobile charged particles in solid conductors (conduction electrons in metals or electrons and holes in semiconductors) under such conditions that the properties of the collection are close to those of a plasma.

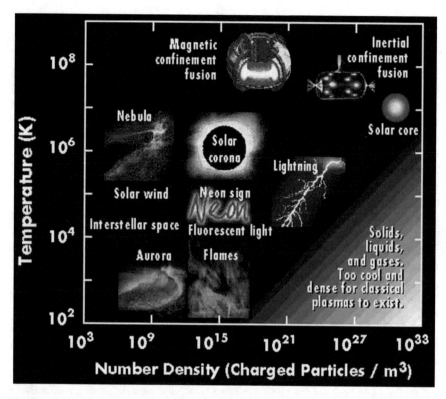

Fig. 2.1 Types of plasma on Earth, including some fusion plasmas, and in the universe (*Image copyright* Contemporary Physics Education Project; www.plasma-universe.com)

component of air, about 14.5 eV is needed for this. We have seen earlier that 1 eV corresponds to about 12,000°, and at that temperature very few particles will have enough energy to knock an electron out of a nitrogen atom. Consequently, for air at room temperature the degree of ionisation is completely negligible. Other atoms, e.g. alkali metals, such as potassium, sodium and caesium, possess much lower ionisation energies, and plasmas may be produced from them already at temperatures of about 3000°. To get appreciable ionisation in air, temperatures of around 100,000° are required, and much higher still for a plasma to be formed. In order for the nuclei in a plasma to undergo fusion with a probability that can keep a fusion reaction going, the plasma temperature must be at least 100 million degrees. For the mere existence of a plasma such high temperatures are not necessary. It is sufficient to have temperatures at which the atoms in the gas break apart into a gas of charged particles (electrons and nuclei) and to sustain this ionisation. Extremely high temperatures are not needed for this. So, plasmas exist in an enormous temperature range from a few thousand degrees to 100 million degrees. It seems obvious that the properties of such plasmas can also differ enormously between such extremes.

For the purpose of this book it is sufficient to consider a plasma as an electrically neutral medium of unbound positively and negatively charged particles, respectively ions and electrons, that move about violently as the particles have a high thermal energy, bump into one another and feel each other's electric and magnetic fields. It is best compared to a boiling hot "soup" of negatively charged electrons and positively charged nuclei. The overall charge of a plasma is roughly zero, but the ionised gaseous substance forming the plasma becomes highly electrically conductive and long-range electric and magnetic fields dominate the behaviour of the matter. In ordinary air the molecules are neutral in charge and do not experience any net electromagnetic forces. The molecule moves undisturbed until it collides with another molecule. In a plasma with its charged particles the situation is totally different. As the particles move around, local concentrations of positively or negatively charged particles can spontaneously be generated, giving rise to electric fields. Motion of charges generates currents and hence magnetic fields, as a current always generates a magnetic field. These fields act on particles far away. Therefore, in a plasma many charged particles interact with each other by long range forces and various collective movements occur, giving rise to many kinds of instabilities and wave phenomena. The plasma can exhibit collective behaviour or coherent motion, such as waves of charged particles that propagate through the plasma without damping or amplification.

At 1 eV (about 12,000°) the electrical conductivity of hydrogen plasma, a plasma consisting of hydrogen nuclei (protons) and electrons, is about equal to that of copper, and at the high temperatures required for fusion it is much higher still. This conductivity is one of the most notable features of a plasma. The conductivity is so great that within the main body of the plasma externally applied electric fields are effectively cancelled by the currents they induce. If two charged balls connected to a battery are inserted inside a neutral medium, an electric potential will arise between the balls. In a plasma with its mix of oppositely charged particles (ions and electrons) the balls will attract these particles. One ball will attract the positive ions and the other the negative electrons, such that the charges on the balls will be shielded off and no electric field will be present in the plasma between the two charged balls (Chen 2016, pp. 2–8).

The fact that this hot gas consists of charged instead of neutral particles is both a blessing and a curse. It is a blessing as charged particles can be influenced and controlled, so it is hoped, by magnetic fields so that the plasma can be confined long enough to allow fusion reactions to occur. It is a curse as the particles themselves also produce electric and magnetic fields, which makes modelling of plasma behaviour extremely complicated.

The interiors of stars like the Sun consist of a fully ionised hydrogen plasma with all the electrons stripped off the atoms; the bare nuclei (protons) and electrons swirl about and electric fields are rampant everywhere. The plasma in the Sun's core (which extends to about a quarter of the Sun's radius, i.e. 175,000 km) is very dense (ten times the density of lead). Its temperature is about 15 million degrees Celsius and the pressure at the centre is estimated at 265 billion bar. It is held together by the vast gravitational forces created by the Sun's mass, which is 300,000 times that of the Earth. Consequently, due to the relatively small mass of a planet like Earth, this force is far too weak to do the same job in terrestrial circumstances. Even in the Sun gravitation is not a perfect confining force as a continuous stream of particles is escaping from the Sun's inner core. Perhaps that should be a warning: if even the colossal gravitational forces in the Sun cannot confine a plasma, how on earth can we be so arrogant as to think that we can do this with some puny magnetic field? It defines the next and main problem for any viable fusion reactor, to be discussed in the next chapters. How to maintain the plasma at the high temperature required and confine it in a vessel or any other contraption that might be able to do that job? Secondly, as we will see, any plasma that will be created on Earth in nuclear fusion machines will have vastly different properties from the plasma in the Sun's interior. The Sun's core is in some sort of dynamic equilibrium with the gravitational forces acting on it. The gravitational pressure on the core is resisted by a gradual increase in the rate at which fusion occurs. This process speeds up over time as the core gradually becomes denser. It is estimated that the Sun has become 30% brighter in the last four and a half billion years.[3] So the common references in nuclear fusion stories to the Sun or other stars, cf. the claim in one of the ITER brochures that "a star is born",[4] are far from true and, if anything, rather misleading and to some extent annoying as they present the wrong picture. We will never (be able to) create a "piece of the Sun" here on Earth. ITER (the International Thermonuclear Experimental Reactor, currently being constructed in the South of France) will not be a star. Just read a book on the Sun, like the one by Lucie Green referred to above, and it will soon be clear that all these references to the Sun are just a load of baloney.

2.2   Plasma in Nuclear Fusion Devices

Research in plasma physics really got going in the early 1950s when proposals were put forward to harness the power of nuclear fusion. At that early stage all proposals involved the application of magnetic fields in order to try to contain the plasma. The second approach to fusion, *inertial confinement fusion*, only became a viable option after the development of lasers in the early 1960s. Inertial confinement has a quite independent history as it always was and still is closely connected to weapons research. It will be dealt with in some detail in a later chapter. In this and the following chapters we will mostly discuss *magnetic confinement fusion*, the use of magnetic fields to contain the plasma.

Once the plasma has been achieved, the next problem is how to maintain it, prevent it from losing energy and heat it in order to increase the likelihood of fusion. For making fusion a reality, we need enough plasma at a high enough temperature that holds its heat for long enough. The plasma however loses energy in various ways and energy is needed to heat it up. The energy required for all this must be less than the

---

[3]http://faculty.wcas.northwestern.edu/~infocom/The%20Website/evolution.html.

[4]A star is born, accessed 28 April 2020, https://www.iter.org/doc/www/content/com/Lists/list_i tems/Attachments/764/Progress_in_Pictures_Dec2017.pdf.

energy that eventually is produced, as there clearly would be little sense and interest in a fusion power plant that produces less energy than needed to operate it. This energy balance must be positive and has come to be denoted by Q: the ratio of the energy gained to the energy lost. This energy balance is a seemingly simple quantity (just the ratio of what goes in to what goes out), but, as we will see in one of the future chapters, fusion scientists have managed to make it a very complicated and muddled business by defining all kinds of Q.

2.2.1 The Lawson Criterion

In the mid-1950s the English physicist John D. Lawson (1923–2008) derived the so-called Lawson criterion (Lawson 1957). It defines the conditions needed for a fusion reactor to reach *ignition*, i.e. the situation in which the heating of the plasma by the products of the fusion reactions (in particular the α-particles (see Chap. 1)) is sufficient to maintain the temperature of the plasma against all losses without any external power input being necessary, in other words to keep the engine running. In that connection we also speak of a *burning* plasma, which is a plasma in which the heat from the fusion reactions is almost or fully sufficient and produced for a sufficiently long time to maintain the temperature in the plasma. Once the fusion has been set in motion, enough energy is produced both for keeping the plasma at the required temperature and for generating energy.

Lawson calculated the requirements for this to happen. He came up with a dependence on three quantities: plasma temperature (T), discussed above, plasma density (n) and confinement time (τ). The Lawson criterion gives a minimum value for the product of these three quantities (called the triple product) required for the fusion reaction to become self-sustaining. The plasma density (usually denoted by n) is the number of fuel ions (nuclei of deuterium and tritium that can fuse) per cubic metre. The *energy confinement time*, usually denoted by the Greek letter τ with subscript E (τ_E), measures the rate at which the system loses energy to its environment, and is defined as the total amount of energy in the plasma divided by the power supplied to heat the plasma:

$$\tau_E = \frac{energy\ in\ plasma}{power\ supplied\ to\ heat\ plasma}$$

It measures how well the magnetic field insulates the plasma. An analogy of this quantity is the rate at which a house cools down when the heating is switched off. If that rate is high, you don't want to be confined long in the house. A plasma loses energy in various ways, some of it by conduction (the transfer of energy by collisions and movement of particles). A second unavoidable loss of energy is heat radiation in the form of X-rays, a consequence of collisions of electrons and ions. The magnetic fields have no effect on such X-rays. For a plasma of pure deuterium and tritium the power of such X-ray radiation is rather small and can be neglected at the temperatures of 100 million degrees we are after. However, as Lawson stressed in his original paper, up to at least 40 million degrees the energy lost from the plasma by radiation is always larger than the amount of heating available from fusion reactions. This implies that some other means of heating remains necessary to reach the conditions for ignition. A further source of unavoidable heat loss is microwave radiation created by the motion of electrons in the magnetic field. Most of such microwave radiation will however be reflected from the vessel walls and be reabsorbed by the plasma.

Energy must be confined long enough for the plasma to reach the temperature of the order of 100 million degrees and the ions must be confined long enough for a considerable fraction to fuse, so it is obvious that the larger τ_E, the more effective a fusion reactor will be as a net source of power. The ideal situation is that no extra power is needed to supply heat to the plasma, in which case τ_E will become infinite and the plasma will burn until all fuel has been consumed.

The general idea is that outside energy is at first used to bring the plasma at the required temperature, and once the fusion reactions get going the α-particles (helium-4 nuclei) produced in these reactions provide more and more of the energy needed to keep the plasma at the right temperature. The Lawson criterion or, the virtually identical, ignition condition states that the "fusion triple product" $nT\,\tau_E$ of density, energy confinement time and plasma temperature must fulfil the condition:

$$nT\tau_E > 3 \times 10^{21} \mathrm{m}^{-3}\,\mathrm{keV\,s},$$

in units of *particles per cubic metre* $\times$ *kiloelectronvolts* $\times$ *seconds*. This is the most important condition in fusion. A more transparent way to express it is to use different units. The product of density and temperature is the pressure of the plasma. The ignition condition then becomes *pressure* $\times$ *energy confinement time* must be greater than 5, in units of *bars* $\times$ *seconds*, bar being the unit of pressure.[5] It implies that a confinement time of 5 s at a pressure of one bar, about the atmospheric pressure on Earth, would be sufficient for reaching ignition (McCracken and Stott 2013, pp. 38–43). As an illustration to show how difficult it is to satisfy the Lawson criterion we note that the biggest fusion machine that has so far been operating, the Joint European Torus (JET) at Culham (UK), achieved around one fifth of this value in its D-T experiments in 1997 (see Chap. 8).

The best route to ignition for a D-T plasma turns out to be at around 10–20 keV (100–200 million degrees). The Lawson condition then gives (with a confinement time of 5 s and a temperature of 10 keV) that the density of the plasma should be at least about 10^{20} particles per cubic metre. This is rather low: a factor of 100,000 less than the density of air, the number of particles in a cubic metre of dry air being about 2.5×10^{25} particles/m^3. If the density were much higher, the outward pressure would also be higher, and it would be more difficult to confine the plasma. On the other hand, if the density is too low, the rate at which fusion reactions would take place and energy can be generated will be too small to be of practical interest. All this

---

[5] 1 bar = 10^5 N/m^2 = 10^5 kg/(m s^2), while 1 keV = 1.6×10^{-16} kg m^2/s^2. A little unit juggling gives the result quoted in the text. The number 5 results from $3 \times 1.6 = 4.8 \approx 5$.

applies for fusion by magnetic confinement. As we will see in Chap. 12, for inertial confinement the confinement times are much smaller (a few tens of a billionth of a second) and consequently, in order to meet the Lawson condition, the density must be much higher (10^{31} nuclei per cubic metre, a factor of about one million higher than the density of air).

The fusion reactions also cause the plasma composition to change appreciably as the products of the reaction (D + T → ^{4}He (3.5 MeV) + n (14.1 MeV)) spread through the plasma. The neutrons, being neutral in charge, but highly energetic, are not affected by the magnetic fields and fly off. They must be absorbed somewhere, and their energy be used for generating electricity, the eventual aim of any power plant. The much less energetic α-particles (^{4}He) carry about 20% of the surplus energy of the reaction. Since they are positively charged, they are trapped by the magnetic field and stay in the plasma. They can be used to further heat the plasma or keep it at the right temperature. But they are also spent fuel, 'waste' products of the fusion processes, and should not become a major fraction of the plasma as it would reduce the thermonuclear reaction rate. This is also happening in the Sun where hydrogen nuclei fuse (via the three-step process explained in Chap. 1) into helium. The helium 'ashes' are left behind and eventually the core becomes so depleted of fuel that the energy production will start to fall.

So, let us summarise what is needed to get controlled fusion going and turn it into a useful energy source (Küppers in Van de Daele et al. 1979, p. 296):

1. a plasma at a temperature of a couple of hundred million degrees;
2. with a density that is not too high (about 1/10,000 of atmospheric density) to withstand thermodynamic pressures at such temperatures and to keep the fusion process under control;
3. a plasma that is very pure (i.e. consists only of deuterium and tritium), since radiation losses strongly increase with impurities (especially metallic impurities from the vacuum vessel wall have high atomic number Z and radiate much more than deuterium and tritium, which have low Z);
4. and that is confined long enough by a magnetic field (the magnetic pressure to compensate the thermodynamic pressure) for a considerable fraction of the deuterium-tritium ions to fuse and net energy gain to become possible; so that
5. after the plasma starts to burn, more fusion energy is produced per unit of time than the total of all losses (radiation, particle loss) and the process will sustain itself;
6. which still leaves the problem to extract the fusion energy from the reactor vessel and turn it into electricity.

Progress towards ignition is being made, but so far this goal has not been reached. Seventy years after Lawson formulated the condition discussed here and in spite of frequent promises by scientists and leaders of experiments, no fusion experiment has reached a situation in which the α-particle heating balances the energy losses, i.e. the point at which a nuclear fusion reaction becomes self-sustaining. Of course, outside factors have been identified on which to shift the blame for these failures. The easy option of lack of funding, foolishly withheld by short-sighted politicians who did not fall head over heels for the fanciful stories of the fusion aficionados, is most prominent among them. See e.g. Dean 2013, who describes the battles for funding with various US administrations in the 1980s and 1990s.

2.2.2 Plasma Heating

Let us now discuss how a plasma can be heated to the temperature at which fusion reactions can start to happen. A plasma can be formed by sending a current through a gas. It will heat up the gas and strip the atoms of their electrons. This type of heating is called *ohmic* heating, *joule* heating or resistance heating, the process by which an electric current through a conductor produces heat due to the resistance of the matter the conductor is made of. A copper wire for instance becomes hot when a current is passed through it, because the electrons in the wire collide with the much heavier ions and transfer heat to them. As we know from secondary school, the amount of heat is proportional to the product of the wire's resistance R and the square of the current I, i.e. proportional to I^2R. In a plasma the number of collisions between ions and electrons is much smaller than in a wire, up to 10 orders of magnitude (10 billion times) smaller. On the other hand, the currents that can be driven through a plasma can be very large, more than 100,000 amperes, even many million amperes.

The first plasma was created by this method in 1947 by two doctoral students, Stan Cousins and Alan Ware, at Imperial College in London in a precursor of the magnetic confinement machines in use today, a doughnut-shaped glass vacuum vessel (Cousins and Ware 1951). They were unable to measure the temperature of the plasma and only saw a bright flash of light in the gas, so their plasma was highly unstable and lasted only for fractions of a second (Herman 1990, p. 40). As the plasma gets hotter and hotter, its electric resistance decreases (as noted above the conductivity of a plasma is one of its most notable features; a plasma is almost a superconductor) and this type of heating becomes less effective.[6] The maximum temperature that can be reached with ohmic heating is less than 50 million degrees, very hot by almost any standards, but insufficient for fusion. Ohmic heating is also limited by the fact that increasing the plasma current beyond a certain limit tends to disrupt the plasma.

Two main heating techniques have been developed to go beyond ohmic heating.

(1) The first one uses powerful beams of highly energetic neutral particles (deuterium atoms or a mixture of deuterium and tritium atoms) that are injected into the plasma. This is called neutral-beam heating. It involves a rather convoluted process whereby the deuterium atoms are first ionized (so the atoms are stripped off their electrons and consequently become positively charged) and then accelerated to high energy by passing them through an electric field. The resulting beam of energetic charged particles cannot be directly injected into the plasma as the particles would be deflected by the magnetic field and harmful

---

[6]The electrical resistance in a plasma is approximately given by the so-called Spitzer conductivity (see Glossary).

plasma instabilities would arise. So, they first have to be (re)neutralised by passing them through a deuterium gas from which they pick up electrons to recombine into neutral atoms. Then they are injected and, since they are now neutral but still very energetic, can penetrate deep into the plasma, collide with many plasma particles and give up some of their energy, further heating the plasma. In the process they are again ionized and become part of the plasma (McCracken and Stott 2013, p. 102). For this method to be useful, the injected particles must have very high velocity in order to penetrate into the centre of the plasma. If they are too slow, they would give up all or most of their energy before getting to the centre. At JET the particles in the neutral beam travel at more than 3,000 km/s, more than 10 million km per hour.

(2) The second approach works rather like the heating of food in a microwave oven. Electromagnetic waves (typically radio waves) are launched into the plasma at different frequencies. It is important to tune the frequency of these waves to some natural resonance frequency of the ions and electrons in the magnetic field holding the plasma together, such as the frequency associated with the motion of the charged electrons or ions in the magnetic field. These resonance frequencies are vastly different for ions and electrons at respectively about 50 MHz and 170 GHz.[7] At such frequencies the energy carried by these waves is transferred to the charged particles, increasing their velocity, and at the same time their temperature and the temperature of the plasma as a whole. The radio waves, typically of wavelengths of several metres in free space (for ions) and a few mm (for electrons), are launched from an antenna or waveguide at the edge of the plasma (Cairns in Dendy 1993, p. 391; McCracken and Stott 2013, pp. 103–104). Such heating is called ion cyclotron resonance heating (ICRH), lower hybrid (LH) (resonance) heating, and electron cyclotron resonance heating (ECRH). In LH heating use is made of longitudinal oscillations of ions and electrons in the magnetised plasma. This type of heating is also called LHCD (lower hybrid current drive).

The use of these techniques was no plain sailing. The temperature rise was less than expected, as energy losses also increased when increasing the heating power. It turned out that the situation was like "a house with central heating where the windows open more and more as the heating is turned up. The room can still be heated, but it takes more energy than if the windows stay closed" (McCracken and Stott 2013, p. 104). As we will see later in this book, a machine like ITER will use a combination of these heating techniques: neutral beam injection and two sources of high-frequency electromagnetic waves, with antennae the size of a bus.[8]

---

[7] https://www.iter.org/mach/Heating.
[8] https://www.iter.org/newsline/225/1193.

2.2.3 *Lorentz Force and Particle Trajectory in a Magnetic Field*

In most of the devices constructed to achieve nuclear fusion, external magnetic fields are applied to confine the hot plasma in a certain region, preventing the particles from making contact with the vessel, as this would rapidly cool the plasma down. In this section we will consider the motion of an individual charged particle in a magnetic field. This is eventually the level of detail needed to be able to make reliable models of a plasma. A charged particle in a magnetic field is subject to a force, called the Lorentz force, after the Dutch physicist Hendrik Anton Lorentz (1853–1928). This force has four main properties: (1) it only acts on charged particles, in opposite directions for negative and positive charges (the force is proportional to the charge of the particle); (2) it only acts on moving particles (the force is also proportional to the velocity of the particle); (3) the force is proportional to the strength of the magnetic field; and (4) the action of the force is perpendicular to the velocity of the particle *and* to the direction of the magnetic field. In a uniform field with no additional forces, a charged particle will gyrate around the magnetic field according to the component of its velocity that is perpendicular to the magnetic field, and drift parallel to the field according to its initial parallel velocity. The result is that the particle describes a helical (corkscrew) motion around a guiding centre in the direction of the magnetic field (Fig. 2.2), which is the basic motion of any charged particle in a magnetically confined plasma. It can be treated as the superposition of the relatively fast circular motion around the guiding centre and the relatively slow parallel drift of this point.

If the speed of the particle is constant, the force is directed inwards and has everywhere the same strength; the particle orbits are circles and the radius ρ is called the gyro-radius or Larmor radius, after Joseph Larmor (1857–1942), an Irish physicist and mathematician, and a contemporary of Lorentz. In a fusion reactor a deuterium nucleus will typically have a gyro-radius of 1 cm, while the plasma itself will have a radius of about one metre. It will complete about 30 orbits per microsecond. An electron with the same energy would have a much smaller radius than a deuterium nucleus. As it is much less massive it will move much faster, which at first sight would increase its radius. The Lorentz force (being proportional to the velocity) is

Fig. 2.2 Corkscrew motion of a charged particle in a magnetic field in de R direction (guiding centre); ρ, the radius of the particle's motion, is called the gyro-radius or Larmor radius

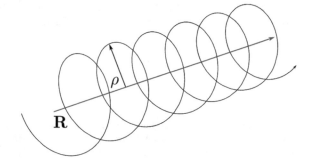

however much stronger for an electron and the result is that the electron gyro-radius is about a factor 60 smaller (the square root of the mass ratio of deuterium and an electron).[9] The gyro-radius is proportional to the particle's mass and velocity, while it is inversely proportional to the particle's electric charge and the magnetic field strength. So, the larger the mass and/or velocity the bigger the circle, and the larger the charge and/or magnetic field the smaller the circle, which implies that increasing the field strength will squeeze the particles around the magnetic field lines.

If wire coils are wound in a helix around a vacuum glass tube and a current is sent through the wire, a magnetic field will be created inside the tube. A charged particle travelling from one end of the tube to the other would then describe such helical motion and be prevented from touching the walls of the tube, if it does not collide with any of the (neutral) particles left behind in the tube. With the proper configuration of magnetic fields this can also be done for a plasma, which is then suspended as it were in "mid air", in this case a vacuum vessel. In this way it can be prevented from touching the walls of the vessel and hopefully from losing too much energy. It was realised from the very beginning of nuclear fusion research that this is *the* essential and fundamental problem this research must solve, and its complexity was grossly underestimated. It is clear that a tube, which is open at both ends, is not suitable as the particles would eventually escape at the ends and fly off. We will see in future chapters which "magnetic bottles" have been proposed for letting the particles go around and around.

However, the magnetic fields themselves add another considerable complication. Was a plasma in itself already an extremely complex system, now we have to face the added complications of external magnetic fields and changes in the composition of the plasma due to the fusion reactions. Energy is stored in the confined plasma, both in the confining magnetic field and in the energy of the plasma particles. As we have already noted above, several types of collective motion (waves) can propagate in a plasma. Under certain conditions they become unstable and such instabilities grow by extracting the energy stored in the plasma. To build proper theories of these instabilities nonlinear analysis is needed, which, as can be imagined, poses exceedingly difficult, almost insurmountable problems (Stringer in Dendy 1993, p. 369). All physical processes are nonlinear to a certain extent, but normally the nonlinearity is rather small. One can then start with a linear analysis and try to calculate nonlinear effects as small perturbations. Such an approach is no longer valid for the turbulent behaviour in a plasma, which is nonlinear in the extreme, a situation in which fluctuations can no longer be treated as small perturbations.

---

[9]Chen (2011), p. 192. A deuterium nucleus consists of two nucleons (a proton and a neutron) and the mass of a nucleon is about 2000 times the mass of an electron. So, the mass of a deuterium nucleus is about 4000 times the mass of an electron (3670 to be exact) and the square root of this number is about 60.

2.2.4 Bohm Diffusion

The first person to throw a spanner in the fusion works was the American physicist David Bohm (1917–1992), a man treated poorly by his country and essentially chased into exile in the early 1950s during the McCarthyism witch-hunts, eventually ending up at the University of London's delightful Birkbeck College. During the war Bohm did some experimental research on the diffusion of plasma particles across a magnetic field. The results of this research, which proved useful for the Manhattan Project, were immediately declared classified and, since Bohm was refused security clearance, it created the preposterous situation that he no longer had access to his own research.

In the case of a plasma in a magnetic field large numbers of particles travel along the corkscrew paths of Fig. 2.2. Depending on the design of the device, the paths they travel can be more or less parallel to each other and particles travelling along adjacent paths may collide and scatter when their electric fields influence each other, resulting eventually in particles diffusing outside the magnetic field, instead of steadfastly continuing to move along a certain path. From experiment Bohm had found that the rate of diffusion was increasing linearly with temperature (increasing the temperature increases the energy of the particles and gives them more room for manoeuvring), and inversely linearly with the strength of the magnetic field applied to the plasma, so increasing the magnetic field would decrease the diffusion (tying the particles tighter to the magnetic field line). This inverse linearity was however much weaker than predicted by classical diffusion, which is inversely proportional to the square of the magnetic field strength. If the classical model is correct, small increases in the field strength would result in much longer confinement times and would offset a possible increase of the diffusion by a temperature raise. If diffusion à la Bohm were correct, confinement by magnetic fields would not be practical. Early fusion machines appeared to behave in accordance with the Bohm model and this so-called Bohm diffusion haunted fusion research for a number of years. It took until the late 1960s before the situation was resolved, when the geometry of the Russian designed tokamak, which will be discussed in extensive detail in the rest of this book, as it is currently by far the most applied geometry for fusion, was shown to be able to overcome the Bohm restriction. It took a while to realise this and throughout the early 1960s the atmosphere in the fusion community was rather pessimistic with many researchers having reached the conclusion that the magnetic field would always destroy the confinement of the hot plasma (Sagdeev 1994, p. 64).

2.2.5 Plasma Modelling

Plasma modelling is necessary and important in order to try to understand the fundamental processes in a plasma. Without such knowledge it will undoubtedly be impossible to ever have a reliably working nuclear fusion reactor. Plasma modelling concerns solving equations of motion that describe the behaviour of the plasma. It will be clear that due to the turbulent motion of the particles in a plasma this is not an easy matter. Neither is there is a single preferred way of modelling for all plasmas and for all circumstances.

We have seen that one of the roots of plasma physics lies in the physics of gas discharges, which was mainly experimental physics. A theoretical explanation of the observed phenomena started from the equations of motion of individual charged particles in external electric fields, at first without considering the effect of external magnetic fields. In the 1920s collective effects (waves of organised motion of plasma particles) were also studied, e.g. plasma oscillations were discovered in 1926. Such collective behaviour is typical for plasmas and a plasma is often defined as a quasi-neutral gas of charged and neutral particles that shows collective behaviour. The word quasi-neutral indicates that locally concentrations of charges can arise, and the collective behaviour refers to the waves and other phenomena of collective behaviour that a plasma can exhibit.

The astrophysicists also needed a theory for the motion of electrically conducting fluids or ionised gases (plasmas) in external magnetic fields. This gave birth to a theory called magnetohydrodynamics (MHD), the study of the magnetic properties and behaviour of electrically conducting fluids, which connected hydrodynamics (gas and fluid dynamics) with electrodynamics. The pioneer of MHD was the Swedish physicist Hannes Alfvén (1908–1995), who developed this theory as early as 1942 and won the 1970 Nobel Prize in physics for this work. In the context of MHD a plasma is described as a magnetohydrodynamic fluid; it is a macroscopic theory that leaves the motion of individual particles out of consideration.

On the other hand, the application of the principles of kinetic gas theory to ionised gases led to a plasma theory that also described microscopic phenomena. The kinetic theory of gases is a simple model of the thermodynamic behaviour of gases and describes a gas as a large number of identical sub-microscopic particles, all of which are in constant, rapid, random motion. An energy (mass × square of the velocity) is associated with that motion. The model assumes that the particles are very small compared to the average distance between the particles. All collisions of particles with each other and with the enclosing walls of the container are assumed to be perfectly elastic. The basic version of the model describes an ideal gas and assumes that the particles have no attractive or repulsive forces between them.

Nowadays there are several types of plasma models, based on a combination of the two basic approaches mentioned above, the main ones being single particle models, kinetic models, single and two fluid models, kinetic/fluid hybrid models, gyrokinetic models and models that treat the plasma as a system of many particles. In a kinetic description a plasma is divided into species consisting of a collection of particles with various velocities, while in a fluid description a continuous medium characterized by macroscopic quantities such as density and temperature is used. In addition, in recent decades numerical simulation of plasmas has become an indispensable addition to plasma diagnostics.

To model the behaviour of gases the ideal gas concept is widely used. In it, point-like particles move randomly and only experience perfectly elastic collisions. A normal ideal gas consists of neutral (non-ionized) particles, so if this concept is to be used for plasma modelling electrostatics has to be added. In this respect the most simple picture for a plasma would be to treat it as an ideal gas with infinite electrical conductivity (the conductivity of a gas is normally very small; air for instance is under normal circumstances an excellent insulator, but as noted before electrical conductivity is one of the most notable features of a plasma). In the early days of nuclear fusion research, it was thought that such an approach could be valid for high-temperature plasmas in a nuclear fusion environment. The ideal gas picture seems however to be contradictory to the whole fusion idea, as elastic collisions can never lead to fusion. A plasma is physically much more complicated than an ideal gas, certainly when an external magnetic field is applied as is the case in magnetic confinement devices.

Also in other ways plasmas are fundamentally different from gases. In a gas all particles move in a similar way influenced by gravity and by collisions with other particles. Two-particle collisions are the rule and there will be very few relatively fast particles. In a plasma negatively charged electrons, positively charged nuclei and neutral neutrons behave independently in many circumstances, giving rise to new phenomena like waves (small amplitude oscillations) and instabilities. Such spontaneously arising organised motion is very important in a plasma and there is in general a significant population of unusually fast particles. The velocity distribution of the particles in a plasma will differ substantially from the Maxwellian velocity distribution of particles in a gas. And so it should, for a system with a Maxwellian velocity distribution would be devoid of any interesting physics. Where a gas could be viewed as an old people's home, a plasma is more like a crowded playground of young kids; both are populated by humans, but the dynamics is rather different. A model based on the former and applied to the latter would probably not produce much useful information.

It is therefore more useful to treat a plasma as a conducting fluid, although its density is much lower, less even than the density of air. The turbulent phenomena in plasmas are however similar to turbulence in fluids (a pattern of motion characterised by chaotic changes in pressure and flow velocity). Turbulence is a very common phenomenon, also in daily life, e.g. smoke rising from a chimney, fast flowing rivers, surf waves, and has been studied for more than 100 years, albeit without a clear understanding, both theoretically and conceptually, of ordinary fluid turbulence. Plasma turbulence is even more complicated than fluid turbulence and it is therefore not surprising that a generally accepted theory of plasma turbulence is still lacking (Haas in Dendy 1993, p. 103). Nonetheless, plasma problems are often handled using a fluid model, describing the plasma by a modified set of fluid equations as it can give a better description of the cooperative motion, like the waves mentioned above (Elliott in Dendy 1993, pp. 29–30). Any modelling that is applied always tries to simplify the situation in one way or another by various approximations (for instance, assuming the medium to be continuous or distributions to be (locally) uniform or averaged over) which causes that effects due to non-equilibrium distribution functions are lost.

Since the particles in a plasma are also charged, in the fluid approach the mathematics governing fluids is combined with the mathematics governing electromagnetism. The resulting equations are coupled partial differential equations. In many situations, the equations cannot be solved directly and computational codes are used to get solutions. As mentioned above, astrophysics came to the rescue here with MHD, which treats the plasma like a single, continuous, conducting fluid. If electric and heat conductivity as well as viscosity can be neglected, the model is called ideal magnetohydrodynamics (*ideal MHD*), still the most commonly studied model. For the case that the fluid cannot be considered a perfect conductor (i.e. zero resistance), but the other conditions for ideal MHD are satisfied, an extended model called *resistive MHD* is used. These models have been used by fusion researchers to help find ways to control a plasma and to build a stable plasma structure.

A further refinement is the two-fluid model in which the electrons and ions (the latter being thousands of times bigger and heavier than the electrons) are treated as separate, charged fluids: an ion fluid and an electron fluid. The plasma is treated as two overlapping fluids, which only interact through their electric and magnetic fields, but the masses flow past each other without touching. It will be clear that in this case the equations become more complicated, and that solving them takes vastly more time, but can still be done by hand. A two-fluid model will capture more effects than a single fluid model, and does a good job mimicking effects due to high concentrations of one kind of plasma, for instance when the ions clump together in a drift wave.

A limitation of MHD (and other fluid theories) is their dependence on the assumption that the plasma is strongly collisional, so that the time scale of collisions is shorter than the other characteristic times in the system, and the particle distributions are Maxwellian. This is usually not the case in fusion plasmas, and it may be necessary to use a kinetic model which properly accounts for such non-Maxwellian distributions. However, because MHD is relatively simple and captures many of the important properties of plasma dynamics, it is often qualitatively accurate and is therefore often the first model tried.

Beyond the two-fluid model computers are needed to solve the equations. In such approaches more details can be accounted for, as single particles, or the fields they generate, can be tracked by computer programs. The reactor or the plasma is subdivided into small cells in which the particles move (particle-in-cell or PIC method). Today numerical simulations of such cellular models form the common approach to plasma modelling. It has been pioneered at Los Alamos in the 1960s. Within each cell, for a given point in time the computer figures out the force that acts on that particular particle, based on its local magnetic and electric environment. The fields acting on the charged particles create a Lorentz force, so inside these little cells, the most common motion of a particle is the corkscrew motion we have seen before. This is fed into Newton's equations of motion, which are then solved. Once these equations have been solved, a step in time is taken, and the computer repeats the whole process for each cell. In this way the motion of the plasma is uncovered. It will be clear that it requires vast amounts of computing time, even on supercomputers, to uncover all this detail and no computer simulation has so far been able to track

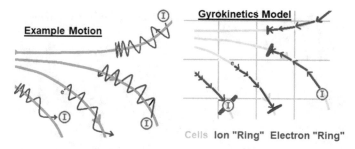

Fig. 2.3 Approximation of the helical motion of charged particles in a magnetic field by ring-like objects moving in straight lines in gyrokinetics

every individual particle in a plasma. Various simplifications have been made to keep the problem under control. The object that forms the basis of the simulation can be simplified particle motion, a so-called macro-particle (representing 1000 or more real ions or electrons) or a blob of matter in a cell.

One of the approaches goes under the name of gyrokinetics in which the corkscrew trajectory of a particle (Fig. 2.2) is decomposed into the relatively slow drift of the guiding centre along the magnetic field line and the fast circular motion, called the gyromotion, around this centre. For most plasma behaviour, this gyromotion is irrelevant and can be averaged over, which reduces the number of dimensions by one (six instead of seven). The particle is then represented by a little ring, with radius equal to the gyro-radius, that moves in a straight line, see Fig. 2.3. Gyrokinetics works and is simple enough to run on desktop computers, but it runs into problems over long periods of time. As you run the simulation longer, the code will stray further and further from reality.

Gyrokinetics is one of the approaches known under the name of kinetic plasma modelling. It is not a well-defined term, but kinetic modelling represents the state-of-the-art of present-day plasma modelling. The basic feature is that it tracks a blob, a simulated chunk of the plasma material, instead of a single particle. The kinetics code tracks the density and velocity distribution inside each blob of plasma, resulting in pictures like the one shown in Fig. 2.4.

Kinetic models are fairly detailed, but still miss part of reality and we are still far from an accurate description of plasma behaviour. Every approach misses out a number of important features. Most importantly and not surprisingly this is the case for plasma instabilities. Moreover, the most promising approaches are also very hard to solve. The ultimate would of course be when we could track everything, every particle at all times. In 2018 the most advanced plasma code can handle a complete model of 1 billion particles, while keeping the energy of the system constant.[10] There exist nowadays quite a number of (commercial) computer codes which can be used for plasma modelling and will undoubtedly become ever more sophisticated in the future. Some of their names are VizGlow, VSim, USim etc. (For a review see Van Dijk et al. 2009).

---

[10]*The Fusion Podcast*, "Interview with Dr. Jaeyoung Park", 27 January 2018, Matt Moynihan.

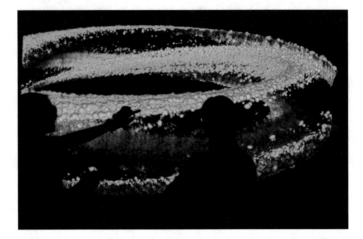

Fig. 2.4 Image of a plasma generated by using a kinetics code

2.3 Plasma Instabilities and Disruptions

2.3.1 *Introduction*

Plasma instabilities or disruptions often lead to catastrophic loss of plasma, which must be avoided at all cost. For a fusion reactor stability is crucial. A disruption event causes the plasma to crash into the wall of the containing vessel. The resulting loss of plasma combined with the potential damage to the wall has led to the realization within the fusion community that in a fusion reactor such instabilities must be avoided. But how to do this as due to the turbulent behaviour of the hot fusion plasma instabilities seem to be part and parcel of it and plasma containment is greatly hampered by hundreds of various instabilities?

Disruptions, still poorly understood, are characterised by very rapid growth and give rise to rather violent effects causing the current to abruptly terminate, which results in loss of temperature and confinement (McCracken and Stott 2013, p. 94). Because of the fast time scale on which the thermal and electromagnetic energy of the plasma is released, large forces act on the plasma containing vacuum vessel, putting it under huge mechanical strain. All the energy is dumped into the structure of the machine and in big machines forces of several hundred tonnes have been measured. Such events have been witnessed both at the JET tokamak in Britain, where it caused the whole giant machine to jump a few centimetres into the air (Seife 2008, p. 111),[11] and in TFTR at Princeton (USA), producing "a noise that sounded like the hammer in hell, followed by echoing thunder" (Clery 2013, p. 182). It is not uncommon that disruptions cause considerable damage, and prevention or mitigation is essential. An eventual power plant will have to be able to withstand forces of at least an order of

---

[11]Private communication with Dr. M. Gryaznevich who witnessed the event.

magnitude higher than estimated to be unleashed by a possible disruption to make sure that the construction is safe, which adds considerably to the construction costs. To a certain extent the chance of disruptions under increase of the current can be diminished by proportionally increasing the external magnetic field, but this will put too much strain on the magnetic field coils.

Disruptions are caused when we push against the limits for a plasma in the device, such as density and pressure. When the plasma is pushed too close to one of these limits, disruptions are more likely to occur. How this occurs is not entirely clear, so the approach to tackle them is avoidance (by staying away from the plasma limits), prediction (by employing sensors) and amelioration (by automatic controls that change the plasma parameters). The problem is a potential showstopper and is receiving a great deal of attention (Chen 2011, p. 290 ff). We will come back to it several times in this book.

Disruptions and plasma instabilities are very closely related, and the latter are often the cause of disruptions. Instabilities have been plaguing fusion research since its early beginning, and it may well be that they will eventually finish off the fusion effort as new and more vicious instabilities keep popping up. It will be a tremendous challenge for fusion research to find sufficiently stable plasma configurations. All instabilities have in common that they are due to exponentially growing small imperfections or noise in the system and are therefore unavoidable since such imperfections are always present. The huge range of instabilities is well illustrated by the list of more than 50 plasma instabilities, from Buneman instability to Weibel instability, in *Wikipedia*.[12] The list is incomplete as one of the most recent and troublesome type of instabilities, the edge-localised modes (ELMs), is not even mentioned. There are many types of ELMs, but the most common is closely related to ballooning instabilities which act like the elongations formed in a long balloon when it is squeezed.

In general, plasma stability is improved by limiting the pressure or current through the device. However, as we have seen when discussing the Lawson criterion, a high pressure is desirable for achieving a high value for the product of the pressure p and the energy confinement time τ_E, and a high current is desirable for increasing τ_E. Stability theory is thus concerned with two basic problems: how can the actual limits on pressure and current be calculated for any given magnetic configuration, and how can the magnetic configuration be optimised so that the pressure and current limits are as high as possible?

Trying to contain a hot plasma in a magnetic field is a hazardous affair. It resembles trying to balance a stick on the top of your finger or to position a ball on the top of a hill. A small displacement will grow with increasing speed and result in the rapid and unstoppable loss of balance or position.

Figure 2.5 shows mechanical analogues of (in)stability situations: (a) a stable equilibrium situation, any small perturbation on the motion of the ball will rapidly die out and the ball will return to its stable position; (b) an unstable situation, any small perturbation will cause the ball to run further away from the top of the hill; (c)

---

[12]https://en.wikipedia.org/wiki/Plasma_stability#Plasma_instabilities, accessed 20 May 2020.

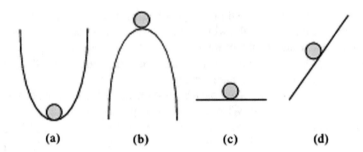

Fig. 2.5 Mechanical analogues of stability: **a** stable, **b** unstable, **c** marginally stable, **d** no equilibrium

neither stable or unstable, but in equilibrium, while (d) represents a situation that is not even in equilibrium.

2.3.2 Classification of Plasma Instabilities

The classification of plasma instabilities does not seem to be standardized. Various authors use different classifications and names for common occurring instabilities, which makes a rather confused field even more confused. A common division into two general groups is (1) (magneto)hydrodynamic instabilities or MHD-instabilities and (2) kinetic instabilities (caused by anisotropies (deviations from a Maxwell distribution) in the velocity distribution function of the plasma particles). They can however also be categorised into different modes, depending on the direction in which the perturbation that starts the instability occurs.

A useful classification of MHD-instabilities into three main classes is the following (most of the following is adapted from Friedberg 2007, pp. 296–306).

2.3.2.1 Internal and External Modes

In the situation of a well-confined plasma in equilibrium separated from the wall of the containment vessel by a vacuum region, instabilities can be classified in internal and external instabilities, based on whether or not the surface of the plasma moves as the instability grows. An internal instability implies that the plasma surface remains in place and the instability occurs purely within the plasma. Often, they do not lead to catastrophic loss of plasma as the plasma remains contained, but can result in important experimental operational limits or enhanced transport. External instabilities, on the other hand, involve motion of the plasma surface, and hence the entire plasma. Such motion may lead to the plasma striking the wall and ending its existence. In a fusion plasma external modes are particularly dangerous.

2.3.2.2 Pressure-Driven and Current-Driven Modes

A second way to classify plasma instabilities is by the source driving the instability. In general, both perpendicular and parallel currents (relative to the magnetic field) occur in a plasma and each of these can drive instabilities. Instabilities driven by perpendicular currents are often called "pressure-driven" modes. They are usually further subclassified into *"interchange modes"* (like the Rayleigh–Taylor instability, see below) and *"ballooning modes."* Pressure-driven instabilities are usually internal modes and set an important limit on some plasma parameters, e.g. the ratio of the thermal plasma pressure to the magnetic pressure, called *beta*,[13] that can be achieved in a fusion plasma. Ballooning modes especially arise in a torus geometry, which is the common geometry of many devices, as we will see later. They do not occur in a straight cylindrical setup.

Instabilities driven by the parallel current are often called "current-driven" modes. A common name for current-driven instabilities is also *"kink modes"* because the plasma deforms into a kink-like shape. Kink modes can be either internal or external.

In certain situations, the parallel and perpendicular currents combine to drive an instability, often referred to as the *"ballooning-kink mode"*. This is usually the most dangerous mode in a fusion plasma. It sets the strictest limits on the achievable pressure and current. Furthermore, it is an external mode, implying that it can lead to a rapid loss of plasma energy and plasma current to the wall of the containment vessel. Another name for it is *"explosive instability"*, the narrow fingers of plasma (ballooning fingers) produced by the instability are capable of accelerating and pushing aside the surrounding magnetic field causing a sudden, explosive release of energy. Other ballooning instabilities go under the names of *peeling-ballooning* and the already mentioned ELMs. There are many types of the latter modes, which are local instabilities that take the form of filamentary eruptions ballooning out from the plasma towards the wall of the reactor vessel. It is a way for the plasma to let off steam when the pressure in the plasma is building up. They constitute a major concern for any tokamak, but especially for ITER (Davidson 2010, pp. 534–535).

2.3.2.3 Conducting Wall Versus No Wall Configurations

The last classification scheme is based on whether or not a perfectly conducting wall (the first wall on the inside of the plasma confining vessel) is required. A close-fitting perfectly conducting wall can greatly improve the stability of a plasma against external ballooning-kink modes. But a real experiment or reactor cannot maintain a superconducting wall close to the plasma. The reason is that in any realistic situation the high temperature of a fusion plasma, coupled with the release of large amounts of neutron power (neutrons from the fusion reactions), does not allow this wall to be superconducting.

---

[13] Since for fusion the plasma pressure has to be kept as high as possible and for cost reasons the external magnetic field should be as small as possible, a high value of this ratio is desirable.

To summarize, MHD-instabilities can be classified by several different methods. In terms of the physics involved one can distinguish internal versus external modes and current-driven versus pressure-driven modes. In terms of experimental practicality, one can distinguish whether a given magnetic configuration does or does not require a perfectly conducting wall to achieve reactor relevant pressures and currents.

2.3.3 Rayleigh–Taylor Instability

In the first fusion devices, so-called pinch devices (see Chap. 3), constructed in the late 1940s to early 1950s, it was soon found that the plasma becomes dreadfully unstable when it is attempted to extend the plasma confinement time beyond a few microseconds. Theoretical analysis showed that several types of pressure and current driven MHD-instabilities occurred, which go under the names (although the use of these terms does not seem to be consistent) of kink instability, representing transverse displacements of the beam cross section, changing only the position of the beam's centre of mass (Fig. 2.6a), sausage instability, resulting from harmonic variations of the beam radius with distance along the beam axis (the plasma was pinched unevenly at one or several places along its length and broken up into a series of blobs like a string of sausages) (Fig. 2.6b), and Rayleigh-Taylor instability (Fig. 2.6c).

The last one is a classical instability that originated in fluid dynamics and will be discussed here in some detail. The Rayleigh-Taylor instability is named after the

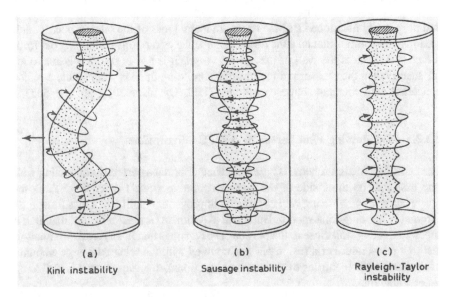

Fig. 2.6 Schematic representation of the simplest MHD instabilities (*from* Pease in Dendy 1993, p. 496)

English physicists Lord Rayleigh (John William Strutt) (1842–1919), who in 1904 earned the Nobel Prize for the discovery of argon and is known for a great number of other discoveries, and Geoffrey Ingram Taylor (1886–1975), who was a major figure in fluid dynamics and wave theory and among other things was part of the British delegation to the Manhattan Project. It concerns the instability of an interface between two media, normally fluids of different densities, which occurs when the light fluid is pushing the heavy fluid.

To explain this instability we can look at two fluids that do not mix, let's say water and oil. Water is denser than oil, so in a normal situation oil will float on water and not the other way round. The situation where the water is on top can briefly be in equilibrium, but the equilibrium is unstable, and any perturbation of the water–oil interface will cause the situation to be reversed. If for some reason a drop of water is displaced downwards with an equal volume of oil displaced upwards, the disturbance will grow very fast, moving the water down and displacing the oil upwards, thus restoring the normal stable configuration. The instability as described here is driven by gravity. There are many of such more or less everyday examples for this instability. The setup studied by Lord Rayleigh was the oil–water system, on which he wrote a paper in 1883 (Rayleigh 1883). The important insight by G. I. Taylor, written down in 1950 (Taylor 1950), was his realisation that this situation is equivalent to the situation when the fluids are accelerated, with the less dense fluid accelerating into the more dense fluid.

What has this to do with a plasma or with nuclear fusion? The instability cannot be driven by gravity as the plasma weighs almost nothing and gravitational forces are rarely of much importance in plasmas. Here pressure on the plasma-vacuum interface is the culprit. The vacuum only contains the magnetic field, while in the plasma (consisting of charged particles) there are both electric and magnetic fields. The magnetic field exerts a magnetic pressure (due to the energy density of the magnetic field) on the plasma, and since the magnetic field is curved (in most doughnut or torus shaped devices) this pressure is not uniform and under its action the plasma and the magnetic field try to exchange position. They are the equivalents of the fluids with different density of the ordinary Rayleigh-Taylor instability.

The Rayleigh-Taylor instability has spawned a number of related instabilities like the flute or interchange instability (which tends to appear whenever the confining magnetic field decreases outwards from the plasma) and the Farley-Buneman instability, which is driven by drift currents, caused by the electric field and always in the direction of the electric field.

Part II
Early Fusion Activity and The Rise of the Tokamak

Chapter 3
Early Years of the Fusion Effort

3.1 How to Make a Plasma and Contain It?

The basic questions that the early pioneers in fusion research had to answer were how to make a plasma and contain it, and what kind of apparatus can achieve this? In this chapter we will deal with the various early attempts by the UK, US and Soviet Union to create plasma containing devices until roughly the end of the 1960s when the tokamak achieved a breakthrough.

3.1.1 The Pinch Effect

The earliest proposals to confine a plasma (i.e. keep it in place for some (very short) time) are so-called pinch devices. A pinch is the compression of an electrically conducting element (e.g. a filament, usually a spiral-shaped wire made of a special metal as in a light bulb) by magnetic forces, but it can also be applied to a (column of) gas. An early version of such compression was described in 1905 by the Australian physicists J. A. Pollock and S. Barraclough (Pollock and Barraclough 1905). They were sent a copper pipe that appeared to have been crushed (Fig. 3.1) by a huge force, while the only thing that had happened was that a large current from a lightning strike had flowed through the pipe. They correctly explained that the current from the lightning strike had generated a magnetic field and that the resulting inward force had crushed the pipe (Clery 2013, p. 46). A modern application of this effect is an industrial device for crushing (tin or steel) cans and compressing other tube-like structures.

Similarly, when a current of sufficient strength is passed through a gas, it strips the electrons from the atoms in the gas (ionizes the gas), raises the temperature (the ohmic heating we discussed before) and produces a circular magnetic field surrounding the current. This is the wonderfully simple idea of the pinch: the plasma current, which heats the plasma, produces its own magnetic field to confine the

© The Author(s), under exclusive license to Springer Nature Switzerland AG 2021
L. J. Reinders, *The Fairy Tale of Nuclear Fusion*,
https://doi.org/10.1007/978-3-030-64344-7_3

Fig. 3.1 The lightning rod crushed by the pinch effect as described by Pollock and Barraclough (*photo Wikimedia*)

plasma. If the current is raised further, the degree of ionization, the temperature and the magnetic field strength also increase. The magnetic field exerts a force on the column of ionized gas (the plasma) and compresses the gas, i.e. "pinches" it into a thin filament (Fig. 3.2). The ionized gas acts like a series of parallel conducting elements which are pulled together by the magnetic field. On further raising the current and the magnetic field, the column of plasma is further compressed, which in turn raises the density and temperature. So, the basic idea of the pinch concept is that the confining magnetic field, which must keep the hot gas from the walls of the vessel, is produced by an internal current in the plasma itself. It is an application of the Lorentz force, the combination of electric and magnetic forces on a charged particle moving in an electromagnetic field.

This "pinch effect" is known as the Z-pinch, originally just called pinch or Bennett pinch, after William Harrison Bennett (1903–1987), an American scientist and inventor who predicted the pinch effect in 1934 and showed that the effect would cause a current to concentrate a plasma into a thin column. Bennett's work went almost unnoticed and another American scientist Lewi Tonks (1897–1971) independently discovered the effect in 1937 and called it the pinch effect. It became the basis of most early plasma confinement systems and of early attempts to produce fusion in the laboratory. The basic idea was rather simple: just pump enough current

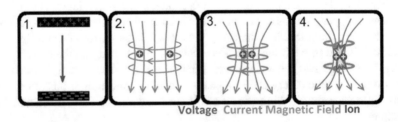

Voltage Current Magnetic Field Ion

Fig. 3.2 Basic explanation of how a pinch works: (1) apply a huge voltage across a tube, filled with a gas, typically deuterium. If the product of voltage and charge is higher than the ionization energy of the gas, the gas ionizes; (2) current jumps across this gap; (3) the current makes a magnetic field which is perpendicular to the current. This magnetic field pulls the material together; (4) the nuclei can get close enough to fuse (*Wikipedia*)

into a plasma, which in a single action would heat and perhaps compress the plasma enough to ignite fusion. Whether it would yield any net energy was a different, although crucial, question that could not yet be answered. The name Z-pinch refers to the direction of the current, running down the axis of the cylinder, which is taken as the Z-axis, while the magnetic field describes circles around the tube (azimuthal).

In plasma physics there are still two other pinches: the θ-pinch and the screw pinch. All of them are cylindrically shaped. The cylinder is symmetric in the axial (z) direction and the azimuthal (θ) direction. The pinches are named for the direction in which the current travels. The θ-pinch has a magnetic field directed in the z direction (along the tube) and a current directed in the θ direction (making circles around the tube). θ-pinches tend to be more resistant to plasma instabilities. The screw pinch is an effort to combine the stability aspects of the θ-pinch and the confinement aspects of the Z-pinch. In this case the magnetic field has a component in both the z and θ direction.

3.2 Early British Efforts

The first research on the pinch effect with nuclear fusion in mind was carried out in the UK by scientists at Imperial College and Oxford University. In the US, as in the Soviet Union, nuclear research remained preoccupied with building bombs and nuclear fusion for energy-generating purposes did not get much attention in the first few years after World War II.

In 1946 the English physicist George Paget Thomson (1892–1975), who in 1937 earned a Nobel Prize for confirming the de Broglie[1] hypothesis about the wave properties of the electron,[2] and the South-African born British theoretical physicist Moses Blackman (1908–1983) were the first to take out a patent for a nuclear fusion power plant based on the Z-pinch concept. Their work was quickly classified as secret, so that no details were made public at the time.

Immediately after the end of the war and stimulated by Fermi's success in building the atomic (fission) pile, Thomson had become enthusiastic about achieving controlled fusion reactions. He was already looking ahead to the possibility that nuclear fusion reactions in light elements, like deuterium, could be used for power generation. Building on the earlier work on particle accelerators at Cambridge, he initiated some work to explore whether fusion reactions could be produced in a gas in a toroidal vessel by means of an electric discharge in the gas, brought about by the ionisation of the gas when a current flows through it. The proper geometric term for such a doughnut shaped device is a torus, from which the word toroidal is derived. It is just a tube in the form of a hollow ring. All currently used magnetic fusion devices

---

[1] Louis-Victor de Broglie (1892–1987), French physicist who hypothesized the dual (particle and wave) nature of the electron in 1924, for which he was awarded the Nobel Prize in 1929.

[2] Thomson's father J. J. Thomson (1856–1940) had won the Nobel Prize in 1906 for the discovery and identification of the electron.

are toroidal in shape. See Intermezzo below. Thomson conceived a ring discharge in deuterium, with the electrons carrying such a large (toroidal) current that they would be held tightly in an area of small cross section by the pinch-effect forces of the poloidal magnetic field created by the current. Energy fed by a radiofrequency source into the discharge would be shared by electrons and ions, and, if they reached 105 eV (1.2 million degrees), fusion of deuterium should occur and the released energy might exceed the radiation loss and the loss of energy by escaping ions and neutrons.

Intermezzo I: toroidal and poloidal

The word 'toroidal', which appears very often in writings about torus-shaped confinement systems, indicates that the magnetic field, current or whatever is called 'toroidal' goes the long way round in the torus (indicated by the blue arrow in the figure below). So, the current sent through the torus in the pinch devices discussed here is a toroidal current, while the magnetic field induced by this current is poloidal, as indicated by the red arrow. According to the Oxford English Dictionary the word 'poloidal' relates to a magnetic field associated with a toroidal electric field, in which each line of force is confined to a radial or meridian plane. The words were first used by Walter Elsasser in discussing the magnetic field of the Earth which has a toroidal component (parallel to the lines of latitude) and a poloidal component (i.e. towards the poles).

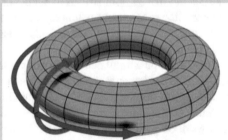

Figure from *Wikipedia*

Simplest, of course, would have been to take a straight cylindrical tube, in which a perfectly uniform magnetic field might be created, but then the particles would escape at the ends. So, the cylinder had to be bent into a torus-like device. However, the magnetic field would then no longer be perfectly uniform, and the charged particles would start to drift. To prevent such outward drift, it was proposed to apply a vertical magnetic field which would interact with the toroidal electron current.

Moses Blackman became part of the enterprise when Thomson realized the need for a theoretical physicist to calculate radiation losses and collision processes. The work was aimed at demonstrating the production of neutrons by the fusion of deuterium ions. For heating the plasma and holding it together Thomson and Blackman's theoretical reactor only used the pinch effect generated by the current passing through the plasma. The consequence was that a tremendous current was required for which he failed to obtain sufficient funding.

By May 1946 they filed the first provisional patent covering a discharge fed by radiofrequency power.[3] The patent application did not say how to ionise the gas, nor how to make it flow around the torus, but suggested several methods. It claimed that a torus 3 m across would be large enough to accelerate particles and to achieve fusion. Thomson and Blackman submitted a second patent application in 1952. The only publication from this early work, which as said was soon classified as secret, was a paper by Blackman on the theory of the pinch effect (Pashley 1987; Moon 1977).

Thomson and Blackman carried out their initial work at Imperial College in London, where Ware and Cousins, two doctoral students, were also working under Thomson's guidance. As mentioned in Chap. 2, they are credited with having produced the first plasma in a doughnut-shaped vacuum vessel.

A parallel effort on pinch devices had by that time started at the Clarendon Laboratory at Oxford University by Peter Thonemann (1917–2017) and James (Jim) Tuck (1910–1980), who apparently had been talking to Ware at Imperial College (Herman 1990, pp. 40–41). Thonemann was born in Australia and came to Oxford in 1946 with a B.Sc. from Melbourne University. At Oxford he started work on his Ph.D. thesis entitled "Study of Gas Discharge Phenomena and their application to Nuclear Sciences". Tuck had just returned from America, where he had been involved in the Manhattan Project. The two of them started work on the magnetic confinement of plasma in small glass tori, an effort that just like the Thomson-Blackman work was soon declared secret. Tuck, who found the post-war conditions in Britain difficult to cope with, returned to the US in 1949. He was invited by Edward Teller to work on the hydrogen bomb and was deeply involved in this until 1951, after which he would return to the pinch idea, as we will see later.

Thonemann had a ring of glass tubing made, pumped out the air and filled it with hydrogen gas at a much lower than atmospheric pressure. This gas was ionised to make a plasma. To get a current flowing through the tube to further heat the plasma, he made use of electromagnetic induction: the induction of a current due to a varying magnetic field. So, he wrapped a coil around the glass tubing, sent a current through it, which induced a magnetic field which in turn induced a current in the tube. For the whole contraption to work the magnetic field had to be changing, so that the increasing magnetic field will induce a growing current in the torus. But it is of course not feasible to keep increasing the current, so Thonemann used an alternating current instead, which flipped the direction of the magnetic field and consequently of the current through the plasma in quick succession. With some adaptations it worked

---

[3] A facsimile of the original patent application by Thomson and Blackman has been published in Haines 1996.

sufficiently well to be able to study the behaviour of a plasma in a magnetic field (Clery 2013, p. 44ff).

In 1946 the Atomic Energy Research Establishment (AERE) was founded at Harwell, just 25 kms from Oxford, and after in 1950 the spy-scandal broke out with the German born Klaus Fuchs (1911–1988), one of the principal theoretical physicists at AERE and a long-time spy for the Soviet Union (see e.g. Rossiter 2014), the need for greater secrecy was realised and Thonemann and his team were moved to AERE, while the Imperial College team went to a secure facility in Aldermaston in Berkshire, about 30 kms to the south of AERE.

At AERE Thonemann continued his work but it soon became clear that his alternating current approach was no good, as it failed to contain the plasma for any length of time. Alternating currents were abandoned and pulsed power supplies were brought in, ramping up the current from zero to a high value with a single pulse in one direction, producing a single burst of pinch effect. The early results were very encouraging: pinched plasma that lasted for one ten-thousandth of a second. They proceeded to scale up by building a number of larger tori, creating ever greater plasma currents. But it was soon found that the plasma was dreadfully unstable and started to wriggle about like a snake until making contact with the vessel wall and being extinguished. It was the appearance of the first plasma instabilities that have been plaguing the fusion enterprise in various forms to the current day: the kink instability. A kink instability shows a wavy behaviour: crosswise displacements of the beam without changing the form of the beam other than the position of its centre of mass (Fig. 3.3). Kink instabilities are a natural consequence of the pinch effect. The magnetic field lines induced by the current can be seen as a series of evenly spaced rings around the current. Any slight kink in the current will disturb this equal spacing, causing the field lines to be more closely packed together on one side and more widely apart on the other. More closely packed field lines mean a stronger magnetic field, disturbing the balance of forces on the current and reinforcing the kink causing the whole thing to get out of control and the pinched plasma to crash into the wall of the containing vessel.

Fig. 3.3 Kink instability of a plasma (*adapted from* McCracken and Stott, p. 49)

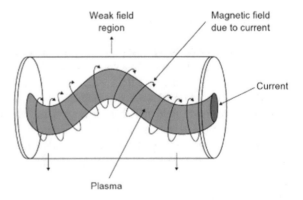

Fig. 3.4 Diagram of one of the first torus-shaped (toroidal) devices, the toroidal pinch. Two magnetic fields are applied: a *poloidal* field generated by the current flowing around in the plasma and a *toroidal* field produced by the external coils. The poloidal field is much stronger than the toroidal field. The combined field twists helically around the torus (*from* McCracken and Stott 2013, p. 49)

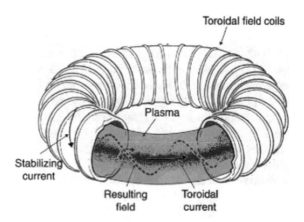

In spite of a lack of understanding of such instabilities, there was at AERE an ever increasing pressure to build larger machines that could produce temperatures necessary for fusion, leading to the design of the ZETA project, Zero Energy Thermonuclear Assembly, an apt name as in its first runs it produced 10^{-12} (one million millionth) of the energy that it used. It was the biggest fusion experiment so far. The problem with the kink instability was solved by applying another magnetic field going around the torus (toroidal field) and exerting a force that would push the kink back into line. The resulting setup has been illustrated in Fig. 3.4. In August 1957 ZETA was finished and ready to produce some fusion (Clery 2013, pp. 54–57).

In parallel to ZETA the Imperial College group at Aldermaston built a smaller machine to test the results coming from ZETA. Two versions, called Sceptre I and II, were constructed, one with a copper-covered quartz tube, 2 in. in bore and 10 in. in diameter, and the other an all-copper version, 2 in. in bore and 18 in. across. They demonstrated temperatures of around 1 million degrees. The system worked as expected, producing clear images of the kink instabilities using high-speed photography. The machine was subsequently changed, re-christened Sceptre III, and experimental runs like those on ZETA were started. By measuring the spectral lines of oxygen, they calculated interior temperatures of 2–3.5 million degrees. Photographs showed the plasma column remaining stable for 300–400 µs, a dramatic improvement on previous efforts. Sceptre III was followed by Sceptre IV, the last classic pinch device built in the UK.[4]

Ten years had passed since the first work at Oxford and Imperial College, and of course the rest of the world had not sat still either during that time. Britain had exploded its first nuclear weapon in 1952. Both the Americans and the Soviets had followed this up with hydrogen bombs and had now their hands free to also devote some time on power generation by nuclear fusion. A surprise event had just happened in April 1956 when the Soviet leader Nikita Khrushchev visited the UK with an official delegation. The party included Igor Kurchatov (1903–1960), the Soviet Union's leading nuclear scientist, leader of the Soviet effort in building nuclear bombs, and

---

[4]*Wikipedia,* https://en.wikipedia.org/wiki/Sceptre_(fusion_reactor).

director of the Institute of Atomic Energy in Moscow, now named after him as the I.V. Kurchatov Institute of Atomic Energy. Until 1955 it was named Laboratory no. 2 of the USSR Academy of Sciences, which was founded in 1943 to develop the Soviet atomic bomb. Kurchatov's institute was also involved in nuclear fusion experiments and he delivered a lecture at Harwell, in which he set out the complexity of the problem without giving any specific details of what the Soviets were exactly doing, although it became clear during his lecture that Soviet scientists were also tackling the problem with pinch-effect systems. He did not mention the circular tokamak invented by Sakharov and Tamm (see below) and actually managed to give his British counterparts the wrong idea that the Soviets were not yet experimenting with plasmas in circular devices.

One technical point that Kurchatov stressed in particular in his lecture at Harwell was the difficulty of determining whether the neutrons produced by the plasma were indeed the result of nuclear reactions. His exact words were the following: "In 1952, soon after experiments with pulsed discharges were started, it was found that at sufficiently high currents the discharge in deuterium becomes a source of neutrons. (…) In the early stages of the investigation it was quite natural to assume that the neutrons resulted from thermonuclear reactions in the high temperature plasma. This was exactly what was expected from the beginning; and the fact that the phenomenon was detected under conditions which completely corresponded to the a priori theoretical predictions seemed to speak in favour of this interpretation. The behaviour of the neutron radiation (its dependence on pressure and current) observed in the first experiments qualitatively concorded with the assumption that the phenomenon was due to a thermonuclear mechanism. However, very soon, serious doubt about the correctness of this assumption began to appear" (Kurchatov 1957). There were all kinds of other processes that could produce neutrons and one should be extremely cautious in claiming to see neutrons from fusion. The Russians had apparently been fooled by neutrons coming from deuterium in a pinch device which they first thought were due to fusion reactions. The Americans had also seen such neutrons but had not reported them to the British for reasons of secrecy (Bromberg 1982, pp. 68–69).

Although Kurchatov's warning must have been heard by his British audience, they let themselves be carried away and be tricked when ZETA was fired up for the first time in August 1957. It soon started to register neutrons, up to a million per plasma pulse. Such neutrons were also observed in Sceptre III run by the Imperial College group at Aldermaston in roughly the same numbers as at ZETA. The neutrons disappeared when the deuterium plasma was replaced by hydrogen. This suggested that deuterium was important for the neutron production and that they resulted from D-D-fusion processes, as it is much harder to fuse hydrogen atoms (see Chap. 1). The temperature reached was at least 1 million degrees, and probably 5 million. The director of AERE, John Cockcroft, in spite of all his experience, couldn't resist the temptation to write to the chairman of the Atomic Energy Authority about the neutrons, but that he was not 100% sure that the neutrons were thermonuclear, clearly implying that the probability they were was very high. The press got wind of this and soon the newspapers were full of wild, overblown stories about Britain's fusion success. An official statement from the Harwell laboratory was however not forthcoming, partly also as the Americans within the framework of a secrecy arrangement with Britain refused to give consent for making the results public, as it would look as if the British were ahead in the thermonuclear field (Bromberg 1982, p. 81). But more time was also needed to check the origin of the neutrons. In the end they were overtaken by events that were beyond their control.

Early in October 1957 the Soviet Union took the entire world by surprise by the launch of the Sputnik, the first artificial satellite, in an orbit around the Earth. Less than a week later a nuclear accident, the worst in Britain's history, occurred with a nuclear fission reactor at Windscale on the northwest coast in Cumberland, releasing radioactive contamination that spread across the UK and Europe.[5] The Sputnik launch and the Windscale fire were two devastating blows to the prestige of poor, old Britain, which desperately needed some good news to restore the public's confidence in the greatness of the nation. In January 1958 John Cockcroft gave a press conference announcing that ZETA had produced temperatures of 5 million degrees, had held these temperatures for up to three thousandths of a second and that he was 90% sure that fusion reactions had taken place. Later, on television he declared that "To Britain this discovery is greater than Sputnik" and Britain would have a working fusion reactor within twenty years' time (Herman 1990, pp. 50–51).[6]

The next day the newspapers exploded with a chauvinistic fanfare of the British fusion triumph (Fig. 3.5) that would usher in a new age of limitless power, not only for Britain but for the whole of mankind of course. Let's be generous. Why not?

The triumph, which was not universally accepted and doubted by many scientists working in the field around the world,[7] proved to be brief. If the neutrons originated from thermonuclear reactions, they would be flying out equally in all directions and with similar energies. The ZETA team had postponed measuring the neutrons' directions on the ground that there were too few of them. Others made the measurements and showed that the majority of the neutrons were of spurious origin. The neutrons produced in the Sceptre III device at Aldermaston were likewise shown to be spurious. They had all been fooled and this while being warned by Kurchatov and, a most poignant aspect, the Americans had already shown such neutrons to be spurious in 1955 by measuring their directions (Bromberg 1982, p. 69). The "mighty ZETA", as the *Daily Mail* had called it, had crashed. It was a disaster of their own making as the instruments that eventually did the measurements had also been available to them. Cockcroft issued a sober press release in May 1958. The neutrons were not from fusion reactions, but a by-product of plasma heating. Neither had the plasma

---

[5]Compared to the Chernobyl and Fukushima disasters the release of radioactive material was negligible.

[6]It is always within 'twenty years' that working fusion reactors are predicted. It is unclear what is so magical about this number of years, probably just long enough to be completely forgotten when it does not come true and still within the predictor's lifetime to reap the glory when it happens to be correct. Cockcroft was 60 years old when he made a fool of himself here.

[7]Noteworthy in this respect is a note by Lyman Spitzer in the same issue of *Nature* that published the ZETA story, in which he pointed out a contradiction between theory and the rate of particle heating reported by the ZETA scientists.

Fig. 3.5 Frontpage of the *Daily Sketch* newspaper after Cockcroft's announcement on ZETA, creating in the bargain a new English verb "to sputnik"; it didn't catch on (*Wikipedia*)

reached the temperature necessary for fusion. Cockcroft did not update his prediction for a working fusion reactor. For the next ten years ZETA remained a useful
research machine, but it would always be remembered as a debacle. The disaster did
not prevent Thonemann from stating at the 2nd Geneva Conference on the Peaceful
Uses of Atomic Energy a little later in the year that "it is still impossible to answer
the question, 'Can electrical power be generated using the light elements as fuel by
themselves?' I believe that this question will be answered in the next decade. If the
answer is yes, a further ten years will be required to answer the next question, 'Is
such a power source economically valuable?'" (Thonemann 1958). How a serious
scientist, just after the pitiable adventure with ZETA, can first state that a question is
impossible to answer, and then immediately continues to answer it in this fashion, is
beyond me. For some reason, many fusion scientists, Thonemann is certainly not the
only one, had difficulty distinguishing between what they wanted to happen and what
they thought was going to happen based on the available facts. The ZETA failure in
any case caused the British to occupy a humbler place in the fusion effort for the
next three decades.

3.3 Developments in the US: Project Sherwood

3.3.1 Introduction

The very first people who made a practical attempt to generate power from nuclear
fusion in the US, and probably in the world, were two American engineers, called
Arthur Kantrowitz (1913–2008)[8] and Eastman Jacobs (1902–1987), working at the
Langley Memorial Aeronautical Laboratory, now the Langley Research Center, in
Hampton (Virginia). In 1938, without their employers knowing anything about it, as
their experiment had nothing to do with aerodynamics, and after having read Bethe's
paper on stellar energy generation, they secretly attempted to heat a hydrogen[9] plasma
by high-power radio waves in a toroidal magnetic field in a dough-nut shaped device
the size of a big truck tire. According to Kantrowitz's calculations, at the resonance
frequency of the electrons in the plasma, enormous amounts of energy would be
built up heating the plasma to ten million degrees. It did not take them long to realise
that only a magnetic field would be able to force the particles away from the walls
of the torus and confine the plasma. To generate such a field, they wound water-
cooled electric cables around their torus, making a magnetic coil. The magnetic
fields would crowd the hydrogen nuclei closer together, increasing the frequency of
atomic collisions and thus the chances of fusion reactions to occur. They dubbed
their machine, which came very close to the later Russian tokamak, the "Diffusion

---

[8]Edward Teller was Kantrowitz's Ph.D. supervisor. Kantrowitz is credited with being the founder
of laser propulsion.

[9]They would have preferred deuterium, but this was not readily available, having only been
discovered in 1932.

Inhibitor". Because of the secrecy they had to maintain, the testing took place at night. And although the plasma began to heat up and glow blue, the X rays they had hoped would be emitted from the fusion processes were not seen. They believed that their plasma was leaking away, so they tried to crank up the power, but to no avail. The diagnosis was right, plasma was indeed leaking away, but the remedy chosen was not the proper one. As we now know, wrapping a simple coil of wire around a torus, does not create a perfect magnetic trap. The plasma will be drifting out of the confining magnetic field. Kantrowitz and Jacobs did not understand this very well, nor did they know anything about plasma instabilities. No one at the time knew about these things. Had they continued to pursue their experiments, they might have discovered the source of their problems. Before they could learn anything more, they were themselves discovered and had to cancel their project. There were no adverse consequences for them, but it was the last and only time they were involved in a nuclear fusion experiment (Hansen 1992; Braams and Stott 2002, p. 16).

A couple of years later, within the framework of building the hydrogen bomb fusion energy was from time to time discussed in the US and even some preliminary studies were made in the early 1940s, but all these activities came to an end by late 1946 when many of the scientists working at Los Alamos left New Mexico to return to their former civilian careers.

It lasted until the very early 1950s that three separate groups, each pursuing a different route to fusion, sprang up almost simultaneously at three laboratories in the country. James Tuck, back from his sojourn in Britain referred to above, assembled a group at Los Alamos, Lyman Spitzer (1914–1997) did so at Princeton University, and Herbert York (1921–2009) and Richard Post (1918–2015) at the Livermore branch[10] of the University of California's Radiation Laboratory. Following the experience he had gained in the UK, Tuck had put his trust in pinch devices, Spitzer, as we will discuss later in this chapter, had developed his own device, which he named the stellarator, while York and Post were favouring so-called magnetic mirrors.

At first the projects fell under different divisions of the Atomic Energy Commission (AEC), the US agency founded in 1946 to promote and control the peacetime development of atomic science and technology. Because of the research nature of the projects it was decided to bring them under in one division, the Research Division, and bundle the efforts into one project which was given the name Sherwood. Tuck's earlier funding for controlled nuclear research had been diverted from Project Lincoln, in the Hood Laboratory at Los Alamos. The coincidence of names (Hood with Robin Hood and Tuck with Friar Tuck, one of Robin Hood's companions) prompted the cover name "Project Sherwood" (Sherwood Forest was Robin Hood's principal hideout).[11]

One striking feature of the programme's first years was the astonishingly short period considered necessary for establishing the feasibility of fusion as an energy

---

[10]This laboratory was founded in 1952 and evolved into the Lawrence Livermore National Laboratory.

[11]J. Tuck, Curriculum Vitae and Autobiography, https://bayesrules.net/JamesTuckVitaeAndBiography.pdf.

source. The director of the AEC's Research Division estimated that 3–4 years and 1 million dollars would suffice to find out if a hot plasma could be contained (Bromberg 1982, p. 33), and the scientists did not contradict him. Today it is clear that this estimate was vastly wrong, so wrong that one wonders how such a huge mistake could be made. When in 1961 President Kennedy announced that "this Nation should commit itself to achieving the goal, before this decade is out, of landing a man on the Moon and returning him safely to the Earth", he was obviously acting on more sound advice in a matter that looked perhaps even more dubious than just proving the feasibility of controlled nuclear fusion in the laboratory. Of course, in 1952 in the matter of fusion there was room for wide extremes of optimism and scepticism. Almost nothing was known about the behaviour of plasma at the extreme temperatures required for fusion. A scientific basis for fusion hardly existed, while the situation with respect to technology was even more wide open. Magnetic fields of the strength the scientists needed had never before been created, nor had plasmas of millions of degrees ever been made in the laboratory. No one really knew how to make them. This of course contrasted sharply with landing a man on the moon, which was mainly a matter of technology. The science for it had all been well worked out. There were no unknowns in that respect. The fusion scientists all took the stance of optimism, but just for their own particular design and on what basis they did so is not all that clear. Spitzer did not think that pinches would work, and Tuck doubted the viability of the stellarator, while York, following an advice of his mentor Ernest Lawrence, just wanted to do things differently for the sake of being different (Bromberg 1982, p. 27). This reeks a little of irresponsibility, scientists seeking public funds for chasing after a wild, poorly thought-out idea. Social factors probably also played a role. Post-war exuberance had created an atmosphere that everything is possible, that anything can be done if only the decision is made to make it a matter of national priority. Overconfidence perhaps after the successful completion of the Manhattan project, in which several of the fusion pioneers had participated or been in contact with. Another factor is that there was no one who knew all the bits and pieces of scientific and engineering knowledge underlying fusion generators. No one had an overview. There were no established disciplines in this field. There was no fusion community yet. There were no nuclear engineers. The participants came from various other fields of physics, e.g. Spitzer was an astrophysicist, Tuck was an accelerator specialist, Post had studied microwaves at Stanford. But one also would have thought that this lack of established practice would have made them more cautious in making predictions and claims in uncharted territory.

The grossly unwarranted optimism is also reflected in the phrase "too cheap to meter" used by Lewis Strauss (1896–1974), a former businessman, not a scientist, who was appointed chairman of the AEC by president Eisenhower, in a speech he gave in 1954.[12] He made it in the context of nuclear fission, but according to his son actually meant fusion, as at the time he was not allowed to reveal anything of that

---

[12]The full text of his speech can be found at https://www.nrc.gov/docs/ML1613/ML16131A120.pdf; see also https://public-blog.nrc-gateway.gov/2016/06/03/too-cheap-to-meter-a-history-of-the-phrase/.

research or to refer to Project Sherwood, which he strongly championed. Whether it was for fission or fusion does not really matter, as it is clear that "too cheap to meter" is a myth for both. The phrase however started a life of its own and is often quoted to deride fusion.

Not everyone expressed the same optimism though. Edwin McMillan (1907–1991), an American physicist and Nobel laureate credited with being the first ever to produce a transuranium element,[13] for instance, did not work on fusion, but pointed out in 1952 that "to confine a plasma with a magnetic field was to place it in a distinctly unnatural situation, somewhat like the situation that would result from trying to push all the water in one's bath to one side of the tub and keep it there". The plasma, like the water, would try every permissible mechanism to escape. (Bromberg 1982, pp. 33–34.) And Edward Teller, in a now famous talk on fusion at a Sherwood conference in Princeton in 1954, pointed out that any device with magnetic field lines that are convex towards the plasma would likely be unstable. All the devices considered by the American groups had such convex magnetic field lines and would suffer from such instabilities.

In any case, the conclusion must be that the scientists were not clear to the administrators about the prospects for fusion while the latter apparently also failed to listen carefully. For instance, in a thoughtfully written review article of the fusion effort in 1956 Richard Post wanted to have it both ways by stating on the one hand that "the technical problems to be solved seem great indeed. When made aware of these, some physicists would not hesitate to pronounce the problem impossible of solution," and on the other hand that "[i]t is the firm belief of many of the physicists actively engaged in controlled fusion research in this country that all of the scientific and technological problems of controlled fusion will be mastered—perhaps in the next few years" (Post 1956).

The end result is a situation in which large amounts of money continue to be spent, while until the present day the same kind of promises and predictions are being made as in the early 1950s.

3.3.2 Pinch Devices

The first experiments in the US were carried out in 1951 with a Z-pinch machine built by James Tuck. It consisted of a doughnut-shaped glass tube about one metre across surrounded by magnets (Fig. 3.6). A capacitor (an electrical device that is designed to store an electric charge and release it instantaneously, see Glossary) was discharged through a coil coupled to this tube, inducing a large pulse of current (tens of thousands of amperes) into the gas in the tube. The purpose of the experiment, which started to run in the fall of 1952, was to create a highly ionised pinched discharge and study its properties.

---

[13] An element with atomic number larger than 92, the atomic number of uranium.

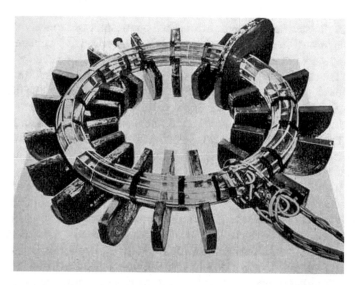

Fig. 3.6 The Perhapsatron the first US fusion experiment at Los Alamos: a simple circular cylinder (doughnut) filled with deuterium gas and surrounded by magnets. Only the bottom half of the magnets is shown in the figure

Tuck coined the name Perhapsatron for his brainchild, a name that was chosen to express the doubts he apparently had about the approach: "Perhaps it will work and perhaps it won't" (Clery 2013, p. 83). Well it didn't, as his venture encountered the same difficulties with kink instabilities as Thonemann had (see above). The plasma remained stable for only a few millionths of a second before breaking up. The kink instability had been predicted theoretically already before the machine had been started up by Martin Kruskal (1925–2006) and Martin Schwartzschild (1912–1997) at Princeton University (Kruskal and Schwarzschild 1954). It was independently discovered by the Soviet physicist Vitaly Shafranov (1929–2014), who worked at the Kurchatov Institute in Moscow, and is therefore also known as the Kruskal-Shafranov instability. Tuck observed the pinch and its subsequent rapid disintegration, so his work can be seen as a textbook confirmation of a theoretical prediction. The conclusion from this work must be that pinches are inherently unstable.

The next idea to be tried was the fast Z-pinch. It was suggested that a very strong electric field could form a pinch so fast that the heating of the plasma by its inward motion would initiate thermonuclear burn before the plasma had a chance to disperse. In other words, the hope was that, if the pinch could be made strong and fast enough, fusion could get going before the instability destroyed the pinch. Strong electric fields were easier to get from linear pinch tubes than from torus-like Perhapsatrons, so such linear machines, known as Columbus, were built at Los Alamos. The stability of the Z-pinch could be improved by adding a longitudinal magnetic field, and a series of experiments were done over the next few years incorporating such a field. Early in 1955 the researchers at Los Alamos actually saw a burst of neutrons every time the plasma was pinched hard enough. After initial optimism it was quickly shown that they were spurious, like the ones seen later on the Harwell ZETA machine, as related above. They were not coming out in all directions with the same energy, as should have been the case if they were truly fusion neutrons. Thermonuclear temperatures had not been reached. The neutrons were the work of another instability: the sausage instability (see Chap. 2; Fig. 2.6b). If the plasma is not pinched evenly along its length, but a small section gets pinched a little more than the rest, the plasma becomes unstable. The tighter pinch grows progressively; the plasma gets wasp-waisted and pinches itself off (Seife 2008, p. 94). The Columbus was unfortunately not the egg of the same name.

The first experiment in which thermonuclear fusion was achieved in any laboratory was actually carried out at Los Alamos in 1958 with the Scylla I machine, the first in a series of θ-pinch machines. The pinch was produced by a very short, intense pulse of current in a coil outside the tube. The measured energy distributions of the neutrons, protons, and tritium nuclei (all products from D-D fusion) from the Scylla I experiments gave definitive evidence that the plasma had reached a temperature of about 15 million degrees Celsius and that the neutrons were the result of thermonuclear fusion reactions. The reason that this result did not cause much upheaval at the time was that the achievement of fusion reactions was no longer seen as a breakthrough. The long-time confinement of the plasma in a device of reasonable size was now rightly viewed as the important intermediate goal on the route towards a reactor (Bromberg 1982, p. 88).

Over the next decade attempts were made to scale up the θ-pinch experiment to improve the confinement times. In 1964 plasma temperatures of approximately 80 million degrees Celsius at plasma densities of 2×10^{22} particles per m^3 and a few billion D-D fusion reactions per discharge were achieved with the Scylla IV, but the plasma confinement times were less than 10 millionths of a second. The largest θ-pinch machine was Scyllac, the fifth in the Scylla series and completed in 1974. Experiments with Scyllac demonstrated the behaviour of high-density pinches in a torus. However, during this time the US national fusion research programme began, rather prematurely, to examine the technologies required for fusion reactors. The characteristics of a Scyllac type θ-pinch machine did not project to an attractive reactor. Work on Scyllac was discontinued in 1977 (Phillips 1983; Bromberg 1982, p. 143).

Stability remained the overriding problem, not only for the pinch devices but also for the other two approaches to be discussed below, and for that matter for present-day devices. Pinch devices were eventually found not to bring any effective results. It turned out not to be possible to keep the plasma stable as the current passing through the plasma readily results in the instabilities mentioned, while there were further problems with the plasma slowly leaking away.

3.3.3 Early Stellarator

The second approach pioneered in the US, and the first to be funded by the AEC, was the so-called stellarator (a name chosen to refer to stellar processes) invented by Lyman Spitzer, a professor of astronomy at Princeton University, where for several years he had been studying very hot rarefied gases in interstellar space and was also working on the electric and thermal conductivity of ionised gases. Legend has it that a story in the newspapers in March 1951 about the fake results of the German physicist Ronald Richter (1909–1991) in Juan Perón's Argentina set Spitzer thinking about plasma confinement while on a skiing trip, which led him to the stellarator. (See e.g. Dean 2013, p. 4; Bromberg 1982, p. 13; Bishop 1958, p. 33.) The story of Richter's false claims has been told in detail elsewhere[14] and will not be repeated here. It is one of these unfortunate stories, which crop up now and again, in which people hot on publicity and short-lived fame waste other people's time and effort with their cranky ideas and results. News media, often averse to taking the trouble to do a proper investigation, tend to stir up the fever, forcing serious scientists to go to extremes to debunk such stories.

Spitzer quickly realised that only a magnetic field could do the confinement trick, but did not believe that the pinch effect was the answer to magnetic confinement, so he took a different route. A pinch device, if it were to work, would deliver neutrons originating from fusion reactions in fast pulses, providing energy in rapidly succeeding bursts, while Spitzer rightly thought that only a steady-state device, which would be able to provide energy continuously, would be practical. He discarded the idea of sending a current through the plasma to generate a confining magnetic field, but envisaged a single coil system to produce a magnetic field to confine the plasma.

Like many others in the field, he started with a linear tube, in which a uniform magnetic field was created by winding a wire round the tube along its full length, and sent a current through the wire. Any charged particles in the tube, instead of flying around in random directions, would start to move along the magnetic field lines while spiralling around them (as we have seen in Chap. 2). As long as the field stays away from the walls of the tube, so will the particles, Spitzer reasoned. A linear tube of finite length is obviously no good, as others also found out, since the particles will drift out at the end, so he bent it into the dough-nut shape (torus) we are already familiar with from various pinch devices. This bending however means that the magnetic field is no longer truly uniform, resulting in particles drifting upwards and downwards depending on their charge (a problem common to all torus-like devices). It results in a separation of positively and negatively charged particles. This separation gives rise to a strong electric field that neutralises the confining effect of the magnetic field and sweeps the ionised gas towards the wall of the tube before having even once circled the torus. So, instead of a torus he built a figure-8 shaped device consisting of two crossing straight sections joined by curved ends (Fig. 3.7), essentially a double torus, with magnetic fields in opposite directions, linked to form a kind of pretzel.

---

[14]E.g. Seife 2008, p. 75 ff. Another similar story is the cold fusion excitement of the late 1980s; see for this and also for Richter's saga Close 1990.

Fig. 3.7 Simplest stellarator design (*from* Bishop 1958, p. 37)

With this modification a particle that tends to drift upwards at one curved end will tend to drift downwards at the other end, cancelling out the net drift.

Let us discuss this in somewhat more detail as it is the basic feature for confining a plasma by a magnetic field in any device, not just in a stellarator. Consider a cross section through the tube at some point, perpendicular to the magnetic field. Assuming that the magnetic field is nonzero at all points in this cross-sectional plane, a magnetic field line will pass through each point of the plane. Each time a magnetic field line has gone around the tube, the point at which it will go through the plane will be slightly displaced, slowly rotating about the magnetic axis of the tube and forming an entire toroidal surface, called a "magnetic surface". Ideally the magnetic field lines will describe closed circles. In this configuration this is only true for the field line going through the magnetic axis, the centre of this cross-sectional plane. All other lines will rotate around this axis and when the parameters are chosen properly will form closed magnetic surfaces that completely stay within the tube, preventing the particles from crashing into the wall. For the gas to remain confined inside the tube and not to crash into the wall, the tube must enclose an entire family of such magnetic surfaces (Spitzer 1958, p. 255). The magnitude of the rotation is determined by the twist of the figure-8 configuration and it is possible to design the tube such that any rotational angle is obtained. The essential idea of the stellarator is that the helical twisting of the magnetic field lines is achieved solely with external coils; no current is sent through the plasma.

One of the other main differences between stellarators and pinches is the plasma density, which is low for stellarators (and tokamaks) and high for pinches. For high-density plasmas equilibrium was the major concern, while for low-density plasmas this is the shape of the orbits of the particles. This made that there was often little common ground between the various approaches to fusion (Bromberg 1982, p. 143).

Spitzer had a complete, at first rather optimistic development plan in mind, starting with a table-top device (Model A; Fig. 3.8) to show that the plasma could be created and confined and could heat the electrons in the plasma to 1 million degrees. This would be followed by a larger Model B which would also heat the plasma ions to 1 million degrees. The final Model C would then be a virtual prototype power reactor reaching thermonuclear temperatures of 100 million degrees. He had calculated that such temperatures were needed to get a positive energy balance Q, whereby the energy released by the fusion processes would exceed the energy lost. He took the

Fig. 3.8 Spitzer next to his first stellarator: the table-top model A

energy radiated by electrons that are decelerated by collisions with other particles in
the plasma (*Bremsstrahlung*) as the dominant source of energy loss. At 100 million
degrees the fusion energy would exceed the *Bremsstrahlung* losses by a factor of
100, so his Q would be 100! History would be rather different, as even now $Q = 1$
has not even been achieved. Spitzer also thought that the whole project would take
about a decade (Clery 2013, p. 80). He was aware though that there could be many
difficulties ahead, in the form of instabilities and other unwanted behaviour. In one
of his papers he wrote: "Little is known about the instability of plasma oscillations
in otherwise quiescent plasma" and "In a system with so many particles and with so
few collisions as a stellarator plasma, almost any type of behaviour would not be too
surprising" (Stix 1998, p. 4).

Spitzer was quick off the mark, for already in May 1951, just two months after
Richter's publicity stunt, he managed to convince the AEC of the usefulness of
his design and to obtain funding for it ($50,000) as part of the Project Matterhorn.
This project, based at Princeton University, was part of the top-secret programme to
develop thermonuclear weapons and supposed to be concerned with the theoretical
aspects of the H-bomb. Spitzer had been one of the people recruited to work on
this because of his expertise in interstellar hydrogen plasmas. He was head of the

controlled thermonuclear research section. The classified status of Project Matterhorn implied that Spitzer's stellarator work was also to be secret.

The Model-A stellarator was ready for work in the course of 1952. Its construction was extremely simple as Fig. 3.8 shows. Its vacuum chamber was made from sections of 5 cm diameter Pyrex glass tube comprising a figure-8 shape about 350 cm in length. Magnet coils to produce a 1000 gauss (0.1 T) steady-state magnetic field were wound directly onto the Pyrex tubing and were energized by a motor-generator set.

What were the first results like on this table-top size machine? The aim was to use ohmic heating to create a fully ionised plasma and heat its electrons to 1 million degrees. At these temperatures, Spitzer reasoned, the ionisation of hydrogen will be nearly complete and the plasma may be regarded as a classical assembly of free charged particles. The device was too small to allow the electrons to heat the rest of the plasma to this temperature, but it proved that the figure-8 geometry could indeed make plasmas more easily (at lower voltage and magnetic field) than was possible in a simple torus. Model B, which was not much larger than Model A, was designed for a considerably larger magnetic field strength, about a factor of 50 higher. Spitzer hoped that it would confine the ions long enough (according to his calculations about 30 ms) to also reach one million degrees (Bromberg 1958, p. 25 and 45ff).

Practical complications make the original figure-8 device less than ideal. This led to alternative designs and additions, which were tested in Model B. Difficulties with the original of this model and new developments resulted in the construction of various improved versions of the machine (B1, B2,[15] B64, B65). The B1 model demonstrated that impurities in the plasma caused large X-ray emissions that rapidly cooled the plasma. In the stellarators B64 and B65 the cross-over in the figure-8 was eliminated and the device flattened into an oval, or as they called it, a racetrack. The models all suffered from loss of plasma at rates far worse than theoretical predictions. This, i.e. theoretical predictions not being borne out by experiment, turns out be a common theme in fusion research.

Planning for Model C began already in 1954, even before Model B had produced results, and consequently was far too optimistic. As Spitzer later wrote: it was designed to "serve partly as a research facility, partly as a prototype or pilot plant for a full-scale power producing reactor." And, still more optimistic, such a full-scale reactor was also already on the drawing boards (Spitzer et al. 1954), called Model D, its design being ready at the end of 1954: a 165 m long machine in the form of a large figure-8 operating at a temperature of about 200 million degrees. It would require an investment of close to 1 billion dollars (1950s dollars!) and was designed to be capable of producing a net electric power output of 5 GW, enough for 5 million households and about 10 times the output of large power plants in operation at that time (Bromberg 1958, p. 47). All this happened before Teller gave his famous talk predicting instabilities in devices like the stellarator (see above) and work on the smaller models and on other toroidal systems established that several of the physics and technology assumptions of the Model-D design were untenable. The work on

---

[15]This model was presented at the second Geneva Conference in 1958 (Fig. 3.18), together with models of Model C.

Model D was stopped, before it had even started, and it was sensibly decided to first gain further experience with the smaller models to find ways to stabilise the plasma (Kragh 1999, p. 291). Model D has never risen from its slumber, although its design set a precedent for future stellarator design.

Construction of Model C started in 1957 and operation of the machine in 1961. By this time information collected from the previous machines was making it clear that Model C would not be able to produce any large-scale fusion. Ion transport across the magnetic field lines was much higher than theory suggested. The greatly increased magnetic fields of the Model-B machines did little to address this, and confinement times were simply not improving. The machine now was in the form of a racetrack 1200 cm in length with a 5–7.5 cm tube radius for the plasma. The typical toroidal magnetic field was 3.5 T. A principal finding over a broad range of experiments on Model C was confinement consistent with Bohm scaling, i.e. a confinement time inversely proportional to the temperature and linearly proportional to the applied magnetic field strength (Stix 1998, p. 7), which as explained in Chap. 2 would make confinement impractical.

Model C continued in operation until the facility went under in the tokamak stampede in 1969 and stellarator research at Princeton was halted. By this time Spitzer had already abandoned his brainchild. In 1966 he resigned his post at the Princeton lab and returned to astrophysics. When he designed his first stellarator in the very early 1950s, he considered it a ten-year project, which in his naivety he thought sufficient for building a prototype fusion power reactor. Now fifteen years later his stellarator still ran into the same type of difficulties and a prototype reactor was not in sight.

It did not mean though that the stellarator idea was abandoned everywhere, although research on it greatly diminished. Soon after the 1958 Geneva conference, which declassified nuclear fusion research worldwide, Princeton had already lost its monopoly on the stellarator. The Max Planck Institute in Garching and the Lebedev Institute in Moscow were next to build stellarators, with especially Garching continuing this research until the present day, while also in the US at Oak Ridge, in Japan and surprisingly enough in Kharkov in Soviet Ukraine research continued. In the early twenty-first century the stellarator experienced a considerable revival. Important facilities include the Wendelstein 7-X in Germany, the Helically Symmetric Experiment (HSX) in the USA, and the Large Helical Device in Japan. Chap. 13 will deal with these developments in some detail.

3.3.4 Mirror Devices

Magnetic mirrors, called magnetic traps in Russia and pyrotrons in the early days in the US, are open-ended linear confinement systems. They are the simplest geometry for plasma confinement one can think of. The basic design is shown in Fig. 3.9. It is essentially a bunch of magnetic field lines, bundled together at each end of a straight tube, hence the magnetic density is higher at these end points than in the middle.

Fig. 3.9 Basic magnetic mirror, the black wiggly line shows the motion of a charged particle along a magnetic field line. The (optional) rings in the centre extend the confinement area horizontally (*Wikipedia*)

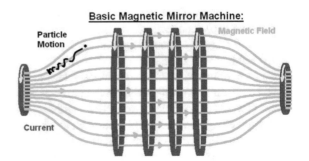

When plasma moves from lower to higher density fields, it can reflect. This reflection is known as a plasma mirror. The mirror machine was supposed to have two plasma mirrors facing each other. The goal was to have plasma ricochet back and forth until it slammed together in the centre and fused.

A charged particle will travel in a helical motion around a magnetic field line, as shown by the black wiggly line in Fig. 3.9, and try to flow out of the system at the end. When approaching such end, it will encounter an increasingly stronger magnetic field and the radius of its motion will become ever smaller. It will start to gyrate faster and faster, increasing its rotational energy. Since energy is conserved, the particle has to take the extra rotational energy out of its translational motion, and it will consequently slow down its efforts to escape from the end. This will eventually cause it to reverse direction (as if reflecting from a mirror) and return to the confinement area. Such a magnetic mirror field was known from cosmic ray physics (Fig. 3.10) and was actually used by Fermi to explain how cosmic rays

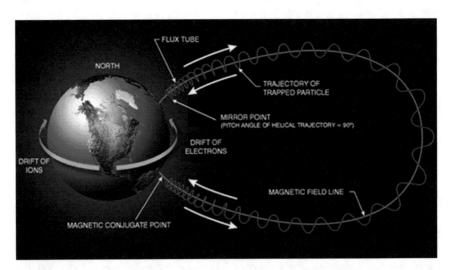

Fig. 3.10 Illustration of the motion of charged particles in a magnetic field (*based on Fig. 5–10* in Jursa 1985)

acquire their large energy (Fermi acceleration) (Fermi 1949). If charged particles spiral along a magnetic field that is not constant (like the Earth's magnetic field), but is stronger at the ends than at the centre, some of them will be reflected. A condition for this to happen is that the velocity component perpendicular to the magnetic field, which creates the circular motion around the field line, is sufficiently large compared to the parallel component.[16]

So, the idea of mirror confinement is to confine the particles in a magnetic field in a kind of bottle by making the field stronger at the ends. Magnetohydrodynamics (MHD) can easily model this system. The boundary conditions can be written down and the equations solved. The math will tell you that everything will work out fine. Except it doesn't. In real life, mirrors have problems. The plasma leaks out at the ends and it is very hard to ensure that no plasma is leaking away. MHD mathematics does not capture the flaws in a mirror machine. On paper, everything looks fine, but in reality there are problems. This highlights an important deficiency of plasma modelling. The simpler the math, the more it deviates from real life.

The mirror device is essentially a leaking magnetic bottle. This problem would eventually be the death knell for these systems. Nevertheless, in spite of this drawback, studies on magnetic mirrors were conducted in a large number of laboratories and in the early days were at least as important in overall fusion research as the work on closed systems. Important considerations for the choice of such a device were technical simplicity and the possibility of conducting a wide range of experiments on them.

Pioneers of mirror confinement devices are the Soviet physicist Gersh Budker (1918–1977) and the American physicist Richard Post, already mentioned above. In 1952 at the Livermore branch of the University of California's Radiation laboratory Herbert York decided to set up some fusion research and do it differently from the teams working at Los Alamos (pinch) and Princeton (stellarator): an externally produced magnetic field (not induced via a current as in Los Alamos) and a straight cylinder (not a torus or figure-8 device as in Princeton). The choice of a straight cylinder would make the magnetic field coils much simpler than for Spitzer's pretzel-shaped device. Richard Post, who had just joined the laboratory, was invited by York to set up a controlled thermonuclear research group. He would remain the leader of this group at Lawrence Livermore National Laboratory (LLNL) for 23 years.

Post started his work by developing a small device to test the mirror configuration, consisting of a linear Pyrex tube with magnets around it, small ones in the middle and big ones at the endpoints to prevent the particles from leaking away. To ensure that the perpendicular velocity component be large enough, radiofrequency fields were used to enhance the mirror effect. Almost simultaneously Budker suggested a similar machine and the first small-scale mirror (for which he coined the name *probkotron*) was built in 1959 in Novosibirsk at what is now called the Budker Institute of Nuclear Physics (BINP). The confined plasmas in these devices are spindle-shaped blobs typically 20 cm long and 2–4 cm thick (Bromberg 1982, p. 57).

---

[16]See for this https://pluto.space.swri.edu/image/glossary/pitch.html.

As recalled above, in his talk at Princeton in 1954, Edward Teller had noted that any device with magnetic field lines that are convex towards the plasma were likely to be unstable. Today this instability is known as the interchange or flute instability, a type of Rayleigh-Taylor instability. The mirror has precisely such a configuration. According to Teller, the instability could be traced to the fact that the magnetic field lines curve away from the plasma. Conversely, in configurations in which the magnetic field lines are convex towards the plasma, stability would intuitively be possible. Teller had only made preliminary calculations and still he had been able to point out a real problem. At that time better than preliminary calculations were hardly possible, as making calculations on hot plasmas was only possible for greatly simplified models. The mathematical tools (computers) and a proper physical understanding were simply not available. The relation between ideal MHD models, which picture the plasma as a simple fluid, and actual plasmas was completely unclear.

The Post group did not see the problem for its so-called Toy Top machines and even declared that the prediction of instability was probably not applicable to its experiments. The Russians however immediately saw what Teller warned about and fixed it by making the magnetic field concave by adding a series of additional bars (so called Ioffe bars after their inventor, the Russian physicist Mikhail Ioffe (1917–1996)).

The Ioffe bars consist of four extra conductors parallel to the axis of the device. When current was sent in the proper directions through these bars and the original coils, the resulting combined magnetic field created by the two types of conductors took the shape of a twisted bow-tie or a peppermint candy, called a *minimum-B* configuration (Fig. 3.11). The magnetic field B at the centre is finite, but at a minimum and the strength of the field increases in every direction outwards from that minimum (it is a sort of "magnetic well"). Plasma in a minimum-B configuration feels a stronger magnetic field strength in all directions and its confinement is stable, so it greatly improved the confinement times.

Fig. 3.11 Magnetic mirror with Ioffe bars (*picture from* LLNL)

At the Salzburg Conference in September 1961, where both the Americans and Russians presented their results, Lev Artsimovich (1909–1973), who headed the controlled nuclear fusion research programme in the Soviet Union, pointed out to the Americans that they had overlooked the problem and misinterpreted some of the neutrons that were seen coming out. He caustically remarked "that Ioffe's results are in sharp contradiction with the attractive picture of a thermonuclear Eldorado … drawn by Dr. Post." And he drew the devastating conclusion that "we now do not have a single experimental fact indicating long and stable confinement of plasma with hot ions within a simple magnetic mirror geometry." (Bromberg 1982, p. 110–111.) This was ten years after the start of the Livermore programme and caused a shift of its strategy towards emphasis on basic research instead of towards designing a real fusion reactor, which was not to the undivided liking of the administrators and lawmakers in Washington.

At Livermore the ALICE (Adiabatic Low-Energy Injection and Capture Experiment) machine, no longer a toy or table top machine, but a sizeable apparatus at a cost of several million dollars, was under construction in the early 1960s and in 1963, after more than a year of deliberation, it was decided to redesign the machine to incorporate Ioffe bars, and not just four, but at first six and later even twelve.

Although the physics and the mathematics were wanting, the artistic skills of the scientists were beyond doubt as shown by Fig. 3.12, which combines the Ioffe bars with the circular magnetic coils around the tube into a single coil. It is a pure piece of art and, what is more, was providing the stability one was looking for. It was designed in 1965 at the Culham Centre for Fusion Energy in England, picked up in the US and renamed "baseball coil" after the stitching of the seam on a baseball. Baseball

Fig. 3.12 A baseball coil (*picture from* LLNL)

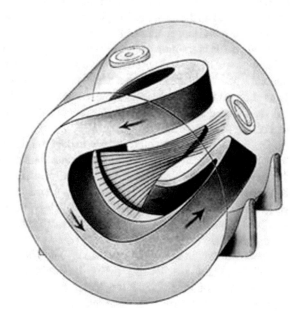

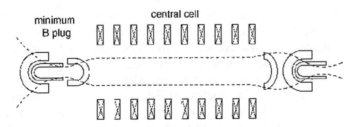

Fig. 3.13 Principle of tandem mirror configuration with long central cell and two mirror end coils (*from* Kawabe 1983, p. 63)

coils had the great advantage that they left the internal volume of the reactor open, allowing easy access for diagnostic instruments.

Post later introduced a further improvement, the "yin-yang coils" (Fig. 3.13), which used two horseshoe-shaped magnets to produce the same field configuration, but in a smaller volume. The bars indeed did the job and suppressed the flute insta- bility, but it was now discovered that the leakiness of the design was far higher than expected. This was traced to a host of newly discovered "micro-instabilities" (a hall- mark of which was the emission of high-frequency radio waves) that were hitherto masked by the more virulent flute stability. They caused plasma to flow out at the ends of the mirror. Suppressing these new problems filled much of the 1960s and created the need for ever larger machines.

How did this leakage occur? At first there was a rapid leakage of electrons, which were much lighter and much more mobile than ions. This produced an excess number of ions in the plasma, resulting in the build-up of charge in the plasma. This 'space charge' pushed on the ions, repelling them and soon causing them to escape as rapidly as the electrons (Heppenheimer 1984, p. 87).

In the 1970s, a solution was developed, independently by a group in Novosibirsk (Dimov et al. 1976) and by Post's group at LLNL, by arranging three mirrors in tandem whereby the outer mirrors help confine the plasma within the central mirror. In its simplest form it consists of a long central solenoid (the central mirror or central cell) terminated at each end by two mirror coils that form a short mirror system, the end cell or plug, as illustrated in Fig. 3.13.

The space charge that would develop in the central mirror would then be balanced by the space charges in the end mirrors that flanked the centre. This led to the "tandem mirror" machines (TMX (Tandem Mirror Experiment) and TMX-U), which operated from 1979 to 1983. The basic idea consisted of a tube formed from ring- shaped magnets, closed at the end with baseball-type magnetic mirrors, as shown in Fig. 3.14. The end coils became more and more complex as ever more difficulties were overcome.

The TMX, which was built to test the layout to be used in a planned large device, the Mirror Fusion Test Facility (MFTF), demonstrated a new series of problems and also suggested that MFTF would not be able to reach its performance goals. Calculations had already made clear that the best it could obtain was a power output

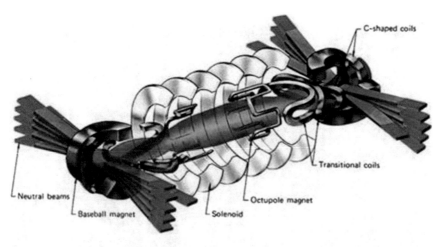

Fig. 3.14 Diagram of the TMX. The flat bars are neutral beams heating the plasma in the stabilizing baseball coils (Original Livermore drawing)

equal to 3% of the input, while the Tokamak Fusion Test Reactor, at that time under construction at Princeton, was supposed to achieve breakeven. It didn't, but that is another story which will be told in Chap. 8. However, theoretical calculations had also shown that the maximum amount of energy the best yin-yang reactor conceivable could produce would be about the same as the energy needed to run the magnets (Heppenheimer 1984, p. 86).

The TMX was upgraded to TMX-U and the MFTF design was modified during construction to MFTF-B (Fig. 3.15a, b) to incorporate the tandem idea.

The cost of the MFTF-B was 372 million dollars and, at that time, the most expensive project in Livermore history. It opened on 21 February 1986 and was promptly shut down. The reason given was to balance the United States federal budget, but it had already become clear that as a power-producing machine, the mirror appeared to be a dead end and would never be able to beat the tokamak. There was no point in continuing with such devices, which soon meant the end of mirror research in the US.

3.3.5 Other US Programmes

3.3.5.1 Industrial Programmes

Industrial fusion programmes in the US date from the mid 1950s when it was made possible for industry, in particular General Electric and a new firm called General Atomic, to gain access to information from Project Sherwood, even though most of that information was still secret. At that time industry was not yet convinced of the

a

b

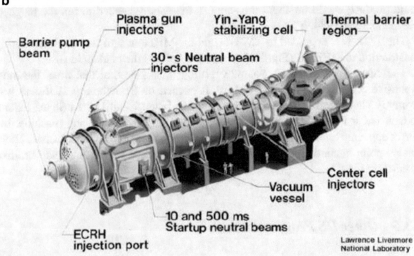

Fig. 3.15 a Mirror Fusion Test Facility at Livermore. One of the two yin-yang mirrors (the size of a two-story house) arrives at LLNL. The plasma was confined in the small area between the two magnets. **b** The MFTF tandem mirror configuration, showing the position of the yin-yang coils of Fig. 3.15 and giving an idea of the total size of the colossus

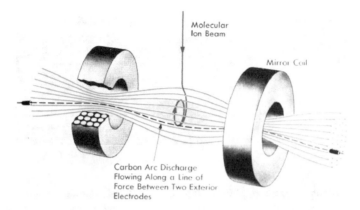

Fig. 3.16 The DCX experiment. A carbon arc discharge is used to cause a high-energy molecular ion beam to dissociate. The resulting deuterium nuclei are then trapped in a magnetic field (*from* Bishop 1958, p. 137)

need for energy production from nuclear fission, let alone nuclear fusion, because fossil fuel supplies were still adequate and waste disposal from nuclear fission plants were thought to pose major problems. The first generation of nuclear fission reactors would only come into operation in the 1960s and in the 1950s it was not clear whether they would be commercially attractive. Fusion scientists were optimistic that controlled nuclear fusion could be demonstrated within a few years, and that a demonstration fusion power plant could be running in 1975. So, the interest of utility companies in nuclear fission reactors was limited, as they might perhaps be able to skip the fission stage altogether and proceed immediately to fusion. With that prospect in mind, electricity companies could of course not ignore the fusion option.

In 1956 when industry got access to Sherwood information, General Electric decided to start a substantial fusion research programme. Other companies, among them Westinghouse, did not start their own programmes, but had contacts on scientific and engineering problems with various government labs and universities. The only firm apart from General Electric that came with its own fusion programme was General Atomic, founded in 1955 in California with assistance from the physicists Edward Teller and Freeman Dyson (1923–1920) and originally part of the General Atomic division of General Dynamics, a major American multinational aerospace and defence corporation.[17] In 1957 General Atomic started a fusion research programme, from 1959 partly financed by TAERF (Texas Atomic Energy Research Foundation), a consortium of Texan utility companies. The stabilised toroidal pinch was chosen as the first major experimental device at General Atomic. It was built in 1958 and followed in the years to come by many other devices.

---

[17]General Atomic was acquired by Gulf Oil in 1967, from 1973 jointly owned by Gulf Oil and Royal Dutch Shell, from 1984 wholly owned by Chevron following its merger with Gulf Oil, and in 1986 sold to the Blue brothers (James Neal and Linden Stanley Blue) after which it assumed the name General Atomics.

Fig. 3.17 Formation of
pattern of closed field lines
in Astron (*from* Bromberg
1982, p. 121)

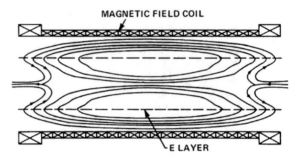

In spite of being a commercial company, the attitude at General Atomic was more academic than industrial, in contrast to General Electric where fusion was supposed to contribute to profitable commercial production. When that did not work out, General Electric wound down its nuclear research programme in the doldrum years of the 1960s. General Atomic did not follow suit and, when in 1967 TAERF withdrew its funds from the fusion programme, General Atomic sought AEC funds to continue its fusion research and its project became gradually absorbed into the AEC programme (Bromberg 1982, pp. 101–104, 149).

3.3.5.2   DCX

In addition to the three major projects, at Princeton, Livermore and Los Alamos, described above, a small group at the Oak Ridge National Laboratory also became interested in fusion research. Within the framework of the Sherwood programme, they started experiments to clear up the discrepancy found by Bohm, and carried out other studies for in particular the Livermore group. In the mid 1950s this led to the Direct Current Experiment (DCX), a facility in which a beam of molecular ions (D_2^+, singly ionised molecules of deuterium, so consisting of two nuclei of deuterium and a single electron) was injected into a carbon arc,[18] an electric arc between two carbon electrodes. The molecules would partly break up into their constituent nuclei and the latter were trapped in the magnetic field. The setup, which in fact is a type of mirror device, is shown in Fig. 3.16. The purpose of this first experiment was to determine the degree of break-up and trapping of the beam and the possibility of building up a plasma in this way. The preliminary results were encouraging (Bishop 1958, pp. 136–137; Bromberg 1982, p. 79).

The deuterium nuclei were however catching free electrons from the carbon discharge and after neutralization flew out of the containment vessel. The result was that the plasma density remained very low. No improvement worked. Various new designs were tried in a whole series of DCX machines, but the plasma density

---

[18]An arc is an electrical breakdown of a gas that produces an ongoing electrical discharge. It originates when an electric charge is led between two electrodes. Since a plasma is formed, the gas between the electrodes lights up.

remained far too low (10^{15} particles per m^3). All the advances made showed that the same type of micro-instabilities that had plagued the Livermore mirror machines were also wreaking havoc here. The situation already reached crisis point in 1965 and, late in the 1960s, just before everybody joined the tokamak bandwagon, the Oak Ridge fusion programme abandoned mirror devices and switched to tokamaks.

3.3.5.3 Astron

A final machine in the Sherwood programme worth mentioning is the Astron, a device pioneered by the Greek-American physicist and engineer Nicholas Christofilos (1916–1972) at the Lawrence Livermore National Laboratory. In 1953, without being aware of the then secret controlled thermonuclear research programme going on in the US, he went to Washington with his own scheme for a fusion reactor, which was positively received.

Astron was a novel approach to controlled fusion with, as its basic principle, the use of circulating beams of charged particles rather than solid coils to generate the magnetic field for plasma confinement. It was strikingly different from any other concept under study at the time. First, an axial magnetic field is produced in a cylindrical tube equipped with magnetic mirror coils, as in a mirror machine. In Christofilos' design high-speed electrons were then injected into this tube; they were trapped in the magnetic field and their orbits bent into a cylindrical shell, the E-layer. This rotating cylindrical shell of very high-energy electrons (20 MeV, about 1000 times higher than the 20 keV of ions in a reactor plasma) was the key feature of his approach. The E-layer would act as a solenoid and create its own magnetic field opposite in direction to the field of the external coils. The injection of electrons would continue until the current in the E-layer would exceed the external current flowing in the mirror coils, at which point the overall magnetic field produced by both the E-layer and the coils would reverse its direction to oppose the fields that would have been created by the mirror coils alone.

So, by means of the E-layer it should be possible to form a pattern of closed magnetic field lines in which plasma could be well contained (Fig. 3.17). In addition to being a unique means of producing confinement, the E-layer also boasted a revolutionary confinement geometry of closed field lines in the form of a torus within a *linear* machine, which implied that plasma loss at the ends, along outgoing field lines, as in other linear machines would not be an issue (apart from the outermost portion of the chamber as can be seen in the schematic picture of Fig. 3.17).

Once the E-layer had been formed a neutral gas of deuterium and tritium would be injected into the chamber. The E-layer would ionise the gas into a plasma and heat it further to thermonuclear temperatures. Since the electrons of the E-layer would continually lose energy to ionise and heat the plasma, new electrons would have to be continually fed in. This implied that an electron accelerator had to be part of the installation. It will be clear from this that the Astron was a complex and challenging design, both scientifically and technologically.

Work on the Astron started early in 1957 at Livermore and construction continued until the early 1960s, when the first experiments were conducted. The results were promising, but after several years of adjustments and experiments it failed to achieve an E-layer to a density sufficient to produce field reversal and the resulting configuration of closed field lines. At various instances in the late sixties and early seventies the Astron project was threatened with termination, especially Robert Hirsch, who took over the AEC fusion programme in 1972 (see Chap. 4), was very much against the project (Heppenheimer 1984, Chap. 4), but it continued until its final shutdown after Christofilos' untimely death in 1972 from a massive heart attack.

With the demise of the Christofilos' project the trial-and-error period of early fusion research came to an end (Bishop 1958, p. 148ff; Bromberg 1982, pp. 119ff, 201–202; Coleman et al. 2011).

Before continuing with the early developments in the Soviet Union and the first tokamaks, it seems appropriate to first relate the story of the declassification of fusion research.

3.4 Declassification

After World War II America greatly tightened security around all nuclear efforts, irrespective of whether it had to do with fission or fusion. The Atomic Energy Act (or McMahon Act after its sponsor senator Brian McMahon) was passed in 1946 and provided for the establishment of the AEC. The most significant stipulation in the Act was that nuclear weapons development and nuclear power management would be under civilian, not military control. The AEC was in charge of developing the US nuclear arsenal, taking over these responsibilities from the war-time Manhattan Project.

However, in spite of its civilian nature the Act had rather draconian secrecy regulations. At first it did not look that way as in the bill put before Congress one of the purposes of the Act was stated as "A program for the free dissemination of basic scientific information and for maximum liberality in dissemination of related technical information." And section 9 of the bill "Dissemination of Information" acknowledged that the government had accumulated secret information on nuclear technology during the war, and called for release of that information "with the utmost liberality as freely as may be consistent with the foreign and domestic policies established by the President." For national security and public safety purposes, the bill emphasized control of nuclear materials, not control of information. However, after having gone through the motions in Congress, in the eventual Act the new second purpose had become "(2) A program for the *control* of scientific and technical information...," and section 9 was replaced by a new section 10, "Control of Information." Slightly modified in 1954 to allow for the development of commercial nuclear power, the Act is still in force today. When the Atomic Energy Act became law, it defined a new legal term "restricted data" as "all data concerning the manufacture or utilization of atomic weapons, the production of fissionable material, or the use of fissionable material in the production of power," unless the information has been declassified. The phrase "all data" included every suggestion, speculation, scenario, or rumour—past, present, or future, regardless of its source, or even of its accuracy—unless it was declassified. All such data was "born secret" and belonged to the government (Morland 2005). From then onwards the words "born secret" started a life of their own.

So, the fact that in the US the fusion programme came under the AEC automatically implied that it was born secret, not only the details of the programme, but even the fact that the AEC was sponsoring such work and the names of the sites where the work was being carried out (Bromberg 1983, pp. 30–31). The Atomic Energy Act prevented any foreigners, including the British and Canadians who had participated in the Manhattan project, from having access to American nuclear secrets, and virtually ended the collaboration between Britain and the US. The early efforts, both in the UK and the US, described in the first part of this chapter were classified with very limited exchange of information between their respective programmes. It lasted until 1957, more than a decade after the British had been barred from access to the American programme, that declassification between the two sides was finally worked out in a declassification guide. This resulted in a common secrecy policy whereby neither side would publish anything without the other's consent (Clery 2013, p. 61). All information was to be made available except for information on devices exhibiting a net power gain, and since there was no prospect of the latter it actually meant total declassification. As reported earlier, after some bickering it allowed the British to publish the spurious ZETA data, causing considerable embarrassment.

The first calls for declassification in the US date from 1952 when people were arguing that secrecy hampered the recruitment of capable people, the free flow of information and ideas, and the publication of papers that could be read, discussed and criticized by others working outside or inside the field. As far as recruitment is concerned, it is striking in this respect that none of the great American theoretical and experimental physicists of the 1950s, who would harvest Nobel Prizes by the dozen in the decades to come, even those who had worked on the Manhattan Project, were actively involved in any of the fusion work, either in its technical problems, or in the theory of plasma physics, the only exception perhaps being Edward Teller.[19] Secrecy moreover was against the open research atmosphere at universities, so especially at Princeton University, where the stellarator was explored within the framework of the classified Project Matterhorn, there was opposition to having such secret research programmes on its campus, while MIT even refused to set up a plasma physics/nuclear fusion programme before secrecy was lifted. For the national laboratories (Oak Ridge, Livermore, Los Alamos) this was less of a problem. Since they were basically weapons laboratories, the fusion facilities were located within fenced-in security areas and remained so even after declassification, which continued to hamper access for foreign fusion scientists. Although most scientists were in general

[19]Teller also seemed to have a more realistic view of the possibilities for achieving nuclear fusion, judging from his 1958 statement: "Fusion technology is very complex. It is almost impossible to build a fusion reactor in this century".

strongly in favour of declassification, the AEC decided against it. The mood for declassification also fluctuated with the perceived state of affairs. The early period of fusion research from 1952 to 1958 is characterised by wild swings from optimism to pessimism about imminent success and back. Fusion and plasma physics were not (yet) a normal scientific discipline and at least in the US it worked on a trial-and-error invention cycle often based on back-of-the-envelope calculations. Scientists and administrators were convinced that the nuclear fusion nut could be cracked in a matter of a few years. In such a situation declassification would of course not be opportune as it might enable others, perhaps even foreigners, to reap (part of) the glory. So, in June 1956 when the fad of the day, the so-called stabilised pinch with an extra external magnetic field to tame the kink instability, suddenly on the basis of yet another new theoretical prediction became the hot favourite in the fusion race, Jim Tuck of Perhapsatron fame wrote to the AEC that "it seems a particularly unsuitable time to take hasty action like complete declassification just now, since we surely ought to wait and see whether the very new ideas on stabilization constitute a breakthrough." For, according to Tuck, "once a stellarator or pinch is found to be stable, the rest is mere slugging at detail" (Bromberg 1982, pp. 70–71). It was again one of those instances where scientists were much too quick in pronouncing new ideas 'promising' or a 'breakthrough' while they actually did not have a clue how plasma behaved, apart from that it was very turbulent.

Declassification became of course also a topic in the discussion on the peaceful uses of atomic energy, either from fission or from fusion. After all, apart from producing weapons nuclear power can also be used to generate electricity and all nations in the world were entitled to benefit from such peaceful use. So far, the world's governments had failed to establish international control over atomic energy and ever more countries were starting their own programmes to harness the power of the atomic nucleus. The discussion on the peaceful uses of atomic energy was started by President Eisenhower in his Atoms for Peace address to the United Nations General Assembly in December 1953. He called for an international conference on such peaceful uses to be held in Geneva and proposed that an international agency be established for making information and nuclear material available under strict guidelines to other nations wishing to engage in peaceful applications of nuclear technology. "Thus, the contributing powers would be dedicating some of their strength, to serve the needs rather than the fears of mankind," he said.

The Russians initially rejected Eisenhower's proposals as empty propaganda, but a few months later the new leadership of the Soviet Union showed some willingness to start negotiations with the Americans and to participate in some international organisation. They came round when they realised the propaganda value of such atoms for peace. It gave the Soviet Union the opportunity to show the world that they had become a leading industrial and scientific power, although of course the launch of the Sputnik satellite in 1957, followed by the first man orbiting the Earth in 1961, would do this to a still greater extent and to the utter stupefaction of the Western world. It resulted in 1955 in the first United Nations International Conference on the Peaceful Uses of Atomic Energy and in 1957 in the creation of the International

Atomic Energy Agency (IAEA) (Josephson 2000, p. 172ff). Apart from this conference it was not directly followed up by any concrete action towards declassification
on the part of the Americans, but ever widening cracks were appearing in the secrecy
surrounding nuclear fusion work. At a press conference associated with this conference in August 1955, the AEC confirmed that the US was actively engaged in a
controlled thermonuclear programme. And that was all it was prepared to reveal for
the time being. By prior arrangement, Homi Bhabha, the chairman of the conference,
had mentioned fusion in his opening address and prophesied on its radiant future,
but he cannot have known much about it, as all major programmes (US, UK and
SU) run in the world at the time were secret. He was working in India on the Indian
nuclear (weapons) programme and was not involved in any fusion research.[20] The
main topics of the conference were, for that matter, not on nuclear fusion and/or
nuclear fusion reactors, as it was much too early for that, but on power generation in
nuclear fission reactors.

The lifting of secrecy of nuclear fusion research was finally decided on in
April 1956 within the framework of the Canadian-British-American declassification conference and in July 1956 the first review article on the subject appeared (Post
1956 already cited above).

In the Soviet Union research in all nuclear fields was of course also secret, if
possible even more secret, and in the West not much was known about the Soviet
efforts, while all kinds of rumours were circulating that the Russians had solved
the problem of plasma heating and were well on their way to a working fusion
reactor. Much of this was debunked by Kurchatov's celebrated speech at Harwell,
which showed that the Russians had the same problems (short plasma confinement
times, low temperatures, instabilities and Bohm diffusion) and were trying the same
solutions as others. Kurchatov's speech, which was undoubtedly also inspired by
Eisenhower's speech mentioned above, but went quite a bit further than the AEC's
statement at the 1955 conference, has been depicted by some Russian/Soviet authors
as the first open step towards declassifying worldwide research on nuclear fusion,
and that Kurchatov was the driving force behind this. But why would the Soviets
work towards declassification apart from the propaganda value mentioned above?
Did they expect to hear from their Western colleagues developments that they were
not aware of? The latter is quite unlikely as the flow of information from west to
east through Soviet agents must have remained considerable and they must have
known that in the West no progress was being made either. It is perfectly possible
that Kurchatov was genuinely of the opinion that collaboration with the West would
be advantageous to both parties, but his opinion, although valued by the Soviet
authorities, would by no means have been decisive in this respect. The arguments
advanced by scientists in the US in favour of declassification as mentioned above
(recruitment, free flow of information and suchlike) did in any case not hold for the
Soviet Union where there were no difficulties in recruiting the best and brightest

---

[20]Bhabha is known as the father of Indian nuclear power, including nuclear weapons, and it is a
little ironic that he, an aggressive and somewhat cynical promotor of nuclear weapons for India,
acted as chairman of the very first conference on the peaceful uses of atomic energy.

(Sakharov, Tamm, Artsimovich) for work on nuclear fusion. Nonetheless, in his Memoirs Sakharov also credits Kurchatov with the initiative in arguing for openness and international collaboration. Sagdeev suggests that Kurchatov had a kind of guilt complex from his many years of working on atomic bombs, and wanted to develop something peaceful, like a nuclear reactor, for which he could be remembered. He also suggests though that the Soviet fusion project was troubled (because of unavoidable and uncontrollable instabilities) and that Kurchatov needed a new impetus for it (Sagdeev 1994, pp. 66, 101). Already in March 1954, just one year after Stalin's death and after the removal of Beria, who so far had been the boss of everything nuclear in the Soviet Union, Kurchatov and a few colleagues had written to Georgy Malenkov, at the time one of the pretenders for the prime position in the Soviet power pyramid, about the dangers of atmospheric nuclear testing because of radioactive fallout and urged the declassification of fusion research. Apparently Kurchatov saw this as a good start for trying to stop the madness of a nuclear arms race (Josephson 2000, p. 172). Although Malenkov lost the power struggle to Khrushchev, the latter apparently also became convinced that there was no military excuse for secrecy in this field and allowed Kurchatov to give his speech at Harwell. Just before he went to Britain, declassification on a national Soviet level had been decided at an all-Union conference in December 1955 and confirmed in February 1956 at the 20th Congress of the Soviet Communist Party. It typified the confident mood the Soviet leadership was in and was soon to culminate in the launch of Sputnik a year later, which gave the feeling in the West that they were losing the space race and the fusion race as well. It is often said that Kurchatov did not say anything that most people in his audience did not already know. That may be true, but it was also true that the people in his audience were themselves not allowed to be as open about their own research or the research of others as Kurchatov had been about Soviet research, as became clear to the chagrin of the audience when in the summer of 1956 Edward Teller gave a talk on nuclear fusion to about 1000 scientists of the American Nuclear Society in which he said much less than Kurchatov had told the British (Lapp 1956).

In spite of the fact that Kurchatov's talk did not give away much, it was quite radical, even revolutionary, in the sense that the Soviets were making the first step in lifting the secrecy surrounding nuclear fusion, implying that there was no overriding military use for controlled nuclear fusion and that the peaceful uses of the atom could be pursued by international cooperation.

The British responded by publishing a number of papers in the January 1957 issues of the *Philosophical Magazine* and the *Proceedings of the Physical Society*, e.g. Lawson's paper with his criterion for a fusion reactor to reach ignition (see Chap. 2), and were also present at the IAU Symposium on Electromagnetic Phenomena in Cosmical Physics in August 1956, four months after Kurchatov's visit to Harwell, where the later director of the Culham Laboratory for Plasma Physics and Nuclear Fusion Bas Pease (1922–2002) gave a talk on pinch phenomena. At this same conference Artsimovich and Igor Golovin more or less summarised the Soviet efforts on nuclear fusion so far. (See Braams and Stott 2002, p. 24ff, which contains a rather detailed description of the slow crumbling of the secrecy regime.) Lyman Spitzer also spoke at this symposium, but not on the stellarators that were keeping him busy at home, but on "Theoretical problems of stellar magnetism", fully in line with the secrecy policy of the United States.

Declassification culminated in the 1958 Geneva Atoms for Peace Conference which by the way was not a conference solely on nuclear fusion, but covered everything nuclear (physics, controlled fusion, fission, nuclear materials, fission reactors, applications in medicine and biology, chemistry). As far as nuclear fusion was concerned it became more an exercise in boasting and propaganda than a serious conference.[21] The conference was open to the general public and it included a giant exhibition hall where the various countries could vie with each other for the public's attention with their fancy exhibits. None of the teams working on fusion had essentially anything to show for, so it made sense to try to impress the public, which the Russians did by showing a full-scale model of one of their sputnik satellites, although it had of course nothing to do with fusion. As far as nuclear fusion was concerned, the Americans stole the limelight with a smartly polished exhibit of the Spitzer stellarator (Fig. 3.18), a much fancier specimen than the cobbled-together design of Fig. 3.8 and for which no expense had been spared, but was nonetheless not working. There were about 100 exhibits at the conference, 62 of which originated from the USA, 9 from Britain, 5 from France and only 13 from the Soviet Union.

After the conference nuclear fusion could be shrouded in the aura of the 'peaceful atom' and be declared a triumph of Western civilization. The US stellarator made a big impression on the Russians, and in particular on Kurchatov, so much so that for some time he wanted to abandon the tokamak (the very T-3 tokamak that was going to triumph in 1968!) and build a stellarator in Moscow (Clery 2013, pp. 117–118). And indeed, when back in the SU Kurchatov almost immediately set to work and started construction of a huge stellarator at Kharkov in Ukraine (see Chap. 13). It was the stellarator's stationary magnetic system for plasma confinement, which was ideal for controlled nuclear fusion, that in Kurchatov's eyes made it superior to the tokamak, which by necessity was a pulsed system. This argument is still valid today.

International contact also led to an increase in Soviet efforts in the field of fusion. Kurchatov had no difficulty in persuading the Soviet authorities that more money and facilities were needed in order not to lose the race in fusion with the Americans and the British (Josephson 2000, pp. 176–177).

In spite of the suspicion, bravura and rivalry the conference was also a celebration of openness. Scientists could openly talk with each other and exchange ideas. They became aware, as was already clear to some extent from Kurchatov's Harwell speech, that in spite of the almost complete isolation the various parties had worked in they all had invented circular and linear devices using magnetic fields and all had encountered the same difficulties and seemingly insurmountable problems, especially with instabilities (Herman 1990, p. 54ff).

The whole exercise showed how immature the field still was and the atmosphere of showmanship, exaggeration and hullabaloo present at the 1958 conference has remained one of the hallmarks of the nuclear fusion enterprise ever since. After the

---

[21] A short American propaganda film about the conference can be viewed at https://go.nature.com/2LP10FD.

Fig. 3.18 The stellarator presented at the 1958 Geneva conference

conference everybody went home and continued doing what they had done before. Any direct collaboration between east and west was not one of its results.

3.5   Early Developments in the Soviet Union

3.5.1   The Birth of the Tokamak

The tokamak is currently the most common device that uses magnetic fields to confine plasma. The word is derived from the Russian acronym for toroidal chamber with magnetic coils (*toroidal'naja kamera s magnitnymi katushkami*) or toroidal chamber with an axial magnetic field (*toroidal'naja kamera s aksial'nym magnitnym polem*). The latter would result in tokamag, which easily evolves into tokamak. The first design stems from work by Igor Tamm (1895–1971) and Andrei Sakharov (1921–1989) in the Soviet Union in the very early 1950s when they were studying the problem of thermonuclear reactions within the framework of the Soviet effort to build a hydrogen bomb (Fig. 3.19).

Before going a little more into the details of the tokamak, we will first relate the tale of Oleg Lavrentiev.

  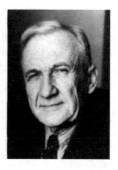

Fig. 3.19 The initiators of the Soviet effort on nuclear fusion: Oleg Lavrentiev, Andrei Sakharov and Igor Tamm

3.5.2 The Story of Oleg Aleksandrovich Lavrentiev

In the summer of 1950 a letter from Oleg Lavrentiev (1926–2011) (Fig. 3.19), at the time a soldier serving in the far east of the Soviet Union, was forwarded to Andrei Sakharov. Lavrentiev was born in Pskov, where he personally experienced the German occupation in the first years of World War II. After the liberation of his hometown in 1944 and before having finished his secondary schooling, he volunteered for army service and was sent to the front in the Baltics. After the war he was posted to the far east on Sakhalin island as a radiotelegraph operator, where he apparently had plenty of time to study nuclear physics. In his letter, which was sent to the Central Committee of the Communist Party and is seen as the beginning of controlled nuclear fusion research in the Soviet Union (Shafranov 2001, p. 837), Lavrentiev made a proposal for a hydrogen bomb, noted the possible importance of controlled thermonuclear reactions for future energy production and sketched a possible approach to the latter problem. Because of this letter Lavrentiev is credited with being the first person in the Soviet Union to formulate the problem of controlled nuclear fusion and to suggest a certain approach to its solution.[22] Knowing where the letter came from, Sakharov was very quick with his reaction (Lavrentiev's letter was received by the Central Committee on July 29 and Sakharov's reaction dates from August 18) and he was very cautious in his reply. Lavrentiev's idea of electrostatic confinement of deuterium nuclei between two spherical grids was no good and Sakharov pointed out some difficulties but concluded very diplomatically that

---

[22]Bondarenko 2001, p. 844; however, in his biography George Gamow mentions that in 1932 Nikolai Bukharin attended a lecture Gamow gave "on thermonuclear reactions and their role in the energy production in the sun and other stars. After this talk, he (Bukharin) suggested that I head a project for the development of controlled thermonuclear reactions (and that in 1932!). I would have at my disposal for a few minutes one night a week the entire electric power of the Moscow industrial district to send it through a very thick copper wire impregnated with small "bubbles" of lithium-hydrogen mixture. I decided to decline that proposal, and I am glad I did because it certainly would not have worked." (Gamow 1970, p. 121). So, actually if Gamow is to be believed, Bukharin, not even a physicist, must be given this credit.

"it cannot be excluded that certain changes in the project may correct this difficulty". He also praised "the creative initiative of the author." In his Memoirs Sakharov writes that the letter did raise vague thoughts in his mind about magnetic confinement, instead of Lavrentiev's electrostatic confinement, in whose setup "there was no way to ensure that the hot plasma would not come into contact with the grids, which would inevitably lead to heat loss and render such means incapable of attaining sufficiently high temperatures for thermonuclear reactions" (Sakharov 1990, p. 139ff). As he wrote in his report, for Sakharov the main new feature in Lavrentiev's idea was the low density of the confined particles ("… a thermonuclear reaction in a high-temperature gas (billions (sic) of degrees) of such low density that the existing materials could withstand the resulting pressure" (Sakharov 2001). So it turns out that all serious scientists (Lavrentiev not being one of them) when thinking about the problem at the time, be it in the UK, US or Soviet Union, all very quickly realised that the solution for the plasma confinement problem had to be looked for in using magnetic fields. The difference in the various approaches is the geometry and strength of the magnetic fields or the way in which they are generated. The essential feature is that magnetic field lines can form closed loops or closed surfaces and keep the charged particles, which follow these field lines, within a closed region, as explained above when describing Spitzer's stellarator. This can in principle solve the contact problem with the external walls of the vessel containing the experimental setup. "The appearance of closed magnetic force lines is particularly evident", Sakharov writes, "within the interior of a torus when the current passes through the toroidal windings on its surface". The particles will start to describe the helical motion along the field lines, as we have discussed earlier. In view of all this it goes much too far in my view to claim that Lavrentiev's letter was a catalyst for the Soviet programme of controlled nuclear fusion, or that he invented fusion as some Russian scientists still assert. Lavrentiev's letter has become part of the folklore of nuclear fusion, but his contribution was in fact not all that significant. If his letter had not been passed on by the feared Soviet organs, Sakharov might not even have bothered to answer it.

Before Sakharov sent in his report, Lavrentiev had already arrived in Moscow on 8 August, just a week after the Central Committee had received his letter. What necessitated the incredible haste to have the young man transferred to Moscow and who had taken the initiative in this is unclear. In any case he passed the entrance exams to Moscow State University (MGU) and started to study there in September 1950 (with a scholarship, superior housing and special tutoring). Lavrentiev's letter and later elaborations continued to play a role in the higher circles of government. He was also received by leading scientists, among whom Kurchatov. It is not clear if they took his ideas all that seriously (as far as I know Sakharov is the only scientist who has put his thoughts about Lavrentiev's ideas on paper, and as mentioned he was very cautious in his writing), but Lavrentiev's casual mentioning that he had been to see Beria (for unclear reasons Lavrentiev had very good relations with the notorious People's Commissar Lavrentiy Beria, who headed the Soviet secret service KGB and was also the leader of the atomic bomb project) worked like magic for him and got him a position in Kurchatov's institute in spite of just being a first-year physics student (Bondarenko 2001, pp. 846–847). It appears to me that Lavrentiev was either naïve or very cunning; I am inclined to the latter. Sakharov was just kind and cautious as he was very much aware of Beria's infamy. (On Beria see Knight 1993.)

The acquaintance with Beria made Lavrentiev an object of suspicion among his fellow students and scientists and after Beria's fall in 1953 Lavrentiev's luck also ran out. His privileges evaporated, he was barred from Kurchatov's institute and sent to Kharkov where he continued work on plasma confinement with an electromagnetic trap. Without success. (Clery 2013, pp. 113–114; for the story of Lavrentiev from a contemporary, although not correct in all details, see Sagdeev 1994, pp. 33–36.)

3.5.3 The Magnetic Thermonuclear Reactor

Whatever Lavrentiev's contribution may have been, it remains a fact that he jolted other physicists into action. In particular Sakharov and Tamm started to study the problem in 1950 almost immediately after Lavrentiev's letter arrived, although Sakharov also says in his Memoirs that these questions were already on his mind in 1949. And they must have been, as is clear from the compilation of papers in volume I of *"Plasma Physics and the Problem of Controlled Thermonuclear Reactions"*.[23] It is hardly possible that all this work could have been done in the few months in 1950 after Lavrentiev's letter. At the time Tamm and Sakharov were both kept busy with their work on the development of the Soviet hydrogen bomb at the secret Arzamas-16 nuclear centre, now called Sarov, a town about 150 km south of Nizhny Novgorod, at the time called Gorky, where Sakharov would spend a few years in exile in the 1980s. Sarov is the Russian centre of nuclear research and for that reason still a closed town that can only be visited by Russians or foreigners with special permission.

Already in October 1950, just a few months after Lavrentiev's letter, they reported the principal design of such a magnetic thermonuclear reactor to a high Soviet official. This first report was followed by further communications in which the construction of a laboratory model of a magnetic nuclear reactor was urged. The announcement by the president of Argentina, Juan Perón, on 25 March 1951 that Ronald Richter had achieved thermonuclear fusion in his country (which also inspired Spitzer as we have seen above) resulted in further frantic activity by Kurchatov and others, culminating in Stalin signing in May 1951 the top-secret resolution of the USSR Council of Ministers "On conducting research and experimental work to clarify the feasibility of building a magnetic thermonuclear reactor" (Goncharov 2001, p. 854; Shafranov 2001, p. 837ff).

---

[23] A four-volume compilation of papers on *Plasma Physics and the Problem of Controlled Thermonuclear Reactions,* edited by M. A. Leontovich and published in 1958 just before the second Geneva conference within the framework of the declassification of nuclear fusion research (English translation published in 1959 by Pergamon Press).

Fig. 3.20 Leading Soviet fusion researchers: upper row left to right: Igor Kurchatov (1903–1960), Lev Artsimovich (1909–1973), Mikhail Leontovich (1903–1981), Igor Golovin (1913–1997); bottom row left to right: Natan Yavlinsky (1912–1962), Vitaly Shafranov (1919–2014), Boris Kadomtsev (1928–1998), Evgeny Velikhov (b. 1935)

It was proposed that Lev Andreevich Artsimovich would head the controlled nuclear fusion research programme (with Mikhail Alexandrovich Leontovich (1903–1981) as head of the theoretical division[24] and in the early days with Tamm and Sakharov[25] as permanent consultants). Artsimovich would remain the driving force behind the tokamak design until his death. He managed to restore confidence in fusion research when he showed in the late 1960s that the tokamak design could overcome the syndrome of Bohm diffusion. Figures 3.19 and 3.20 show pictures of some of the most important players in Soviet fusion research in the first few decades.

Most of the work was carried out at Kurchatov's Institute of Atomic Energy in Moscow, but Kharkov (see Chap. 13), Leningrad, Sukhumi in Georgia, and later Novosibirsk were also drawn into this work.

The term 'magnetic thermonuclear reactor' (MTR) was coined by Tamm. In 1957 a more specific version was given the name tokamak. Igor Golovin is credited with being the creator of this name. In 1951 Tamm and Sakharov wrote three astonishingly rich papers, two by Tamm and one by Sakharov, all three with the title "The theory of a

---

[24]For some more details on the lives and background of both Artsimovich and Leontovich, see Paul R. Josephson 2000.

[25]Sakharov says in his Memoirs (p. 147) that he took no active part in later work, i.e. after 1950–1951, on fusion reactors.

Fig. 3.21 Schematic view of a torus, showing the z-axis (major axis), the major radius R (the distance from the centre of the torus to the centre of the tube) and the minor radius r (the radius of the tube). The ratio of R divided by r is called the aspect ratio

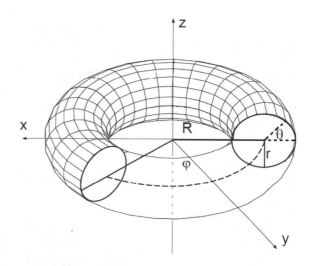

magnetic thermonuclear reactor".[26] In his contribution Sakharov proposed to confine the plasma in a doughnut-formed device (torus) by means of a toroidal magnetic field. He already drew the outlines of a D-D fusion reactor with a power capacity of 900 megawatt, a major radius of 12 m, a minor radius of 2 m (see Fig. 3.21), a magnetic field of 5 T, a plasma density of 3×10^{20} particles per m^3 and an ion temperature of 100 keV, not far from the dimensions of modern devices. He was also aware of the drift of the plasma particles due to the nonuniformity of the magnetic field and knew that a toroidal field alone, hence a field going around the torus the long way, cannot confine a plasma. If we take the schematic representation of Fig. 3.21, with the major axis as indicated called the z-axis, the magnetic field generated by the coils wound around the torus (the toroidal field) will create lines of force going around the torus and charged particles will travel in circles around the z-axis (describing their corkscrew trajectories along the field lines in the torus). Due to the curvature of the torus the magnetic field is however not completely uniform and the particles will at the same time slowly drift in the z direction; this drift is called the toroidal drift (cf. Spitzer, who also had to deal with this drift and designed his stellarator in the form of a figure-8 to cancel it). The positively charged ions and the negatively charged electrons will drift in opposite directions. The resulting separation of charges will induce an electric field causing the particles to drift further and further apart and eventually hit the wall of the torus. Therefore, this setup cannot confine the plasma unless the charge separation is cancelled (see e.g. Miyamoto 2011, p. 28).

Sakharov suggested two possible solutions to stabilize this toroidal drift, either send a current through the torus, which would produce a poloidal magnetic field with its field lines going the short away around the torus, or apply a direct external poloidal field, which will then induce a toroidal current, i.e. a current that goes round

---

[26]Published as the first three papers in the first volume of *Plasma Physics and the Problem of Controlled Thermonuclear Reactions*. Sakharov's paper has been reprinted in *Physics-Uspekhi* 34 (1991) 378.

and round in the torus (Shafranov 2001, p. 840) (Fig. 3.22). The poloidal field would combine with the toroidal field from the coils on the outside of the torus (cf. Fig. 3.4 for the pinch device) and the resulting combined field would be twisted into a helix, so that any given particle would find itself alternately on the outside and the inside of the torus, and the drift would be cancelled out. This is essentially how a tokamak works, not very different from the pinch devices we already met. In the next chapter the working of the tokamak, which would soon come to dominate the field, will be described in more detail.

Fig. 3.22 Magnetic fields and current in a tokamak. In the top figure indicated in blue are the toroidal field coils that produce a field that goes round the torus. The middle figure shows the plasma current and the resulting poloidal field. The bottom figure combines the two fields in the resultant helical (twisted) field, cancelling out any drift. Figure from *Wikipedia*

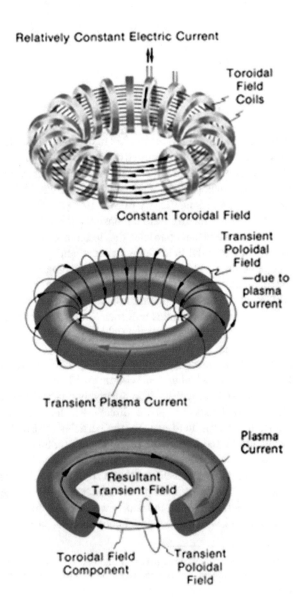

However, Sakharov still missed an essential point, which was supplied by the stability criterion formulated by Vitaly Shafranov, which makes the history of the tokamak a little more complicated. In 1929 in his book *Osnovy Teorii Elektrichestva* (Leningrad 1929) Tamm had already invented nested toroidal magnetic surfaces, and, as recalled here, in 1950–1951 he and Sakharov developed toroidal plasma confinement with compensation of particle drift. However, the tokamak as a toroidal plasma device, stabilised by strong toroidal magnetic fields, also needs Shafranov's stability criterion (known as the Kruskal-Shafranov limit) which imposes a condition on the strength of the toroidal field relative to the poloidal field as we will see in the next chapter. Shafranov was the one who eventually convinced Natan Yavlinsky to build the first tokamak in 1957.

3.5.4 Soviet Experiments from 1955–1969

At the Kurchatov institute, where research with toroidal and linear fusion systems started in 1951 following the aforementioned resolution of the USSR Council of Ministers, the need for a toroidal current, producing a poloidal field, led at first to the proposal to just do without a toroidal field altogether and send a current through the plasma in both linear and toroidal devices to create a pinch effect, the same technique as used in the UK and US with pinch devices and magnetic mirrors. At about the same time as the DCX was constructed at Oak Ridge, the Russians built a similar machine of rather gigantic dimensions (a 20-m-long linear pipe). The machine was called OGRA which stands for *odin gram neitronov v sutki* (one gram of neutrons per day), which if actually achieved would have been a momentous achievement. It also employed injection of molecular ions (as shown for the DCX in Fig. 3.16). The OGRA machine was shown at the 1958 Geneva Conference, where Artsimovich explained that it had encountered the same problems as the DCX.

The Soviet Union explored a whole range of devices (so not only tokamaks) with as its main objective to create high temperatures (the temperature was supposed to increase in proportion to the square of the current). As in the UK and the US, 'false' neutrons were observed (already in 1952) (as mentioned by Kurchatov in his Harwell talk) and the kink/sausage instability threw a similar spanner in the works. Its stabilization would still require a separate toroidal magnetic field, but it also seemed impossible to raise the temperature above 10–30 eV (100,000–300,000 degrees) by increasing the current. These problems again diverted the attention to the original Sakharov-Tamm idea of using both a toroidal magnetic field and a toroidal current. As explained above, the function of the current was to provide equilibrium and plasma confinement, while the magnetic field was supposed to create stability, i.e. keep the kink instability in check as long as the current is not too high. This led to the construction of the first tokamak-like device, called TMP, which just stands for 'torus with magnetic field' (*tor s magnitnym polem*). The name tokamak was not yet in use. The TMP consisted of a porcelain torus (the porcelain turned out to evaporate as shown by silicon lines in the plasma radiation spectra) with a major

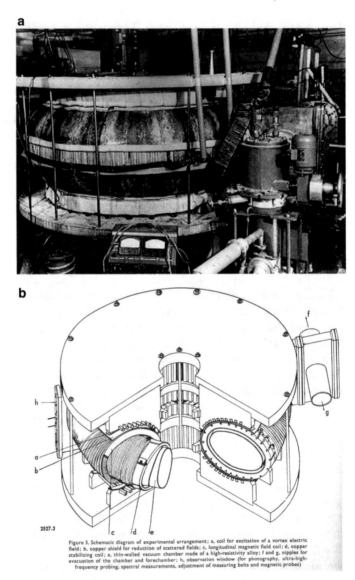

Figure 3. Schematic diagram of experimental arrangement; a, coil for excitation of a vortex electric field; b, copper shield for reduction of scattered fields; c, longitudinal magnetic field coil; d, copper stabilizing coil; e, thin-walled vacuum chamber made of a high-resistivity alloy; f and g, nipples for evacuation of the chamber and forechamber; h, observation window (for photography, ultra-high-frequency probing, spectral measurements, adjustment of measuring belts and magnetic probes)

Fig. 3.23 a Picture of the first tokamak in the world the T-1 at the Kurchatov Institute with major radius $R = 0.67$ m, minor radius $r = 0.17$ m. Operations started in 1958 (*from* Smirnov 2010). **b** A cut-out of the tokamak shown under a, as presented at the 1958 Geneva conference

radius of 80 cm and a minor radius of 13 cm. It was followed by a whole generation of Soviet tokamaks. The T-1 (Fig. 3.23), which started to operate in 1957, was the first proper tokamak in the world, a device with a stainless-steel liner inside a copper vacuum chamber. From 1955 to 1965 eight facilities of this type (TMF, T-1, T-2, T-3, T-5, TM-1, TM-2, and TM-3) were built. The last one in the series and in the Soviet

Union was the T-15, which started to operate in 1988. Nowadays Russia is part of the ITER project, which consumes most of its fusion research budget.

The physical layout of the tokamak is essentially identical to a toroidal pinch device. Figure 3.23a shows a picture of the T-1 and Fig. 3.23b a cut-out of this machine. The essential difference with a toroidal pinch is the relative strengths of the magnetic fields. In a pinch device the poloidal field (responsible for the plasma current and the pinch) is much stronger than the toroidal field, which is almost completely captured within the plasma. In the tokamak it is just the other way round. The toroidal field in a tokamak is typically about 10 times stronger (1–1.5 T in the Soviet designs) than the poloidal field and is present outside as well as inside the plasma, with nearly the same value in both regions. The superposition of the two fields (poloidal and toroidal) provides the helical shape (a gentle helix for the tokamak and a tightly twisted one for a pinch), while the outside toroidal field surrounding the plasma acts as a sort of straitjacket and provides greater stability than the geometries explored in the West (McCracken and Stott 2013, p. 92; Bromberg 1982, p. 134).

The various devices were successful in remedying various 'childhood diseases' of tokamaks, like the influx of impurities into the plasma, radiation losses and shifting of the plasma column (which required the installation of correcting and controlling coils (Azizov 2012, p. 192)), but in spite of these successes it cannot be said that an obviously healthy 'adolescent' was quickly emerging. The Bohm diffusion restriction (see Chap. 2), which limited confinement times, remained a persistent problem. The whole business was in the doldrums for quite a while and for five years no significant progress was made both with pinch and toroidal systems (Shafranov 2001, p. 840; Bromberg 1982, p. 130ff).

The T-3 tokamak[27] was the first machine that managed to beat the Bohm restriction by at least a factor of ten, showing that not all magnetic confinement geometries necessarily had to suffer from this. In the now declassified era, Artsimovich reported the Russian results in 1965 at the 2nd IAEA Conference in Culham (UK), but they were not fully appreciated by a somewhat incredulous Western audience. Especially Lyman Spitzer, who reviewed the experimental work submitted to the conference, was sceptical about the results reported by Artsimovich and very pessimistic about the possibility of taming the instabilities. He stressed the in his view general feature of all devices in operation at that time "that anomalous particle loss is present and that it is roughly within an order of magnitude of that predicted by the Bohm formula" and that "instabilities seem to be an intrinsic property of a heated plasma in a toroidal chamber" (Spitzer 1965). As reported above, less than a year later Spitzer, the inventor of the stellarator, apparently disillusioned, left the field altogether and returned to astrophysics.

The Russians went home and quietly continued improving their results, so that at the next conference in Novosibirsk in August 1968 Artsimovich could report dramatically better values for temperature, density and confinement than measured on any other device. This tale with its dramatic consequences will be told in the next chapter.

---

[27]Natan Yavlinsky was the leader of the T-3 experiment; after he died in a plane crash in 1962, Artsimovich placed the machine under his own direct supervision.

Chapter 4
The Tokamak Takes Over

4.1 Introduction

If something had been learned from the early period described in the previous chapter, it was that fusion was not that easy to control as initially thought, and that some basic research was needed to understand the physics better. Various devices had been constructed, with the emphasis on toroidal structures (stellarators, tokamaks or toroidal pinch devices) as they seemed to be the only ones to have a remote chance of avoiding catastrophic particle losses. Secrecy was abandoned and nothing, apart from politics, stood in the way of international collaboration, but for the time being it did not really get off the ground, although it allowed for the freer exchange of information.

The problems facing nuclear fusion were clear. At the 1958 Geneva conference Thonemann stated that "the problem of stability is of paramount importance. Unless the rate at which charged particles cross magnetic lines of force can be reduced to that given by classical diffusion theory (i.e. overcoming the Bohm diffusion restriction (LJR)), the loss of energy to the walls will prevent fusion reactions from becoming a practical power source" (Thonemann 1958, p. 37).

Where Bhabha had given the world twenty years in 1955 to show that fusion energy could be liberated in a controlled way, Thonemann, as we have seen, was already a little more cautious as twenty years would now be required to only show the economic value of such power, but in 2020, while this book is being written, we still don't know the answer, although many of the technical problems which he mentions have been solved. Edward Teller, who also spoke at the conference and always was invariably optimistic about even the most improbable technological schemes (Park 2000, p. 185), was much more cautious and did not share Thonemann's optimism: "If we want to shoot for the jackpot, for energy production, I think that it can be done, but do not believe that in this century it will be a thing of practical importance" (Teller 1958, p. 32), and indeed it hasn't been, but Teller at least allowed for more than half a century.

L. J. Reinders, *The Fairy Tale of Nuclear Fusion*,
https://doi.org/10.1007/978-3-030-64344-7_4

At the same conference Artsimovich from Russia was even more cautious, as fitting for a Soviet scientist who knew that rash predictions not proving to be correct could cost you an extended time in the Gulag (at least in the days just passed), but also showing a greater sense of reality:

"Three years have passed since this prophecy (*i.e. Bhabha's (LJR)*) and now, before our eyes, there begins to emerge a rough outline of the scientific foundation on which the methods of solving the problem of controlled fusion reactions will probably rest. This foundation has been laid by the numerous experimental and theoretical results obtained in recent years in Great Britain, the USA, the USSR and other countries. For the first time these results will be discussed on an international scale, and this is probably the most important step which has been made towards the solution of this problem. The importance of this fact is greater than that of the separate investigations, which as yet have not brought us very much nearer to our ultimate goal.

We do not wish to be pessimistic in appraising the future of our work, yet we must not underestimate the difficulties which will have to be overcome before we learn to master thermonuclear fusion. In the long run, the main difficulty lies in the fact that in such a light substance as rarefied plasma, any manifestation of instability develops at an enormous rate," reflecting Spitzer's earlier remark quoted in Chap. 3 that "[i]n a system with so many particles and with so few collisions as a (…) plasma, almost any type of behaviour would not be too surprising."

Artsimovich also, in vain as it would turn out, made an appeal for closer international collaboration as: "[a] most important factor in ensuring success in these investigations is the continuation and further development of the international cooperation initiated by our conference." (Artsimovich 1958, p. 20.)

In spite of the cautious remarks by Artsimovich and others, the general atmosphere during and after the 1958 Geneva Conference was one of optimism, bordering on naivety, which is borne out by the summary of the conference in the UN Review[1] of which the first sentence reads: "Secrecy in civilian science has ended", followed by an overly rosy view of the future of nuclear power and cooperation in this field. Although in the 1970s there were regular exchanges of scientists between the Soviet Union and the US (and at a smaller scale with other countries), it seems that this international collaboration is only now being realised with ITER.

In my view Braams and Stott hit the nail on the head by saying that "it became clear that the principal outcome of the [1958] Conference had been the insight that first the forgotten chapter of plasma physics needed to be written, before the question of the scientific feasibility of controlled fusion could be addressed" (Braams and Stott 2002, p. 55). In other words, the nuclear fusion physicists did not understand (and in my view still do not understand) the stuff, i.e. the plasma, they are trying to control in their devices. Plasma physics, Braams and Stott say, had been shunned by mainstream physicists as being 'dirty' physics, although in fact all the ingredients to come to a fuller understanding of the behaviour of plasmas (Maxwell's theory of electromagnetism, Boltzmann's statistical mechanics and the principles of fluid

---

[1]Free Exchange of Information, *UN Review* 5 (1958) 22–27.

dynamics) date from the nineteenth century and just needed combining. But the behaviour of plasma is still largely a mystery even today and, when you put it to nuclear fusion scientists that the understanding of plasma physics is insufficient, you may get an answer like Alvin W. Trivelpiece, one of the principal actors in US nuclear fusion research and a former director of the Oak Ridge National Laboratory, gave as late as 2013: "There are some who say that we don't know enough of the basic science and technology to proceed and that we should focus on them until we have a better understanding of all of the fundamental processes." In an attempt to refute this clearly reasonable argument he continues: "Suppose that when someone noticed that we could make steam with fire, it was insisted by some that we should understand the physics of flame before proceeding to make a steam locomotive. That would have possibly stopped the development of steam engines." (Trivelpiece in Dean 2013, p. 232.) In this deeply flawed comparison Trivelpiece forgot to mention that flames and the use of boiling water (steam) for producing mechanical motion had been already in use by mankind for centuries before a steam engine was invented around 1700. Once it was realised how to transform the pushing force of the steam into a rotational force for work, the development went very fast. Knowledge of the physics of flames is completely unnecessary for this.

In other words, Trivelpiece essentially admits that basic understanding is still missing and he and others claim that we do not have to know how a plasma behaves; at least not in great detail, a nuclear fusion power station might still work and, if it does, who cares what the details are. A rather unscientific attitude.

But let us return to our story. In the 1960s things were not going well in US experiments, the Model-C stellarator was apparently lost to Bohm diffusion, with more rapid plasma loss rates than expected, the so-called pumpout. The magnetic mirror at LLNL had similar problems etc., and so did experiments in other countries. A bewildering variety of collective plasma effects and instabilities plagued fusion experiments and resulted in disappointingly short confinement times. This caused the frontier of plasma research to become the experimental identification and theoretical analysis of such plasma instabilities. From 1956 to 1970, in a series of review articles Richard Post of Livermore expressed among other things the sentiment that the enthusiasm of the 1950s had given way to a more cautious and thorough study of the physics of plasmas. He suggested that, during the 1950s, "the lure of the final objective was so great that some of the traditional scientific precautions were by-passed in hopes of leapfrogging into an early solution." However, the difficulty of the fusion problem led the community to "now see that a more fundamental attack on the problem ... would probably have put us farther along the way of understanding." (Post 1959). Harold Fürth, at the time also at Livermore, and Post criticized the course of fusion research in the 1950s as having been dependent on "short-term technological pressure." As a result, the community failed to allow sufficient time to interpret its experimental results theoretically and so "the importance of stability considerations was poorly understood." They suggested that, instead of the "technology-oriented research effort" of the 1950s, a more "physics-oriented research effort" be followed (Weisel 2001, p. 212–213).

A bright spot was that superconducting magnets were entering the field and promised the ability to work in the near future with larger magnetic fields that would consume much less energy than copper coils. With the latter any commercially viable fusion energy production would be out of the question, as the amount of power consumed by the magnets would preclude a favourable energy balance. Apart from this, the overall situation seemed very bleak. The whole sector was in the doldrums and outside the fusion community itself there was in the early 1960s hardly any support for fusion. Industry began to reconsider its commitment, now that the breakthrough they had hoped for had not materialised. The success of fusion was uncertain and, at best, far off (Bromberg 1982, p. 135–136, and previous chapter). Fission energy, not fusion, was the craze of the early to mid 1960s.

The fact that the fusion enterprise failed to yield the results that were promised was not lost either on the politicians in the US Congress who oversaw the US fusion programme. After the tentative funding of Tuck and Spitzer in the very early fifties, the budget for fusion had rapidly grown. Project Sherwood was so much pampered with money by the AEC that researchers hardly knew how to spend it all. For 1951–1953 the budget amounted to $1 million, an amount just plucked out of the air by the director of AEC's Research Division, then increased from $1.7 million for 1954 to $10.7 million in 1957 and even $29 million in 1958. (Clery 2013, p. 89.) From a small-scale research project involving just a few dozen people it had grown into a dynamic development effort employing hundreds of people. The launch of the Sputnik by the Soviet Union on 4 October 1957 had caused a shock to the United States. It had always been assumed that the US had an unassailable lead in technology and that the Soviet Union would forever just be scrambling to catch up. This second surprise sprung on them by the Soviets, after the unexpectedly early explosion of the Soviet atomic bomb in 1949, led to a sense of urgency in the US and prompted a huge hike in government spending on science and technology, almost tripling the fusion budget for 1958 compared to 1957. In the years thereafter until 1962 the budget remained fairly stable, oscillating around $25 million.[2]

In the early 1960s the mood began to change. Already in 1962, the House Appropriations Committee made a first attempt to trim the fusion budget for the fiscal year 1963. Only $300,000 was cut off, on a total of $24.2 million. In 1963 for fiscal year 1964 the CTR (controlled thermonuclear research) programme was cut by $3.75 million, leaving $20 million to spend. When a scientific "breakthrough" was promised, $2 million was restored. But throughout the 1960s, the AEC never attained a large enough budget to increase its support of university research in plasma physics and related subjects, and concentrated on maintaining fusion research at the four major laboratories (Los Alamos, Livermore, Princeton, and Oak Ridge).

The basic cause of congressional opposition had its roots in part in a general attitude in the early 1960s of questioning the post-Sputnik science spending spree. But very specific questions were also raised about the CTR programme. A dozen years had passed since the programme had started. Not only had no reactor been developed, but no knowledgeable scientist was able to assert that such a reactor is

---

[2]For the precise figures, see https://aries.ucsd.edu/FPA/OFESbudget.shtml.

possible (a situation, by the way, that continues to the present day). Was fusion a wild-goose chase? Democratic Senator John Pastore (1907–2000), serving as senator for Rhode Island from 1950 to 1976 and a member of the Joint Committee, said at the hearings in the spring of 1964: "how long do you have to beat a dead horse over the head to know that he is dead?", essentially setting the tone for the fusion debate in the US Congress for the decades to come. Well, the answer is that the beating is still continuing and the answer wanting.

Moreover, the money that earlier was allocated for technology and spent in a trial-and-error method (think e.g. about the rapid succession of stellarator models whereby the next model was already being developed before the results of the preceding model were known), which clearly did not work, was now being diverted to a purely scientific enterprise: the physics of high-temperature plasmas. This was not to Congress's liking. They were more interested in technological development than research, I guess partly also because of the overblown expectations that the scientists had raised about nuclear fusion. After all, it was from the lips of the fusion scientists themselves, 5–6 years earlier, that Congress had heard the promise that the programme would be at the stage of prototype reactors by the early sixties. This earlier optimism was now returning to haunt the fusion scientists. They now belatedly saw the political dangers of their own premature, foolish and unwarranted promises.

From 1960 to 1964 Sherwood had spent between $24 and $32 million a year. It was again Pastore who asked "Is this not indeed a very expensive way of getting this basic knowledge? We can build these machines until the cows come home. Somewhere along the line somebody has to think that this is a lot of money and maybe we ought to be putting it into some other place where it may be more productive."

Congress also questioned the large number of avenues that were pursued: in the 1963 budget these were the stellarator, the mirror, the direct current mirror machine (DCX) of Oak Ridge, the Astron at Livermore, and various forms of pinches. After so many years, it must be possible to sift the wheat from the chaff (Bromberg 1982, p. 117–119).

This called of course for committees to thoroughly review the fusion effort, a recurring phenomenon over the years in US politics.[3] The divide was (and probably still is) between those who want to view the efforts in nuclear fusion as a scientific research project or physics experiment (trying to understand plasma behaviour, for instance), although such voices are no longer audible, it seems, and those who see it as a march towards the construction of power generating plants that would solve all our energy problems, whose voices are certainly the loudest. This latter attitude of viewing very preliminary machines already as prototype or candidate reactors, is one of the problems with nuclear fusion and has led to grossly inflated expectations, which when the inevitable failure occurred led in turn to a negative over-reaction. The divide became more and more a partisan affair in Congress, the republicans favouring the former viewpoint and the democrats the latter. The amount of money spent on fusion research in the US did not go up in the 1960s and even went down when figures

---

[3] See Dean (2013) for an extensive discussion of the bickering in US political circles about nuclear fusion.

adjusted for inflation are considered. The Johnson administration was short of money because of the Vietnam war and the funding needed for the President's Great Society programmes. The arguments put forward for increasing the fusion budget did not make a big impression, and understandably so.

4.2 The Triumph of the Tokamak

Then came the third international conference on plasma physics and controlled thermonuclear fusion, held in August 1968 in Akademgorodok ('Academic City') near Novosibirsk, in far-away Siberia. Here Lev Artsimovich reported on the latest results with the Soviet T-3 and TM-3 tokamaks (Fig. 4.1).

The values of temperature (over 10 million degrees, where before 1 million had been typical for closed systems like tokamaks), confinement time (of tens of milliseconds) and density (10^{12} cm^{-3} s for the product $n\tau$ of density and confinement time, a

Fig. 4.1 Scientist monitoring the T-3 tokamak at the Kurchatov Institute in Moscow. Built in 1960, it was the third reactor of this type. In 1968 it generated plasma temperatures of around 10 million degrees and kept plasma confined for tens of milliseconds (https://alltheworldstokamaks.wordpr ess.com)

factor 10 higher than before) were dramatically better than so far recorded. In particular the Bohm trauma, which had held up progress for almost a decade, seemed finally to have been overcome. The energy confinement times exceeded the Bohm time by a factor of 50.

At Culham in 1965 the Soviets had already reported confinement times exceeding the Bohm time by a factor of 10, but had been met with incredulity and, as stated in the previous chapter, Spitzer had just lumped them together with the other failing devices. Since then data, some from stellarators (among them the Proto-CLEO machine at Culham and early Wendelstein devices at Garching, near Munich, but not the Princeton stellarators), but mostly from other devices, had piled up to finally discredit Bohm's empirical rule, so the results on the energy confinement time reported by Artsimovich were not such a surprise, but the temperature and density values were. Most scientists at the conference and elsewhere were impressed, but there still was doubt, especially with the Princeton scientists, since the measurement of the temperature was indirect. The Soviets had inferred the plasma pressure from measurements of the magnetic fields and then used the resulting pressure to derive the value of the electron temperature, assuming that the electron velocities obeyed a Maxwellian velocity distribution (see Chap. 2). American critics were questioning this assumption and argued that the Russians might have been measuring the pressure caused by some very high velocity (runaway) electrons. Most electrons are accelerated and then collide with an ion, lose speed and accelerate again. Some manage to avoid colliding and just keep being accelerated. These 'runaway' electrons could have been responsible for the observations. Direct temperature measurements were possible by shining light on the plasma and observe the way in which it scattered. A very intense light source, i.e. a laser, was needed for this, as the intensity of the scattered light is very low and the plasma itself is highly luminous. Such lasers had just entered the field as diagnostic tools and were not yet in use in Russia (Bromberg 1982, p. 152–153). This resulted in the exceptional event (in view of the Cold War; the recent crackdown on dissidents and the suppression of the Prague Spring; science as always had a higher calling and could not be distracted by such trivial events) that a British team went to Moscow to carry out these measurements and confirm the temperature values reported by Artsimovich (Fig. 4.2).[4] Both the temperature and density measurements of the Soviets were correct, and there was no evidence of a runaway electron population. The results were reported a year later at Dubna in the Soviet Union where Derek Robinson (1941–2002), the leader of the British team, told his audience that Artsimovich had been right, the temperature was indeed 10 million degrees. The tokamak, at that time still a device unique to the Soviet Union, had triumphed, and swept away the negative, pessimistic mood that was prevailing in the West. The tokamak stampede was about to begin.

Just a couple of months before this, in the spring of 1969, Artsimovich had gone abroad, to Western Europe and the US, culminating in a series of lectures at the MIT National Magnet Laboratory, where he had presented the strong case of the Russian tokamak. Braams and Stott (Braams and Stott 2002, p. 152) call his visit

---

[4] A short film about this mission to Moscow can be viewed at https://go.nature.com/2ykwP5W.

Fig. 4.2 The experimental setup of the Culham group at the T-3 tokamak (https://alltheworldstok amaks.wordpress.com)

a triumphant success, while Herman (Herman 1990, p. 92) quotes Bruno Coppi (b. 1935), an Italian–American plasma physicist working at MIT, who recalled that after one lecture Artsimovich came into his office with tears in his eyes, complaining that people did not seem to be listening to him. So, although triumphant, for him it was at times also a frustrating experience. And he was wrong, for people were listening, hanging on every word he was saying.

All this, the confirmation by the Culham group and Artsimovich's lectures, had a profound effect and indeed rightly so as it was undoubtedly the most important event in the history of fusion research so far, an event that can with justice be called a "breakthrough", not one of the phoney breakthroughs the history of fusion is littered with. The reaction, though, of people all over the world changing their research programmes overnight and scrambling to jump on the tokamak bandwagon was a little precipitate, to say it mildly. We will discuss this development in detail, starting in the next chapter, but before that let us look a little more in detail at the tokamak.

4.3 Tokamak Fundamentals

4.3.1 Basic Design

Now that the tokamak has become the basic nuclear fusion device, the prime candidate for a future reactor, with emphasis on the word 'future', and its development will take up most of the rest of this book, it seems a good idea to summarise its underlying basic design and list and discuss the fundamental parameters of both the tokamak and the plasma it is supposed to confine. This is unavoidably a little technical, after all the tokamak is a technical device and for understanding its basic workings knowledge of some technical details is required. As is the case with any machine there are some parameters, numbers and concepts that are important to know in order to understand how it works, that characterise how well it works and what its potential is for further development towards an actual reactor, or something that comes close to it.

A definition of the tokamak might read (Braams and Stotts 2002, p. 131): "The tokamak is an axially symmetric [magnetic] field configuration with closed magnetic surfaces, in which a toroidal field is produced by currents in external coils and a poloidal field by a current in the plasma. The weaker poloidal field determines the plasma confinement and the toroidal field provides the stability."

Let us first consider an (ideal) transformer as depicted in Fig. 4.3. A transformer can be used to transfer electrical energy between two coils (the primary and secondary windings in the figure) without a metallic connection between the two circuits. A current sent through the primary winding induces a magnetic flux in the transformer

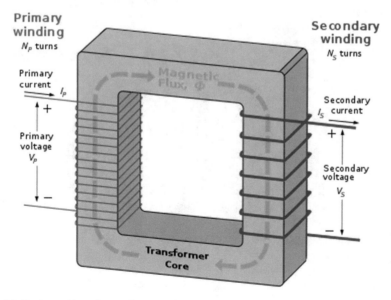

Fig. 4.3 Basic working of a transformer (*from Wikipedia*)

iron core, which in turn induces a current in the secondary winding. This is a consequence of the law of induction, also called Faraday's law, discovered in 1831 by the
English physicist Michael Faraday (1791–1867). Transformers are for instance used
for increasing or decreasing alternating voltages in electric power applications. Since
for an ideal transformer the same magnetic flux passes through both the primary and
secondary windings, a voltage is induced in each coil proportional to the number of
its windings. In case of a non-ideal transformer, which of course applies to the real
world, there will be some losses. For our purpose this is however not of relevance.

The basic design of the tokamak, including a transformer for inducing a current
through the plasma, is depicted in Fig. 4.4. A crucial point here is that the plasma
functions as the secondary winding and that the toroidal current through the plasma
(the plasma current) is generated in this way. In this connection the primary windings
are, perhaps confusingly, sometimes also called ohmic heating coils, as they are
responsible for generating the current which (ohmically) heats the plasma. Apart
from heating the plasma, this current will in turn generate a poloidal magnetic field
around the plasma current. The combination of this latter field with the toroidal field
generated by the toroidal-field coils will give the resultant helical (twisted) field.

How these magnetic fields come about is once more elucidated in the next figure
(Fig. 4.5).

The figure explains how a tokamak magnetic field configuration is created. The
drawing under **a** shows that a flowing current (through a wire for instance) generates
a magnetic field around the current, while the figure under **b** shows how the circular

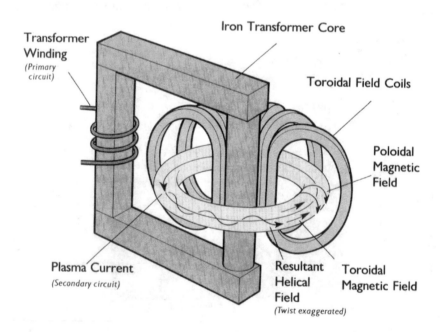

Fig. 4.4 Basic setup for a tokamak

a) Magnetic field around the current   **b) Magnetic field by cylindrical curcular coils**

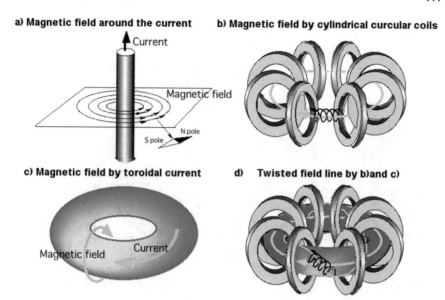

c) Magnetic field by toroidal current   **d)   Twisted field line by b)and c)**

Fig. 4.5 How to create a tokamak configuration (*from* Kikuchi (2010))

coils produce a toroidal magnetic field that forces charged particles to travel in spirals around the torus. The figure under **c** is the analogue of figure **a** for a current flowing in a circle, like the plasma current in a tokamak, and figure **d** combines **b** and **c** to create the twisted magnetic field that we have in a tokamak. So, a tokamak is nothing else than a doughnut-shaped object (also called a toroid or torus and symmetric around the axis passing through the middle (axially symmetric)) with in it a nested set of magnetic surfaces that are produced by coils and currents. We only consider here setups with a circular cross section, as in the figures. Later we will see that current-day tokamaks are constructed with a D-shaped cross section.

Once we know that a transformer is involved in generating the plasma current, we can draw a more encompassing schematic picture of a basic tokamak setup as has been done in Fig. 4.6.

In this figure we have included some other essential elements of the tokamak; the central solenoid acts as the primary winding of the transformer. Solenoids are lengths of coiled wire that generate magnetic fields when electric current is passed through them and are responsible for driving the plasma current in all tokamak devices. The central solenoid is a long straight line of circular coils. It is the heart of the tokamak. By continuously increasing the current through the central solenoid, electromagnetic induction drives a current through the plasma. This current produces the poloidal field and heats the plasma. The current through the central solenoid can of course not be increased indefinitely. This is the essential reason of the inherent pulsed nature of the tokamak design. No matter how big the central solenoid coils are, eventually a limit will be reached, induction will stop, the particle drifts will jump into action and destroy the plasma confinement. A current can only be driven for a short time in this

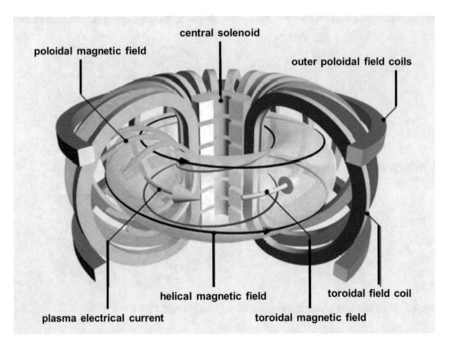

central solenoid

poloidal magnetic field

outer poloidal field coils

helical magnetic field   **toroidal field coil**

plasma electrical current   **toroidal magnetic field**

Fig. 4.6 Schematic figure of a tokamak with the various fields and currents. *Source* EFDA-JET (now EUROfusion)

way. The conclusion from this is that to achieve steady-state operation in a tokamak the plasma current, once driven by a central solenoid, must at least be maintained non-inductively, i.e. without this central solenoid. Or alternatively, instead of inductive current drive we need some sort of non-inductive current drive.

In this respect it is also important to realise that there is a difference between pulse length and confinement time. The **pulse length** is the overall duration of the plasma before it is being destroyed for whatever reason, usually technical limits on the magnetic field, the plasma current and/or the plasma heating systems. On the other hand, the **confinement time**, introduced in Chap. 2, is a measure of the average time that particles (ions and electrons) (or energy, so in Chap. 2 we also spoke of energy confinement time) spend in the plasma. The confinement time is generally shorter than the pulse length, and the particle confinement time is generally longer than the energy confinement time as energy is always lost by both thermal conduction and convection (McCracken and Stott 2013, p. 112). It is the energy confinement time that enters into the Lawson criterion, as we have seen.

The magnetic field coils that generate the twisted magnetic field in the torus-shaped plasma chamber must be arranged such that the magnetic field lines close within the chamber and form closed magnetic surfaces to prevent particles from colliding with the walls of the vessel or from escaping in any other way. This is a delicate matter and some ring-shaped outer poloidal field coils are added in Fig. 4.6

for generating an extra vertical field needed for shaping and positioning the plasma and contributing to its stability by "pinching" it away from the walls.

As already noted in the previous chapter, the physical layout of the tokamak is not that different from the toroidal pinch devices we have seen before. The essential difference is the relative strengths of the magnetic fields. In a pinch device the poloidal field (responsible for the plasma current and the pinch) is much stronger than the toroidal field. In the tokamak it is just the other way round. The toroidal field in a tokamak is typically about 10 times stronger than the poloidal field. The outside toroidal field surrounding the plasma acts as a sort of straitjacket and provides greater stability than a pinch (McCracken and Stott 2013, p. 92; Bromberg 1982, p. 134).

4.3.2 Fundamental Tokamak Parameters

Let us now introduce some parameters that naturally follow from such a setup.

In the schematic view of the torus in Fig. 3.21 we have already defined the major radius R and the minor radius r of the torus. The latter is usually denoted by a and the ratio R/a is called the aspect ratio of the torus. For the early tokamaks aspect ratios were quite large with a being fairly small and R something like 1 m, e.g. for the famous Soviet T-3 it was 8.3 ($R = 1$ m, $a = 0.12$ m) and for the equally famous T-10 it was 4.2 ($R = 1.5$ m, $a = 0.36$ m). The hole in the middle must be large enough for the magnetic windings (the toroidal-field coils) and the central solenoid to fit in without touching, which limits the aspect ratio for conventional tokamaks to about 2.5. In the 1980s it was discovered that tokamaks with low aspect ratios were inherently more stable. The closer a is to R the 'fatter' the torus will start to look, like a sphere with a tube bored right through it, i.e. like a cored apple (Fig. 4.7, which has aspect ratio close to 1). The spherical tokamaks, which use an alternative arrangement of the magnetic coils, as will be discussed in Chap. 11, very much

Fig. 4.7 Cored apple, torus shape preferred for small spherical tokamaks

resemble the cored apple of Fig. 4.7 and allow for an aspect ratio as low as 1.2. For ITER, not a spherical tokamak of course, it will be 3.1 ($R = 6.2$ m and $a = 2$ m), while for JET, the Joint European Torus, it is 2.4 ($R = 2.96$ m, $a = 1.25$ m).

An obvious important quantity is also the strength of the plasma current I_p (measured in megaamperes, millions of amperes) which produces the poloidal magnetic field (Fig. 4.5c), denoted by B_{pol}. This poloidal field depends linearly on the plasma current and, as said before, is needed to cancel the drift of the particles due to the non-uniformity of the toroidal field B_T. It is in a plane at right angles to the toroidal field and shifts the toroidal field by a certain angle each time B_T goes around the torus the long way (i.e. toroidally). The greater the plasma current, the larger this angle.

For tokamaks the ratio of the number of times a toroidal magnetic field line goes around the torus the long way to the number of times a poloidal magnetic field line goes around the short way is defined as the **safety factor q**, which plays a role in the suppression of the kink instability. The onset of the instability depends on the strength of the poloidal field relative to the main magnetic field, the toroidal field. The toroidal field has a stabilizing influence on the instability and can keep it in check if the current induced by the poloidal field is not too strong. The resulting limit to the plasma current is called the Kruskal-Shafranov limit, after Martin Kruskal and Vitaly Shafranov, whom we already met in Chap. 3 in connection with the kink instability. The critical current corresponds to a combination of toroidal and poloidal fields whereby the plasma twisted by the magnetic fields comes back to exactly the same spot after having travelled once around the torus, i.e. a combination of fields for which the safety factor $q = 1$. In terms of magnetic field lines $q = 1$ corresponds to the situation whereby a poloidal field line goes around the torus the short way exactly once for each time a toroidal field line goes around the long way. If the poloidal field line goes around more often it will twist the toroidal field too much, and hence the plasma too, and the situation becomes unstable ($q < 1$). If it goes around less often, no kink will develop ($q > 1$) (Chen 2011, p. 221, 222). So, q is a measure of the twist of the magnetic field and, at the same time, of the stability of the setup. By arranging the reactor such that this q is always greater than 1, tokamaks strongly suppress the instabilities that plagued earlier designs. It was in the T-1 experiment at the Kurchatov Institute in Moscow that an abrupt change in stability was observed when this fundamental parameter became larger than 1 (Strelkov 1985, p. 1189). A fundamental and trail-blazing discovery that has drastically changed the course of fusion research! The strengths of the poloidal and toroidal magnetic fields (and the strength of the plasma current) are closely related. In a tokamak the toroidal field B_T must be strong enough to satisfy the Kruskal-Shafranov condition.

In older tokamaks the plasma current was less than 1 MA and has since been increased to 7 MA for JET. And for ITER plasma currents to a maximum of 15 MA are envisaged. (See e.g. Miyamoto 2011, p. 215 for precise numbers for various devices.)

Table 4.1 Summary of main tokamak and plasma parameters

Tokamak parameters	Plasma parameters
Major radius R	*beta*, ratio of thermal to magnetic pressure
Minor radius a	Density n
Poloidal field B_{pol}	Electron temperature T_e
Toroidal field B_T	Ion temperature T_i
Safety factor q	Plasma pressure p
Plasma current I_p	

The toroidal magnetic field B_T is typically of the order of 5 T,[5] and has not changed very much from the older tokamaks like the Soviet T-10 with 5 T to JET with 3.45 T. It is typically ten times stronger than the poloidal magnetic field. For ITER the toroidal field coils are designed to produce a maximum magnetic field of 11.8 T.

The next fundamental plasma parameter is a quantity called **beta** which is equal to the ratio of the thermal plasma pressure to the magnetic pressure. It is a measure of the efficiency with which the magnetic field confines the plasma. High *beta* is desirable, as it would require less energy to generate the magnetic fields at any given plasma pressure, but is difficult to achieve experimentally because of various plasma instabilities. For a confined plasma *beta* is always smaller than 1 (else it would collapse) and is normally expressed as a percentage. Ideally, a magnetic confinement fusion device would want to have *beta* as close as possible to 1, as this would imply the minimum amount of magnetic force needed for confinement. Low *beta* is easier to achieve but represents a lower confinement efficiency (Friedberg 2007, p. 87).

For conventional tokamaks the record for *beta* is rather low at just over 12%, and it is expected that practical designs would need to operate with *beta* values as high as 20%. The low achievable *beta* is due to instabilities generated through the interaction of the fields and the motion of the particles. As the amount of current is increased in relation to the external field, the instabilities become uncontrollable. Increasing the relative strength of the external magnetic field, the simple instabilities damp out, but at a critical field other instabilities will invariably appear, notably the ballooning mode. For any given fusion reactor design, there is a limit to the *beta* it can sustain. A practical tokamak-based fusion reactor must be able to sustain a *beta* above some critical value, which is calculated to be around 5%. One of the advantages of low-aspect ratio tokamaks, i.e. spherical tokamaks, is that they can operate at *beta* values that are significantly higher, with a record of 40% set by the START device at Culham (see Chap. 11).

Since there are two magnetic fields (poloidal and toroidal) it is possible to define a separate *beta* for each of these fields. We will not further consider this complication in this book. (See e.g. Miyamoto (2011) for more details.)

The parameters discussed above have been summarised in Table 4.1.

---

[5]The former unit for magnetic field strength was gauss; 1 T being 10,000 G.

As nuclear fusion research proceeded over the past decades, many more parameters were introduced and existing ones refined. We will encounter some of them when discussing certain aspects; others will not be discussed as they are not essential for a basic understanding of what is going on in fusion.[6]

4.3.3   Bootstrap Current

In the 1970s, it was predicted theoretically that in a toroidal plasma, the difference in pressure between the hot plasma in the central core and the cooler edge region actually drives a spontaneously arising toroidal plasma current, due to collisions between trapped particles and passing particles. This current, called "bootstrap current", since it is self-generated within the plasma, was experimentally confirmed in tokamaks in the 1980s. One goal of advanced tokamak designs is to maximize the bootstrap current, and thereby reduce or eliminate the need for an external current driver, i.e. a central solenoid. This could dramatically reduce the cost and complexity of the device.

When a plasma is heated by neutral beam injection (NBI) (see Chap. 2), it creates a population of energetic ions which can circulate many times around the torus before they slow down by collisions with the background plasma. This can drive a similar current that can amount to several hundreds of kA, even though the ion current of the input beam is only of order 50 A. In some devices the combination of NBI-driven and bootstrap current has been increased to about 60% of the total plasma current, thereby relaxing the requirement for induction to sustain the current.

4.3.4   Evolution of the D-Shape[7]

Twisted field lines in toroidal devices were needed, as we have seen, to cancel the drift into opposite directions of electrons and ions. This drift is due to the non-uniformity of the magnetic field in the torus, the fact that it is necessarily weaker on the outside of the torus compared to the inside, near the hole in the doughnut. An obvious idea to increase the plasma volume without changing the drifts is simply to vertically extend the torus into an ellipsoid, without actually changing its radius. In 1972 Artsimovich and Shafranov (Artsimovich and Shafranov 1972) were among the first to propose a tokamak with a vertically elongated cross section (the so-called finger-ring) and discuss its advantage. It was realised in the Soviet T-9 tokamak. The important point of such a cross section is that *beta* will be higher. As we have seen above, a high *beta* value is desirable, as it would require less energy to generate the magnetic fields at any given plasma pressure. In the past, various shapes have been considered,

---

[6]Readers who want to know more about (fundamental) plasma parameters may consult the *Wikipedia* page https://en.wikipedia.org/wiki/Plasma_parameters and proceed from there.

[7]Chen 2011, p. 248ff.

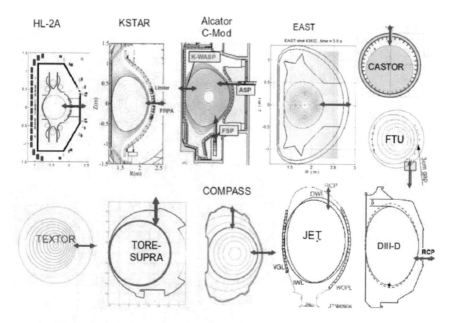

Fig. 4.8 Plasma shapes for various tokamaks (*from* Horacek et al. 2016)

e.g. a bean-shaped cross section (the PDX at Princeton) and a kidney-shaped cross section, which looked like two merged tokamaks (aptly called the Doublet) and was developed at the end of the sixties by the Japanese–American physicist Tihiro Ohkawa (1928–2014) at General Atomic in the US. A D-shaped plasma is currently the most common form and in modern projects for new devices or future reactors only D-shaped tokamaks are considered. Figure 4.8 shows the plasma shapes in some tokamaks, most of which are mentioned in this book, and Fig. 4.9 the D-shape of the plasma vessel of the German ASDEX-Upgrade tokamak.

The D-shape is not all roses though, the curvature at the corners of the D is very sharp, but fortunately occurs in only a small part of the total surface, and this part of the D can actually be used for removing waste material from the plasma, the helium particles (the 'ash') of the fusion reactions. They have to be drained off as they would use up the magnetic confinement capability reserved for the deuterium and tritium. Escaping plasma is channelled into the corners of the D where special devices, called divertors, are placed that can handle the heavy heat load and allow for the online removal of waste material from the plasma while the reactor is still operating. We will discuss these devices in greater detail below, but first have to introduce some extra parameters to describe the shape of the plasma.

When the cross section is no longer circular, we have to redefine what we mean by the minor radius and introduce other parameters, like the elongation and triangularity, to fully describe the geometry of the plasma. To know the exact form of a circle we only need to know its radius, for an ellipse we need two parameters and for a D-shape at least three. We will illustrate this with the aid of Fig. 4.10. In the figure *R*

Fig. 4.9 The D-shape of the plasma vessel of the German ASDEX-Upgrade tokamak (*from* Tiia Monto, https://commons.wikimedia.org/w/index.php?curid=34943904)

is the major radius and *a* the minor radius, so not much has changed as far as these two parameters are concerned. The **elongation**, the extent in which the plasma is lengthened in the vertical direction, is indicated by the Greek letter κ and defined as the ratio of *b* and *a*: $\kappa = b/a$, where *b*, as indicated in Fig. 4.10, is half the height of the D. For a circular cross section $\kappa = 1$, so the elongation measures how much the cross section deviates from a circular form, how ellipse-like the cross section is. In the figure closed field lines or surfaces (the red lines that go round in the D) and open field lines or surfaces (lines that do not close in the D) are also drawn. The "last closed flux surface" (LCFS) is called the **separatrix** (as it separates closed surfaces from open surfaces). A point in the figure at which the poloidal field has zero magnitude is called an 'X-point'. The separatrix is the magnetic flux surface that intersects with such an X-point. The figure drawn here has two separatrices. All flux surfaces external to this surface are unconfined and all flux surfaces internal to this point are closed.

The **triangularity**, indicated by the Greek letter δ, measures the D-ness of the plasma shape and is defined as the horizontal distance between the plasma major radius *R* and the X-point divided by the minor radius *a* ($\delta = 0$ corresponds to a fully elliptic cross section and $\delta \sim 0.5$ to the shape of JET or General Atomic's DIII-D tokamak (see Chap. 5)). In general (when there is no up-down symmetry), there can

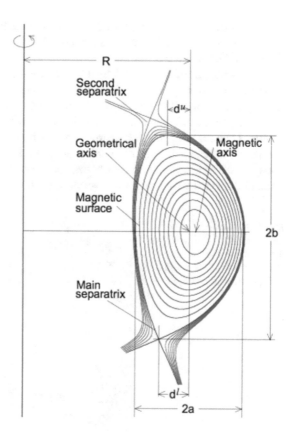

Fig. 4.10 Sketch of tokamak geometry, including separatrix (*Wikifusion*)

be an upper-triangularity and a lower-triangularity as in Fig. 4.10. In this figure the distances related to the triangularity are indicated as d^u and d^l.

The triangularity in Fig. 4.10 is positive. When the D is flipped with the 'belly' of the D pointing inwards, the triangularity is negative. Negative triangularity plasma has some favourable magnetohydrodynamic properties as regards edge-localised modes (ELM), the disruptive instabilities that were introduced in Chap. 2 and pose a major challenge to magnetic fusion. For that reason, there is currently an increased (theoretical) interest in such negative triangularity tokamaks; their disadvantage is that the *beta* limit is relatively low.

4.3.5 Divertors

A divertor is a device that acts as an exhaust, a sort of large ashtray that collects the 'ash' from the fusion reactions and other non-hydrogen particles. Its concept was introduced by Spitzer for the stellarator. A special magnet that pulled off the very outer layer of the plasma was placed around the tube. In the Model-C stellarator for instance it was installed in one of its straight sections to remove the ions before they drifted too far and hit the walls.

It is a vital part of modern tokamaks, one of its main functions being to remove impurities, e.g. from plasma-surface interactions, and to prohibit them from entering the confined plasma. It further serves to extract heat produced by the fusion reactions, minimize plasma contamination, and protect the surrounding walls from thermal and neutron loads. It removes alpha-particle power by transferring heat to a fluid, and pumps out helium ash to avoid dilution of fusion fuel. In short, it is a jack-of-all-trades as far as cleaning and heat extraction is concerned. Figure 4.11 shows the setup of the ITER vessel with the divertor at the bottom.

Fig. 4.11 Tokamak cross section for ITER, showing the divertor at the bottom (*from* Frederici, Talk at Erice 2004; Fusion for Energy)

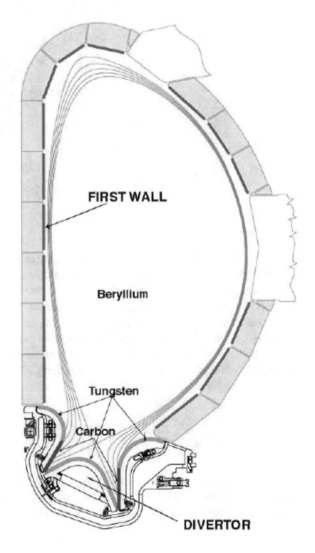

Early tokamak designs only very rarely included a divertor, as it was considered to be required only for operational reactors. When long-shot reactors started to appear in the 1970s, a serious practical problem emerged. Since plasma confinement is never perfect, plasma continues to leak out of the main confinement area, strikes the walls of the reactor vessel and causes all sorts of problems. A major concern in this respect was sputtering, i.e. the ejection of microscopic particles from the wall surface, which caused ions of the wall metal to flow into the fuel and cool it. The heat load of the plasma on the wall is considerable, and although there are materials that apparently can handle this load, they are generally made of expensive heavy metals whose particles you don't want to contaminate the plasma. In the 1980s it therefore became common for devices to include a feature known as a **limiter**, which is a small ring, like a sort of washer, of a light metal that projects a short distance into the outer edge of the main plasma confinement area. It reduces the diameter of the confinement area and restricts the cross section of the plasma column. The plasma column has to squeeze through this slightly narrower part of the tube, which keeps the plasma away from the wall. It also scrapes off the outer layer of plasma which contains most impurities. Any plasma particles leaking out will hit and erode this limiter, instead of the wall of the vessel. These erosion atoms will then mix with the fuel, and although the lighter materials of the limiter cause less problems than atoms from the vessel wall material, material is still being deposited into the fuel, and the more so when temperatures become higher because of external heating. The limiter simply changed the source from which the contaminating material was coming and provided only a partial solution.

When D-shaped plasmas came into use, it was quickly noticed that the flux of particles escaping from the plasma could also be shaped. This resulted in the idea of using the magnetic fields to create an internal divertor that flings (*diverts*) the heavier elements out of the fuel, typically towards the bottom of the reactor. There, a pool of liquid lithium metal is used as a sort of limiter. Magnets pull at the lower edge of the plasma to create a small region where the outer edge of the plasma, the "Scrape-Off Layer" (SOL), hits the lithium metal pool. The particles hit this lithium, are rapidly cooled, and remain in the lithium. This internal pool is much easier to cool, due to its location, and, although some lithium atoms are released into the plasma, lithium's very low atomic mass makes it a much smaller problem than even the lightest metals used previously.

So, the divertor is also installed for protecting the tokamak vessel itself. The divertor improves on the limiter in several ways, but, because modern tokamaks try to create plasmas with D-shaped cross sections, the lower edge of the D is a natural location for the divertor.

In ITER (Fig. 4.11) and the latest configuration of JET, the lowest region of the torus is configured as a divertor, while Alcator C-Mod at MIT (see Chap. 5) was built with divertor channels at both top and bottom.

A tokamak featuring a divertor is known as a divertor tokamak or divertor config-uration tokamak. In this configuration, the particles escape through a magnetic "gap" (determined by the separatrix), which allows the energy absorbing part of the divertor

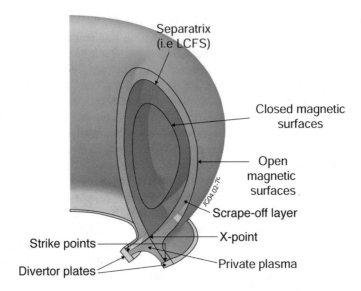

Separatrix
(i.e LCFS)

Closed magnetic
surfaces

Open
magnetic
surfaces

Scrape-off layer

X-point

Private plasma

Strike points

Divertor plates

Fig. 4.12 Schematic picture of a D-shaped plasma with separatrix and SOL (scrape-off layer) (*from* https://www.igvp.uni-stuttgart.de/forschung/projekte-pd/openclose.en.html)

to be placed outside the plasma. The schematic illustration of Fig. 4.12 elucidates the various concepts introduced in this section.

4.4   The Glory Years: The 1970s and a Change of Direction

With the positive Soviet tokamak results the decade of despondency was suddenly over, in spite of the fact that none of the fundamental problems, such as plasma instabilities, had been solved. Everything seemed to come together in the early 1970s with Nixon's US energy independence policy, the end of the Vietnam war, and the troubles with the free supply of oil from an embargo in 1973 by the Arab countries aimed at countries, including the United States, perceived to have supported Israel in the Yom Kippur war. By the end of the embargo in March 1974, the price of oil had risen from US$3 per barrel to nearly $12 globally, with US prices still significantly higher. This first oil crisis was followed later in the decade by the second oil crisis due to the decreased output in the wake if the Iranian revolution, pushing the price of oil up even further to about US$40. In addition, the growing concern for the environment and the clamour for clean energy, or at least energy with minimal environmental impact, made that fusion's claim of abundant, clean energy was increasingly being heard. The Washington Post for instance wrote on 8 March 1971: "*Fusion power … is a far cleaner and safer power source. It won't have the troublesome waste heat discharge that atomic power plants have. … It would produce a fraction of the radioactive by-products. Assuming disaster—a war, an earthquake or a collision of some kind—a fusion plant would be thousands of times safer than an atomic power plant."* Other newspapers too were urging the Nixon administration to support nuclear fusion research and other clean energy sources, all in the completely mistaken notion that a nuclear fusion reactor was just around the corner. All this was a boon to the nuclear fusion effort. Even though fusion had not much to show for, the atmosphere was such that every proposal was likely to get funding and funding for fusion went up dramatically in the 1970s. Very little was actually known about the environmental impact, biological hazards and safety issues of fusion reactors, for the simple reason that there were no specific designs of them yet. Fusion was not even at a stage where such reactor design studies made much sense, so long as it was not clear that a well-behaved plasma could be created and confined long enough for net power to be produced. In spite of this, reactor study groups and panels sprang up at various places in the US. September 1969 is seen by the fusion community as the beginning of serious interest in fusion reactor design studies. In that month the first conference on such reactors was held at the Culham fusion site in Oxfordshire in Britain, with all contributions (except two) coming from British and American scientists, showing that the appetite for such far-fetched enterprises were very much an Anglo-Saxon affair.

A further change happening in the American fusion effort was that strategic decision-making was taken away from the individual laboratories. Funding had always come from the AEC, but the lab directors at Princeton, Livermore, Oak Ridge, Los Alamos, General Atomic and MIT decided which problems to study and which machines to build, and they had a tendency to go ahead with their own ideas even if the AEC did not approve, confident that things could be put right later. It meant of course that there was considerable duplication in the machines proposed and constructed. An example is the approval of five very similar tokamak proposals, which also raised eyebrows in Congress.

The AEC fusion programme had been directed since 1953 by Amasa Bishop (1921–1997)[8] who was used to giving a free hand to the various laboratories, acting as their spokesman and resolving their conflicts. In 1970 he was succeeded by Roy Gould (b. 1927), who pursued the same policy. Gould, however, left after a year to return to Caltech and his successor Robert L. Hirsch (b. 1935), who had been deputy to both Bishop and Gould, pursued a much more aggressive policy during his term of office. Hirsch was a nuclear fusion engineer who had worked at the Fort Wayne Indiana Laboratory of International Telephone and Telegraph on an electrostatic confinement device designed by television inventor Philo T. Farnsworth (1906–1971). A proposal for funding of the device, called the fusor and to be discussed in Chap. 14, was submitted to the AEC, but rejected, after which Hirsch was hired by the AEC.

Hirsch reduced the independence of the lab directors and moved the centre of control and policy-making to Washington, greatly bureaucratizing the system. He committed the fusion community to a programme of milestones and ever larger

---

[8]Bishop is also the author of the first book on the subject: *Project Sherwood: The U.S. Program in Controlled Fusion* (Addison-Wesley, 1958).

machines, steering the community away from the fundamental research programme of the 1960s into plasma instabilities and suchlike, towards the practical goal of fusion energy. Above we have seen that this was identified by Richard Post of LLNL as one of the errors of the fifties; an error that was going to be repeated now.

This major change of direction from mainly pure plasma physics to the practical goal of constructing a working fusion reactor had a considerable impact on the research topics studied in the various research groups. The autonomous development of a scientific field, in this case plasma physics, is disturbed when a practical goal as a fusion reactor is formulated. It steers the research into a certain direction and restricts the problems that must be solved and the order in which this should be done. It determines when a certain development must be broken off, namely when it is considered to distract too much from the practical goal; and it also determines what must be considered as a solution and when the solution is considered adequate. It implied that pure plasma physics had to align itself with the practicalities of constructing a fusion reactor based on the tokamak design, and this at the time when plasma physics was still struggling to make any sensible predictions on the behaviour of plasmas in general, let alone in such a complicated geometry as a toroidal system. The goal now being formulated for plasma physics, within the wider setting of building a reactor which involves in addition a multitude of other problems, was the physical realisation of a system in which fusion processes take place in a controlled way and deliver more energy than is needed for their ignition, and this specifically for the tokamak geometry. (See about this Küppers in Van den Daele et al. 1979.)

Hirsch took the distribution of money in hand, put pressure on labs to do his bidding, and sparked off competition between the various laboratories for the funds needed to build ever larger machines, as we will see in the next chapter when discussing the Tokamak Fusion Test Reactor. He was a good salesman and in 1971 told Congress that with vigorous funding a demonstration fusion reactor could be ready by 1995.[9] This was close to the magical prediction span of two decades, but again totally unfounded as the recent Russian tokamak results in no way justified such a pronouncement and not a single tokamak was yet running in the US or anywhere else outside the Soviet Union. In a later report the claim was even upped: "*an orderly aggressive program might provide commercial fusion power about the year 2000, so that fusion could then have a significant impact on electrical power production by the year 2020*".[10]

Hirsch very much exemplifies what is wrong with the approach to fusion. Heppenheimer, who wrote his book before the term of Hirsch's prediction had expired, quotes Hirsch as saying: "*I had inherited a collection of very bright people, very good physicists, who in many respects were not aggressive* (being aggressive apparently is a virtue (LJR)) *but who were very conservative, and wanted to do a lot more of basic*

---

[9]It was not his last wrong prediction, for Hirsch is also one of the authors of the 2005 Hirsch Report (*Peaking of World Oil Production: Impacts, Mitigation, and Risk Management*, updated in 2007) on peak-oil scenarios, (wrongly) predicting a fall in oil production within five years. The opposite actually happened.

[10]*Fusion power: an assessment of ultimate potential*, Division of Controlled Thermonuclear Research, Atomic Energy Commission (1973).

physics than I thought appropriate. They wanted to solve all the problems before they would take the next step. I had a bunch of people who were generally timid and only weakly committed to making practical fusion power." Hirsch apparently did not understand that commitment is not sufficient for solving a problem. He was of the opinion that increasing the size of the tokamak would move the walls further away from the plasma, which then would take longer to leak away and that all problems would be solved. Without further ado he wanted to build large tokamaks and straight-away burn deuterium and tritium, with all the complexities of handling radioactive tritium and without proper preparations. Not the way to do science, one would think. But all this bravado got him the money he wanted. In 1973 President Nixon had announced that he would dramatically increase spending on energy research and Hirsch managed to get a sizeable slice of this, and of course had long departed the fusion scene (he left the AEC in 1976) before his wrong prediction might come to haunt him. (Heppenheimer 1984, p. 41–44; Bromberg 1982, p. 173ff; see also Dean 2013, p. 40ff; and Weisel 2001.) His detailed and aggressive plan for fusion research specified programme steps between the demonstration of scientific feasibility of a fusion concept and its eventual use as a reactor. Such thorough and optimistic planning helped renew the attractiveness of the AEC programme to Congress, and it greatly helped that he had an influential ally in Congressman Mike McCormack (1921–2020), who held a degree in chemistry and was a strong supporter of fusion. During Hirsch's stewardship, from 1972 to 1976, funding for fusion leaped from $30 to $200 million.

The heading of this section has been called "The Glory Years: 1970s", following Stephen Dean, who used the same heading for Chap. 4 of his book (Dean 2013, p. 27). Those years were glorious because of the increased enthusiasm among scientists after the perceived breakthrough with the tokamak, which enthusiasm was matched by increasingly generous government funding. They were not glorious for any great strides towards nuclear fusion, as we will see in the next chapters. In them we will discuss the various tokamak experiments that were carried out from the early 1970s onwards, first paying attention to the activity at America's foremost fusion laboratory, the Princeton Plasma Physics Laboratory, then discussing the other efforts in the US, before turning to other countries.

Table 4.2 lists the great variety of tokamaks and their most important parameters built around the world in the decade following the 1968 Novosibirsk conference, some of which will be discussed in some detail in the next few chapters.

Since research with tokamaks greatly expanded, with ups and downs, in the decades to come, we will devote a separate chapter to the big tokamaks, TFTR at Princeton, JET at Culham in the UK, and the Japanese JT-60, which paved the way for ITER, the giant device currently under construction at Cadarache in France.

Table 4.2 Tokamaks built in the first decade after the Novosibirsk breakthrough

Location	Name	Operation	R (m)	a (m)	B_T (T)	I_p (kA)
USA						
Princeton	ST	1970–1974	1.09	0.14	4.4	130
	ATC	1972–1976	0.88	0.11	2.0	50
			0.38	0.17	4.7	118
	PLT	1975–1986	1.32	0.42	3.4	700
	PDX	1978–1983	1.4	0.4	2.4	500
MIT	Alcator A	1972–1978	0.54	0.10	9.0	300
	Alcator C	1978–1986	0.64	0.16	13	800
Oak Ridge	ORMAK	1971–1976	0.8	0.23	2.6	230
	ISX-B	1977–1984	0.93	0.27	1.6	250
GA, San Diego	Doublet II	1972–1974	0.63	0.08	0.8–0.95	90–210
	Doublet IIA	1974–1979	0.66	0.15	0.76	<350
	Doublet III	1978–1985	1.45	0.45	2.6	0.61
Europe						
Fontenay, Fr	TFR	1973–1978	0.98	0.20	6.0	400
	TFR-600	1978–1986	0.98	0.22	6.0	600
Grenoble, Fr	WEGA	1975–1978	0.72	0.15	2.2	80
	Petula-B	1974–1986	0.72	0.18	2.7	230
Garching, Ger	Pulsator	1973	0.7	0.12	2.7	125
Frascati, It	TTF	1973	0.3	0.04	1.0	5
	FT	1978	0.83	0.20	10.0	800
Culham, UK	CLEO	1972–1973	0.90	0.18	2.0	120
	Tosca	1974	0.3	0.1	1.0	20
	DITE	1975–1989	1.17	0.27	2.7	260
Japan						
Naka	JFT-2	1972–1982	0.9	0.25	1.8	170
	JFT-2a/DIVA	1974–1979	0.60	0.10	2.0	70
Nagoya	JIPP-T2	1976	0.91	0.17	3.0	160
Tokyo	TNT-A	1976	0.4	0.09	0.42	20
Soviet Union						
Moscow	T-4	1974–1978	0.90	0.16	5	300
	T-7	1979–1982	1.22	0.35	2.4	390
	T-5	1962–1970	0.625	0.15	1.2	60
	T-6	1970–1974	0.7	0.25	1.5	220
	T-11	1975–1984	0.7	0.25	1.0	170
	T-9	1972–1977	0.36	0.07	1.0	

(continued)

Table 4.2 (continued)

Location	Name	Operation	R (m)	a (m)	B_T (T)	I_p (kA)
	T-12	1978–1985	0.36	0.08	1.0	30
	T-10	1975–date	1.5	0.39	5.0	
	TO-1	1972–1978	0.6	0.13	1.5	70
St. Petersburg	TUMAN-2	1971–1975	0.4	0.08	0.4–1.2	8
	TUMAN-2A	1977–1985	0.4	0.08	0.7–1.5	12
	FT-1	1972–2002	0.625	0.15	0.7–1.2	30–50

Adapted from Braams and Stott 2002, p. 154–155

Chapter 5
The Tokamak Stampede, Part 1

5.1 Introduction

The Novosibirsk results unleashed a veritable tokamak stampede in the West. The US saw a change of direction, so drastic as seen never before or after in any scientific endeavour. A variety of research directions were abandoned in the course of a few months and years of research experience thrown out of the window in favour of the blinkered pursuit of the tokamak concept, on the flimsy basis of a single Russian result, that had not yet been independently confirmed on any other device. And that was not all. The character of the exercise was also changed from a scientific research project into an energy generating project.

The Soviet Union of course did not have to change much. It continued its research programme, as will be set out in the final part of this chapter.

As we will see in the next chapter, the French took an equally bold step, by starting without further ado with the construction of the TFR (Tokamak de Fontenay-aux-Roses), similar in size to the Soviet T-3, but with a higher toroidal magnetic field. Other countries started at a smaller scale, like Germany with its Pulsator tokamak, before committing themselves to larger machines. The Italians too started building a small machine at Frascati. The British at Culham, who were in the middle of the construction of their CLEO (Closed Line Electron Orbit) stellarator, now decided to leave out the helical windings and start to operate it (in 1972–1973) as a tokamak. This was later followed up by a purpose-built tokamak, named DITE (Divertor Injection Tokamak Experiment).

The countries in Asia were either not yet involved in fusion research (China, South Korea, India) or, like Japan, had altogether been more cautious and more sensible by concentrating in the early years on basic plasma physics and a few small-scale pinch devices. But, when they entered the scene, they also started to build tokamaks (Sekiguchi 1983).

In this and the following two chapters we will review the response of the various countries to the tokamak triumph and/or describe their later entry in the fusion race, e.g. for China and other Asian countries, until the start of the age of the big tokamaks

© The Author(s), under exclusive license to Springer Nature Switzerland AG 2021   129
L. J. Reinders, *The Fairy Tale of Nuclear Fusion*,
https://doi.org/10.1007/978-3-030-64344-7_5

and its culmination in the ITER project. We will start with the United States and the Soviet Union, and turn to Europe and Asia in Chap. 6, respectively Chap. 7.

5.2   United States

Early research on fusion in Princeton—the pioneering stellarator work by Lyman Spitzer and his collaborators—was carried out within the framework of Project Sherwood. After the declassification of fusion research, the name of the laboratory was changed into the Princeton Plasma Physics Laboratory (PPPL) and throughout the 1960s research on stellarators was vigorously pursued, as has been discussed in Chap. 3.

In December 1969, within two months after confirmation of the Russian results, a drastic switch to tokamaks was made. It was decided to cease operation of the Model-C stellarator and convert it into a tokamak. Within a three-year period, from 1970 to 1973, the Princeton team entirely shifted its focus away from the stellarator and built a series of ever larger tokamaks, culminating in the $300 million TFTR.

It was a surprising decision and almost certainly premature, as Model C was still a promising machine and numerous experiments had yet to be completed. The stellarator had been studied for 18 years and a lot of expertise and knowledge had been gained, which would now for the most part go to waste. The American scientists were also convinced that the stellarator was a better experimental machine. The absence of a large plasma current, which in the tokamak produced an essential part of the confining field and of the heating of the plasma, made it possible to study confinement in isolation of the heating process. The disadvantage was of course that the heating had to be provided by other means; a problem they were convinced could be solved. The stellarator design was better suited as a future reactor as it could be run in a steady state, while the tokamak had to be run in pulses, alternately heating and cooling its plasma (Bromberg 1982, p. 165). But they were unable to figure out why it could not beat the Bohm restriction, while others could,[1] and decided to abandon stellarator research and convert the Model-C stellarator into a tokamak, renaming it ST (Symmetric Tokamak).

It would have been more sensible to continue the stellarator experiments for some time to see if the problems could be solved and the tokamak results be matched. Tokamak research might be concentrated for the time being at Oak Ridge, which

---

[1]The Princeton group believed the results of their calculations of the magnetic system of Model C, without backing up their calculations with experiment. The various components of the Model-C magnetic system had to be tuned to each other with very high accuracy. The PPPL people ignored this step, although it was absolutely necessary for a new machine, and after poor results on plasma confinement they decided to convert their stellarator into a tokamak. But before the conversion took place, they actually did make some measurements and found out that the magnetic surfaces were not closed, which caused particles to leak out. This could have been remedied, but the decision had been made, and there was no way back (V. S. Voitsenya, Kharkov Institute of Physics and Technology, private communication, March 2019).

had been even quicker in abandoning its research in mirror devices and already had a tokamak programme in place, and/or at one or several of the other places that had put in proposals for tokamaks, especially those that would advance beyond existing Russian machines, which was not the case for both Oak Ridge and Princeton.

Pressure from the AEC was however also considerable and the PPPL director cut through the knot, probably anxious that Princeton would lose out on the generous funding that was about to be released.

The fusion programme that now emerged in the US was completely unbalanced, with the various laboratories competing for the, albeit soon to be increased, funds. Duplication was inevitable, and the great emphasis on tokamak research created sky-high expectations for a major breakthrough. After all, if everyone in the fusion community suddenly caught tokamak fever, then surely that must pay off. Or else, what were they doing?

There was also a grave political aspect to this business and causing great worries in Washington. The Soviets were planning a whole series of tokamaks, culminating in the very large T-10 tokamak, which they expected might be capable of reaching full reactor plasma conditions. That would once again (after the Sputnik and the first man in space) snatch the glory away from the United States of being the first to demonstrate the scientific feasibility of controlled nuclear fusion. They needn't have worried. Even today that time has not yet come, and the Soviet Union has in the meantime already been buried. For science there was a great and rather unexpected advantage in all this. Now that national pride was at stake, the members of the Congressional oversight committee were suddenly 'unusually receptive' and even asked what the level of support for the fusion programme should be (Bromberg 1982, p. 164).

The proposals for tokamaks were coming in fast, even before the Soviet results had been confirmed: General Atomic's Doublet (a tokamak with a kidney shaped cross section, soon becoming the craze), Oak Ridge's ORMAK, MIT's Alcator, the Texas Turbulent Tokamak at Austin, and the conversion of the Model-C stellarator at Princeton. Funding was first provided to Oak Ridge and Princeton, although their proposals did not go further than a replication of the Soviet results. The others that would advance beyond existing Soviet machines, had first been rejected because of lack of funds, but were now approved by the AEC. Congress was astonished and in the Johnson/Nixon transitionary period was at first reluctant to just increase the fusion budget for the construction of five similar devices (Bromberg 1982, p. 168). After having become president, Nixon was also at first deciding on more austerity, but later became interested in US 'energy independence' and prepared to fund even such distant prospects as energy from nuclear fusion. Emphasizing the right words with the ignorant helps, but mostly only for a limited time before they feel cheated.

5.2.1   Princeton Symmetric Tokamak (ST)

First experiments on the Symmetric Tokamak (ST), as the converted stellarator was called, started on 1 May 1970 and the first US tokamak results, comparable to the T-3 as regards confinement times and temperatures, but not remarkable in any other respect, were presented a few months later in July 1970.

The fact that the tokamak and the stellarator were similar in concept presented a special opportunity for Princeton University, and paid off in the sense that it was the first laboratory (in the US it always pays to be the first, if only for the chance to indulge in extensive self-congratulation) to have a working machine although it was one of the last laboratories to enter the tokamak race (also a favourite American pastime to turn everything into a race). Oak Ridge and MIT took until late 1971 and early 1972, respectively, to produce their first results.

5.2.2   Adiabatic Toroidal Compressor (ATC) and Princeton Large Torus (PLT)

As a second step, funds were set aside for the Adiabatic Toroidal Compressor (ATC), a small tokamak specially designed to study heating by both adiabatic toroidal compression (i.e. heating the plasma by compression without heat exchange in a magnetic field) and neutral beams (discussed in Chap. 2), and for the Princeton Large Torus (PLT), first called the Proto-Large Torus, which began construction during 1971 and came into operation in 1975. The PLT was an ambitious experiment that was designed to produce higher plasma currents than the Russian T-3 tokamak. It was largely a copy of the Russian T-10, but with novel external heating systems in the form of neutral beam injection and lower hybrid current drive (LHCD), another method for heating the plasma by exciting ions and electrons. The PLT was the first tokamak which achieved a plasma current of 1 megaampere (MA) and thanks to the neutral beam heating it managed to push the ion temperature above 5 keV (60 million degrees). The latter achievement, accomplished in 1978, is important as it was the first time that the ion temperature exceeded the critical threshold for a "burning plasma", also called ignition, i.e. the temperature at which the nuclear fusion reaction can become self-sustaining. With the overstatement so common in fusion research, the American fusion community considered this the most significant achievement in the then nearly 30-year history of fusion research (Dean 2013, p. 55), but for some reason forgot to call it a breakthrough. Ignition was however not achieved, nor was this a goal of the device. A worrying concern was that confinement times seemed to get worse when the neutral beam heating was further increased, which cast a small dark cloud over the future of the big TFTR that was being planned at that time and will be discussed with the other 'big' tokamaks in Chap. 8 (Clery 2013, p. 155).

An experiment at Princeton that obtained (and obtains) very little attention is the PDX (Poloidal Divertor Experiment). It was probably completely eclipsed by the talk about TFTR at the time, whose construction soon started. PDX can mostly be found in lists of abbreviations, but even the meaning of the abbreviation is uncertain as it is said to stand for both the Princeton Divertor Experiment and the Poloidal Divertor Experiment, whereby the latter is probably the correct one. It was later converted into the Princeton Beta Experiment (PBX), and still later into the Princeton Beta Experiment Modification (PBX-M). For some mysterious reason, it and its successors are completely ignored, not even deemed worthy of a mention by Stephen Dean in his history of the US fusion energy programme (Dean 2013). It was by no means a small machine, but of similar size as the PLT and it had the novelty of bean-shaped plasmas. The problems it intended to study were also far from trivial, and included, as its name suggests, divertors, the all important 'ash tray' we discussed in the previous chapter. It was one of the first tokamaks to be equipped with such a device, with the German ASDEX, General Atomic's Doublet III (both to be discussed below) and the Japanese JT-60. The latter is one of the big tokamaks and will be discussed in Chap. 8. PDX's main objectives were to develop so-called poloidal divertors and other techniques for controlling impurities in large, high-temperature collision-less tokamak plasmas and optimise plasma cross sections. A poloidal divertor is a divertor in which the weaker poloidal field lines (going the short way around the torus) are diverted, which because of their weakness are easier to divert than toroidal field lines.[2] The PDX experimented with up to four of such poloidal divertors. Apart from its divertor a further outstanding feature of PDX was its neutral beam injection heating system, which was the most advanced in the world and a joint production of Oak Ridge and Princeton.

PDX operated only for four years from 1979 to 1983 and in its final year it revealed a new instability, called the fishbone instability, that barred the way to higher plasma pressure.[3] The instability directly deteriorates the confinement of the fast ions in the plasma. The poloidal magnetic field fluctuations associated with this instability have a characteristic skeletal signature that suggested its name.

The Princeton Beta Experiment (1984–1985), and its upgrade the PBX-M (1987–1993), both also fairly obscure, were designed to test the extreme shaping of the plasma cross section—in this case indentation of the plasma on the inboard side,[4] so a bean-shaped plasma like PDX—for increasing the stability of the plasma.

The reconfiguration of the Poloidal Divertor Experiment (PDX) into the Princeton Beta Experiment (PBX) resulted in a doubling of the achievable *beta* value: stable tokamak plasmas with average *beta* values above 5% were achieved for the first time. The results of these two experiments were apparently not such that bean-shaped plasmas or otherwise indented plasmas became a common feature of tokamaks.

In 1981, Spitzer's successor Mel Gottlieb (1917–2000) resigned as PPPL director and was succeeded by the Austrian-American physicist Harold Fürth (1930–2002).

---

[2]In the US there was at the time also strong interest in so-called bundle divertors in which a small bundle of the stronger toroidal field lines is diverted (Stacey 2010, p. 68).

[3]https://www.iter.org/newsline/283/1672.

[4]Contrary to the Doublet series of General Atomic in which the plasma had a kidney-shaped cross section with an indentation on the outside (see below).

Ten years had passed since the tokamak stampede had begun and apart from a few 'world' records in temperature, and 'firsts' in the deployment of various techniques not much had been achieved. According to the timeline on the PPPL website, PLT experiments had been expected "… to give a clear indication whether the tokamak concept plus auxiliary heating can form a basis for a future fusion reactor." The website fails to answer whether the expectation was fulfilled. Well, it wasn't, but in spite of this PLT is still considered an extremely successful experiment and led to a machine allegedly capable of reaching breakeven, TFTR, which ushered in the age of the big tokamaks, as we will see in a separate chapter.

5.2.3   Oak Ridge National Laboratory

The second important fusion research laboratory and the first to switch to tokamaks was Oak Ridge National Laboratory (ORNL). Its fusion programme with open mirror systems, the Direct Current Experiments (DCX), did not get anywhere, especially due to problems with low particle density, as we have seen in Chap. 3, and in the fall of 1968, after some of their scientists had returned from the conference in Novosibirsk, it was decided after ample consultation and deliberation to switch to tokamaks, just before everybody else joined the tokamak stampede. The mirror programme was abandoned and a first tokamak constructed, called ORMAK.

For Oak Ridge the conversion to tokamaks involved much more than it did for Princeton. There was at Oak Ridge almost no experience with toroidal devices (the 'mirrors' they had been working on were all straight solenoids) and, although they started before the British confirmed the T-3 tokamak results at the Dubna conference in 1969, they still needed until late 1972 before useful physics results were produced. In the beginning a two-phased project had been proposed: starting with the ORMAK-I which would match the existing Russian experiment TM-3 (a small device run in parallel with the T-3 and oriented at plasma physics rather than fusion physics), but with more attention paid to symmetry, and then continuing with the ORMAK-II, in which the plasma would be heated by injecting neutral beams of very energetic particles[5] and the magnetic field would be considerably increased, going beyond what the Russians already had achieved. A little further in the future, after ORMAK-I and -II would have been successful, a feasibility reactor, ORMAK-III was planned, envisaged by Alvin Weinberg (1915–2006), Oak Ridge's director from 1955 to 1973,[6] as the fusion equivalent to Fermi's Chicago Pile-1 reactor for fission.

---

[5] Oak Ridge had ample expertise with the technology of particle beams although at very low beam intensities, which would have to be increased by a factor of thousand to heat a plasma to 100 million degrees. Such a high temperature was not yet aimed for though (Bromberg 1982, p. 161).

[6] Weinberg was fired by the Nixon administration because of his continued advocacy for nuclear safety and molten salt reactors, instead of the liquid metal fast-breeder reactor, favoured by the AEC and currently the only type of large-scale fast breeder reactor in operation. His firing effectively halted the development of the molten salt reactor. In 1974 Herman Postma (1933–2004) was appointed as Weinberg's successor.

The novel aspect of ORMAK was that instead of having coils placed around the plasma vessel, it deployed a solid copper shell. To preserve the symmetry of the torus about the vertical axis through the centre of the doughnut hole, these coils must be placed with great precision. By deploying a solid copper shell for producing the toroidal magnetic fields, this symmetry would be even better assured, but for the rest it did not struck new ground compared to the Soviet machines (Bromberg 1982, p. 154ff). The plasma itself was surrounded by a thin, vacuum-tight, stainless steel liner; gold plated to minimize impurities. This liner was enclosed in a cooled aluminium shell.

In the last days of July 1970, the Oak Ridge scientists were in Princeton listening with their jaws clenched together to the results of the first US tokamak experiments. They had been quick off the mark with their first ORMAK and had thought to enjoy a tokamak monopoly in the US for some time, a dream that was shattered by the speed of the Model-C stellarator conversion at Princeton and by the presentation of its early results. They were beaten by Princeton in the tokamak 'race' and had to scrap some of the novel features of the ORMAK. There was no longer a need for duplicating the Russian results; that had already been done by the ST. New ground had to be covered. It forced them to discard altogether the single-piece copper coil that was the pride and novelty of ORMAK, in part also because of technical problems with uniformly feeding the current (Bromberg 1982, p. 170). A new ORMAK was designed and constructed, involving a rearrangement of some of the parts and the building of some new ones to produce a fatter, i.e. low-aspect ratio, torus at a moderate magnetic field.[7] The Russians had predicted, but in no way proved, that a low aspect ratio would improve tokamak performance, a point that was taken up much later in the development of spherical tokamaks. ORMAK's aspect ratio was 3.5, as can be calculated from the numbers given in Table 4.2, while the Princeton ST had an aspect ratio of 7.8, even larger than the Russian T-3. ORMAK operated from 1971–1976 and became the first machine where heating by using neutral beams exceeded ohmic heating power. It demonstrated the feasibility of this method and was the first tokamak to reach a temperature of 20 million degrees, but was soon surpassed by Princeton's ATC (see above).[8]

ORMAK-III was never built, but Oak Ridge remained involved in plasma physics experiments, especially the testing of high-temperature superconducting magnets with its International Fusion Superconducting Magnet Test Facility (IFSMTF), whose construction started in 1977, and novel ways of heating. The lab also constructed a slightly larger tokamak than ORMAK, the Impurity Study Experiment (ISX), in collaboration with General Atomic, specifically designed for studying how to control impurities in thermonuclear plasmas, i.e. such basic questions as how best to keep surfaces clean. Its first version ISX-A was soon upgraded to the ISX-B version, which operated from 1978 to 1984. It was technically important for *beta* studies and impurity transport studies, and demonstrated the existence of a *beta* limit before theory had predicted it. It was also one of the few machines that studied the

---

[7]For a rather lively exposé of the construction of this machine, see Roberts 1974.

[8]ORMAK also features in J. Bronowski's renowned BBC TV series, "The Ascent of Man".

beneficial effects of beryllium as a plasma facing material instead of e.g. stainless steel.

ORNL competed with PPPL for the construction of the Tokamak Fusion Test Reactor, which was so large that only one could be built in the US, but they lost the battle as will be described in Chap. 8.

As the energy crisis eased in the 1980s, funding for the US fusion programme declined. ORNL's last large-scale fusion experiment was a stellarator, called the Advanced Toroidal Facility, which operated from 1987 to 1994.

Currently the US ITER office, coordinating the US contribution to the ITER project in France, is at ORNL with partner locations at PPPL and Savannah River National Laboratory.

5.2.4   Massachusetts Institute of Technology (MIT)

MIT had been running some table-top magnetic fusion experiments in the 1960s at its Francis Bitter Magnet Laboratory.[9] The Plasma Fusion Center (now the Plasma Science and Fusion Center) was spun off from the Magnet Laboratory in 1976. It came into fusion in a big way only after Artsimovich's visit in the spring of 1969, especially due to the efforts of Bruno Coppi, which set MIT on the road of constructing its own tokamak, the first one in the Alcator series (Alcator stands for Alto Campo Toro, which is Italian for High Field Torus, revealing Coppi's involvement).

In the previous chapter we have discussed the Kruskal-Shafranov limit, which demands the safety factor q to be larger than 1, but also puts a limit on the plasma current that produces the poloidal magnetic field. The strength of this current is proportional to the ratio of the toroidal magnetic field and the major radius (B_T/R), while the power delivered to the plasma by ohmic heating depends on the resistivity times the square of the current (as for an ordinary wire). Coppi's idea was to create high-temperature plasmas by building a small fat tokamak torus with small R, jacking up the current without exceeding the Kruskal-Shafranov limit (Bromberg 1982, p. 163). In this way ohmic heating could be exploited to the full. In the course of 1969, a tokamak proposal was sent to the AEC. Its major features were an exceptionally high B_T and a small major radius R, to make the ratio B_T/R as large as possible. The design and construction of this first Alcator tokamak were approved in January 1970, but soon ran into trouble, due to contamination of the plasma and a faulty design. Bruno Coppi, who had been in charge in spite of being first and foremost a theoretical physicist, was replaced; the first device was dismantled and a new one built, which operated from 1974 to 1982 (Dean 2013, p. 39). From the start its plasma was of unparalleled purity. Because of its high toroidal field, a feature of all Alcator tokamaks, the plasma density was an order of magnitude larger than in other similar-size tokamaks. The energy confinement time was found to increase almost linearly

---

[9]Named after Francis Bitter (1902–1967) a pioneer in the production of intense magnetic fields. He established a magnet laboratory at MIT in 1938.

with the plasma density over more than a factor of ten, and to be proportional to the product of the plasma density and the square of the minor radius a, a scaling which is now called Alcator scaling. Confirmation that other ohmically heated tokamak experiments satisfied the same energy confinement scaling reinforced the view that this 'law' was in some sense universal for all tokamaks and that bigger tokamaks would provide larger confinement times.

The success of Alcator A led to the design of the second, Alcator-B tokamak, which however was never built because of difficulties with the power supply. When these were resolved, the design was changed and renamed Alcator C. It was a larger machine ($R = 0.64$ m) and its toroidal magnetic field was even higher than Alcator A, up to 12 T. In this new experiment, the energy confinement was at first lower than had been predicted by Alcator scaling, especially at high density. Injecting frozen pellets of "fuel" into the core region restored good confinement. Later experiments with the TFTR tokamak reinforced the conclusion that the density profile can impact confinement. Alcator C, which ran from 1978 to 1987, reached electron temperatures of over 3 keV and values for the product $n\tau$ of over 0.8×10^{20} s per cubic metre.

Several ideas for new devices and upgrades were proposed, but never funded. From 1978 to 1980, design activity was carried out for Alcator D, a larger version of Alcator C, that would allow for more heating power, and possibly even deuterium–tritium (D–T) operation. It was never formally proposed to the Department of Energy, but continued to evolve under Coppi's direction, eventually becoming the Italian–Russian IGNITOR device planned for construction at the Troitsk Institute of Innovative and Thermonuclear Research (TRINITI), a branch of the Kurchatov Institute, located near Troitsk in Russia (about 20 km from Moscow). As efforts are currently concentrated on ITER, it will probably never be constructed.

The final MIT facility to be mentioned here is Alcator C-Mod (Fig. 5.1), which for the purpose of obtaining funding was presented as an upgrade of Alcator C, although it is virtually a new machine. The older machines all had the usual circular plasma cross section, while the C-Mod incorporated a D-shaped non-circular cross section, as well as other modern tokamak design features.

The conceptual design was completed and Alcator C-Mod was formally proposed to DOE in late 1985. The project was approved and construction authorized in 1986. It was the largest fusion reactor operated by any university. It was in use up to 2016 when, after having obtained a few years respite from Congress in the years before, it became a primary casualty of the decrease in the domestic US fusion activity, due to the increase in US ITER spending (Dean 2013, p. 183).

In its final year the device set a world record for plasma pressure in a magnetically confined fusion device, reaching 2.05 atmospheres, a 15% jump over the previous record of 1.77 atmospheres (also held by Alcator C-Mod). This record plasma had a temperature of 35 million degrees, lasted for 2 s, and yielded 600 trillion fusion reactions. It was presented of course as "getting us closer to fusion energy" (Mihai 2016 (retrieved 13.5.2019)), which strictly speaking is true, but if every mm progress on a 200 km stretch or each brick laid for a three-storey house is presented as such, we will in the end be thoroughly dizzy from success, I am afraid.

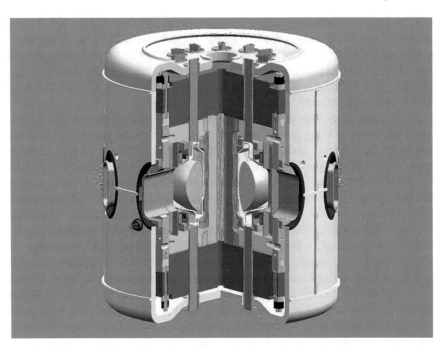

Fig. 5.1 An engineering diagram of the Alcator C-Mod tokamak showing the central portion of the machine without auxiliary systems. The toroidal plasma (red) is inside the vacuum chamber (mustard) with many vacuum ports. Magnetic coils (green, purple) profile the magnetic field. A strong superstructure (grey) surrounds the magnets. The entire assembly is encased in a liquid nitrogen cooled cryostat (light grey)

Following completion of operations at the end of September 2016, the facility has been placed into safe shutdown, with no additional experiments planned at this time. With the Alcator C-Mod gone, the United States was only left with two large tokamaks (the DIII-D, see below, and the National Spherical Torus Experiment (NSTX), which we will discuss in the chapter on spherical tokamaks).

In Chap. 2 we have mentioned the phenomenon of disruption, an event in which all the energy is suddenly in a few milliseconds dumped into the structure of the machine. It is a really serious problem and a major hazard for any tokamak. Unless it can be solved the structure of the tokamak, and especially the divertors have to be beefed up to be able to absorb all the energy, adding considerably to construction costs. Normally the plasma energy escapes slowly into the divertors, which are designed to handle such heat load. The MIT group working with Alcator C-Mod has captured what happens to a plasma in a disruption. This has been illustrated in Fig. 5.2. In such a typical elongated D-shaped tokamak, specially shaped coils must prevent the plasma from drifting up or down. When an instability causes a disruption, the plasma moves vertically, shrinking as it loses its energy. In the figure it moves downward towards the divertor, but it could also move upwards. The time scale shows that the whole event took 6 ms (Chen 2011, p. 290; Granetz et al. 1996).

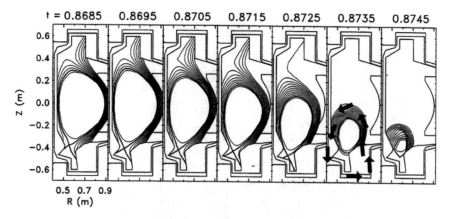

Fig. 5.2 Vertical motion of the plasma in a disruption (*from* Granetz et al. 1996)

In spite of Alcator C-Mod being scrapped, the MIT Plasma Science and Fusion Center has not given up on fusion research. On the contrary, it has an extensive programme that includes collaborations with institutions in the US and around the world on the German ASDEX upgrade, the DIII-D tokamak, the Chinese Experimental Advanced Superconducting Tokamak (EAST), the European JET, the Korean KSTAR, the National Ignition Facility (see Chap. 12), the National Spherical Torus Experiment (NSTX), the OMEGA laser system at the Laboratory for Laser Energetics of the University of Rochester, the Wendelstein 7-X stellarator in Germany (see Chap. 13) and the Z Pulsed Power Facility.

5.2.5 General Atomic

In March 1969 a proposal was submitted to the AEC for an extremely elongated tokamak with a strongly indented (kidney-shaped) cross section (Fig. 5.3) instead of the usual circular cross section.

It was called the Doublet, as it looked like two merged tokamaks with circular cross sections on top of each other. It was the brainchild of the Japanese physicist Tihiro Ohkawa (Fig. 5.4) who had joined the General Atomic fusion group in 1960 and taken charge of the group in 1967.

He had earlier proposed using plasma shapes of non-circular form, but only applied it to the tokamak design after the Russians reported their results in 1968. A theoretical study at General Atomic had shown that a Doublet had the potential to contain a hotter and denser stable plasma than a circular cross-section tokamak. The first device of this type, the Doublet I, was a table-top device and was funded internally by General Atomic. It showed that a Doublet-shaped plasma could be created in a plasma chamber surrounded by an hourglass-shaped conductor (copper) shell.

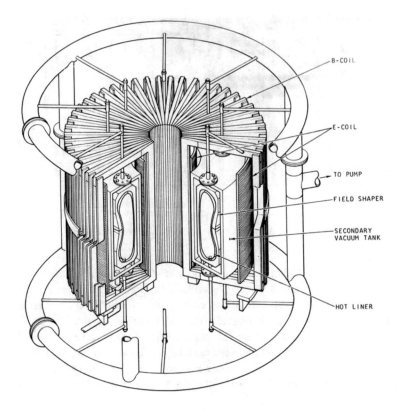

B-COIL

E-COIL

TO PUMP

FIELD SHAPER

SECONDARY
VACUUM TANK

HOT LINER

Fig. 5.3 Schematic presentation of the Doublet II tokamak with its extremely elongated plasma vessel (https://commons.wikimedia.org/w/index.php?curid=35934923)

Encouraged by these results, a larger Doublet II programme was proposed in 1969 and submitted for funding to the AEC. The size of Doublet II was comparable to that of a typical tokamak in those days (see Table 4.2). The Doublet II programme demonstrated its goals during operation from 1972 to 1974, namely that a stable doublet configuration could be maintained for tens of milliseconds with plasma parameters comparable to those achieved in tokamaks with a circular cross section. In 1974, Doublet II was converted into Doublet IIA with separate field-shaping coils surrounding the plasma chamber to control the plasma shape externally, and the design of an even larger follow-on machine, Doublet III (Fig. 5.5) was started. In May 1974, the AEC selected General Atomic to build the Doublet-III magnetic fusion experiment based on the success of their earlier Doublet experiments. When it became doubtful that sufficient funds would be provided by the US Department of Energy, Ohkawa took advantage of the Japanese eagerness to invest money acquired from the then-burgeoning trade imbalance between the US and Japan in a US energy programme and started a joint US-Japan collaboration on Doublet III, which was laid down in an official agreement between Japan and the US in 1977.

Fig. 5.4 Tihiro Ohkawa displaying a partial model of the copper shell of the Doublet I with his innovative 'hourglass-shape' cross section to enhance the plasma flow in a fusion energy device (*from* https://www.fusion-holy-grail.net/doublet-revolutions)

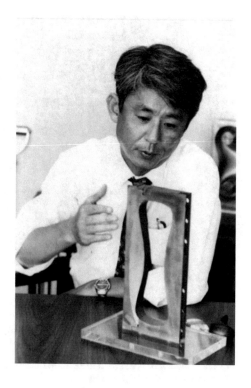

The machine was later upgraded and renamed DIII-D in 1986. It was so named because the plasma is shaped like the letter D, a shape that is now widely used in what are called 'advanced tokamaks'. They are called advanced as they are characterised by operation at high plasma *beta* through strong plasma shaping and active control of various plasma instabilities. The work at General Atomic, later General Atomics, demonstrated in particular that certain shapes of the plasma suppressed a variety of instabilities, which led to higher plasma pressure and performance (Bromberg 1982, p. 164ff, see also https://www.fusion-holy-grail.net/doublet-revolutions).

DIII-D is still operating and the largest tokamak facility currently in use in the US. It is a fairly sizable machine, similar in size to the German ASDEX Upgrade and the Chinese EAST (see next chapters). General Atomics is also manufacturing major components for ITER, including diagnostics systems and the central solenoid, the world's largest pulsed superconducting electromagnet. DIII-D holds the record for *beta* (12.5%) for conventional tokamaks. It has 18 independently controllable shaping coils to produce plasma cross sections ranging from circular to highly triangular D-shapes. An additional, unique set of asymmetric magnetic coils is used to suppress plasma instabilities. DIII-D research is aimed at establishing the scientific basis for future fusion power-producing tokamaks such as ITER, and for developing "advanced" operational regimes with high *beta* and self-sustained plasma current for ITER and tokamaks beyond ITER.

Fig. 5.5 Doublet III being built in the 1970s, expanding the capacity of plasma control experiments and improving stability of plasma

5.2.6   Other Activities in the US

Apart from the four major fusion research centres mentioned above (Princeton, Oak Ridge, MIT and General Atomic), there was and is a whole range of universities and laboratories that at one time or another carried out, and are still carrying out, research in plasma physics or engineering related to fusion. In Chap. 3 we encountered Los Alamos as a major player in early fusion research. Its fusion programme was started by James Tuck in 1951 who utilized the "pinch" effect for confining a fusion plasma. Similar concepts are found in today's Los Alamos work on Magnetized Target Fusion (see Chap. 14). Los Alamos fusion energy projects also include experiments on other fusion energy concepts, and theoretical modelling of fusion energy plasmas. The fusion technology work supports both magnetic fusion and inertial fusion concepts and is closely integrated with related work throughout the US and elsewhere in the world. In the last decade, LANL has largely focused on collaborating with other experiments around the world to build and diagnose the most advanced (and usually quite large) experimental fusion machines.[10]

At the California Institute of Technology (Caltech) a tokamak experiment was run on a fairly small tokamak, known as the Caltech tokamak (sometimes called

---

[10]More information can be found on the laboratory's website https://fusionenergy.lanl.gov/.

ENCORE[11]) from the late 1970s to the early 1990s. Its plasma temperature of 100,000 degrees was far too low to achieve fusion, and the device was mainly used for studying plasma properties.

Another California Institution, the University of California at Los Angeles (UCLA), conducted research with its Continuous Current Tokamak (CCT) from the late 1980s to the mid 1990s. It was a fairly large circular cross section tokamak, but with low toroidal field and low current. As its name suggests it was not a pulsed device but operated on a continuous current. The key to the continuous current operation is the use of radio-frequency waves to accelerate and drive the particles around the torus, so abandoning altogether ohmic heating. The results were not encouraging. In 2000 an even bigger machine was constructed, the Electric Tokamak, also known as the TSX. It claims to be the "largest tokamak" in the world and indeed has the largest major radius ($R = 5$ m and $a = 1$ m) and plasma volume, although not the largest torus volume (190 m^3 compared with 200 m^3 for JET). But like the CCT the plasma current is low. It has mainly been used for $beta$-limit studies. Operations were suspended about 2006 when funding ran out.

At Columbia University a tokamak, called HBT-EP (High Beta Tokamak), has been operating since 1993. It specialises in the study of resistive wall modes (instabilities related to the presence of a resistive wall that surrounds the plasma). It has an adjustable segmented conducting wall close to the plasma.

There are still quite a few more which all have conducted useful research of one type or another, but have not in any big way contributed to bringing the goal of nuclear fusion closer. The website www.tokamak.info gives an (incomplete) list of some 185 conventional tokamaks,[12] not only in the US, but all over the world. It obviously goes too far to discuss them all in this book, nor would it contribute much to the public understanding of nuclear fusion and whether we are ever getting there. It just serves to illustrate that much research has been and still is being carried out, which is a good thing, but all these large and small machines, still in operation or already shut down and the fact that they all think to be able to contribute vital data to the creation of nuclear fusion on Earth, make perhaps clear that it is far too early to start building huge machines like ITER. There are still too many unclear points that need clarifying before such large devices should be built.

5.3   The Soviet Programme

The Soviets of course did not have to change course and continued their programme at pace with the T-4, T-5, T-6, etc. ending with the T-15, whose construction was started just when the Soviet Union was in its death throes in 1988. In their eagerness to be the first to prove scientific feasibility the Soviets designed ever larger tokamaks,

---

[11]This name was chosen because of the device's high repetition rate, a unique design feature that facilitates study of some of the basic physics of tokamaks.

[12]So not spherical tokamaks for which there is a separate list.

culminating in the design of the giant T-20, a genuine test reactor, but they sensibly settled for a more modest, but superconducting physics experiment, the T-15. The T-20 has never been mentioned again. It was all based on the idea that larger is better: when you scale up the reactor size, e.g. increase the plasma radius by a factor of two, confinement times would improve fourfold. Physicists call this scaling laws, like Alcator scaling encountered above. This is also the idea behind building the giant ITER facility. The trouble is that these scaling laws, being just experimentally observed patterns or regularities, do not always work or the scaling just stops at a certain point. Other designs, like stellarators, were not completely ignored (see Chap. 13), but the Soviets understandably concentrated on tokamaks.

Notable in the Soviet series is the T-10, which started operating in 1975 and was at the time the largest tokamak in the world. The Princeton Large Torus (PLT) was largely a copy of this. The Soviets also operated a series of smaller TM tokamaks, like the TM-3 mentioned above, and a few TO tokamaks which were characterised by an O-shaped magnetic core. Judging from the list of notable discoveries and progress with tokamaks from www.tokamak.info they lost their edge on the rest of the world fairly soon after the stampede began; the last notable discovery or 'first' with a Soviet tokamak dates from the early 1970s with the T-7, which was the first tokamak to use superconducting coils (Braams and Stott 2002, p. 153ff).

In general Soviet research gives the impression as being conducted in a careful and well-thought-out manner within the framework of a long-term programme, more so than in other countries, in particular the US, and in the way you expect science to be conducted, although some rigidity is also apparent. No competition for funds between the various institutions, although there must have been some rivalry; no jumping from one extreme to the other, cutting funding one year and increasing it again the next. And although they took notice of what was going on in other countries, it did not divert them from their course. In the view of Lev Artsimovich, the leader of the Soviet fusion effort, the true danger of international contacts for the domestic Soviet programme was that physicists would pay too much attention to foreign programmes and would mistakenly go down the wrong path, chosen by others, rather than finding their own way (Josephson 2000, p. 179). That is why he stubbornly stuck to tokamaks up to his premature death in 1973 and opposed attempts by Kurchatov, although without success, to start an additional stellarator programme.

The Soviets were also putting a lot of money and manpower into nuclear fusion. According to the Report released in 1966 by the US Atomic Energy Commission (AEC) on the status of fusion research in the world, as far as manpower involved in the sector the Soviet Union leads the world. "Their effort is twice the US effort. In plasma theory the Soviets are preeminent and at this time their effort in theory is about four times the US effort. In number and variety of major experimental devices the Soviets also lead the world".[13]

---

[13]USAEC, *AEC and Action Paper on Controlled Thermonuclear Research*, June 1966, III-32, https://fire.pppl.gov/US_AEC_Fusion_Policy_1966.pdf, last accessed 26 July 2019.

By far the most important research was carried out at the Kurchatov Institute in Moscow, where from the early 1950s to the 1990s a huge series of tokamaks numbered T-1–T-15 was constructed with a few side branches called TM (small tokamak, *malyi tokamak*) and TO. In 1961 the Troitsk Institute for Innovation and Fusion Research (TRINITI) at Troitsk (also in the Moscow region) became a division of the Kurchatov Institute. Some of the fusion research was and still is being carried out at TRINITI. From the early 1970s the Ioffe Institute in Leningrad/St Petersburg joined in with the much smaller series of TUMAN and FT tokamaks, some of which were moved from the Kurchatov Institute to Leningrad.

Some tokamak (and stellarator) research was conducted in Sukhumi (at the time in Georgia, but now in the breakaway republic of Abkhazia). In Moscow a very small tokamak called TV-1 was operated from the late 1970s to 1983 at the Institute of High Temperatures of the Academy of Sciences (IVTAN). In 1984 it was transferred to Bishkek (currently the capital of Kyrgyzstan) where it remained in operation until 1998. Research with stellarators was also conducted at the Kharkov Institute for Physics and Technology in Ukraine (see Chap. 13).

Each of the machines they built solved one problem or another (an equilibrium, stability or impurity problem) and culminated in the success with the T-3 and its smaller companion the TM-3 in 1968, resulting in the world-wide tokamak stampede. While the whole world was following their example and building tokamaks all over the place in all kinds of sizes and with various strengths of magnetic fields and currents, the Soviets quietly continued to carry out their programme of experiments, but they never achieved any other major 'breakthrough'.

It has been suggested that the main obstacle to further Soviet success was the big lag in computing power, diagnostics and modelling (Josephson 2000, p. 192). That may well be true, but another reason may have been that their greatest specialist in the field and after Kurchatov's death in 1960 the undisputed sole leader of the Soviet programme Lev Andreevich Artsimovich died prematurely at the age of 64 in 1973. Evgeny Velikhov (b. 1935) (see picture in Fig. 3.20), who upon Artsimovich's death took over the leadership of the Soviet fusion programme, was a less dynamic leader.[14] Moreover he was a theoretical plasma physicist and it is in general not a good idea to place a theoretical physicist at the helm of a largely experimental and engineering project. Unless the direction of the project is clear, which is by no means the case for nuclear fusion, it is bound to veer off in one way or another or get into a rut. As an active Party member and important apparatchik in Soviet scientific life with numerous positions in science and government he was certainly less involved in daily research work (although his name is on more than 1500 scientific publications, a number most scientists not even manage to read, let alone understand in their entire scientific life). He not only was an apparatchik, but apparently also had an

---

[14]That there was at first nobody to take over Artsimovich's role and the fusion program slowed down is confirmed to some extent by Sagdeev (Sagdeev 2018, p. 387). Velikhov is known among other things for the Velikhov-Chandrasekhar instability (a fluid instability that causes an accretion disk orbiting a massive object to become turbulent, first noticed by Velikhov in a non-astrophysical context and generalized later by Chandrasekhar) and the Velikhov instability (a magnetohydrodynamic instability occurring in magnetised cold plasmas).

apparatchik's mind, believing that a top-down approach, initiated by government resolutions and multi-year plans, were the panacea to any technological problem. He espoused a mechanical view of scientific progress through exhortation and slogans, not reality. Social receptivity was less important in his mind (Josephson 2000, p. 193). Velikhov remained the major figure of the Soviet fusion programme until the end of the Soviet empire and even beyond, spanning a period of close to fifty years at the helm, which is too long. It was in general a problem of Soviet society, not only in science, but also in politics, for instance, that people tended to grow old in their jobs, which is bound to lead to stagnation.

Velikhov claims to having persuaded Gorbachev in 1985 that the next generation fusion device needed to be a joint international effort, so by some he is considered ITER's godfather.[15] He was farsighted in that respect, as without ITER Russia would probably no longer play any role of significance in fusion research. He was ITER Council Chair during the technical design phase for ITER and again at the start of ITER construction from 2010–2012.

Another possibility of the lack of any major results from Soviet/Russian tokamaks is of course that the tokamak is after all not the ideal design for a fusion reactor, but it seems that this possibility is not really in the minds of the fusion enthusiasts.

In the following some details of the Soviet research programme, its achievements and prospects from 1968 will be discussed. The basic parameters of most of the Soviet tokamaks can be found in Table 4.2. For earlier Soviet research we refer to Chap. 3, where the early Soviet efforts and the birth of the tokamak have been described.

As recalled in Chap. 4, Artsimovich and Shafranov had noted, independent from Tihiro Ohkawa at General Atomic, that in a tokamak with a vertically elongated plasma cross section (a D-shaped plasma) the stability conditions are preserved at higher current density values than would be the case with a circular cross section, and that consequently it should be possible to obtain higher plasma temperatures in these conditions (Artsimovich and Shafranov 1972). Moreover, the elongated cross section seemed to promise higher *beta* values. It was put into practice in 1973 on the relatively small T-9 tokamak which had an extremely elongated cross section and demonstrated the possibility of creating a stable plasma with such a design. It was given the name "Finger Ring" as the shape of the torus was reminiscent of a ring. It was also shown that the current and safety factor values corresponded with theoretical predictions. The T-9 was later converted to become the T-12. A second device, also built in 1973 and a little smaller than the T-9, with a non-circular cross section but not as extremely elongated as the T-9, was the T-8.

A few years earlier in 1970, while the effectiveness of ohmic heating was studied with the T-6 device, an unusual type of discharge was discovered (a so-called runaway mode in which virtually all the current is carried by a small fraction of accelerated electrons with very high energies from a few tens to hundreds of keV). It led to the discovery of a new instability (the Parail-Pogutse instability (Parail and Pogutse 1976, 1978) or runaway electron beam instability) characterised by an abrupt increase

---

in the transverse plasma energy. In larger devices such as ITER, such beams of runaway electrons can contain enough energy to jeopardize internal components of the machine and therefore must be kept under control to avoid such damage (Strelkov 1985, p. 1191). It should be noted that such instabilities, reminiscent of the disruptions mentioned earlier, have so far only been observed in plasmas that are still far from ignition. It is not clear whether ignited fusion devices can survive such events and the ability to do so is in any case a major engineering design challenge.

In 1975 regular experiments on the T-10 machine, a device pushed by Artsimovich until his last breath, started at the Kurchatov Institute. It is one of the few larger tokamaks of that era still in use. The T-10 was supposed to be shut down around 1985, but when it became clear in 1983–1984 that the construction of the new big tokamak T-15 would be delayed by four to five years, it was decided to preserve the T-10 as else there would be no sizable tokamak left to do experiments with. It has been around ever since.

The physics for the project had already been formulated in 1968. As stated above, the Princeton Large Torus was largely a copy of the T-10, but had superior peripherals, like neutral beam injection, lower hybrid current drive and computers for handling data. At the time, T-10 was the biggest tokamak in the world, but it was poorly conceived from the start and remained a machine of limited capabilities. In comparison with the T-3 it had a high plasma current (10 times as large), but only a modest increase in linear dimensions (less than twice as large). The parameters were optimized to achieve the highest possible ion temperature by ohmic heating, so pushing the current to the maximum value, according to the Kruskal-Shafranov criterion, as determined by the stabilizing toroidal magnetic field. An analysis of data from various tokamaks showed that the limiting temperature not only increased with the toroidal magnetic field, but also with dimensions (scaling), which has resulted in the construction of ever larger devices.

The second stage of the T-10 experimental programme involved the use of external plasma heating by high-frequency and super-high frequency waves. The structural features of the device made neutral beam injection inappropriate and it was necessary to choose between radio-frequency heating at the ion cyclotron resonance frequency and microwave heating at the electron cyclotron resonance frequency, of which after some preliminary testing the latter was chosen. With it an electron temperature of 10 keV was reached, which apparently was a record at the time. It showed the widespread opinion that the electron temperature in tokamaks was limited to 5–7 keV to be erroneous (Dnestrovskij 2001, p. 827). The PLT at Princeton achieved an electron temperature of 8 keV and in general better results than the T-10. The record is now surpassed in larger tokamaks and, although for ITER electron temperatures of 40 keV are predicted, average electron temperature will just be 8.8 keV.

The T-10 further studied transport mechanisms in the plasma, behaviour of impurities and the makeup of the plasma. Its research was complemented by experiments on the TM-3, the T-4, which obtained high parameters of deuterium plasma, the T-6, which demonstrated that a copper shell had a stabilizing effect on some magnetohydrodynamic instabilities, and the TO-1, which studied the equilibrium of the plasma (Josephson 2000, p. 189).

The T-10 met with a major disaster when its toroidal field coils overheated, resulting in the destruction of the coil insulation and requiring extensive repairs. The work on the T-10 was done at the Kurchatov Institute in Moscow, but with the participation of groups from a number of institutes in Russia and abroad (Strelkov 2001).

1975 was a busy year at the Kurchatov Institute, since apart from the T-10 the T-11 and T-12 also came online in that year. The T-11 is equipped with neutral beam injectors for supplementary plasma heating. The T-12 device was built on the basis of T-9 for studying divertors.

As noted several times in the previous chapters, the pulsed regime in which present tokamaks operate is a serious obstacle to the creation of a fusion power reactor. The first step to remedy this is the use of superconducting magnets to set up a constant magnetic field (non-superconducting coils have to be cooled between pulses), and the second step is to develop methods of creating and maintaining a steady-state current in the plasma loop. Work in this direction is carried out on the T-7 tokamak, which went into operation in 1979 and was the world's first tokamak with a superconducting toroidal field winding. The coils of the T-7 later went to China to become part of the HT-7 in Heifei.

In the course of the 1970s and 1980s the time for funding fusion became increasingly unfavourable, because of financial pressure connected with the construction of nuclear fission reactors. Moreover, as in the US, the optimistic, but never fulfilled promises of the fusion scientists made government agencies sceptical. In spite of this, in June 1983 the Politburo still increased funding in line with the Soviet Union's long-term energy plan that envisaged industrial tokamaks to be built early in the twenty-first century (Josephson 2000, p. 190). That vision was at least a century off.

In line with this, tokamak research at the Kurchatov Institute developed further with the introduction of the T-15, designed for the production and analysis of a plasma with thermonuclear parameters, a goal that was not realised either. It was also intended to again secure the Soviet Union's global leadership in fusion research, an intention that likewise miserably failed. Contrary to other large tokamaks designed and built from the early 1980s onwards (like JET), the T-15 still had a circular cross section. It is surprising that of all the tokamaks built at the Kurchatov Institute only the T-8 and T-9 (still built under Artsimovich's supervision in the early 1970s) had a non-circular cross section. The T-15 was optimistically called the first industrial prototype fusion reactor and used superconducting magnets to control the plasma. These enormous magnets confined the plasma but failed to sustain it for more than just a few seconds. The USSR's desire for cheaper energy ensured the continuing progress of the T-15 in the twilight years of the Soviet Union under Mikhail Gorbachev. Reactors based on the T-15 were supposed to replace the country's use of gas and coal as the primary sources of energy. Needless to say that there is no chance of this happening soon. In the Soviet Union too, fusion scientists were utterly devoid of any sense of realism as regards the feasibility of energy from nuclear fusion.

The T-15 achieved its first thermonuclear plasma in 1988 and the reactor remained operational until 1995, but during this entire period it was plagued by technical and budgetary problems, undoubtedly connected with the demise of the Soviet empire.

So, all in all, it can surely be concluded that the T-15 has largely been a waste of time, energy and money.

But, things could have been worse, since the planning in 1983 (it was already talked about in the mid 1970s) had at first foreseen in the construction of the T-20, a very large tokamak with superconducting coils that would greatly surpass the TFTR-sized tokamaks. It would have been a giant machine of close to ITER proportions and twice as large as JET, but (surprisingly) with a circular plasma cross section. It is bewildering to read that *"The parameters of the T-20 are determined by its purpose, i.e. the possibility of prolonged operation with a reacting D-T plasma and the technical possibility of attaining these parameters in the early 1980s"*. (See e.g. Spano 1975.) Now almost fifty years later this situation has still not been achieved with any tokamak, and will not, for considerable time to come. Were the people who wrote this, in this case the Soviet theoretical physicist Boris Kadomtsev, really serious (scientists)? Or was it just hollow Soviet bluff, part of its death rattle, an empty attempt to do better than the West in a country that was already in shambles? Fusion scientists, both in the West and the East, were and perhaps still are incapable of escaping megalomania, it seems.

From its early successes in tokamak research in the 1950s and 1960s the brightly burning flame of Soviet fusion research has become duller and duller and was reduced to a flicker at the collapse of the Soviet Union. Russia took over the T-15 programme from the defunct Soviet Union and the T-15 is expected to remain the main tokamak in Russia to the year 2040, when the country should have been studded with 'industrial tokamaks', if everything had gone according to plan.

An upgrade under the name T-15MD, where MD stands for Modified Divertor, with a D-shaped cross section is currently being assembled at the Kurchatov institute. It will be equipped with auxiliary plasma heating and current drive systems, which will allow the simultaneous achievement of high plasma temperature and plasma density. Start-up of the device is planned for late 2020 and its research will serve both ITER and DEMO.[16] It will be considerably smaller ($R = 1.5$ m) than the original T-15 ($R = 5$ m) and a very modest machine indeed. Its focus will be on carrying out specific tasks in support of ITER, as well as the creation of a fusion neutron source for atomic energy needs. The neutron source can for instance be used to produce nuclear fuel for fission reactors (Melnikov et al. 2015; Khvostenko et al. 2019).

In the middle of the 1960s, the Ioffe Institute in Leningrad also started experiments on tokamaks. The first small toroidal device, TUMAN-1, was put into operation in 1964. In Russian *tuman* means fog, haze or mist and there may be a deeper meaning to this choice of name. Its main aim was to study adiabatic compression. It was followed by a series of TUMANs up to TUMAN-3 which began operation in 1977, was rebuilt in 1990 and renamed TUMAN-3M, and is still in use today (Smirnov 2010, p. 5).

In 1970, the FT-1 tokamak, constructed with the help of the Kurchatov Institute, began operation. The chamber, coils and transformer were sourced from the T-2

---

[16]ITER website https://www.iter.org/sci/tkmkresearch; about the second life of the T-15 see also https://www.iter.org/newsline/152/477.

tokamak. At present in 2020 the institute is still in business with the FT-2. The Ioffe Institute also operates a spherical tokamak, called GLOBUS-M (see Chap. 11).

All these tokamaks are very small, even according to the standards of the 1970s, and although useful research was done with them, there were no major achievements or surprises.

The strategy of the Russian activity in fusion is aimed at the construction of a fusion power plant in about 2050, but as things stand now that deadline is not realistic.

Chapter 6
The Tokamak Stampede in Europe

6.1 Introduction

European nuclear research was coordinated almost from the start by the European Atomic Energy Community (Euratom), established in 1957 by the six founding members of what is now the European Union (EU). It has the same membership as the EU, plus Switzerland as an associated state.

No single European country at the time was able to carry out an exclusively national fusion effort. Fusion seemed ideal as a field of European cooperation and an advantage in this respect was that there were no questions related to industrial and military applications (as in the case of fission). In spite of this there apparently was not enough support for a joint European fusion research institute,[1] so national programmes had to be embedded in some sort of common programme. This would allow to share the costs of research that no single member country would be able to bear individually, in particular in a field still at a very preliminary stage and with experimental results expected only in the distant future.

Lacking a central research institute, all fusion research in Europe (not including Britain, which had started its own programme much earlier and only joined Euratom after its accession to the European Community in 1973) evolved through association agreements with the European Commission, which acts as the representative of Euratom, and Euratom would finance, develop, coordinate and supervise national

---

[1] A joint European program was in part impossible as in some countries, notably France, nuclear fusion and fission research were carried out under the same roof and nuclear fission (still) had very strong ties with the military. An early European Commission had considered the possibility of a European institute and concluded that for the time being, "unless it can be demonstrated that a European laboratory is needed in order to build larger facilities that cannot be built by national groups, the many other advantages of such a centre may prove insufficient to overcome the difficulties in its creation and maintenance" (Curli 2017, p. 69). Of course, as is the case with CERN, when you have such a central laboratory there must be activity in the member countries to feed the laboratory with new ideas and manpower. Such a laboratory without close ties with universities and institutes in the member countries would soon die out, certainly for a long-term project like nuclear fusion.

© The Author(s), under exclusive license to Springer Nature Switzerland AG 2021
L. J. Reinders, *The Fairy Tale of Nuclear Fusion*,
https://doi.org/10.1007/978-3-030-64344-7_6

fusion programmes. In each member state participating in the European Fusion Programme at least one research organisation has such an association agreement. The association agreements were a great success, because of their simplicity and effectiveness. A laboratory that accepted the Community rules on tendering, knowledge sharing and work evaluation could have 25% of its general expenditure paid by EURATOM, without losing any of its autonomy. This was of course a very advantageous arrangement and most countries quickly signed up for it (Claessens 2019, p. 23).

All the fusion research organisations and institutions of a country are connected to the programme through the contracted organisation(s). The groups of fusion research organisations of the member states are called "Associations". This network of association contracts constituted the framework within which all fusion research in Europe was to be developed, and would remain so for many years. The structure was partly modelled on Project Sherwood, where the Sherwood Committee financed and coordinated research in American fusion laboratories. (Curli 2017, p. 70. For Sherwood see Bromberg 1982.) The first Euratom association treaty was concluded in 1959 with the French Atomic Energy Commission (CEA) and its laboratory at Fontenay-aux-Roses, followed in 1960 by the Italian centres at Frascati. Other contracts with German and Dutch institutes followed in 1961 and 1962, and with several others in later years. (See e.g. Palumbo 1987.)

So, contrary to the successful Europeanization of high-energy physics in CERN, controlled nuclear fusion research in Europe, at least until the plans for the Joint European Torus (JET) in the 1970s, remained the domain of national research programmes and institutes. Some see this as a missed chance, while others see it as "probably the only example of a truly "common" European policy" (Curli 2017, p. 59). In view of the high cost of fusion research and in spite of the research being carried out at the national level, all were happy that the association agreements resulted in Euratom becoming a major financier of fusion research and by the mid 1960s, a European fusion community had been established. In those early years activities mainly concentrated on the development of the yet virtually non-existent plasma physics and the tentative exploration of a variety of confinement schemes together with some efforts at heating methods beyond ohmic heating (Palumbo 1988).

While in 1958 experimental research on a large scale was conducted mostly in Britain as far as Europe is concerned, in 1965 France, Germany and some other countries, all working under association treaties with Euratom and coordinated by Joint Committees, had entered the fray and were actively involved in such research. Around that time altogether 43 scientific organizations were working on the problem of controlled thermonuclear reactions in Western Europe. The most important groups were the German Institute of Plasma Physics at Garching; the French centres at Fontenay-aux-Roses and Saclay; the Italian plasma laboratory in Frascati; Dutch Institutes in Jutphaas and Amsterdam; the German centre at Jülich (near Aachen); the Institute of Plasma Physics in Stockholm (Sweden); the Risø National Laboratory at Roskilde (Denmark); and the Plasma Physics Laboratory of the University of Paris at Orsay. An extensive range of experiments with a multitude of device types were carried out at these institutions, with the European Study Group on Fusion,[2] a body set up in 1958 by Euratom and CERN to put forward suggestions for how to coordinate European fusion programmes, acting as the catalyst (Schlüter 1964).

Immediately after the results from the T-3 tokamak became known, various countries started to construct tokamaks, some of which will be dealt with in greater detail below. In total ten (small) tokamaks were constructed in the early 1970s in Germany, France, Italy, Denmark, the Netherlands and Belgium.

In the following the efforts of the various European countries will be discussed in some more detail.

6.2 UK

The early history of fusion research in Britain has been recounted in Chap. 3. Experiments on ZETA continued at the Harwell UKAEA site until 1968 when the machine was shut down. In the meantime, in 1965, the British had opened a dedicated nuclear fusion laboratory at the site of a former airfield in Culham, nearby in Oxfordshire. Fusion activities at Aldermaston and other UK locations were also amalgamated at Culham to form a national centre for fusion research. In the first two decades of its existence Culham built almost 30 different experiments trying out a variety of fusion concepts; among them shock-waves, magnetic mirror machines, stellarators and superconducting levitated rings to trap the plasma, called levitrons.

From Culham the British delegation left for Moscow in 1969 to confirm the Soviet results with the T-3 tokamak and, of course, the excitement of the confirmation forced them to drastically alter their programme. When the Russian results broke, they were in the middle of the construction of their CLEO (Closed Line Electron Orbit) stellarator, but now decided to leave out the helical windings and start to operate it (in 1972, 1973) as a tokamak. This was later followed up by a purpose-built tokamak, named DITE (Divertor Injection Tokamak Experiment). Culham would start to play a pivotal role in European fusion research when it was decided in 1983 to locate JET there.

6.3 Germany

Germany's status as an occupied territory after World War II formally ended when the General Treaty, in German *Deutschlandvertrag*, signed in 1952 by the Federal Republic of Germany and the Western Allies (Britain, France and the US), came

---

[2]In the months leading up to the second Geneva conference, scientists and administrators connected with both CERN and Euratom formed a "European Study Group on Fusion" to consider the possibility that the high-energy physics laboratory might do work on plasma physics. Their main objective was to enable Europe to catch up in fusion research with the US, UK and Soviet Union and perhaps to make presentations at the Geneva conference (Weisel 2001, p. 158).

into force in 1955. It recognised Germany's rights as a sovereign state and made it possible to start preparations for constructing a nuclear reactor. Once restrictions on applied nuclear physics work were lifted, this kind of research really took off in Germany and within a decade West Germany was the world's leading exporter of nuclear technology (Cassidy 1992, p. 530).

In 1955 in a talk at Göttingen the German Nobel Prize winning physicist Werner Heisenberg (1901–1976), who was one of the driving forces behind German nuclear research, also made a statement about nuclear fusion: "Ultimately, one will start to (…) utilize the greatest energy source that nature can provide, which can be found in the interior of the Sun and most stars, for the peaceful development of Earth. This problem can be solved although at first glance it looks almost hopeless. When that goal has been reached, the 'Fire of the Stars' will have been brought to Earth and energy supplies will have become virtually inexhaustible. In terms of daring, this problem is set to surpass anything developed thus far."[3] Heisenberg at least still thought that the problem would be difficult, even seemingly hopeless, but soluble. Others were far more optimistic as reflected in the German foreign policy journal *Außenpolitik: Zeitschrift für Internationale Fragen*, which in 1954 claimed that nuclear fusion would be "practically ready to use in two years," (quoted in Morris 2016) in line with the grossly inflated expectations in other countries. It seems though that in the course of a few years the Germans adopted a more realistic view of the possibilities of achieving nuclear fusion in a power generating reactor.

After having heard about foreign work on controlled nuclear fusion at an astrophysical congress in 1955, a dedicated plasma physics group was formed, but it was in particular Kurchatov's talk at Harwell in 1956 that alerted German scientists about the work in nuclear fusion that was being done abroad. The lecture had an extraordinary effect, which is explained by the fact that for the first time concrete research results were presented to the public. Although the British did not hear anything new, that was not the case for scientists in other countries. Because of the secrecy adhered to in the field, Germany had not been aware of the extent and orientation of the research, and in the climate prevailing at the time in the German Federal Republic, where contact with international research was feverishly sought, such reports fell on particularly fertile ground (Boenke 1991, p. 88). The result was that by the end of 1957 six university fusion programmes of varying size already existed in Germany (Weisel 2001, p. 156–158).

In 1958, shortly before the 2nd Geneva Conference, the relevant German ministries had established that there was general agreement among scientists that for the next 15–20 years the construction of technical fusion installations would still be out of the question, and that it was not even known if nuclear fusion would actually provide a cheaper way of generating electricity than nuclear fission (Boenke 1991, p. 111; see also Eckert 1989). It showed that they were not fooled for long by Bhabha's rash prediction at the first Geneva conference.

---

[3]Quoted in Boenke 1991, p. 88. This PhD thesis studies the early developments of German research in nuclear fusion in detail and can be found at https://pure.mpg.de/rest/items/item_2472138_4/component/file_2472409/content.

At the 2nd Geneva conference Ludwig Biermann (1907–1986), who since 1948 had been heading the astrophysics group at Göttingen University, already presented a paper on "Recent Work on Controlled Thermonuclear Fusion in Germany (Federal Republic)" (Biermann 1958), bearing witness of the fact that German research in this area had already reached a fairly high level, much higher for instance than in France that in 1956 was still in the stage of building up a theory group. Experiments on toroidal discharges were being carried out in Göttingen and Aachen and on high current linear discharges at Aachen, Munich, Hanover, Kiel and Stuttgart, not unlike the experiments performed at that time in Princeton and Harwell. The experiments in Göttingen involved a very small Pyrex torus with a major radius of 50 cm and a minor radius of 2.5 cm. These experiments originated in 1956 in connection with research in theoretical astrophysics. Already at that time three methods of heating were studied: the normal ohmic heating, which was concluded to be effective up to 1 million degrees, but also heating through high-frequency radiation, near the resonance frequencies of the plasma, and through forced oscillations of the magnetic field, which were found to be particularly effective if the oscillation frequency of the magnetic field is near to or larger than the collision frequency of the ions. Stability problems and particle losses through drift motion were also studied. Although the Germans were certainly behind the Americans and British as regards large-scale experimental setups and technology, their research in theoretical areas was on par with foreign efforts.

The 2nd Geneva Conference had a catalysing effect, not only in Germany for that matter, and the German effort really took off in 1960 with the foundation[4] of the Max Planck Institute for Plasma Physics (IPP) by the Max Planck Society and Werner Heisenberg as shareholders[5] in Garching, near Munich. A few years later the institute was integrated into the European Fusion Programme through an association agreement concluded with Euratom. The Germans were at first hesitant to conclude such an agreement as they feared the dominance of the French CEA, which had been the first to conclude an agreement in 1959. This hesitation is rather strange as a far-reaching cooperation between European countries, or at least the original six signatories of the 1957 Euratom Treaty, would be a logical consequence of that treaty, one would think. In the end the cooperation between Germany and Euratom has never given any problems. With a staff of more than 1000 people IPP is currently one of the biggest fusion research institutes in Europe.

The objective of the institute was to carry out research in plasma physics and adjacent fields, as well as the development of methods and tools needed for such research. As in its name, no allusion was made to the eventual goal of building a fusion reactor; even fusion research was not mentioned. The early phase of German nuclear policy has been described as an "imitation phase". At first sight, from the

---

[4]It actually concerned a move of the institute from Göttingen to Munich.

[5]A curious arrangement to have a private person as a shareholder of a national institute. The initial idea was to have three private individuals as shareholders, namely Werner Heisenberg, Ludwig Biermann and Ernst Telschow (1889–1988). The latter had been an administrator of the Kaiser-Wilhelm-Gesellschaft during the Nazi-time, having himself become a member of the Nazi Party in 1933, immediately after Hitler's takeover of power.

repeated reference to foreign role models, one might conclude that the foundation of a separate dedicated institute for fusion research was indeed a consequence of imitating foreign developments. On closer inspection, however, this statement must be put into perspective: fundamental considerations on harnessing fusion as an energy source were made at the Max-Planck Institute at an early stage when there was no further specific information from abroad, and the numerous publications from before 1957 bear witness of the thorough theoretical research in this field (Boenke 1991, p. 145).

In the 1960s plasma behaviour was studied with linear and ring-shaped pinch devices, especially theta-pinches. A special battery was built which could produce a current of about 20 million amperes in about 10^{-5} s, and in early 1965 in the first large experiment, called ISAR-I, after the tributary of the Danube that flows past Munich, a record temperature of 60 million degrees was achieved. As we have seen in earlier chapters, there is in general no lack of records in fusion/plasma physics. Of course, they all show something, if only, that something is possible, but do not really solve anything substantial as regards the realization of nuclear fusion as a power source.

Throughout its early history IPP took the developments at Princeton as its guideline, and so the institute's foundation year already saw the launching of its first stellarator, the Wendelstein 1-A, which would eventually culminate in the Wendelstein 7-X. Research on stellarators has always remained part of the core business of IPP and German efforts in stellarator research will be discussed in greater detail in Chap. 13. At the end of the 1960s IPP broadened its research into experiments in inertial fusion with lasers and in relativistic plasmas for the acceleration of electrons and ions.

After the Soviet results with the T-3 tokamak were reported at the 1968 Novosibirsk conference, IPP also caught tokamak fever, but started, at a smaller scale than France, with its first tokamak, Pulsator, in 1970, before committing itself to larger machines. Stellarator research was however not abandoned. IPP now concentrated its fusion research in three areas: tokamaks, stellarators and high *beta* experiments.[6] Since that time IPP is the world's only institute pursuing comparative investigations of both tokamaks and stellarators.

The Pulsator device operated at IPP Garching from 1973 to 1979. The plasma of this small tokamak was just heated by the plasma current (ohmic heating). It made two important contributions to international tokamak research. Pulsator was provided with external helical windings and depending on the current through these windings drastic effects were observed. For the first time a tokamak plasma was influenced locally by an external helical field, which provided the starting point for the experimental investigation of the disruptive instability and its theoretical description. As we have seen earlier, a disruptive instability or disruption is a violent event that terminates a magnetically confined plasma, usually as a consequence of a rapidly growing instability. In a disruption, the temperature drops drastically, and

---

[6]Max Planck Institute for Plasma Physics, *50 years of research for the energy of the future*, https://www.ipp.mpg.de/17194/geschichte; Boenke 1991, p. 173 ff.

heat and particles are released from confinement on a short timescale and dumped on the vessel wall, causing damage in proportion to the stored energy. The loss of confinement is associated with the production of runaway electrons, which may also cause damage. In Pulsator the cause of the disruption could be traced to the spontaneous occurrence of magnetic field disturbances.

Pulsator was followed by ASDEX (Axially Symmetric Divertor Experiment) (in operation from 1980–1990) and upgraded into ASDEX Upgrade (from 1991 onwards). Instead of having a material limiter, as in Pulsator, the ASDEX plasma was limited to the outside with an auxiliary magnetic field. This worked as a divertor.

6.3.1 High-Confinement Mode (H-Mode)

As we saw above when discussing the experiments at Princeton, the most important property of magnetic confinement, viz. thermal insulation of the plasma, decreased whenever the plasma temperature was increased by external heating. Confinement times seemed to get worse when neutral beam or high-frequency heating was further increased. The problem was completely unexpected, and scientists were at a loss about its cause. According to new calculations gigantic amounts of power would be needed to heat the plasma. Under such circumstances it seemed impossible to ever achieve a burning plasma. The solution, as unexpected as the problem itself, came in 1982 with IPP's ASDEX tokamak (Wagner et al. 1982). It was found completely accidentally ('serendipity' is pushed to the brink in fusion science, it seems, but then as Nietzsche said *"Das Wesentliche an jeder Erfindung tut der Zufall, aber den meisten Menschen begegnet dieser Zufall nicht"*[7]) that when increasing the heating power slightly from 1.6 to 1.9 MW there was an abrupt transition in the plasma characteristics and energy and particle confinement improved; the plasma snapped into a new mode as it were. Plasma confinement improved by about a factor of 2, which seems little considering that confinement times have increased millionfold since the beginning of fusion research, but it may be the last push needed to get to the regime of a proper reactor. Plasma pressure was improved by about 60%. The discovery of this high-confinement regime in ASDEX, for short H-regime or H-mode, versus L-mode or low-confinement mode which denotes the normal confinement in tokamaks, doubled the thermal insulation attainable and is one of the most important discoveries in fusion research. Essentially only two achievements in fusion are worth the word "breakthrough": the defeat of Bohm diffusion by the tokamak in 1968 and this H-mode discovery. The L-mode is characterized by relatively large amounts of turbulence in the edge region of the plasma, which allows energy to escape the confined plasma. The energy confinement time in H-mode is roughly twice that of L mode.

---

[7]The essential part of every invention is the work of chance, but most men never encounter this chance (Friedrich Nietzsche, *Daybreak: Thoughts on the Prejudices of Morality* (Cambridge University Press, 1997), translation by Maudemairie Clark and Brian Leiter).

H-mode performance is achieved via the spontaneous formation of a transport barrier in the outer few percent of the confined plasma, as the result of suppression of the turbulence in the few centimetres of plasma, just inside the last closed flux surface. This narrow insulating layer, referred to as a 'pedestal' or 'pedestal region' typically results in the pressure increasing more than 30 times across a 0.4–5 cm layer, which has a major impact on fusion performance (Bowman et al. 2018).

The pioneering achievement of H-mode had to do with the use of a divertor, or actually with the presence of a separatrix. The separatrix signals the transition from the confined plasma with closed field lines to the Scrape-Off Layer (SOL) with open field lines (see Chap. 4). The outermost boundary layer of the plasma is diverted off into side chambers, where the plasma particles are pumped off, removing impurities from the plasma. This also affects the confinement properties: the boundary layer produced by the divertor induces a transport barrier at the plasma edge, creating good thermal insulation. Divertor operation and the H-regime have now become the basis of modern tokamaks, right up to ITER.

In the following years, H-mode was observed in the American machines PDX and DIII-D, in JET and in the Japanese JT-60, and subsequently in several other tokamaks. On DIII-D an even higher performance mode, dubbed VH-mode or very high mode, was discovered with energy confinement times again up by a factor of two compared to H-mode (Jackson et al. 1991; Osborne 1995). As we will see in Chap. 8, the H-mode was vital for JET in achieving its record results in the 1990s. In 1993 H-mode was also achieved in the German W7-AS stellarator, thus demonstrating that it was a "generic feature" of all toroidal configurations and it can indeed be reproduced in any tokamak or stellarator. For the H-mode to arise it seems that only two requirements have to be met: (1) that the input power be high enough (the actual level varies from device to device); and (2) that the plasma be led out by a divertor into an external chamber rather than be allowed to strike the wall (Chen 2011, p. 265).

The physics behind the H-mode is still not completely understood (Llewellyn Smith and Cowley 2010, p. 1098), in spite of the fact that a large fraction of fusion physicists have been occupied with this problem for over 20 years and a special annual conference is devoted to this topic.[8] However, such lack of understanding is also a recurring feature in fusion research and has never greatly bothered fusion scientists. The value of H-mode is unquestioned. All present-day tokamaks are designed to run in H-mode and without it, ITER would need to be twice as large and, consequently, twice as expensive.

6.3.2 Edge-Localised Modes (ELMs)

But, as usual, nothing comes for free, and while the plasma characteristics in the H-mode are better than in L-mode a new type of instability emerged. These are the

---

[8]The *International Workshop on H-mode Physics and Transport Barriers*, the 17th of which took place in Shanghai in October 2019.

edge-localised modes (ELMs) we have met before. They occur most violently in a thin layer of the plasma in the edge region of the torus with its large radial gradients in temperature and pressure. These instabilities pose a major challenge for nuclear fusion with tokamaks (Davidson 2010, p. 534ff).

Not surprisingly, ASDEX was also the first tokamak on which the very troublesome ELMs have been seen. The ELM is a periodic distortion of the plasma boundary, which rotates with the velocity of several kilometres per second and exists for about half a millisecond, ejecting bundled plasma particles and energies outwards to the vessel wall. This instability (or 'mode') appears in the standard H-mode operation regime and leads to a crash ('ELM crash'), causing plasma to eject the particles in bursts onto the chamber wall. Up to one-tenth of the total energy content can thus be ejected. In bigger future devices the particle bursts caused by the ELMs may result in the destruction of the plasma facing components (PFCs) and can cause overloading of, in particular, the divertor. Their mitigation is therefore of paramount importance.

ASDEX was shut down in 1990 and five years later, the device was passed on to the Southwestern Institute of Plasma Physics in Leshan, China, where its components were used in the construction of a Chinese tokamak, the HL-2A, which was put into operation in 2002.

ASDEX's follow-up in 1991 was ASDEX Upgrade, a brand-new machine, not just an upgrade of ASDEX. It is a midsize tokamak experiment, with a plasma volume of 13 m^3. Figure 4.9 shows the D-shape of the plasma vessel of this tokamak and Fig. 6.1 installation work in the vessel. The aim of ASDEX Upgrade is to prepare

Fig. 6.1 Installation work in the ASDEX Upgrade plasma vessel. The collector plates of the divertor are seen at the bottom (*from* https://www.ipp.mpg.de/16208/einfuehrung)

the physics base for ITER and the follow-up demonstration plant. For this purpose, essential plasma properties, primarily plasma density, plasma pressure and wall load, are matched to the conditions in a future fusion power plant.

How foresighted the planning of the device was is shown by comparing it with the construction plans for ITER, which were completed ten years later in 2001 and essentially look like an enlarged copy of the device at Garching. In particular, the divertor research on ASDEX Upgrade has largely been incorporated in the conception of the ITER divertor.[9]

One innovative feature of the ASDEX Upgrade experiment is its all-tungsten first wall. The first wall is the wall of the vessel facing the plasma. Tungsten (wolfram) has been chosen for the first wall because of its very high melting point of more than 3000 °C (the highest for any metal), which enables it to endure the very high heat fluxes emanating from the hot plasma at the heart of the tokamak. However, tungsten also has the tendency to ionise at high temperatures, which can lead to pollution of the plasma. Furthermore, as a material with a high atomic number Z (number of protons in the nucleus), radiation from fully ionized tungsten in the plasma is several orders of magnitude higher than that of other proposed first wall components such as carbon fibre composites (CFCs) or beryllium. It will therefore more easily "contaminate" a proposed breakeven plasma. ASDEX Upgrade has shown that this problem can be avoided by the appropriate injection of impurity atoms at the plasma edge and in the divertor, which is good news for the construction of ITER's first wall, but a rather complex solution to the problem.

One of its major successes, announced in January 2011, has been the reduction of the perturbing edge-localised modes, by fitting eight magnetic control coils on the wall of the plasma vessel.

ASDEX Upgrade plays an important role in investigations in preparation for ITER.[10] In general, it must be concluded that German research is rather impressive, with a solid and sound programme.

6.4 France

France's nuclear fusion programme, like in Japan and Sweden, started with the formation of a working group for theoretical research after Bhabha's prediction at the 1955 Geneva meeting. Proper research into fusion began in 1957 with a torus-like device, called the Tore TA 2000 at an old fort in Fontenay-aux-Roses (15 kms from Saclay, just outside Paris, home of the Commissariat à l'énergie atomique et aux énergies alternatives (CEA)). Some papers were submitted to the 1958 Geneva Conference, mainly on theoretical topics, but also some experimental work on rectilinear and

---

[9]Fifty years of research for energy of the future, https://www.ipp.mpg.de/4239000/50_years_en.pdf, p. 4.

[10]For more information the reader can visit the excellent and very transparent IPP website: (https://www.ipp.mpg.de/ippcms/de/pr/forschung/asdex/index).

ring-shaped pinch devices. The work apparently did not warrant a review talk at the plenary session like the one given by Biermann referred to above about German research. Saclay was chosen as the site of the CEA laboratory and after the foundation of Euratom it was the first in 1959 to conclude an association agreement. In these first years, mirror devices (discussed in Chap. 3) had the special attention of French efforts, as evidenced by the papers submitted to the Culham conference in 1965.

Like in other countries, 1968 was a watershed year for the French who suddenly knew what to do and immediately without further ado embarked on the construction of a for that time large tokamak, TFR (Tokamak de Fontenay-aux- Roses). The first version, called TFR 400, operated from 1973 to 1978 and was similar in size to the Soviet T-3, but with a higher toroidal magnetic field. For a few years it was the highest performing tokamak in the world reaching temperatures of 2 keV and achieving important results as regards plasma duration. With the Russian T-4 it was the most powerful tokamak in operation in the early 1970s with maximum plasma currents of about 400,000 amperes. Initially it was damaged by runaway electrons and had a thick copper shell for stability, which was removed in the 1978 follow-up version TFR 600. This upgrade was shut down in 1984.

TFR was followed by Tore Supra at Cadarache in the south of France, the future site of ITER. Its name comes from the words torus and superconducting and for a long time it was the world's largest tokamak with superconducting toroidal magnets. Superconductors have zero resistivity, so once a current has been started in them, it will keep running (almost) forever. It must provide the solution to the still elusive steady-state operation of the tokamak. The drawback is that these superconductors, at least the ones used in Tore Supra, have to be cooled to below 4.2 °K (about −269 °C). This cooling consumes enormous amounts of energy and is a serious impediment to making the generation of fusion energy, if ever possible, commercially attractive. An important development, to be discussed in a later chapter, is that present-day superconducting technology allows magnets to be made out of materials that become superconducting at higher temperatures, reducing energy consumption for cooling.

Tore Supra operated from 1988 to 2010. Its basic parameters are $R = 2.25$ m, $a = 0.7$ m and a toroidal magnetic field of 4.5 T (see Fig. 6.2 for a cut-open view which gives an impression of the complexity of such a device), making it one of the bigger tokamaks. Thanks to the superconducting magnets, it managed to achieve very long plasma pulses, reaching a record of 6.5 min in 2003, with a 500 kA plasma current. It runs fully non-inductive pulses, i.e. it can do completely without ohmic heating and fully relies on various resonance heating methods. Most in-vessel components are actively cooled. Due to its particular configuration, including this active cooling of the components, Tore Supra formed a unique opportunity for experimenting with various plasma-facing components (Jacquinot 2005).

When Tore Supra had fulfilled its mission in 2011, it was upgraded to a device called WEST (which stands for Tungsten (chemical symbol 'W' for wolfram) Environment in Steady-state Tokamak. The acronym WEST was specifically chosen to set it against the Chinese tokamak EAST. Both WEST and EAST are part of SIFFER, the SIno-French Fusion Energy center (see below). The Tore Supra upgrade (WEST)

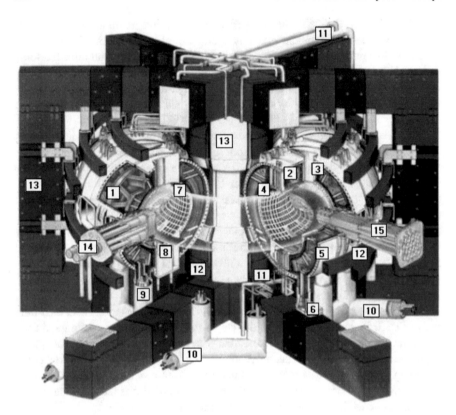

Fig. 6.2 Cross section of Tore Supra: 1. 4 °K mechanical structure of coils; 2. Superconducting winding 1.8 °K; 3. 80 °K thermal screen; 4. Cryostat, internal area at −220 °C; 5. Cryostat, external area at 20 °C; 6. Supporting pedestal for the cryostat and screens; 7. First wall, actively cooled to − 220 °C; 8. Toroidal pumping limiter; 9. Cryogenic supply, 1.8 and 80 °K; 10. Towards exchanger, water pressurised at 40 bar; 11. and 12. Poloidal field coils; 13. Magnetic circuit; 14. Heating antenna at ion cyclotronic frequency; 15. Heating antenna at lower hybrid frequency (*from* https:// www-fusion-magnetique.cea.fr)

involved conversion to a D-shaped cross section and the installation of tungsten walls and a divertor (Tora Supra only had a limiter). It has been in operation since 2016.

With its new magnetic configuration, specific equipment and all-metal environment, WEST will test tungsten components, identical to those to be installed on ITER, in particular its full ITER-grade tungsten divertor. Although the temperature and the density at the heart of the WEST plasma are lower than in ITER, the conditions at the periphery of the plasma are very similar, with heat and particle flows in the divertor of up to 20 MW/m^2—ten times more intense than what the shield of a space shuttle has to withstand when entering the atmosphere. In the summer of 2021 it will enter Phase II of its operational program, extending plasma durations up to 1000 s. Its research plan and progress can be followed on the WEST website.[11]

CEA, as part of the Euratom community, contributes to the European fusion programme and is fully committed to ITER, currently being built at the CEA Cadarache site.

France also participates in the JET project at Culham (UK). Contrary to Germany, France's fusion programme is fully integrated into its various international collaborations either with JET or with ITER.

6.5 Italy

The start of fusion research in Italy dates back to May 1957 with the creation of a research group on ionized gasses in Rome. In 1960 the Laboratorio Gas Ionizzati (LGI), consisting of a theoretical and an experimental group, was established and moved to Frascati, where the Laboratori Nazionali had just been set up. In the same year the association contract between LGI, through the Comitato Nazionale per l'Energia Nucleare (CNEN), and Euratom was signed.

Like in many other countries that just entered the scene, the Italian fusion programme was launched in the optimistic climate following the 1958 Geneva Conference. The Italian effort, although limited, showed that Italy, although a late-comer, intended to make up for lost time. In this early phase, research in Frascati developed along two main lines: a Programme A which included the theta-pinch "Cariddi" (the Italian word for the mythological sea monster Charybdis, pared with Scylla) and the "Hot Ice" experiment (for studies of the properties of a dense plasma, through irradiating a cylinder of solid deuterium by an intense laser beam); and an early inertial confinement Programme B consisting of MIRAPI (MInimum RAdius PInch), whereby a cylindrical plasma shell was imploded towards the rotational axis, and MAFIN (MAgnetic Field Intensification), producing the high magnetic fields necessary for fusion, by means of implosions induced by conventional explosives.

The second phase of the association contract would soon face a series of difficulties and shortcomings, related to the progressive bureaucratization of Euratom and its early "crisis", reflected in a drastic financial cut of the Community's second five-year plan, and related to the crisis of the Joint Research Centre at Ispra. Cuts in the fusion programme affected all fusion programmes in the various centres where association contracts were in operation, not only in Italy, but also in France, Germany, the Netherlands and Belgium.

In the second half of the 1960s the general economic and monetary difficulties, the reconsideration of national fission programmes (e.g. the French shift to light water reactors); and social and political unrest in 1968 also affected national nuclear programmes, including fusion programmes. In the Italian case, in particular, the crisis

---

[11] https://irfm.cea.fr/en/west/index.php.

of the Frascati centre took place within the framework of a more general crisis of the Italian nuclear programme.

This situation soon required a re-launching of the European fusion programme as a whole, which took place following the 1968 "tokamak revolution". A new phase of European fusion history was set in motion (Curli 2017, p. 72ff).

In this dawning tokamak age, the Italians started with the construction of a small machine at Frascati, TTF (Turbulent Tokamak Frascati also called "torello"), used for studying turbulent plasma heating in a toroidal configuration. It was built in 1973, followed in 1977 by the high-field Frascati Tokamak (FT)—a compact machine, characterised by a high (10 T) magnetic field, high current and a special heating method by means of electromagnetic waves in the radiofrequency range. The Frascati Tokamak Upgrade (FTU) has been operating since 1989 until the present day. It is similar to FT, but has a much larger surface facing the plasma. FTU has three additional heating systems and two pellet injection systems; this allows substantial heating of the plasma by means of radiofrequency power injection systems. With this method it is expected to be able to increase the temperature of the high-density plasma to 50–100 million degrees.[12] Within the framework of the use of liquid metals as plasma facing materials in fusion devices, a cooled liquid-lithium limiter was installed some years ago on FTU in order to investigate the effectiveness of liquid lithium in tokamaks and in particular its role in plasma-wall interactions. The limiter was recently replaced by a liquid-tin limiter.

For completeness we also mention here the THOR tokamak which operated from 1978 to 1989 at the University of Milan for ECRH studies of supra-thermal plasmas (Argenti et al. 1981), and the reversed field pinch device RFX that ran from 1991–1999 at the University of Padua. Reversed field pinch systems will be discussed in Chap. 14.

Italy is currently a member of the EUROfusion consortium through its National Agency for New Technologies, Energy and Sustainable Economic Development (ENEA). ENEA's Fusion Division participates in the design of ITER. Within this framework ENEA has made substantial contributions in the fields of superconductivity, plasma-facing components, neutron transport, safety and remote handling (needed for working with the radioactive tritium).

One of the main challenges in realizing a fusion power plant is the adequate control of heat exhaust through a divertor, as explained in Chap. 4. Already demanding for ITER, the problem is amplified for a real power plant or for a DEMO plant, the successor of ITER, if the latter turns out to be successful. The assumed linear dimensions of DEMO are about 50% larger and the fusion power output at least 3 times higher than of ITER. ITER will only generate energy in bursts of a few minutes. The DEMO reactor, which would have to show that fusion can supply electricity to the grid, would need to operate more or less continuously. It is not clear whether conventional divertors can deal with the resulting greater heat load. Within this framework ENEA has run ahead of things by making a proposal in 2015 for a US$570 million Divertor Tokamak Test facility (DTT) to test novel divertor

---

[12]https://www.fusione.enea.it/WHO/history.html.en.

designs.[13] DTT would be a tokamak with a plasma volume of only 4% of ITER's, but would have room to test different ways to divert waste heat. EUROfusion, the consortium of national fusion research institutes in the EU, Switzerland and Ukraine, agrees that the project is an important stepping stone to DEMO, but considers funding the expensive new facility now to be premature and has sensibly postponed a decision on this for five years. Other researchers also have reservations about DTT. Physicists at MIT have proposed an as-yet-unfunded Advanced Divertor Experiment. At an estimated US$70 million, it would be far cheaper than DTT because, unlike the Italian design, it would not use superconducting cables to create its magnetic fields and would have a smaller plasma core, but it would still have room for testing alternative divertors (Cartlidge 2017). In Chap. 9 we will come back to DTT, which also involves a lot of politics, when discussing the European roadmap to fusion.

Finally, we mention the joint Russian-Italian project on the IGNITOR reactor, which has evolved out of Bruno Coppi's activities at MIT (see Chap. 5). It is part of the line of research that started with the Alcator machine at MIT in the 1970s, and continued with Alcator C/C-Mod at MIT and the FT/FTU experiments at Frascati. It is so far the first and only experiment proposed and designed to obtain physical conditions in magnetically confined D-T plasmas that sustain the plasma under controllable conditions without the addition of extra heat, i.e. to achieve ignition. It is a compact D-shaped fusion reactor with total plasma volume of just 10 m^3. So far only model calculations have been carried out (Bombarda et al. 2004) and construction of the reactor itself at the TRINITI site in Troisk (Russia) is long overdue.[14]

6.6 Other European Countries

Other European countries that carried out fusion experiments from quite early in the game were the Netherlands and Sweden.

Swedish fusion research originates from immediately after the 2nd Geneva Conference, but is now part of the European effort via a Contract of Association with Euratom and within this framework it contributes to JET and ITER. The Alfvén Laboratory at the Royal Institute of Technology in Stockholm runs the EXTRAP T2R device (Extrap stands for External Ring Trap), which is a medium-sized reversed-field pinch (RFP) and the successor of various other devices. Reversed-field pinches will be briefly discussed in Chap. 14. The experiment received Euratom priority status in 1990 and first plasma was obtained in 1994.

In the Netherlands fusion research was already well under way around 1958. Work on rotating plasmas and plasma-beam (ion and electron) interaction experiments were carried out in Amsterdam already before 1958. Soon the centre of gravity of Dutch fusion work came to lie at the newly founded Institute for Plasma Physics

---

[13]ENEA, Divertor Tokamak Test facility—Project Proposal, July 2015; Crisanti et al. 2017.
[14]IGNITOR website https://www2.lns.mit.edu/ignitorproject/Ignitor@MIT/Home.html and https://www.frascati.enea.it/ignitor/.

in Jutphaas. Aside from the experimental department (work on toroidal pinches), a strong theoretical group was formed that has always covered a broad range of subjects (Braams 1987).

From 1963 toroidal and other pinches became a speciality of the Dutch institute. Like everywhere else, from 1969 they were followed by tokamaks. Two small tokamaks have been operating in the Netherlands. The first one was called TORTURE, which stands for TORoidal TURbulence Experiment. Although perhaps a very apt name for an unruly fusion device, it was felt to call up wrong associations and its name was rapidly changed into TORTUR. Various versions (I to IV) were built and operated from 1974 to 1991. In 1992 it moved to Portugal where it became known as ISTTOK (Instituto Superior Técnico TOKamak). It is currently the oldest operating tokamak in the EU. The second tokamak was RTP (Rijnhuizen Tokamak Project) from 1989 to 1998, named after the Rijnhuizen country estate in Jutphaas (now Nieuwegein), which until 2012 housed the Dutch FOM Institute for Plasma Physics. It was obtained from France where it had operated in Grenoble under the name Petula. RTP was used to study the development of instabilities and the layered structure of the plasma. Heat losses in the plasma were found not to be uniform throughout the plasma, but showed a layered structure, like an onion.

Cursed with the same rosy vision as other 'experts' in the field C. M. (Kees) Braams (1925–2003), the first director of the Dutch plasma institute, said in 1973 that, although all problems of nuclear fusion in the laboratory have not yet been solved, such will be the case by the end of the 1970s. At the same time he expressed the opinion though that it will still take considerable time before nuclear fusion will supply electricity (a few months later made more specific by expressing the hope that this will be the case before the end of the century (i.e. the twentieth century)). The politicians, in particular the then minister of economic affairs Ruud Lubbers, showed more common sense by stating (in 1976) that nuclear fusion is not yet realisable in practice and hence cannot be considered energy research, but must be seen as a pure science research project (which had consequences for (the source of) its funding). In 1982 the Dutch budget for fusion research was frozen for the coming four years at the 1982 level, in view of the fact that since 1976 there had been no change in the prospects for energy generation from nuclear fusion. This was in line with developments in other countries, notably the US, where, after the glory years of the 1970s, fusion had an increasingly hard time in the 1980s.

Developments had stalled by the fact that at higher temperatures the plasma was losing energy faster than calculations had shown and that it turned out to be far more difficult to increase the plasma temperature than envisaged. As we saw earlier in this chapter, confidence was restored when this problem was solved by the accidental discovery of the high-confinement mode (H-mode) on the German ASDEX tokamak.

Since the decommissioning of RTP in 1998, no tokamak has been in operation in the Netherlands. All fusion research carried out since has been in support of international collaborations, first JET at Culham and now ITER. Since 2012, Dutch research is being carried out under the name DIFFER, Dutch Institute for Fundamental Energy Research, which in 2015 moved from Jutphaas to the campus of Eindhoven Technical University. Like all EU fusion efforts, it is a member of the EUROfusion consortium.

Denmark operated a small tokamak called DANTE (Danish Tokamak Experiment) from 1977 to 1989 at Roskilde. It was a testing ground for the Risø pellet injectors, which were later installed on other machines around Europe.

In August 2019 DTU (Technical University of Denmark), which ran the DANTE device at Roskilde, installed a spherical tokamak, the ST25, on permanent loan from the UK start-up company Tokamak Energy (see Chap. 11), at its Kongens Lyngby site and took it into use as a teaching tool. The small spherical tokamak—with a plasma major radius of just 25 cm and a plasma volume of about 100 L—is called NORTH (the NORdic Tokamak Device).[15]

Since 1990 Portugal also has an association agreement with Euratom and runs the small ISTTOK tokamak (a refurbishment of the Dutch TORTUR), carrying out projects within the Euratom fusion programme.

---

[15]https://www.iter.org/newsline/-/3330.

Chapter 7
Tokamak Research in Asia and Rest of the World

7.1 Introduction

In this chapter the efforts in Asia (Japan, China, South Korea, India) and other countries (e.g. Australia) will be briefly reviewed. All these countries were relative latecomers in the nuclear fusion game, starting to participate mostly only after the breakthrough of the tokamak in 1968.

7.2 Japan

The programme of nuclear fusion projects in Japan has been very extensive. Fusion as a future source of energy apparently was and is taken very seriously in Japan, perhaps understandable for a country that has very few fossil fuels of its own (apart from coal). In the early years policy as regards nuclear fusion was rather cautious and sensible by concentrating on basic plasma physics and a few small-scale pinch devices, all for research purposes. Late in the sixties, after the Russian T-3 results, construction of fusion devices really took off and Japan very swiftly jumped on the tokamak bandwagon.

Research in plasma physics and nuclear fusion started in Japan around 1955 with a small-scale linear pinch experiment at Osaka University. Instead of immediately starting the construction of relatively large-scale fusion experiments at the just founded Japan Atomic Energy Research Institute (JAERI),[1] the Japanese decided to first start with basic plasma studies through the construction of small-scale devices and the training of young scientists at universities. For this a new central research institute for plasma and fusion studies, the Institute of Plasma Physics at Nagoya University, was founded in 1961. It has played a central role in Japanese fusion

---

[1] JAERI had been founded in 1956. In 2005 it merged with the Japan Nuclear Cycle Development Institute into the Japan Atomic Energy Agency, which is in charge of Japan's nuclear power plants.

© The Author(s), under exclusive license to Springer Nature Switzerland AG 2021
L. J. Reinders, *The Fairy Tale of Nuclear Fusion*,
https://doi.org/10.1007/978-3-030-64344-7_7

research and was succeeded in 1989 by the National Institute for Fusion Science (NIFS) (Sekiguchi 1983).

Late in the 1960s, Japanese fusion efforts entered their second stage through the launch of a series of toroidal plasma projects at JAERI (the JFT-1 hexapole, followed in the 1970s by the JFT-2 and −2a DIVA tokamaks, and in 1983 the JFT-2 M tokamak which continued to operate until 2004 (JFT standing for Jaeri Fusion Torus)), the HYBTOK-I and—II tokamaks (HYBTOK stands for hybrid tokamak; in spite of this name these are conventional tokamaks; hybrid referring to hybrid wave) at Nagoya (the latter tokamak is still in operation), various pinch experiments at the Electrotechnical Laboratory (ETL) in Tsukuba (Ibaraki), and supporting work on plasma heating by ECRH, plasma diagnostics, and plasma-wall interactions at the Institute of Physical and Chemical Research (IPCR) at RIKEN.[2] This was further expanded in the mid-1970s by planning the construction of a large tokamak (JT-60) at JAERI, for carrying out a scientific feasibility experiment as a milestone towards the realization of fusion energy. Its construction was finally authorized early in 1978. JT-60 is one of the big tokamaks to be discussed with JET and TFTR in the next chapter. In preparation for the eventuality that Japan would have to go it alone on the route towards nuclear fusion, an even bigger machine, the $2 billion Fusion Experimental Reactor (FER), was being planned in the early 1980s, but was eventually absorbed into the ITER programme.

More than 40 conventional tokamaks have been built in Japan at various institutions from the 1970s onwards, more than in any other country, in addition to 10 spherical tokamaks and a large number of devices of alternative design. (See www.tokamak.info.) Of the 40 conventional tokamaks only a couple (two or three) are still in operation, now that most effort is spent on JT-60 and ITER.

From the mid 1970s alternatives to the tokamak were also realised to be important, resulting in a variety of projects including heliotron projects at Kyoto University, inertial confinement fusion projects at Osaka University, and tandem-mirror projects at the University of Tsukuba (Sekiguchi 1983). Some of these developments will be discussed in Chaps. 12 and 13.

The current Japanese nuclear fusion programme can be divided into two parts, one related to ITER, and a part that focuses on higher performance and more challenging options of components and systems.[3] JAERI has been charged with the first part, while universities and the National Institute for Fusion Science (NIFS) mainly cover the latter aspect.

The breeding blanket, to be discussed in greater detail in the chapter on ITER, is an important component of any future reactor. It covers the interior of the fusion chamber and must be capable of sustaining a high heat load and an intense neutron flux. One of its purposes is to assure self-sufficiency of the fusion reactor with regard

---

[2]Riken is short for Kokuritsu Kenkyū Kaihatsu Hōjin **Ri**kagaku **Ken**kyūsho (National Institute of Physical and Chemical Research). It currently has about 3000 scientists on seven campuses across Japan, including the main site at Wakō, and runs several institutes overseas.

[3]This in addition to a fairly extensive program of spherical tokamaks that will be discussed in Chap. 11.

to tritium (by producing (breeding) tritium from lithium contained in the blanket, at least the same amount of tritium as consumed in the nuclear fusion reactions in the plasma). In the division of tasks laid down in the Japanese fusion programme JAERI is responsible for the development of breeding blanket concepts in collaboration with universities and NIFS. (Tanaka 2006.)

Japan apparently has great confidence in the prospects for fusion energy and is already looking beyond ITER. After the 2011 Fukushima disaster it was decided to gradually phase out power generation by nuclear fission. All its nuclear power plants will be shut down and plans to build another plant on the main island of Honshu were cancelled in the face of strong opposition. It is extremely doubtful that this is a good policy to follow, as sizable power generation from nuclear fusion is still but a distant and dim prospect. Fact is though that Japan is now taking a leading role in developing nuclear fusion as a next-generation power source and is closely cooperating in this respect with Europe. To accelerate the development of nuclear fusion, Europe and Japan identified three major projects to be carried out jointly in Japan, complementing the construction and operation of ITER. These three projects are (1) the further development of planning for the construction of a materials test facility (IFMIF), (2) the establishment of an international centre for studies of fusion technology, remote operation and plasma simulation, directed towards the construction of the demonstration electricity generating plant (DEMO) that will follow ITER (IFERC), and (3) the construction and operation of an advanced superconducting tokamak, JT-60SA, to act as a satellite tokamak to ITER during its operation.

JT-60SA, SA standing for Super Advanced, an upgrade of JT-60 Upgrade, which in turn was an upgrade of JT-60 (see next chapter) is designed to support the operation of ITER and to investigate how best to optimise the operation of fusion power plants to be built after ITER. It is a fully superconducting tokamak capable of confining high-temperature (100 million degrees) deuterium plasmas. Its assembly took seven years and has just (April 2020) been completed at Naka in Japan, using infrastructure of the existing JT-60 Upgrade experiment.[4]

Japanese fusion development is now in its third phase, whose targets are ignition and long-pulse burn by ITER and the establishment of technologies towards the DEMO plant. The fourth phase of its programme, assuming that both ITER and JT-60SA will be successful, will then focus on a DEMO project aimed at the technological demonstration and economic feasibility of fusion power. The objective of its DEMO reactor is to realize a steady and stable electric output of over several hundred megawatt, availability sufficient for commercialization and overall tritium breeding that will fulfil self-sufficiency in fuel.

In the timeline depicted in Fig. 7.1 success of ITER and D-T ignition have been assumed, as well as steady-state operation of JT-60SA. Without these assumptions to be fulfilled, there will be no point in starting construction of the DEMO plant. To achieve the goal of starting commercial fusion by 2050, the construction site of DEMO must be decided before 2035, i.e. before the start of ITER D-T operations,

---

[4]http://www.jt60sa.org/b/index.htm.

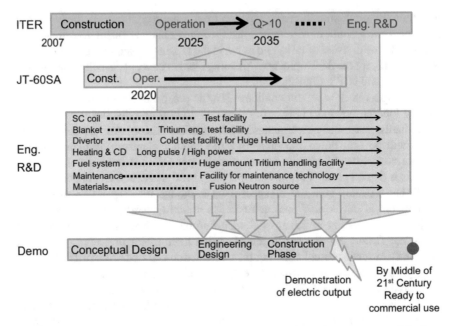

Fig. 7.1 Schematic timeline of Japan's roadmap towards DEMO and commercial plants (*from* Okano 2019)

which are planned for 2035, but will be later as construction of ITER has suffered delays. Such a decision seems difficult, but necessary if DEMO construction has to start before 2040. There is furthermore a multitude of technological issues that have to be resolved before DEMO construction can start. So, the timeline of Fig. 7.1 seems to be overambitious. To have both a DEMO reactor and a commercial power generating device ready in just ten years is completely unrealistic. It is out of the question that nuclear fusion will be ready for commercial use by 2050, just thirty years from now and only a decade after ITER will have provided any results, if everything goes well.

7.3   South Korea

South Korea, which relies on imports for more than 90% of its energy needs, has been a player in nuclear fusion since the oil crisis of the 1970s when it developed its first tokamak, SNUT-79 (Seoul National University Tokamak), which operated from 1982 to 1992. This was followed by small scale fusion devices such as KT-1 in the Korea Atomic Energy Research Institute (KAERI), the KAIST-tokamak from 1992–2002 in the Korea Advanced Institute of Science and Technology (KAIST), and the Hanbit mirror device in the National Fusion Research Institute (NFRI). It was a re-assembly of the PRETEXT tokamak which was shipped from Texas University and used for basic studies of high-temperature hydrogen plasmas.

Fusion research was then concentrated in Daejeon at the National Fusion R&D Center which in 2005 was converted into the National Fusion Research Institute (NFRI). South Korea subsequently joined the ITER consortium in 2006.

In 1995, the Korean government launched KSTAR, the Korea Superconducting Tokamak Advanced Research, the flagship of South Korea's research in nuclear fusion. It was designed in collaboration with the Princeton Plasma Physics Laboratory, which had previously proposed a similar machine in the US, the Tokamak Physics Experiment (TPX), but failed to get the necessary funding from Congress (Dean 2013, p. 136). KSTAR is a medium-sized all superconducting tokamak, with parameters as listed in Fig. 7.2, next to those of ITER. The figure also shows the relative size of KSTAR's plasma compared to other tokamaks. Heating and current drive will be initiated using neutral beam injection, ion cyclotron resonance heating (ICRH), radio frequency heating and electron cyclotron resonance heating (ECRH). KSTAR is one of the first research tokamaks in the world to feature fully superconducting magnets, which will be of great relevance to ITER as the latter will also use superconducting magnets. Otherwise too it will study aspects of magnetic confinement fusion that will be pertinent to the ITER fusion project. First plasma was obtained in June 2008. It expects to produce pulses of up to 2 MA over 300 s. In 2016 KSTAR achieved a high-performance H-mode plasma that was stable for 70s, a world record at the time and since broken by the Chinese EAST.

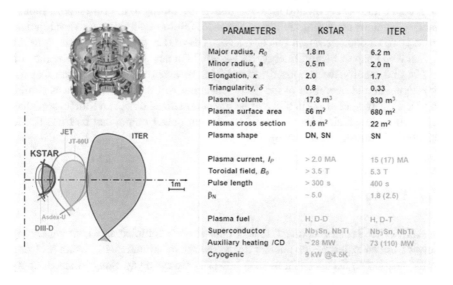

PARAMETERS	KSTAR	ITER
Major radius, R_0	1.8 m	6.2 m
Minor radius, a	0.5 m	2.0 m
Elongation, κ	2.0	1.7
Triangularity, δ	0.8	0.33
Plasma volume	17.8 m³	830 m³
Plasma surface area	56 m²	680 m²
Plasma cross section	1.6 m²	22 m²
Plasma shape	DN, SN	SN
Plasma current, I_p	> 2.0 MA	15 (17) MA
Toroidal field, B_0	> 3.5 T	5.3 T
Pulse length	> 300 s	400 s
β_N	~ 5.0	1.8 (2.5)
Plasma fuel	H, D-D	H, D-T
Superconductor	Nb₃Sn, NbTi	Nb₃Sn, NbTi
Auxiliary heating /CD	~ 28 MW	73 (110) MW
Cryogenic	9 kW @4.5K	

Fig. 7.2 Parameters of the South Korean KSTAR tokamak (*from* https://www.nfri.re.kr/eng/pageView/53)

In 2019 KSTAR maintained high-temperature plasma at a temperature >100 million degrees for 1.5 s, the first time in the world that a nuclear fusion reactor reached such high temperature for its central plasma ions for such a long period.

In the ten years of its operation the Korean machine has built up a valuable database for the future operation of ITER as well as for the design basis of a next-generation DEMO machine.[5] Such a demo plant, called K-DEMO, is currently being planned, again in collaboration with PPPL.

7.4   India

India has a small tokamak, called ADITYA (meaning *Sun* in Hindi), running since 1989 at the Institute for Plasma Research in Gandhinagar. It is the first indigenously designed and fabricated tokamak in India and its heating systems have gradually been upgraded to be used as a testbed for SST-1 (Steady-state Superconducting Tokamak). SST-1 which started in 1994 and achieved first plasma in June 2013.

SST-1 is part of a new generation of advanced tokamaks with a D-shaped plasma and steady-state operation (i.e. pulses up to 1000 s) as its major objective. It has been designed as a medium-sized tokamak with superconducting magnets. Although SST-1 is not yet in full operation, its successor SST-2 is already being designed and a demonstration reactor contemplated. This seems an established practice in fusion research, where even close to 100 years of failing research has not yet instilled a sense of patience and reflection into the community. According to a recent report, India's energy demands are rising rapidly and to meet future requirements, a development strategy for a demonstration power reactor (DEMO) is planned. According to the Indian Roadmap of Fusion Energy, SST-2 will be a new D-T machine with the aim to test and qualify the proposed technologies of a reactor. It is foreseen that the development and operation of the SST-2 machine will provide experiences similar to ITER (Danani et al. 2019). It will be a medium sized device, a little bigger than JET, with a low fusion gain ($Q = 5$) and fusion power output can be from 100 to 300 MW (Srinivasan et al. 2016).

7.5   China

In April 2019 Reuters reported[6] that "China aims to complete and start generating power from an experimental nuclear fusion reactor by around 2040", again reverting to the magic two decades that nuclear power is always away. Song Yuntao, deputy director of the Institute of Plasma Physics at the Hefei Institute of Physical Science,

---

[5]https://www.iter.org/newsline/-/3241.

[6]https://www.reuters.com/article/us-china-nuclearpower-fusion/china-targets-nuclear-fusion-power-generation-by-2040-idUSKCN1RO0NB.

was quoted as saying: "Five years from now, we will start to build our fusion reactor, which will need another 10 years of construction. After that is built, we will construct the power generator and start generating power by around 2040". Nothing can be simpler, it seems: build, construct and build again and everything will be ready. Of course, optimism is a good thing, but failing to learn from the past (and the present) can be fatal. Reuters was not the only one to sing the praises of China's nuclear fusion prowess in April 2019. There were stories about "Earth's Second Sun"—China's Fusion Future, the Holy Grail of Unlimited Energy[7] and even the BBC joined in with the less exalted heading: "Will China beat the world to nuclear fusion and clean energy?" The press too apparently never learns to take with a large pinch of salt the stories of politically astute scientists with long years of experience in playing the media with their fake, but apparently always exciting news. Are they really unaware that scientists are telling us such stories now for close to seventy years, albeit that the epithets used become ever more exalted?

It is of course not surprising that China puts great store on nuclear fusion as a possible future source of energy. Like Japan, the country is heavily dependent on foreign suppliers for oil and gas to meet its voracious energy needs (although it produces a lot of coal, which provides about 60% of Chinese energy, and natural gas of its own). Since the Fukushima disaster power from nuclear fission is also meeting opposition in China. As of March 2019, China has 46 nuclear power stations in operation with a combined installed electric capacity of more than 45 GWe. Nuclear power contributed about 5% of total Chinese electricity production in 2019. Another 11 nuclear reactors with 11 GWe of electric power will be added over the next two years. Additional reactors are planned for a further 36 GWe. China was planning to have 58 GWe of capacity by 2020. However, few plants have commenced construction since 2015, and it is now unlikely that this target will be met.[8]

Its population is huge, 1.3 billion at present and forecast to grow to 1.5–1.6 billion in 2050. While energy consumption per capita in China is at present still fairly low, it is expected to grow rapidly. This however does not necessarily have to be the case as the example of the UK shows. Today the UK consumes less energy than it did in 1970, despite an extra 6.5 million people living there. Per capita its consumption is 2764 kg of oil equivalent, against a figure for China of 2237, so not all that different, surprisingly close actually. In 1970 this figure for the UK was still close to 4000 kg of oil equivalent per capita, stayed almost constant until 1995 after which it rapidly started to decline to the quoted figure (of 2015).[9] It seems that the UK is more efficient both in producing energy and using it. Households use 12% less, while industry uses a massive 60% less. This is largely offset by a 50% rise in energy use in the transport sector.[10] Still the amount of energy used per capita in the UK is 30% less than in 1970! This example shows that China can perhaps gain more by introducing energy

---

[7]https://dailygalaxy.com/2019/04/earths-second-sun-chinas-fusion-future-the-holy-grail-of-unlimited-energy-weekend-feature/.

[8]https://en.wikipedia.org/wiki/Nuclear_power_in_China.

[9]Data from worldbank.org.

[10]https://www.bbc.co.uk/bitesize/guides/zpmmmp3/revision/6.

savings and efficiency schemes than by investing heavily in nuclear fusion monsters. This is also borne out by the fact that the efficiency of turning fossil energy (coal) into economic output in China is very poor, just one seventh of that in Japan. (Li and Wan 2019.)

We will here review China's efforts so far and see if its bold optimism is justified. The fact of the matter is that China is part of the ITER consortium. Everything depends on ITER's success and it probably already needs unparalleled luck to get ITER properly working by 2040.

Experimental plasma physics research in China began in 1958 on linear Z pinches by several groups in Beijing. In the early sixties this work was followed (and replaced) by studies on a magnetic compression mirror and small theta pinches, and since the 1970s Chinese fusion research has developed extensively.

The two major research institutes for controlled nuclear fusion in China are the Southwestern Institute of Physics (SWIP) and the Institute of Plasma Physics of the Chinese Academy of Sciences (ASIPP). SWIP was set up in 1965 at Leshan (Chengdu) and has since become the largest research base for nuclear fusion in China. A large superconducting mirror was built in the 1970s and shut down in the beginning of the 1980s. SWIP designed and constructed a number of tokamaks. After the two very small Pretest Torus (1977) and Mini Torus (1978) experiments and the small, non-circular FY-1 from 1981 to 1988, it followed with the larger HL-1 tokamak in 1984 (shut down in 1992) and the HL-1 M tokamak in 1994 (shut down in 2002). HL stands for Huan-Liuqi which is "Toroidal Current Device" in Chinese. HL-1 was of medium size with a major radius of about 1 m. HL-1 M (M standing for "modified") was the first machine to demonstrate the electron fishbone instability (discovered on PDX at Princeton (see Chap. 5)) excited by RF heating alone.

In 2002 SWIP put into operation the first divertor tokamak in China, HL-2A. The plasma vessel and magnet coils were from the German ASDEX. Its mission is to study advanced tokamak operation, including confinement, MHD instabilities, wall conditioning, heating and current drive. Initially it had a large closed divertor chamber, which is fairly unique in modern tokamak experiments. Supersonic molecular beam injection (SMBI) and deuterium pellet injection are used for plasma core fuelling. Over the years, HL-2A has gained a world reputation due to its significant contributions to fusion technology development and physics research in many fields.

An upgrade of HL-2A, called HL-2 M, is currently under construction at SWIP. Once it is in full operation, with 25 MW of heating power and up to 3 MA of plasma current, it can provide high performance, high *beta*, and high bootstrap current plasma to simulate a fusion relevant environment. It is set to be operational as soon as 2020 and expected to generate plasmas hotter than 200 million degrees.[11]

The HL-2A and HL-2 M devices are part of the Sino-French Fusion Energy Center (SIFFER), a joint venture, established in November 2017, between the French Atomic Energy Commission (CEA) and the Chinese Ministry of Science and Technology to further promote bilateral cooperation on fusion energy research. The purpose of SIFFER is to offer support to the ITER organization and its partners, develop

---

[11] https://www.neimagazine.com/news/newschina-completes-new-tokamak-7531412.

fusion components and technologies, conduct experimental physics research, develop safety codes and technical standards for fusion energy and prepare for the next-step device. The institutes that conduct cooperation in SIFFER include the French Institut de Recherche sur la Fusion par confinement Magnétique (IRFM), the China International Nuclear Fusion Energy Program Execution Center (ITER-CHINA), and the major Chinese institutes ASIPP and SWIP.

ASIPP was established in Hefei, capital of the province Anhui in the east of China, in 1978. Currently China's major contributor to ITER, it has formerly built the small HT-6B and HT-6 M tokamaks, the middle-sized HT-7, the first superconducting tokamak in China, and China's most modern device EAST. HT-6B operated from 1983–1992 and is now installed as IR-T1 in Iran. HT-6 M was in use from 1985–2000 and HT-7 from 1993–2013, using coils from the Russian T-7. EAST at ASIPP and HL-2A at SWIFT are currently the two most important fusion facilities in China, and roughly of the same size.

EAST, the Experimental Advanced Superconducting Tokamak, which also goes under the name of HT-7U (to stress the Hefei origin as HT stands for Hefei Tokamak), is one of the devices brought into the Chinese-French collaboration SIFFER by China (WEST (see previous chapter) being the French contribution). It is however not an upgrade of HT-7, but running in parallel on the site of the former HT-6M. EAST is the first non-circular advanced steady-state fully superconducting tokamak. The project was approved in July 1998. Construction started in October 2000 and was completed in March 2006. So, it took only 5 years to build and the claim is that costs were a mere $37 million, where a comparable reactor would cost 15–20 times more in other countries. Its design, R&D, construction and assembly have been done mainly at ASIPP. EAST is designed on the basis of the latest tokamak achievements. Its mission is to conduct fundamental physics and engineering research on advanced tokamak fusion reactors. Its distinct features are a non-circular cross-section, fully superconducting magnets and fully actively water-cooled plasma-facing components, which will help to explore the advanced steady-state plasma operation modes. Its construction and physics research will provide direct experience for the construction of ITER, and ultimately contribute to the development of ITER and fusion energy. It is smaller than ITER, but similar in shape and equilibrium, yet more flexible. EAST will have a long pulse capability (60–1000 s), a flexible poloidal magnetic field system, and auxiliary heating and current drive systems, and will be able to accommodate divertor heat loads that make it an attractive test facility for the development of advanced tokamak operating modes.

First plasma on EAST was obtained in September 2006. In early 2016 EAST achieved 100 s operation in L-mode with high core-electron temperature. Later in 2016 the machine achieved another milestone: 61 s in fully non-inductive H-mode, with a tungsten ITER-like divertor. (This significant achievement was surpassed by the Korean KSTAR later in 2016, but this is surely not the end of the race.)

Very good results have been obtained, which addressed some very important key issues for the future operation of ITER. HL-2A focused on edge plasma physics and MHD control, especially on mitigation of the very troublesome edge localized modes (ELMs), the instabilities we have encountered before. Large ELMs can lead to rapid erosion of first-wall material, which is not acceptable for DEMO operation. Several ways have been developed towards robust control of ELMs. By using supersonic beam injection, the ELM amplitude could be reduced by a factor of 3 and the ELM frequency increased dramatically. Utilizing its superconducting long-pulse capability, EAST concentrates on exploring advanced high-performance steady-state plasma operations. (Li and Wan 2019, p. 114.)

For completeness we also mention here a few tokamak experiments at other Chinese institutes.

Since 2007 Huazhong University in Wuhan operates J-TEXT (Joint Text), a medium-sized circular tokamak that formerly operated as TEXT-U (Texas Experimental Tokamak) at Austin, but was moved to Wuhan, but is still run jointly with the University of Texas at Austin.

The Department of Modern Physics of the University of Science and Technology in Hefei runs the small KT-5C for studying fluctuations and transport in the plasma edge since 1985 (still in operation in 2004; an upgrade was proposed in 2017).

In addition, the Institute of Physics at Beijing has been operating the small tokamak CT6-B, a modification of an earlier device called CT6-A, from 1995–2002 and a field-reversed pinch; a group at the University of Science and Technology of China is working on the development of some special diagnostics; plasma focus processes are being studied at Qinghua University, and some groups in Beijing and Shanghai are developing special fusion technology such as gyrotrons, microwave systems, etc.

Within the framework of SIFFER, China has put forward a proposal for a successor to ITER: the China Fusion Engineering Test Reactor (CFETR). See Fig. 7.3 for an artist impression. Its construction is planned for the 2020s as a demonstration of the feasibility of large-scale fusion power generation.

So, like Japan, China is already planning its own sequel to the ITER adventure, both countries apparently expecting that after ITER they have to go further on their own, but apparently the Chinese want to start building their demo plant even before ITER is ready! The project would include two phases of operation. The first phase aims to demonstrate steady-state operation and tritium breeding. The second phase would include an update of the system to obtain fusion power production of 1 GW or 1000 MW (compared to ITER's 500 MW) and a fusion gain (Q) higher than 12, with tritium self-sufficiency, so enough tritium breeding to keep the process going.

We finish this section with the rather optimistic Chinese roadmap to nuclear fusion as envisaged now:

(1)   2006–2045: Join and fully support the construction, operation and experiments of ITER, while increasing the support for experiments on EAST, HL-2 M and J-TEXT;
(2)   2020: Start of construction of the Chinese Fusion Engineering Test Reactor (CFETR);
(3)   2030: Completion of the construction of CFETR (fusion power of more than 200 MW and test of steady-state operation and tritium self-sufficiency);
(4)   2040: Completion of the upgrade of CFETR (fusion power of about 1 GW, Q_{eng} > 1);

Fig. 7.3 Artist's impression of the future CFETR (*from* SIFFER)

(5) 2050–2060: Completion of the construction of the Prototype Fusion Power Plant (PFPP) (about 1GWe, power plant validation).

Everything depends on the operation of ITER which as things stand now (2020) will certainly not start operation in 2025 as envisaged in the Chinese roadmap.

The same has been put into a picture in Fig. 7.4.

7.6 Australia

Australia has a connection with nuclear fusion from its very beginning. In 1934, under the guidance of Ernest Rutherford, the Australian physicist Mark Oliphant (1901–2000) and the German physical chemist Paul Harteck (1902–1985) were the first to identify a fusion reaction in their lab by firing deuterons (nuclei of deuterium) into a deuterium gas. In 1946 the Australian Peter Thonemann (see Chap. 3) and a British colleague, pioneered studies of plasma magnetic confinement in a toroidal configuration at Oxford University. In 1953, Australia entered the nuclear science arena, when its Atomic Energy Act came into effect. The Australian Atomic Energy Commission (AAEC) followed and in 1987 the AAEC evolved into the Australian Nuclear Science and Technology Organisation (ANSTO), as it is known today.

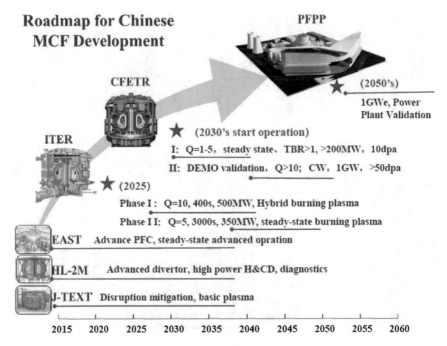

Fig. 7.4 Roadmap for Chinese magnetic controlled fusion (MCF) development (*from* Li and Wan 2019, p. 116)

From 1964 to 1978 the Australian National University (ANU) at Canberra built and operated the first tokamak outside the Soviet Union. It was called LT-3, Liley Tokamak, and was followed from 1978 to 1984 by LT-4. A tokamak, called TORTUS, worked from 1981–1992 at the University of Sydney, while in recent years ANU has decided to pursue advanced stellarators.

Australian researchers invented the Rotamak, a near-spherical device that uses a rotating magnetic field to drive a steady toroidal current in a compact torus device. Such devices are still in use today at various places in the world, e.g. the Prairie View (PV) Rotamak at Prairie View A&M University in Texas, which can either be used as a spherical tokamak or as a field-reversed configuration.

The first heliac, H-1, was made at ANU in 1985 (see Chap. 13) and the first experimental demonstration of a spherical torus configuration was achieved in 1988. Until its decommissioning in 2017 and transfer to China, the H-1 Heliac stellarator at the Australian Plasma Fusion Research Facility (APFRF) at ANU was Australia's focus of basic experimental research on magnetically confined plasma. It provided a plasma confinement environment suitable for developing and testing the proposed ITER diagnostic system.

In 2011 MAGPEI (Magnetised Plasma Interaction Experiment), a helicon linear magnetic mirror plasma device at APFRF, began operations and is used to produce a plasma environment suitable for the study of plasma-material interactions with candidate materials for future fusion reactors, including ITER. The environment inside the ITER fusion reactor will be extremely challenging for materials, in particular the reactor wall. Advanced materials are required that can withstand extreme radiation, extreme heat, plasma chemistry and thermo-mechanical stresses.

Australia takes part in ITER through a cooperation agreement formalised late in 2016, to enable the country to engage with and benefit from participation in the world's largest engineering project to create fusion energy.

7.7 Miscellaneous

There are various other countries that had and/or still have a tokamak programme, with small to very small tokamaks (typically less than 0.5 m for the major radius, apart from the Swiss TCV and the Canadian TdeV). Most devices have been listed on the website www.tokamak.info. Noteworthy among them are:

Czech Republic. From 1985–2007 it operated a small tokamak CASTOR, which was the former Russian TM1-VCH (an upgrade of TM-1). This has been replaced by COMPASS, received from Culham (UK) in 2007. With its ITER-like plasma shape, flexible neutral beam heating system, and ability to achieve H-mode, COMPASS contributes to ITER-relevant plasma physics studies. An upgrade COMPASS-U has been approved and scheduled to start operations in 2021. Its research must serve both ITER and its DEMO successor.

Iran runs at least three small tokamaks: Alvand (from 1976, named after one of Iran's highest mountains (3580 m), followed by several upgrades), Damavand (from 1993, named after Iran's highest mountain (5609 m), obtained from Russia where it was called TVD) and IR-T1 (a former Chinese machine, HT-6B, which also benefitted from Russian and German components; since 1994 operating in Teheran at Azad University). In addition, design and construction of the Alborz tokamak have been studied since 2012 at Amirkabir University of Technology. Recently, this device has been assembled and initial tests are currently in process. (Amrollahi *et al.* 2019.)

South Africa operated the Tokoloshe tokamak in Pretoria from 1980–1990, which had a low aspect ratio of 2 (spherical tokamak (see Chap. 11)) and was used for studying fuel recycling.

In **Spain** TJ-1, a very small tokamak, operated from 1983 to 1994 in Madrid; in 1994 it started operating as the TJ-1U torsatron and was subsequently transferred to Kiel in Germany to become TJ-K. About the stellarator TJ-2 see Chap. 13.

Switzerland operated TCA (Tokamak Chauffage Alfven), which was later sent to Brazil, and TCV (Tokamak à Configuration Variable) since 1992 in Lausanne. TCV's distinguishing feature over other tokamaks is that its torus cross section is three times higher than wide. It achieved 100% bootstrap current in 2006 and holds the world record of fully non-inductive operation (i.e. no ohmic heating) with electron cyclotron current drive (ECCD), i.e. the current is driven by electron cyclotron waves. It is the only machine capable of negative triangularity.

Other countries that operated tokamaks in the past include **Brazil**, **Canada**, **Egypt**, **Hungary**, **Libya**, **Mexico** and **Ukraine.**

Chapter 8
The Big Tokamaks: TFTR, JET, JT-60

8.1 Introduction

Tokamaks are getting bigger and bigger. Devices have become progressively larger
in terms of the torus' major radius R and also use an increasingly stronger confining
magnetic field. Why is that necessary? The simple reason is that the bigger the
machine the more time the particles spend inside the plasma before leaving it, and
the more time they have to fuse and produce energy. Particles in a plasma gyrate
around the magnetic field lines and sometimes collide. When colliding, the guiding
centre of their motion will jump two gyro-radii (the gyro-radius is the radius of their
motion around the magnetic field line, see Fig. 2.2), so every collision causes these
particles to jump towards the wall. Such collisions are necessary as particles that do
not collide will never fuse. In a fusion reactor a deuterium nucleus will typically have
a gyro-radius of 1 cm. It takes more than 1000 collisions before a fusion reaction
occurs. In these 1000 collisions the particle comes ever closer to the wall, unless the
machine is big enough to allow for so many collisions. In other words, the bigger
the plasma volume the more energy will be generated, while losses are proportional
to the surface area. In the following we will look at this a little more in detail.

   Confinement was and still is poorly understood theoretically, e.g. there is no proper
theory to predict the dependence of the confinement time on the plasma density or
other parameters. This in itself is not surprising as the orbits of charged particles in a
torus are very complex. Different regimes of turbulence are responsible for different
aspects of plasma transport, e.g. particle diffusion and heat conduction, in different
regions of the plasma, whereby these various turbulence regimes interact with and
influence each other. A theory of a single turbulence regime is not sufficient to provide
a realistic model for the whole plasma. To address the problem theoretically a set
of highly nonlinear equations describing the plasma in non-uniform magnetic fields
needs to be solved. Such sets of equations are extremely hard to crack, and although
some qualitative aspects can be modelled, reliable quantitative estimates are much
more difficult to obtain (see also the section on plasma modelling in Chap. 2).

© The Author(s), under exclusive license to Springer Nature Switzerland AG 2021   183
L. J. Reinders, *The Fairy Tale of Nuclear Fusion*,
https://doi.org/10.1007/978-3-030-64344-7_8

Due to the plasma turbulence, it is still impossible to calculate energy and particle transport in tokamaks from first principles. Energy losses through electrons for instance exceed those predicted by neoclassical theory by a factor of ten. Many real mechanisms of losses are not clear. In the absence of adequate theoretical understanding, empirical methods have to be relied upon, i.e. collect data from many tokamaks over a range of different conditions and parameters and try to establish from them how the confinement depends on the plasma parameters, and then subsequently extrapolate the results to higher values of these parameters. From such extrapolations it is then derived how a new machine should be constructed. As mentioned before, such extrapolation relationships are generally called 'scaling laws', although the use of the word 'law' is a little pretentious in this respect. They are no more than empirical rules of thumb, which for instance connect confinement times with machine and plasma parameters. Figure 8.1 shows the empirical scaling of energy confinement time from present day experiments to ITER, which due the logarithmic scales chosen looks better than it in reality is.

In Chap. 5 we have already met such scaling behaviour, as discovered in the Alcator tokamaks. Energy confinement time was found to be proportional to the product of the plasma density and the square of the minor radius a and to increase almost linearly with the plasma density over more than a factor of ten (Braams and

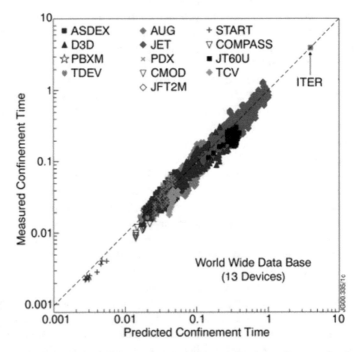

Fig. 8.1 Showing the agreement between experimental results for the confinement time from a large number of different machines (ordinate) and the result of the scaling law (abscissa) extrapolated to ITER (from http://www-fusion-magnetique.cea.fr/gb/fusion/physique/modesconfinement.htm)

Stott 2002, p. 176). In the 1980s it was established and confirmed on a large number of tokamaks that the energy confinement time of tokamak plasmas scales positively with plasma size (meaning: increases with plasma size). It is consequently generally expected that Lawson's triple product will also increase with plasma size. This has been part of the motivation for building ever larger devices, up to ITER with a plasma volume of over 800 m^3 (the volume of a sizable house).

However, tokamak plasmas are also subject to operational limits and two important limits are a density limit and a *beta* limit, *beta* being the ratio of the plasma pressure to the magnetic pressure. Current and density limits are usually determined by disruptions, while other instabilities set the *beta* limit (Chap. 2). A disruption, as we have seen, is the dramatic event in which the plasma current abruptly terminates and confinement is lost. The scaling for the confinement time can be combined with the density limit into an expression for the triple product showing that the triple product scales like $a^2 B_T^3$. From this it follows that increasing the minor radius, i.e. plasma radius, and working at the highest possible magnetic field are key steps in reaching ignition (Costley 2016).

In this chapter we will discuss three big tokamaks: the American Tokamak Fusion Test Reactor (TFTR) at PPPL in Princeton, the European Joint European Torus (JET) at Culham in the UK and the Japanese JT-60 (and its upgrades JT-60U and JT-60SA) that link the small and medium-sized tokamaks of the 1970s and 1980s to a future thermonuclear experimental reactor like ITER. Figure 8.2 compares the cross sections of the three devices. The Soviet T-15 was of similar size as the three tokamaks discussed here. One of the novelties of the T-15 was that it used superconducting magnets to control the plasma. Since it has never worked properly and was shutdown in 1995 due to lack of funds, we do not discuss it here. For some more details about the T-15, see Chap. 5.

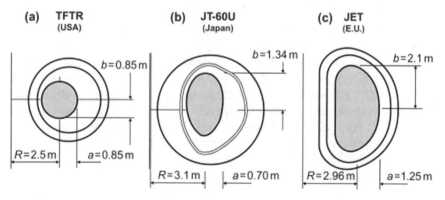

Fig. 8.2 Comparison of the three large tokamaks TFTR, JT-60U and JET. The size of the fusion chamber and the sizes of the plasma volume for each device (*from* McCracken and Stott 2013, p. 111)

8.2   Tokamak Fusion Test Reactor (TFTR)

The first tokamak that deserves to be called big is the Tokamak Fusion Test Reactor (TFTR) at Princeton. The idea for TFTR went back to the very early 1970s. The fact that it was conceived at precisely this early, not to say premature, stage in tokamak research, was mainly due to Robert Hirsch, and perhaps also due to pressure from the decision taken by the European Union in 1971 in favour of a robust European fusion programme. From the onset of his term in office starting in 1971, Hirsch was bent on building a tokamak feasibility experiment, i.e. an experiment that would produce and contain a high temperature plasma exceeding the Lawson criterion. He did not want another machine for studying plasma physics, but wanted to make a major step towards future power generation, so he insisted on a machine that would burn deuterium and tritium,[1] which as we have seen is the most promising fuel combination for achieving fusion. Of course, there was wide-spread expectation in the fusion community that one day that step would have to be made, but the general consensus was that the time for this had not yet come. Hirsch had other ideas though. He saw several advantages in a D-T experiment. In the D-T reaction

$$D(np) + T(nnp) \rightarrow {}^{4}He(nnpp) \quad (3.5\,MeV) + n\,(14.1\,MeV)$$

the energy of the charged helium nucleus ^{4}He (3.5 MeV or 3500 keV) is roughly 300 times the 10 keV (in temperature about 100 million degrees) of the deuterium and tritium fuel ions. They would be captured by the magnetic fields, remain in the plasma and substantially alter its properties. Hirsch felt that this had to be studied as it would be a typical situation for a functioning fusion reactor. In principle there was not much wrong with this idea; the only snag was that at that time nobody knew how to get a burning plasma that would produce these helium nuclei. In his view this lack of knowledge was essentially due to the attitude of the fusion scientists. He wanted them to change that attitude from a scientific plasma physics interest (to find out how it might work) to the practical goal of energy generation (it doesn't matter how it works, so long as it works), in other words he wanted to turn them from scientists into engineers, which probably also had to do with his own background as a nuclear fusion engineer. A D-T experiment would provide the opportunity to automatically shift the attention to engineering problems, as Hirsch was of the opinion that that was were the real problems lay. And finally, there would be the political advantage, it would bring home the importance of fusion to the public and the politicians by demonstrating the actual production of power. The amazing thing in all this is that, although most physicists and engineers did not like his suggestion, nobody managed to stop him. The physicists objected as the physics of tokamaks, so early in the game, was still very poorly understood (especially the role of impurities[2] and diffusion losses),

---

[1]Hirsch had used a deuterium-tritium mix in his work at International Telephone and Telegraph.

[2]Energetic plasma ions diffuse outwards towards the walls and bombard the walls and other structures of the vessel. This releases heavy atoms from the wall material into the plasma and pollutes it.

while the engineers abhorred a D-T burning tokamak as it would be plagued with problems of radioactivity (tritium is radioactive and easily absorbed in the human body, implying that much of the work must be done by remote control) and hazardous neutron fluxes (the high-energy neutrons released in the D-T reactions would also cause induced radioactivity in the apparatus). But the fusion community was divided, a lot of money was being offered for doing this experiment, and especially Oak Ridge, which had a long history of dealing with radioactive reactors, was positive. The scientists working on ORMAK, the Oak Ridge tokamak, had actually opened Hirsch's eyes to the possibility of starting D-T experiments. They thought they could build such a tokamak experiment at reasonable size and expense. By feeding them money Hirsch got them to design the tokamak he wanted and went even further by plumping for a device that would not just achieve breakeven, but even ignition (Bromberg 1982, p. 206).

Princeton, also a prime candidate for building a big tokamak, had great misgivings because of the use of radioactive fuel. It was not a federal laboratory, but a university where they felt radioactivity had no place. The Princeton laboratory saw itself purely as a plasma physics laboratory. This contrary to Oak Ridge which was and always had been a reactor laboratory. The scientists at Princeton and many others in the fusion community were not convinced either that the scientific problems connected with the burning of deuterium and tritium were the important ones. The behaviour of a plasma containing helium nuclei from fusion reactions could be reliably predicted by theory, they thought, supplemented if needed by much cheaper experiments. At this point in the fusion game there were far more important problems to study than the burning of deuterium and tritium, such as impurities and diffusion losses.

Apart from the radioactivity bit, Princeton was as eager as Oak Ridge to build a tokamak feasibility experiment, the successor to their PLT experiment, for which funding had just been obtained. Hirsch at first favoured Oak Ridge as he was very encouraged by recent results on neutral beam heating with ORMAK and, due to his engineering background, the technical orientation of their work was in general more to his taste than Princeton's physics approach. In order to soften up the higher echelons of the AEC for providing the necessary funds, he even suggested to describe the results of the neutral beam heating as a 'major breakthrough', setting a precedent for the amazing number of breakthroughs that have been achieved over the years in fusion research, most of which have to be taken with a large grain of salt or, alternatively, change your definition of 'breakthrough'. However, the expected doubling or tripling of ion temperatures at Oak Ridge with neutral beam heating turned out to be elusive and Princeton scooped them also in this respect with its Adiabatic Toroidal Compression experiment, which pushed the temperature up to 60 million degrees, as we have seen in Chap. 5. On top of that, the Oak Ridge design became too costly at several times the hundred million dollars budgeted for the experiment.

The competition between the two laboratories went on for some time and Hirsch skilfully played them off against each other. Both laboratories started to design a D-T burning experiment (Princeton with the help of Westinghouse Electric Corporation, which was a major player in building nuclear fission reactors). Although Princeton still had misgivings about housing a radioactive experiment, it did not want to lose the feasibility experiment either. If it refused such an experiment, its future might be in jeopardy. The outcome of the rivalry was such that the prospects dramatically reversed in the course of a single year. In September 1973 Oak Ridge was still favoured to house the D-T burning experiment and the scientists at Princeton were worrying about the future of their jobs, but less than a year later, in July 1974, Princeton was actually awarded the construction of TFTR and on the path of a major expansion. Hirsch chose for the more conservative Princeton proposal, which used for instance ordinary magnets where Oak Ridge proposed to use superconducting magnets. The Oak Ridge design could also be extrapolated to an ignition device, which was not the case for the Princeton proposal. In the end the decisive factor was the projected cost of the device which for Oak Ridge was three times what Princeton was proposing. Although the design was Princeton's, Hirsch at first still wanted to site the experiment at Oak Ridge, since as a nuclear engineering laboratory it was more suitable for such an engineering project and its possible follow-up than Princeton's plasma laboratory. In the end he came round to Princeton as the site for TFTR, the main argument being that a Princeton team would be more committed than an ORNL team running someone else's design.

In this period each of the four major fusion laboratories (Oak Ridge, Princeton, Los Alamos and Livermore) lived with the constant threat of their main-line experiments being terminated. Each of them tried to find a secure niche that would be essential to the overall programme progress. Oak Ridge was historically the most vulnerable of the four and hence it zealously sought the role of host of the tokamak feasibility experiment. This posed the danger for Princeton of ceding its pre-eminent position in fusion research to Oak Ridge and forced it to submit its own proposal against its own prior best judgement. It accepted a D-T experiment which it actually did not want but was not in a position to refuse either (Bromberg 1982, p. 204ff; Heppenheimer 1984, p. 44ff; Bromberg, TFTR 1982). I have dwelt at length on this episode to show how political and scientific forces interact in such a government sponsored programme and that in the competition for resources scientific judgement often has to take a backseat to political and pecuniary considerations.

The great danger in all this was that a failure of TFTR might jeopardize the entire magnetic fusion programme. But when would it have failed? What were its objectives? The 1976 TFTR Technical Requirements Document described the objectives of TFTR as: (1) demonstrate fusion energy production from the burning, on a pulsed basis, of deuterium and tritium in a magnetically confined toroidal plasma system; (2) study the plasma physics of large tokamaks; and (3) gain experience in the solution of engineering problems associated with large fusion systems that approach the size of planned experimental reactors (Williams 1997). The first objective was to be satisfied by "the production of 1–10 megajoule (MJ)[3] of thermonuclear energy (per pulse) in a D-T tokamak with neutral beam injection under plasma conditions approximating those of an experimental fusion power reactor" (Dean 2013, p. 44). A plasma temperature of 5–10 keV, a density approaching 10^{20} m^{-3}, and an energy

---

[3] 1 MJ is not a great deal of energy; a 100-watt lightbulb will burn it away in less than 3 h (1 kWh = 3.6 MJ).

confinement time of 0.1 s would be required (roughly a factor 100 below the Lawson criterion). The device was expected to exploit a special approach in which beams of high-energy neutral deuterium were injected into a mainly tritium plasma. The neutral deuterium atoms would be stripped of their electrons and form a 50/50 D-T plasma (Bell in Neilson 2016).

In July 1974 the price tag for this experiment stood at $228 million dollars (Clery 2013, p. 154), which soon went up to $300 million, and had ballooned well beyond that at the end of the construction period in 1982. The eventual costs amounted to almost $1 billion (Herman 1990, p. 216). One reason for this was that, due to the positive results with PLT, the objectives were amended in 1978 and 1979 and additional funding added to reach the goal of scientific breakeven for the D-T plasma (Meade 1988). To meet this latter requirement, the heating system was upgraded to 50 MW.

8.3   Breakeven

At this point it may be useful to dwell in some detail on the term **breakeven**. It is a word that has a seemingly simple meaning, one would think: breakeven has been achieved when the power released by the nuclear fusion reactions is equal to the power used to get the fusion reactions going, in other words when what goes in is equal to what comes out. With some creativity in the choice of adjectives and by being misleading about what is meant by "what goes in" or where it goes in, the fusion community has managed to make it a very confusing term.

We first define the **fusion energy gain factor** (expressed by the symbol Q) as the ratio of the fusion power produced in a nuclear fusion device to the power required to maintain the plasma heated (P_{fus}/P_{heat}). This latter power should not be taken to mean all the power needed to bring the plasma into the state in which fusion reactions can start to occur. In fusion parlance it is only a fraction of that power, namely just that part of the power that actually goes into the plasma to keep it at the required temperature, i.e. to heat it while fusion reactions are actually going on. All losses and power used for the magnets, cooling etc. are not included. The power spent in running the reactor itself is never taken into account in any of the calculations, nor the power needed to keep the device on standby.

With this definition of Q, until at least $Q = 5$ self-heating of the plasma (by the α-particles produced in the fusion reactions) is not expected to be sufficient to make up for losses that are inevitable like neutrons produced in the fusion reactions that fly off unimpeded by the magnetic field. External heating still remains necessary.

If Q increases past this point, increasing self-heating eventually removes the need for external heating. The reaction then becomes self-sustaining, a condition called **ignition**, and no further external energy input will be needed. From the definition of Q it follows that ignition corresponds to infinite Q as P_{heat} is then zero. This is generally regarded as highly desirable for a practical reactor design. This does not mean that such a device would produce an infinite amount of energy, far from it, only that the fusion reactions continue for some (short) time without external heating of the plasma being required. ITER is designed to give Q of at least 10 and $Q > 15$ would be needed in a power station.

For fusion power reactors several types of breakeven have been defined:

The first one is **scientific breakeven** (a term mainly used in inertial confinement fusion, but also popping up in magnetic confinement). It is normally just called breakeven and corresponds to $Q = 1$. It is the minimal quantitative challenge. This is what fusion devices have been aiming to achieve for the past 40 years. It is the situation in which the power released by the fusion reactions is equal to the power required to heat the plasma or just fulfilling the Lawson criterion (a system at $Q = 1$ will still cool without external heating). As said, the power required to heat the plasma P_{heat} only includes the power that actually goes into the plasma; any losses in this heating process are not included. When really talking about breakeven, one should of course include all the power used in running the reactor itself and powering it up, including all the losses suffered in the process. This is not done here. Only that (tiny) part of the external power is taken that actually goes directly into the plasma. And amazingly it can nonetheless be called breakeven.

Next, we have **engineering breakeven** which is when sufficient electric power can be generated from the fusion power output to feed back into the heating system for keeping the reactor going (recirculation) (up to $Q = 5$).

Economic breakeven applies to a machine that can sell enough electricity to cover its operating costs, estimated to require at least $Q = 20$, while **commercial breakeven** is the situation when sufficient power can be converted into electric power to cover the costs of the power plant at economically competitive rates and any net energy left over is enough to finance the construction of the reactor. This is eventually the only breakeven that matters; a power plant that cannot achieve this will not be economically viable.

Finally, we have **extrapolated breakeven** which is when (scientific) breakeven is projected for a reactor hypothetically using D-T fuel from experimental results using only deuterium as fuel, by scaling the reaction rates for the two fuels. (Basu 2001, p. 22; see also fusion energy gain factor in *Wikipedia*.)

Within this framework it is amusing, but at the same time rather sad, to read APS News of August/September 1995 (volume 4, no 8)[4] in which the Chair and Vice-Chair of the APS Plasma Division lambaste Congressman Rohrabacher for stating that there is no and will not be any breakeven in fusion for many years to come. They call his statement "fantastically incorrect", and continue by saying that "[i]n fact, breakeven conditions have already been achieved in the JET tokamak operating with deuterium plasmas. In 1996, when tritium will be introduced into JET, it is expected to generate more fusion power than is put into the plasma. (…)" Here they talk about extrapolated breakeven. From the results obtained on JET with

---

[4]https://www.aps.org/publications/apsnews/199508/letters.cfm, accessed 2 June 2020. There are plenty of other instances where such misleading statements have been made by leading members of the fusion community, which is very regrettable, if not outright dishonest. The website of *New Energy Times* has exposed several of such statements in respect of ITER.

purely deuterium plasma they confidently state that breakeven ($Q = 1$) will be a certainty when the plasma is changed to D-T. An objectionable statement by the APS Plasma Division, and very close to a straight lie. It is without doubt an untruth and in any case "fantastically incorrect", and nobody in the science community has taken the trouble to point this out or correct it. In his statement Rohrabacher probably meant *commercial* breakeven as there is no reason for him to be interested in any other of the more dressed-down forms of breakeven. The concept of breakeven as used by the APS includes only a fraction of the external power used. But even that kind of breakeven was not achieved by JET. In 1997, as we will see, the JET tokamak in the UK in its tritium operation achieved $Q = (16\ \mathrm{MW})/(24\ \mathrm{MW}) \approx 0.67$, meaning that just 67% of the plasma power input came out as heat. And this of course only in a pulsed operation, not steady state. (McCracken and Stott 2013), p. 115, Fig. 10.6.) That is the best that has ever been achieved, as openly stated for instance on the ITER website.[5] As we will see below, the TFTR result was even worse. So when the same publication says: "The fact is that the fusion community no longer considers breakeven an important scientific challenge: several years ago it began looking beyond breakeven at fusion self-sustainment (ignition) and issues that affect fusion power plant size, cost and complexity", they are again not telling the truth, but just complete bullshit. None of that is even close to being realised today, let alone in 1995. I am certain that when ITER achieves the most limited possible form of breakeven ($Q = 1$) in quite a number of years to come (I am tempted to say 20 of course), the noise of popping champagne bottles will be deafening. It seems that even the experts are confused or in some cases annoyed about the juggling with the breakeven concept, and that attempts are or have been made to make it even more confusing (see Meade 1998).

8.4   TFTR Continued

Conceptual design studies for TFTR had begun at PPPL in January 1974 before the start-up of the Princeton Large Torus (PLT). The dimensions chosen were about a factor of two larger than those of PLT. As can be seen in the cutaway model of Fig. 8.3 (and in Fig. 8.2), the cross section of the vessel is circular, this contrary to many other tokamaks built around the same time and later, like JET and JT-60U, which had D-shaped cross sections. Princeton stuck to the tried and tested circular design to keep their device as simple as possible. This undoubtedly was a mistake as it would make it difficult to achieve high-confinement mode and/or to install a divertor. TFTR had to do with a limiter and one of the outstanding issues in the design was the choice of limiter material. PLT had a tungsten limiter, which had caused a substantial concentration of tungsten impurities, resulting in collapse of the plasma core. Subsequent experiments with graphite limiters on PLT showed acceptable impurity influxes, and graphite was incorporated in the TFTR design.

---

[5]https://www.iter.org/sci/iterandbeyond/.

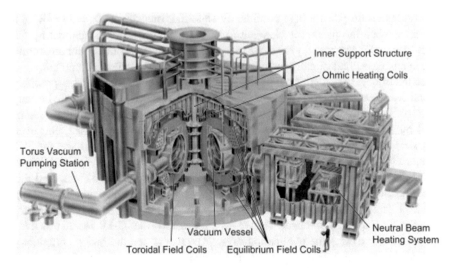

Fig. 8.3 Cutaway model of TFTR, showing the magnetic field coils, the vacuum vessel, and the neutral beam heating equipment (*from* McCracken and Stott 2013, p. 110)

Heating was also a major issue on tokamaks of this size as ohmic heating alone was no longer sufficient. Because plasma resistivity falls with increasing temperature, ohmic heating has its limit and becomes negligible at the temperatures to be achieved in these tokamaks. All these tokamaks start operation without any additional heating, which is then added later and upgraded over the years. In TFTR heating by neutral beam injection (NBI) and by waves in the ion cyclotron range of frequencies (ICRF) was used. The development of high-power neutral beams was a major technological undertaking. It had been pioneered on mirror machines and developed further with ORMAK at Oak Ridge, but in 1976 the maximum injected power into a tokamak was a few hundred kW, while the neutral beam system eventually deployed on TFTR injected a maximum of 40 MW into a D-T plasma. In addition to heating the plasma, the neutral beams were also used for fuelling the plasma (Hawryluk, Batha et al. 1998).

The design parameters are shown in Table 8.1, together with the actually achieved values, where different. Its major radius was considerably larger than any tokamak built so far. The largest ones had been Princeton's PDX with $R = 1.4$ m and the Soviet T-10 with $R = 1.5$ m. The plasma volume of TFTR was about 38 m^3.

Construction of TFTR started in April 1976 and it produced its first (hydrogen) plasma in December 1982, nearly nine years since its conceptual design study in 1974, showing how the construction time increases with each new generation of fusion device. This was followed by deuterium plasmas in the mid 1980s. Experiments with plasmas containing tritium (50% tritium and 50% deuterium) were planned to start a few years later but were eventually delayed until 1993. The process of optimization with deuterium plasmas took much longer than expected. In the end TFTR was beaten by the European JET device at Culham in the UK which used a D-T mixture for the

Table 8.1 Basic design values of TFTR and actually achieved values, where different

	Design value	Achieved value (where different)
Vacuum vessel major radius (m)	2.65	
Vacuum vessel minor radius (m)	1.1	
Toroidal field on axis (tesla)	5.1	5.9
Duration (in seconds) of maximum toroidal field	3	
Plasma current (MA)	2.5	3.0
Duration (in seconds) of maximum plasma current	2	
NBI heating power (MW)	20	40
Ion cyclotron power (MW)		10
NBI pulse duration at maximum power (s)	0.5	0.8
NBI accelerating voltage (kV)	120	

first time in November 1991, albeit with only 10% tritium. (McCracken and Stott 2013, p. 114; Clery 2013, p. 180). TFTR and JET were the only two tokamaks in the world designed to work with a real D-T fuel mix. Tritium had been avoided as it is radioactive (with a half-life of eleven years). Special licences would have to be obtained and special shielding has to be in place, as well as provisions for remote handling and maintenance.

From the outset it was clear that TFTR would not reach breakeven (i.e. the very limited form of breakeven ($Q = 1$) as discussed above), but experiments went ahead anyway. The ability to reliably operate a large tokamak device with a D-T plasma was a major issue during the design of TFTR. Despite the complications introduced by the use of tritium, TFTR operated routinely and reliably from the start of D-T operations in November 1993 to the completion of experiments in April 1997. More than 1000 shots of D-T experiments were carried out with a peak fusion power of 10.7 MW, which when achieved in November 1994 was a world record and the first time that any real fusion power was produced in any device. It was presented as being enough to meet the power needs of more than 3000 homes, although just for a few seconds, or 'transient' as they tend to call it, which of course sounds much posher.[6] This value should be compared to the more than 30 MW of input power used to heat the plasma (i.e. just the power that went into the plasma; the neutral beam power alone, which only partly went into the plasma, was already 39.5 MW (Table 8.2)), indicating how far short of scientific breakeven the TFTR has remained (Q was just 0.27). The total electric power consumed for the production of this 10 MW of thermal fusion energy amounted to 950 MW, so a factor hundred would still be needed to

---

[6]It is a sly way to present the result in this way: "enough to power 3000 homes for a few seconds"; the use of 3000 makes it impressive, as 3000 is a fairly big number, but it is actually the same as "enough to power a single home for just a few hours", which will not impress anybody.

Table 8.2 Values of TFTR parameters for the D-T operation

TFTR D-T plasma parameters for shot 80539	
Parameter	Value
Plasma major radius (R)	2.52 m
Plasma minor radius (a)	0.87 m
Plasma current (I_p)	2.7 MA
Toroidal Magnetic Field (B_T)	5.6 T
Neutral Beam Power (P_{NBI})	39.5 MW
Electron density (N_e)	1.0×10^{20} m^{-3}
Ion Temperature (T_i)	32 keV
Electron temperature (T_e)	13.5 keV
Stored Energy (W_{tot})	6.9 MJ
Energy Confinement Time (τ_E)	0.21 s
Generated Fusion Power ($P_{D\text{-}T}$)	10.7 MW

approach any useful sort of breakeven. The values of the TFTR parameters in the D-T operation have been listed in Table 8.2.

It may seem surprising, but TFTR has so far been the only device that actually has attempted to reach breakeven. To reach breakeven, the system would have to meet several goals at the same time, a combination of temperature, density and confinement time, i.e. not only for the triple fusion product discussed in Chap. 2, but also for the individual components in the triple product, e.g. the temperature alone had to be around 150 million degrees. In spite of considerable effort, the system could at any given time only demonstrate any one of the required values. In April 1986, TFTR experiments with a purely deuterium plasma produced 1.5 $\times$ 10^{20} seconds per cubic metre for the product of density and confinement time, while the value for the triple product remained a factor 20 short of the required value. One of the problems was that the use of neutral beams for the heating caused a relative reduction in confinement time as the plasma got hotter, and it remained in this predicament, dubbed the "low mode" until, quite by accident, a so-called "supershot regime" was discovered. (Herman 1990, p. 216; Clery 2013, p. 170.) It refers to an operating regime in which the number of fusion reactions among deuterium nuclei is up to 25 times higher than previously observed. The supershot regime was an experimental discovery, a sort of H-mode (see Chap. 6) and apparently particular to TFTR as it has not been reproduced anywhere else. It has to do with decreasing the influx of impurities from the vacuum vessel walls. These supershots, as measured by the Lawson triple product, enhanced performance by a factor of about 20 over comparable L-mode plasmas, or a factor of about 5 over standard H-mode plasmas. (McGuire et al. 1995; Hawryluk 1998). From Table 8.2 it can be calculated that the value of the triple product remains about a factor 10 short of the required value of 3 $\times$ 10^{21} m^{-3} keV s quoted in Chap. 2.

Although at one of the PPPL web pages it is bluntly stated that TFTR accomplished all its research goals (both its physics and hardware goals)[7], breakeven ($Q = 1$), which definitely was one of its goals (Meade 1988), and arguably the most important one, was never achieved, but scientists managed to sell other achievements of the device as 'breakthroughs' of one kind or another.

Overall, 99 grams of tritium was processed. TFTR set a world record plasma temperature of 510 million degrees (about 30 times the temperature in the middle of the Sun). Such records are announced with great fanfare but are not such a great deal. After all, at the time TFTR was the biggest machine around and it would have been rather peculiar if it had not reached the highest temperature or other record values. It remains unclear what the significance of such a record temperature actually is, as it is well beyond the 150 million degrees or so required for (commercial) fusion.

Apart from achieving record pressures (the plasma pressure at the centre of TFTR reached 6 atmospheres, comparable to that needed for a commercial fusion reactor) and temperatures, TFTR was also the machine on which so-called "ballooning modes" were observed for the first time. When heating the plasma, variations in plasma pressure can cause the plasma to become unstable. A particular type of instability called "ballooning mode" had been predicted, a prediction that was confirmed by TFTR experiments. Ballooning modes take place in regions where the magnetic field (which must confine the plasma) is weakest. Then, if the variations in plasma pressure are large enough, the plasma breaks out, much like an aneurism or a balloon. The region expands until the plasma ruptures and collapses. In any reactor design, the ballooning instability limits the fusion power that can be produced.

The device also provided the first experimental confirmation of the bootstrap current, the self-generated non-inductive current (as opposed to the inductive current induced by the central solenoid into the plasma), which we discussed in Chap. 4. Researchers were able to drive the entire plasma current in TFTR by means of the bootstrap effect and neutral-beam injection, which is important, for if the bootstrap current can replace the external current drive, it could be used to realise steady-state operation.

TFTR was also the first tokamak, together with JET, to demonstrate tritium handling technology. Following plasma operations, this tritium was routinely reclaimed, converted to tritiated water (HTO) and safely stored for ultimate reprocessing or disposal.

Finally, it demonstrated reliable plasma heating technology. Its neutral-beam heating systems provided reliable plasma heating power at levels routinely exceeding thirty megawatts. The Ion Cyclotron Radio Frequency (ICRF) heating systems demonstrated the first radio-frequency heating of D-T plasmas.

TFTR remained in use until 1997 and was dismantled in September 2002, after 15 years of operation. Its shutdown was due to Congressional budget cuts in the US Fusion Research budget. It was clear that the world would not have an operating fusion power plant by the year 2000, as envisaged in the 1976 plan of the US Energy Research and Development Administration (also published in Dean 1998b), as Stephen Dean

---

[7] https://www.pppl.gov/Tokamak%20Fusion%20Test%20Reactor.

laments (Dean 2013, p. 146–147), placing the blame for this squarely with the US government: it was all due to the government's failure to provide the necessary funds, its failure to commit to construction of the necessary facilities, such as the fusion engineering test facility that had been designed as the successor to TFTR, and its failure to manage the programme so as to achieve its avowed practical purpose. It is a lame and misguided excuse, as the fusion community itself must take most of the blame for its failure to deliver its promises and fulfil the expectations it had raised. A single result on a Russian machine was enough to rush head over heels into tokamaks, completely reverse direction and throw overboard most of what had so far been learned. The saying 'one swallow does not make a summer' was wasted on Hirsch and his mates.

There can be no doubt though that from a physics point of view TFTR was a successful experiment, but it was also clear to everybody that, in spite of Hirsch's pronouncements, commercial nuclear fusion power was still very far from being in sight, and no amount of funding could have remedied that. Every experiment teaches you something, solves some problem or perhaps even several problems, but the question is whether TFTR has solved any of the problems that needed solving for bringing controlled fusion closer to its realisation.

8.5 Joint European Torus (JET)

- **Introduction**

In Chap. 6 we saw that European nuclear fusion research was coordinated by Euratom, establishing a network of institutions in various countries of the then European Community. The fusion research programme of the Community was adopted for periods not exceeding five years, as part of a long-term cooperative project embracing all the work carried out in nuclear fusion. It aims to achieve in due course the joint construction of fusion power-producing prototype reactors, with a view to industrial production and marketing.

After the publication of the Russian results with the T-3 tokamak, the construction of tokamaks was well under way in many countries, also in Europe (see Table 4.2). First discussions on constructing a big tokamak in Europe started in the summer of 1970 in the context of arrangements for the UK fusion programme to become part of the Euratom programme and culminated in the recommendation that such a device should be built. Britain joined the European Communities in 1972 and their expertise was added to the European pool. The Americans had just given approval for PLT, which began construction during 1971 and came into operation in 1975. It was the biggest device so far (with the Soviet T-10), so the Europeans set their sight on an even bigger machine that would go beyond 1 MA for the plasma current. The cost of such a machine would be beyond the means of any individual country and a European collaboration was proposed.

The objective of the Joint European Torus (JET) in the European programme, as stated in its progress reports, was held somewhat vague by stating that its essential objective is to obtain and study plasma in conditions and with dimensions approaching those needed in a fusion reactor. This is further worked out, but still not made very specific, by elaborating that the realization of this objective involves four main, rather general and obvious areas of work, like investigating plasma processes and scaling laws, examining plasma-wall interactions and impurity influxes, demonstrating effective heating techniques, and studying alpha-particle production and confinement. There was no mention of net production of fusion power or of reaching breakeven or ignition, nor how close to reactor conditions the results had to be, although some fusion reactions in a tritium-deuterium plasma were envisaged. (See JET Joint Undertaking Progress Reports.) It was all sufficiently vague for it to be declared a success whatever the outcome would be. From the design report, a document of close to 700 pages published in 1976 (JET Project 1976), it is clear though that the aim was breakeven.

The vagueness in JET's objectives is not all that surprising as very little was actually known about the confinement properties of the extremely hot plasmas that would be required and it was necessary to discover the rules governing confinement in plasmas closer to reactor conditions. Moreover, there were other equally or even more important uncertainties. Perhaps the most fundamental one was the question of the so-called *beta* limit (see Chap. 4). Another subject surrounded by mystery was "disruptions". It was known that increasing the plasma density too far or, for a given toroidal magnetic field, increasing the plasma current too much would cause a fatal disruption of the plasma. Since for a high thermonuclear reaction rate high density was to be desired, and larger currents were expected to improve confinement, the disruption problem was crucial, but unfortunately not understood. In spite of being faced with these uncertainties the JET team had to come up with a design (Wesson 1999, p. 21–22).

- **Design**

The UK offered to host the design team at Culham (near Oxford), although it was not yet, but would soon become a member of the European Community. The offer was accepted and the multinational 'Design Team of the JET Joint Undertaking' started work in September 1973. I will describe some of the problems with the design for JET in some detail, as it shows which mountains had to be climbed in designing a tokamak of such gigantic size.

The largest tokamak actually running at that time had a plasma current of 400 kA and the largest in construction would have about 1 MA (PLT), so the original specification of 3.8 MA for JET's plasma current, with the possibility of extension to 4.8 MA was very ambitious (and higher even than the design value for TFTR). With some adaptations it later even managed to reach 7 MA, bearing witness to its excellent design. This was in stark contrast to many earlier fusion experiments that had failed to reach their design values (Braams and Stott 2002, p. 204).

One of the key decisions in the design was the geometry. Should the plasma vessel be circular or elongated, and what should be the aspect ratio, R/a? For the toroidal field coils a D-shaped profile was chosen as in that configuration they experience the lowest mechanical stress, exerted by the confining magnetic fields which push the coils towards the central column of the tokamak. Major structural reinforcement would be required to support the toroidal field coils against these forces if a circular profile was insisted on, which would add considerably to the cost. So, the coils were allowed to be moulded by the magnetic forces, to find their own equilibrium. This naturally resulted in D-shaped coils and consequently a D-shaped vacuum vessel inside them, which was 60% taller than it was wide. It implied that both a circular and non-circular (elongated) plasma was possible. From earlier calculations by Artsimovich and Shafranov it was known that a D-shaped plasma might give better performance as it would give a higher *beta* (see Chap. 4). The Russians were still testing these ideas and there was little proof yet, but it was believed that a D-shape would allow JET to go beyond the 3 MA for the plasma current. The plasma current was considered to be the main element defining plasma performance and alpha-particle confinement. Therefore, JET was designed to have a D shaped plasma cross-section which allowed, at the elongation chosen ($b/a = 1.7$), a 1.7-fold increase in plasma current over that of a circular plasma. If this choice proved to be wrong, the plasma could still be forced to be circular and the initially planned plasma current could still be achieved.

Apart from a bigger volume and increased current, a more important aspect of the elongated shape, although not yet known at the design stage, was that it would later allow for the installation of a divertor.

The aspect ratio was decided on the basis of minimising costs. The optimum value of R/a was found to be between 2 and 3, and the value chosen was 2.4. These values would make the volume of JET more than 100 m^3, a factor 100 bigger than the largest European tokamak so far, the French TFR. JET implied a gigantic step forwards, at any rate in size.

With the values of the parameters given above we can now see how the physical size of JET was determined. From earlier experiments it appeared that a safety factor q of about 3 was needed to provide an adequate margin of stability[8]. In Chap. 4 we have seen that q can be expressed as $q = (B_T/B_{pol})(a/R)$. With the geometric ratios given above, and a toroidal field of 3 T, this requirement on q means that the poloidal magnetic field, B_{pol}, should not be greater than 0.5 T.[9] With the given geometry, and the value $B_{pol} = 0.5$ T, a current of 4 MA requires a minor radius of just over 1 m. With aspect ratio $R/a = 2.4$ and elongation $b/a = 1.7$, the conveniently rounded dimensions $R = 3$ m and $b = 2$ m give a satisfactory minor radius of $a = 1.25$ m, and these were the plasma dimensions chosen for the JET design. The exact figures can be found in Table 8.3. (Wesson 1999, p. 21–23.)

---

[8] A condition for stability is $q > 1$ everywhere; in practice q at the edge of a large aspect ratio ($R/a \gg 1$) plasma should be 3 in order to fulfil this condition.

[9] The size of the plasma then follows from Ampère's law, which relates the integrated magnetic field around a closed loop to the electric current passing through the loop, so here it relates the total

Table 8.3 Basic design values of JET and actually achieved values, where different

	Value foreseen in design report	Achieved value (where different) (Rebut 2018, p. 473)
Vacuum vessel major radius R	2.96 m	
Vacuum vessel minor radius a (horizontal)	1.25 m	0.96 (divertor version)
Vacuum vessel minor radius b (vertical)	2.10 m	
Aspect ratio R/a	2.37	
Elongation b/a	1.7	1.8
Plasma volume (m^3)	100	100
Toroidal field on axis	3.4 tesla	4.0 tesla
Plasma current (limiter mode)	3.2 MA	
Plasma current (in D-shapeshape)	4.8 MA	7 MA
Duration (in seconds) of maximum plasma current	10 s	60 s
NBI heating power	10 MW, 25 possible	24 MW
Ion cyclotron power	0	32 MW
Lower hybrid power	possible	12 MW

The next problem that had to be addressed was the heating. At the temperatures required for thermonuclear plasmas, ohmic heating in JET would be negligible (heating the plasma only to 2–3 keV where at least 10 keV were needed). So, some of the other heating schemes, briefly discussed before in Chap. 2, should be used here. But little experience had so far been gained with these methods, apart from the injection of high-energy beams, which already was an established technique, as was radio-frequency heating (RF heating). A further method which was seriously considered by the design team was compressional heating. Just as a gas can be heated by compressing it, so can a plasma.

How much heating would be needed to reach interesting temperatures could not be answered. The fundamental uncertainty in this respect was the energy confinement. In TFR the energy confinement time was 20 ms at a temperature of about 1 keV. How the confinement time depends on the size of the plasma and on the magnitude of the current and applied magnetic field was however unknown. Confinement would also depend on the density and temperature of the plasma, and these, of course, would depend on the confinement itself.

A possible goal for JET was to achieve breakeven conditions, in which the generated thermonuclear power would be equal to the power supplied for heating the

---

plasma current to B_{pol} and the minor radius. For a given plasma current, a too small plasma gives too high B_{pol} and a too small safety factor.

plasma. External heating enhances the turbulence in the plasma resulting in the temperature increasing proportionally to the square root of the heating power. The design team decided pragmatically to begin with 3 MW of heating and to increase this in stages to 10 and 25 MW. A plasma current of about 3 MA would give a plasma confinement time which, for the most optimistic theory, allowed a margin of about a factor of 4 for reaching thermonuclear conditions. Furthermore, such a plasma current also ensures that many of the 3.5 MeV alpha particles generated in fusion reactions will remain within the plasma while they give up their energy. Table 8.3 shows that the eventual additional heating amounted to 68 MW, using neutral beam injection, ion cyclotron heating and lower hybrid heating systems.

A further problem was the stability of the plasma, especially the magnetohydro-dynamic stability, whereby plasma is treated as a fluid. The theories then available were hardly relevant to real tokamak plasmas, so studies were commissioned to try to improve the situation. This resulted in the development of the first numerical code for calculating stability in a fully toroidal geometry, making it possible to explore the stability of any proposed JET plasma. One of the instabilities was due to the choice of a vertically elongated plasma for JET. Without stabilisation this would lead to an extremely fast instability in the vertical direction. However, the simulations demon-strated that the growth rate of the instability was slow enough for it to be counteracted by using additional magnets and an electronic feedback system.

Figure 8.4. shows a schematic cross section of the JET design with the basic elements, like transformer core, vacuum vessel and magnet coils, and the dimensions of the device, while Fig. 8.5 shows a more dressed-up cut-away model of the tokamak. The size of the device is apparent from the figure of a man drawn in the picture.

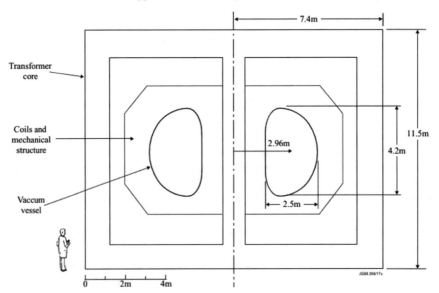

Fig. 8.4 Cross section showing the basic elements and dimensions of the JET design (*from* Wesson 1999)

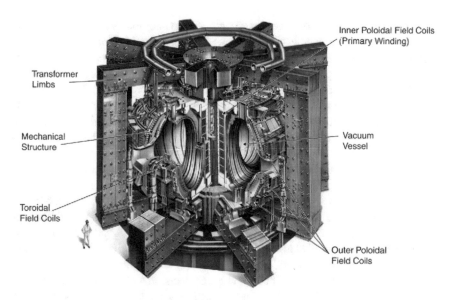

Fig. 8.5 Cutaway model of the JET tokamak. Heating systems and the many diagnostic systems are not shown in this view (*from* Rebut 2018, p. 465)

- **Politics** (Clery 2013, p. 143ff.; Herman 1990, p. 120ff.)

When the proposal for the design was presented in September 1975, there was considerable debate about some of the choices made. The project as presented by the design team was very different from what most laboratories had in mind. Eventually most was agreed with and the final result was published in 1976 in the bulky design report (JET Project 1976).

In the meantime, the political bickering about where the experiment should be sited and what kind of legal form it should take had already started. It would bring the whole project to the brink of disaster. JET was a high-profile international project and many EU nations were keen to have it in their country. Initially there were five nations vying for the prize: Belgium, Italy, France, Germany, which offered two sites, one of which was its plasma physics laboratory in Garching near Munich, and Britain, which naturally wanted to have it at Culham. The European Council, consisting of the President of the European Commission and the heads of state or government of the EU member states,[10] had to take a decision about where JET was to be sited. It met several times late in 1975 and in 1976, but in spite of hours long debates it was unable to reach a decision. The Research ministers and Foreign ministers of the EU countries also had a say in the decision, which was made the more complicated as action on

---

[10]Nowadays the President of the European Council would have to be added to this. From 1975 to 2009 this was not a separate function but held by the head of state or government of the member state holding the semi-annually rotating presidency of the Council.

major questions like this one was taken only after unanimous votes by the various councils, which was much easier than nowadays, one would think, for at the time the European Community had still only nine members. The European Commission was asked to make a recommendation and opted for Ispra, where since 1961 its Joint Research Centre had been sited. This choice however was vetoed by Britain, France and Germany. For two years they were locked in debate, during which time half of the design team left, moved back to their native countries or accepted jobs elsewhere and the JET project came within a hair's breadth of its demise. Pleas to various government leaders were ineffective in moving the JET decision any closer, and this while no one had argued against building JET. Everyone wanted it to go ahead, but only after the pie had been sliced properly and to everyone's satisfaction. British prime-minister James Callaghan got personally involved and started to lobby the German chancellor Helmut Schmidt, the French president Valery Giscard D'Estaing and everybody else who had a say in the matter. He even threatened that, if the European Community failed to build JET, the UK would make its own arrangements and build it at Culham anyway, perhaps with help from other countries (Iran, for instance, showed an interest). It did not help; no decision was taken in 1976. By the summer of 1977, after a lot of talk, it was decided to favour sites with fusion expertise as compared to nuclear expertise, which whittled the competition down to a choice between Culham in the UK and Garching in Germany, but further progress seemed impossible.

Then help, although uncalled for, came from an unexpected quarter. In October 1977 a Lufthansa plane was hijacked over the Mediterranean by Palestinian terrorists. The objective of the hijacking was to secure the release of imprisoned leaders of the German Rote Armee Fraktion (Red Army Faction, also known as the Baader-Meinhof Group). After several intermediate stops the plane ended up in Mogadishu in Somalia. Five days of negotiations with the terrorists followed after which the plane was stormed by German commandos who shot the four terrorists and freed all passengers unharmed.[11] The German commandos had been accompanied by two members of the British Special Air Services (SAS) who blinded the terrorists for six seconds with special magnesium-flash grenades allowing the Germans to storm the plane successfully. So, when Callaghan visited Schmidt one day after the successful ending of the hijacking, he was emotionally greeted and profusely thanked by Schmidt. In the generous atmosphere of the meetings that followed the two leaders were able to resolve some of their nations' differences over European Community matters, with Schmidt acquiescing to the siting of JET at Culham. In return it was implied that a German would become director of the project. The Council of Minsters met just a week later, agreed to a majority vote and officially selected Culham as the site for JET. Four years after the design team had started its work, the issue had finally been settled, whereby the last two years had been spent in a frightful deadlock. The delay had put them firmly in second place to TFTR, whose ground breaking ceremony had been held in October 1977, a few days after the end of the hijack of the Lufthansa

---

[11]Earlier at Aden International Airport, one of the intermediate stops, the pilot Jürgen Schumann had been killed by the terrorists.

plane. One wonders what would have happened with fusion research if it had become necessary to jettison the JET project because of the failure of European leaders to agree on a site.

The legal status of JET also generated deep controversies. Three considerably different solutions were put forward. The Culham laboratory preferred integrating JET into a national laboratory (i.e. their own if it were awarded the siting) that would take responsibility towards the other partners. This was also the situation with TFTR which was part of the Princeton Plasma Physics Laboratory. The other two solutions differed by the control the European Commission would have: one with direct control by the Commission and the other an independent organisation without interference from the Commission. After lengthy debate a final compromise was reached by making JET an autonomous organisation as defined in the Rome Treaty[12], a so called "Joint Undertaking", serving the European Community interests. A JET Council, with representatives of the Commission and the various Associations, or of the states themselves, would be running JET (Rebut 2018, p. 466). No large contributions towards the project financing were required from the Associations, as about 80% of JET was financed by the Commission from European Community funds.

In line with the gentlemen's agreement between Callaghan and Schmidt, a German, Hans-Otto Wüster (1927–1985), a deputy director of CERN at the time, was chosen as JET's first director, bypassing the Frenchman Paul-Henri Rebut (b. 1935) who had headed the design team and had wanted the job for himself, but had to be satisfied with being Wüster's deputy. When Wüster died prematurely in 1985, Rebut as yet became director of JET (until 1992).

- **Construction** (Wesson 1999, p. 29–36)

Finally, construction could start and on 18 May 1979 the foundation stone of the Torus Hall was laid. The participation in the project had by now been extended to 11 nations with Sweden and Switzerland, which had concluded association agreements with Euratom, also becoming partners, while Greece joined in 1983. Construction of JET itself began immediately after the completion of the Torus Hall. The Torus Hall had 3 m thick walls to stop the 14 MeV neutrons freed by the fusion reactions and their secondary radiation. Figure 8.5 presents a cutaway view of the tokamak without the heating and diagnostic systems. The latter are shown in the drawing of Fig. 8.6.

The figure shows the layout of the device with its major components. The inner-most element is the vacuum vessel, which had to hold a vacuum in which the pressure was less than one millionth of atmospheric pressure. This meant that it had to be able to bear the force of atmospheric pressure over the whole of its outside surface, 10 tonnes per square metre over an area of 200 square metres. The magnetic field coils

---

[12]This can now be found in Article 187 of the Treaty on the Functioning of the European Union. The members of these Joint Undertakings are typically the European Union (represented by the European Commission) and industry-led association(s), as well as other partners. Joint Undertakings adopt their own research agenda and award funding mainly on the basis of open calls for proposals.

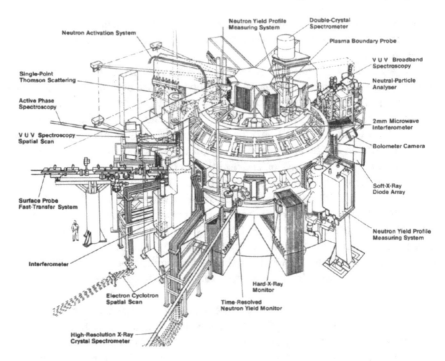

Fig. 8.6 Layout of the diagnostics of the JET device (*from* Rebut (1985))

for producing the toroidal magnetic field consisted of 32 D-shaped coils enclosing the vacuum vessel. Each coil was wound with copper wire and weighed 12 tonnes. The coils were to carry currents for several tens of seconds, and consequently had to be provided with a cooling system, for which water was used. The magnetic field exerts a tensile force on each coil, i.e. a force acting on the material in opposite directions trying to stretch it, of up to 600 tonnes, to be borne by the tensile strength of the copper. The total force on each coil would be almost 2000 tonnes, directed towards the major axis of the torus. A further force arises from the interaction of the currents in the coils with the poloidal magnetic field. The current in the toroidal field coils crosses the vertical component of the poloidal field in opposite directions in the upper and lower halves. This produces a twisting force which, in the JET design, is borne by an outer mechanical structure.

The poloidal field coils are horizontal circular coils. If these coils were placed inside the toroidal field coils, the two sets of coils would be linked, which would cause assembly problems. The poloidal field coils are therefore placed outside the toroidal field coils. The main poloidal field coil is the inner coil (inner poloidal field coils in Fig. 8.5) wound around the central solenoid of the transformer, to act as the primary of the transformer. The other six coils are optimally placed to provide control of the plasma shape and position. The largest of the coils is 11 metres in diameter.

The massive structure of the laminated iron transformer core, weighing 2,600 tonnes, dominates the appearance of JET with its 8 limbs (two of which have been cut away in Fig. 8.5) enveloping the other components. The overall dimensions of the device are about 15 m in diameter and 12 m in height.

Electric power supplied the currents in both the toroidal and poloidal field coils, a similar power being required for each. JET was designed to allow a pulse repetition rate of once every 15 min, each pulse requiring a total power of up to 800 MW—the output of a medium sized power station.

As for TFTR, the plasma heating systems were a major issue. And similar to TFTR, it was decided to provide the main heating by neutral beam injection and ion cyclotron resonance heating, with the further possibility of lower hybrid resonance heating (LHRH). The power could be increased over time by increasing the number of injectors and antennae. The plan was for an initial installation of a few megawatts in each system, as recalled above when discussing the design, to be increased later to 25 MW of neutral beams and an ion cyclotron power of around 15 MW. The RF waves are emitted into the plasma by antennae placed inside the vacuum vessel.

Figure 8.6 shows the layout of the diagnostic systems the JET device was equipped with to record the basic quantities characterising the plasma and other data made available when the device was in operation. Some of this data was required to control the plasma and the auxiliary systems, or to diagnose the plasma behaviour. It was originally envisaged that a hundred thousand readings per pulse would have to be recorded, but ultimately it turned out to be more than a hundred million readings per pulse (a factor 1000 more). The diagnostic systems for providing the information on the plasma had to be integrated into the complex JET structure, and a consistent reliability had to be achieved even when subject to high levels of radiation. Small coils placed around the plasma would detect changes in the magnetic field and determine the plasma current. Larger loops encircling the plasma toroidally would give the voltage around the plasma.

Diagnosis of the plasma itself also involved measuring the emission of varies forms of radiation. The rapid cyclotron motion of the electrons in the magnetic field causes them to emit radiation, and by measuring this electron cyclotron emission the electron temperature can be determined. The full power radiated from the plasma in chosen directions was to be detected by the temperature rise produced in a set of bolometers[13]. Neutron detectors were installed to measure the thermonuclear reaction rate when deuterium and, later, D-T plasmas were produced. The ion temperature would be measured from the small number of neutral particles in the plasma, and by analysing those which escape. These neutral particles were formed by the neutralisation of plasma ions, and their energy, which is characteristic of the ion temperature, can be measured by external detectors. The plasma also emits soft X-rays, which can be followed with very high time resolution. The ultra-violet radiation from the plasma

---

[13] A bolometer is a device for measuring the power of incident electromagnetic radiation via the heating of a material with a temperature-dependent electrical resistance.

Fig. 8.7 View of JET after completion (*from* Rebut 2018, p. 472)

was to be analysed to extract information on the impurity content of the plasma. The average density of the plasma is measured by interferometry[14].

The construction of JET was completed on time, providing Rebut and his colleagues from the design team with the first reward for their efforts which had begun ten years earlier. Figure 8.7 shows a photograph of the completed device. The detail in the photograph gives some indication of the complexity of the construction – in contrast to the simplified drawings in the figures above.

The construction was carried out without any overrun of the budget defined in the original design report and completed in the time of 5 years as foreseen. The total construction cost amounted to 438 million in 2014 US dollars (so considerably less than TFTR whose bill eventually reached 1 billion (more expensive) dollars)[15].

---

[14]Interferometry involves a number of techniques in which waves, usually electromagnetic waves, are superimposed, causing interference, which is used to extract information.

[15]Comparison is however made difficult by the fact that the European project is accounted for in the nebulous Unit of Account (UC), a basket of European currencies, in 1979 changed into the European Currency Unit (ECU) and finally in 1999 into the euro. The total construction cost is estimated to have been about 200 million ECU. The design report states a total construction cost including commissioning, buildings and staff of 135 million UC at March 1975 prices (JET Project 1976, p. 49). The 1994 JET Annual Report states that the construction phase of the project, from

- **Experimental results and upgrade**[16]

Formally, the 'Operation Phase' of the JET Project started on 1 June 1983, following completion and commissioning of the machine in its basic performance configuration. JET's experimental programme started on 23 June 1983, when plasma was successfully produced at the first attempt. The plasma current soon reached 1 MA and routine 3 MA operation for studies of ohmic plasmas (i.e. without additional heating) was established within half a year. These studies took most of the period from 1984 to 1986. The first additional heating system, ICRH, came into operation in early 1985 and the first NBI system in early 1986.

In the first few years most experiments were carried out first with hydrogen, then with deuterium, with both nearly circular ($b/a \approx 1.2$) and elongated plasmas ($b/a = 1.7$), to test the machine and optimize the plasma. Disruptions were suffered (in total 2309 over the last decade of JET operations); instabilities occurred that had to be stabilized, e.g. so-called sawtooth oscillations, tearing modes and ELMs. The events that lead to a disruption are often a complex combination of several destabilizing factors. Disruptions in JET dropped markedly in later experimental campaigns to only 3.4% in 2007, from a high of 27% in 1992. Technical (e.g. insufficient impurity or density control) and physics problems (e.g. MHD instabilities) but also operator (and human) mistakes were found to be common sources of disruptions (De Vries et al. 2011).

Impurities from the limiters (carbon) and from the vessel walls (iron, chromium and nickel) were studied. The vessel walls were also the source of oxygen impurities. This caused 70–100% of the input power to be radiated off. Covering the vessel wall with a layer of lower-Z material (beryllium in 1989) reduced the radiated power to 40%, but impurity levels remained a problem.

Pulse lengths of up to 20s were achieved and confinement times of up to 0.8 s, with the highest values for deuterium plasmas, more than twice the values obtained with TFTR, and found to increase with density. High electron temperatures up to 5 keV and ion temperatures up to 3 keV were reached. In general JET behaved in a way similar to smaller tokamaks. As expected, the size advantage was seen in the record energy confinement time (Rebut et al. 1985).

As we have seen, it was discovered at other tokamaks that confinement times went down when the heating with neutral beam injection was increased and the plasma became hotter, the opposite from what had been expected. A solution to this had been discovered at ASDEX in Munich (see Chap. 6). When increasing the heating slightly, the plasma spontaneously jumped from the lower confinement of L-mode into the improved confinement of H-mode. The H-mode was not understood but, in the magnetic field geometry of ASDEX, seemed to depend on the existence of a separatrix (last closed flux surface), which separates closed magnetic surfaces from open surfaces. Its purpose was to divert plasma escaping from the main plasma

---

1978 to 1983, was completed successfully within the scheduled period and within 8% of projected cost of 184.6 million ECU at January 1977 values.

[16]One of the papers I benefited from in writing this section is Keilhacker et al. 2001.

into a different chamber, the divertor (see Chaps. 4 and 6), before it could do any harm. JET did not have such a divertor, but the question was whether it could still achieve H-mode confinement by adjusting the currents in the external control coils and make the required change in the magnetic geometry. It turned out to be possible (thanks to its D-shaped plasma), but only just, to create a modified geometry with a separatrix and obtain a H-mode plasma. In 1988, halfway through its experimental programme, during such H-mode operation the fusion triple product reached 3 × 10^{20} m^{-3} keV s at temperatures exceeding 5 keV; and a plasma current up to 7 MA (for 2 s) was achieved. This value of the triple product is only a factor of 10 lower than the Lawson criterion stated in Chap. 2. The data showed (by extrapolation as no tritium had yet been used) that a plasma current of around 30 MA would be required for ignition of a D-T plasma. With lower values of the current, the risk of not achieving ignition would be large. A high current value of 30 MA would be possible, but extremely costly. A new device would have to be constructed, which according to a very preliminary estimate would cost more than ten times the cost of JET (Rebut and Lallia 1989). The amazing aspect here is that just halfway through the experimental programme of a brand new device, which had just been operating for a few years, and even before a D-T operation had been started, a new device, which still would be just a further experiment, hence not a real reactor, was already being proposed as JET would not be able to get close enough to reactor conditions.

Both TFTR and JET had originally been designed only with limiters, although the possibility of a divertor in JET had been left open in the original design report, not for achieving H-mode of course, but as a means of impurity control. On TFTR, with its circular cross section, no divertor could be installed and it had to concentrate on enhancing confinement by other means, but JET did allow for the installation of a divertor without any major modifications and this was soon done. Figure 8.8 shows cross sections of the plasma with the divertor, scrape-off-layer etc. The working has already been explained in Chap. 4.

But, still before the installation of a divertor, in November 1991 the world's first D-T fusion experiment was carried out with JET, not yet with 50–50 D-T, but one pulse with 10% tritium and one with 20%, using just five milligrams of tritium that were introduced into the torus by neutral beam injection. The restriction to only two pulses had to do with the limitation of nuclear activation, i.e. the induction of radioactivity in the vacuum vessel. It produced a peak fusion power of 1.7 MW and released 2 MJ of fusion energy. It ended after about 1 s with a bang coinciding with a strong influx of carbon impurities, dubbed 'carbon bloom'.

During the period 1992/1993 a divertor was installed to handle higher levels of exhaust power. It marked the beginning of an extensive series of divertor tokamak studies. The control of the carbon bloom and the improvement of H-mode performance were among the notable achievements.

From July-November 1997 JET conducted a tritium campaign, including a three-month campaign of experiments with 50–50 D-T fuel mixtures. In total 35 grams of tritium were used. The improved plasmas with the new divertor resulted in the release of 14 MJ of fusion energy and the production of 16 MW of peak fusion power, lasting for a few seconds, considerably more than achieved by TFTR. The amount

a) D shaped Plasma section   **b) Scrape Off Layer**

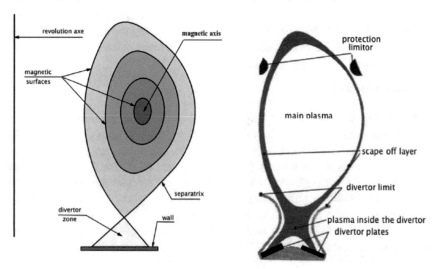

Fig. 8.8 D-shaped plasma with separatrix and Scrape-Off Layer (SOL); (a) represents a plasma cross section, delimitated by a separatrix and showing a few magnetic surfaces. The plasma is D-shaped. The separatrix allows the installation of a divertor; (b) the SOL is the edge of the plasma that touches external materials, the divertor plates are shown where the residual plasma power is evacuated. In a reactor the effective thickness of the SOL could be less than one centimetre (*from* Rebut 2018, p. 470)

of power that had actually gone into heating the plasma was 24 MW, which implied a Q value of 0.67, the highest achieved so far, but still a considerable distance from scientific breakeven. In total JET used 700 million watts of *electricity* to produce the shot that resulted in fusion particles with 16 million watts of *thermal* power, which means that JET lost a staggering 98% of the power it consumed. Only 24 million watts of these 700 million watts actually went into the plasma, and of this power only 16 million watts came out as fusion power. If this were converted into electricity, a considerable part of it, probably 2/3, would further be lost. Looked at in this way the 700 million watts of *electricity* used by JET correspond to a *thermal* power of about 2100 million watts, which makes the gain of 16 million watts in *thermal* power completely insignificant. JET was and is an enormous squanderer of energy as is also apparent from the fact that when it is running plasma shots (typically two per hour) it draws up to 8% of the electricity supplied by the entire UK national grid. That is why fusion experiments on JET are not allowed during the day or at other times when electricity demand is high (Claessens 2019, p. 26).

Tritium is very expensive and both TFTR and JET relied on what was available on the market, and it was used very sparingly in these machines. ITER too will rely on the global supply of tritium, but no sufficient external source of tritium exists for fusion energy development beyond ITER. In a real reactor, if we ever get to that stage, tritium will have to be bred by the reactor itself. A layer of lithium will be part of a structural component called a blanket. This blanket will surround the vacuum vessel and tritium can be produced in it when neutrons escaping the plasma interact with the lithium. This concept of 'breeding' tritium during the fusion reaction is of vital importance for the future needs of a large-scale fusion power plant. ITER must lay the basis for this and we will discuss it in Chap. 10.

- **Conclusion on JET**

In this section we have dwelt at length on JET, its design, construction and exper-imental results as it has been the most important device to date. It can rightly be seen as the precursor of ITER, a test-bed for ITER, and more than 30 years after its construction it is still in operation. In many ways the designs for ITER and JET are very similar. JET is the physics model for ITER and, without the results achieved at JET, ITER would not have been possible. Specific work for ITER has been a prominent part of the JET programme in its final years. The JET Joint Undertaking was established in 1978 for an initial twelve-year period, subsequently extended and concluded at the end of 1999. The JET apparatus itself was, however, still in excellent condition and, with its size and its D-T capability, remained and still is the most powerful fusion device in the world. JET operation continued under the Euro-pean Fusion Development Agreement (EFDA), with experiments being carried out by visiting scientists from the Associated European Fusion Laboratories, and from 2013 under its successor EUROfusion, which is a consortium of national fusion research institutes located in the European Union, Switzerland and Ukraine. All EU countries, except Luxemburg and Malta have national fusion research organisations.

Now that TFTR has been shut down early, JET remains the only tokamak in the world able to work with D-T mixtures. It has a D-shaped plasma with a separatrix and divertor, but its shape can be widely varied. In this configuration, it is possible to reach a 6 MA current that can be maintained for over 10 s; in fact, 7 MA plasma currents have been produced in a slightly different device configuration. These currents suffice to obtain energy confinement times of about a second or even longer. It is designed to work with tritium and intense 14 MeV neutron fluxes, as from the very beginning JET has been equipped with heavy remote handling facilities, like a nuclear machine, which is extremely important for its maintenance (Rebut 2018, p. 472).

JET has not run with D-T mixtures since the groundbreaking campaign in late 1997, when plasmas were obtained with D:T ratios all the way from 99:1 to 10:90. There nevertheless remains the legacy of this campaign in the form of more than a gram of tritium residing in the vessel surfaces, mainly on the inboard side of the divertor structure. This residual tritium presence has dictated that strict procedures of operation and maintenance have always to be followed, due to the radioactive nature of tritium. In 2021 JET will be re-introducing tritium into its vacuum vessel for the first time since 1997, apart from some experiments with small amounts of tritium in 2003 (see EFDA-JET Bulletin, July 2003).

In October 2009, a 15-month shutdown period was started for rebuilding many parts of JET to test concepts from the ITER design, in particular an "ITER-Like Wall" of the vacuum vessel. The first experimental campaign after the installation of this wall is showing stable plasmas and good impurity control. (See e.g. Litaudon et al. 2017.) These upgrades to JET have effectively turned JET into a miniature version of ITER. JET is now as close as any device can get to ITER's operating conditions and until ITER starts operations with tritium (2035 at the earliest), JET experiments with tritium offer the only opportunity for fusion scientists to investigate physics relevant to burning plasmas for ITER and the prototype fusion power plant DEMO.[17]

In July 2014, the European Commission signed a contract worth €283 million for another 5-year extension, so more advanced higher energy research can be performed at JET. Under this plan, a new run of D-T experiments, a "dress rehearsal" for ITER, is planned for 2021.

The British plans for leaving the European Union (Brexit) at first threatened to throw the plans for JET in doubt, as the UK also plans to leave Euratom, which provides the funding for JET. In March 2019 the UK Government and European Commission signed a contract extension for JET, which guaranteed JET operations until the end of 2020. After Brexit Britain will remain part of ITER as part of a Nuclear Cooperation Agreement signed between the UK and Euratom. This agreement also ensures the continued operation of JET.

8.6 JT-60

JT-60, with JT standing for Jaeri Tokamak (although according to others it stands for the rather unimaginative Japan Tokamak or Japan Torus), is the culmination in a long list of Japanese tokamaks constructed since the early 1970s (see Chap. 7). Like the JFT tokamaks, JT-60 is located at JAERI (Japan Atomic Energy Research Institute) in Naka about 120 km north of Tokyo. A few of the earlier tokamaks are still in operation, but JT-60, including its successors JT-60U and JT-60SA, is without doubt the showpiece of the Japanese fusion effort so far. The first plans for building a large tokamak in Japan, essentially skipping the stage of intermediate-size tokamaks as were built in other countries, with the intention to reach conditions in deuterium plasmas that would be equivalent to breakeven in D-T plasmas date from 1975. JT-60 was not intended to operate with tritium, so here we encounter an example of the rather odd concept of extrapolated breakeven, discussed above in this chapter. The rationale of working only with deuterium plasma is that all aspects of plasma physics, except those involving the fusion event itself, can be investigated in such plasmas without producing a high level of neutron activation of the facility, i.e. induced radioactivity in the materials of the facility, especially the vacuum vessel, and without having to work with radioactive tritium. The basic parameters of JT-60 and of its two successors JT-60U and JT-60SA are listed in Table 8.4.

After a seven-year construction period first plasma was obtained in April 1985, lasting about 80 ms. JT-60 only used hydrogen for its plasmas, no deuterium yet. It was the only one of the three big tokamaks that from the outset had been designed with

---

[17]http://www.ccfe.ac.uk/news_detail.aspx?id=486.

Table 8.4 Basic design values of JT-60, JT-60U and JT-60SA; the numbers in brackets denote the energy at the power source; without brackets actually absorbed by the plasma. The JT-60 numbers are from Yoshikawa (1985); the JT-60U ones from Kikuchi (2018) and the JT-60SA numbers from the website

	JT-60	JT-60U	JT-60SA
Vacuum vessel major radius (m)	3.0	3.4	3.16
Vacuum vessel minor radius (m)	1.0	1.0	1.02
Plasma elongation ratio κ	–	1.5	1.95
Plasma volume (m^3)	60	90	140
Toroidal field on axis	4.5	4.2	2.25
Plasma current (MA)	2.1–2.7	6.0	5.5
Pulse length (s)	5–10	28–65	100
NBI heating power (MW)	20	43	34 (100)
Ion cyclotron heating power (MW)	2.5 (6)	10	–
Electron cyclotron heating power (MW)	-	4	7 (30)
Lower hybrid heating power (MW)	7.5 (24)	8-12	–

a divertor. This outer divertor, called a magnetic limiter in some publications, was formed by a set of three poloidal coils inside the vacuum vessel. They were however rather awkwardly placed and took up so much space that the plasma volume was relatively small. It proved effective in controlling impurities, but not so successful in accessing H-mode, which was discovered in 1982 when JT-60 was already in an advanced state of construction. Divertors at the top or bottom of a torus seem to be preferable for achieving H-mode. This was corrected two years later in a first upgrade, whereby a new divertor was installed under the vacuum vessel for H-mode studies. A more extensive upgrade, JT-60U, came online in 1991. The maximum pulse length in JT-60 is 10 s, much longer than the energy confinement time. Long pulse length is one of the characteristic features of JT-60 and its successors (Braams and Stott 2002, p. 209–210).

Like the other two big tokamaks, JT-60 started operation without any additional heating, which however was added in stages and upgraded over a number of years. It was equipped with three additional heating devices: neutral beam injection (NBI), lower hybrid (LH) and ion cyclotron resonance (ICR) heating, providing respectively 20, 24 and 6 MW of power at the power source (total 50 MW), of which respectively 20, 7.5 and 2.5 MW (30 MW in total) is actually used in heating the plasma. Especially lower hybrid heating is apparently very inefficient (Yoshikawa 1985). When the auxiliary NBI heating system came online in 1986 the plasma confinement showed typical L-mode confinement, with the energy confinement time degrading with increase of the auxiliary heating power, as was happening in all tokamaks before H-mode was discovered. They were all inventing the same wheel and stumbling on the same problems.

The first change carried out in 1988 to JT-60 involved the installation of a lower divertor, as mentioned above. The H-mode result obtained with this new divertor was actually disappointing, but a 'serendipitous' phenomenon resembling the TFTR supershot was found. Further changes involved an increase of the toroidal magnetic field and the plasma current, and the installation of a high-speed pellet injector to achieve high central plasma density and performance. With a pellet injector the fuel is replenished by injecting fuel in the form of cryogenic (frozen) pellets into the chamber; such type of injectors has in particular been developed at Oak Ridge National Laboratory since the 1980s. However, the changes made to JT-60 did not really improve results. JT-60 remained far behind JET in its achievements and it was soon decided to go over to a proper upgrade, which resulted in JT-60U. JT-60 was shut down in 1989. The parameter values for the upgraded machine are also given in Table 8.4 (Kikuchi 2018).

Figure 8.9 shows the evolution of the JT-60 configurations and the major results. JT-60SA is not included in the figure since it is not operating yet. JT-60U started from 1991 and used deuterium plasmas (legal requirements had prevented JT-60 from using deuterium). The use of deuterium made it possible to increase the NBI power from 20 to 40 MW.

One of the main design features of JT-60U is high triangularity, a quantity measuring the D-ness of the plasma shape, i.e. how much the shape deviates from an ellipse. An ellipse has triangularity zero (see Chap. 4 for more details).

World records were set in 1996 for ion temperature and for the triple product, which is only a factor of 2 lower than the Lawson criterion. The values are presented in Fig. 8.9. As can also be seen in the figure, a value of $Q = 1.25$ was obtained by extrapolating the actual result with a pure deuterium plasma to a D-T plasma. Although of interest, it remains a theoretical value and it seems that only the Japanese

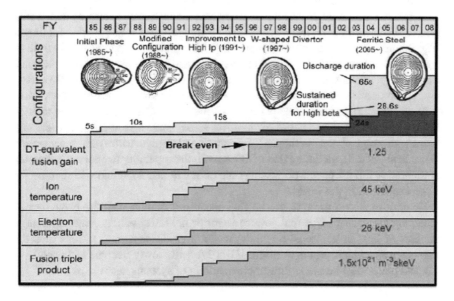

Fig. 8.9 Evolution of configurations and major results of JT-60 (*from* Kikuchi 2018)

are impressed by it. The peculiar thing is that it is not easy to find the actually obtained value of Q in the literature. One would expect that the papers that quote the extrapolated value would also give the value it has been extrapolated from. That is however not the case. Even the original papers do not seem to do this; they do not even report how much fusion power has been produced (See e.g. Fujita et al. 1999). The only thing we know is that the real Q is less than one. Moreover, which is not clear from the way the data are presented in Fig. 8.9, the values of the various record quantities presented in the figure are not obtained simultaneously. For instance, in the regime in which the record value for the triple product was obtained the extrapolated Q value remained below 0.6 (Fujita et al. 1998). One should carefully tread here and cannot just assume any of the values reported at face value; care and suspicion must be your companions. There is a lot of politics involved, due to the fact that certain results have been promised to politicians, who hold the purse strings, and must be presented such that they are suitably impressed, to entice them to further loosen the said strings.

JT-60 and its first upgrade JT-60U operated for 23 years, from 1985 to 2008. JT-60SA, although using, where possible, the infrastructure of JT-60, is essentially a new machine. This final version in the JT-60 series is a fusion experiment designed to support the operation of ITER and to investigate how best to optimise the operation of fusion power plants that are to be built after ITER. It is a joint Japanese-European research and development project (constructed within the framework of the Broader Approach agreement between the EU and Japan), and its construction at Naka has just been completed (April 2020). SA stands for "super advanced", to indicate that the experiment will have superconducting coils and will study advanced modes of plasma operation. Figure 8.10 shows a cutaway view of the tokamak.

JT-60SA is supposed to contribute to the realization of fusion energy by addressing key physics issues for ITER and its DEMO follow-up, if it ever gets that far. It is a fully superconducting tokamak capable of confining high-temperature (100 million degrees) deuterium plasmas, so this machine too will not operate with D-T plasmas. It has a large amount of power available for plasma heating and current drive, from neutral beams and electron cyclotron resonance radio-frequency heating. It will typically operate 100 s pulses once per hour. Its most important novelty in my view is that the machine will be able to explore full steady-state operation. It will operate with a wide range of plasma shapes (elongations and triangularities) and aspect ratios (down to about 2.5), including that of ITER, with the capability to operate in divertor configurations. If all goes well it will come online in late 2020 and will then be the biggest tokamak in the world.

Deuterium is said to be used as a fuel because it mimics well the behaviour of a reacting D-T plasma in a real power reactor or in ITER, without generating large amounts of heat or neutrons. Since the reaction produces some neutrons directly, plus reactions which produce tritium, JT-60SA will slowly become radioactive in use, and plans must be made for remote handling of systems near the plasma.[18]

---

[18]http://www.jt60sa.org/.

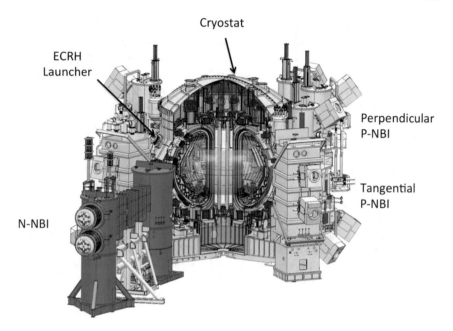

Cryostat

ECRH
Launcher

Perpendicular
P-NBI

Tangential
P-NBI

N-NBI

Fig. 8.10 Cutaway view of the JT-60SA tokamak (*from* jt60sa.org)

Some of the components for the machine are made in Europe. For instance, the D-shaped superconducting toroidal field coils are wound in France and Italy with Niobium-Titanium (NbTi) wires made in Japan. The critical temperature of NbTi, i.e. the temperature at which the superconductivity breaks down, is about 10 °K (10 degrees above absolute zero), which means that they must be kept cooled below this temperature. This will be done with liquid helium ($\sim$4 °K). It is clear that the superconducting magnets must be properly shielded from the vacuum vessel to avoid the magnets from being heated by radiation from the vessel, and for the same purpose must be embedded in a cryostat, which is a vessel that can be evacuated at room temperature. In the cryostat a vacuum will be provided around the cold magnet components to minimise thermal loads. The use of superconducting magnets differs fundamentally from ordinary copper magnets which have to be (water) cooled to prevent overheating and can only provide pulsed operation. With superconducting magnets, it is in principle possible to operate in steady state. One of the objectives of JT-60U was to develop a steady-state advanced tokamak scenario, made possible by the relatively long pulses (up to 65 s) on JT-60U.

The upgraded NBI system for JT-60SA consists both of positive-ion-based NBI (P-NBI) and negative-ion-based NBI (N-NBI). The P-NBI system is modified from that of JT-60U to extend the pulse duration from 10 s to 100 s. Positive-ion based NBI is the traditional method as discussed in the section on Plasma heating in Chap. 2, whereby positive deuterium ions are first accelerated, then neutralised and injected into the plasma, where they are again ionised by collisions with the plasma particles and kept in the plasma by the confining magnetic field. In the case of N-NBI the precursor ions are negatively charged deuterium ions, which are accelerated, neutralised and injected. Negatively charged deuterium ions (whereby a neutral deuterium atom must capture an extra electron) are obviously much more difficult to manufacture than positively charged ions, for which neutral atoms only have to be stripped of their electrons. Impressive progress with negative-ion based NBI systems was made in the 1990s on the Large Helical Device (LHD) (see Chap. 13) and JT-60U, which are to date the only devices that have used N-NBI heating. The power demand for P-NBI at JT-60SA will be 60 MW; the demand for N-NBI will be 40 MW and for ECRF 30 MW. So, the total power demand for auxiliary heating will be 130 MW of which 41 MW (24 P-NBI, 10 N-NBI, and 7 ECRF) will actually be used in heating the plasma (here too a lot of power is lost). A lot of attention is paid to this feature as NBI for ITER is a substantial challenge.

The assembly of the JT-60SA tokamak started in 2013 and was completed in April 2020. Its parameters have been included in Table 8.4. First plasma is planned for September 2020, about two years behind schedule, but it seems that the current Covid-19 pandemic will result in some further delay. All in all it is a very impressive undertaking, the progress of which can be closely followed on the website http://www.jt60sa.org.

8.7   Conclusion

The TFTR story related at the start of this chapter gives the inescapable impression that it just came too early, that it was too hastily embarked on. Before such an experiment should have been considered, more fundamental plasma physics problems, e.g. instabilities, should have been solved.

The decommissioning of TFTR left the US without its major facility. Plans for a successor at Princeton (the Compact Ignition Tokamak (CIT) and the Burning Plasma Experiment (BPX)) failed to win approval, partly since by that time the US was involved in ITER (withdrawing in 1999 and re-joining in 2003) and shifted its domestic focus from reactor development to plasma science. A proposal for an experiment studying plasmas dominated by α-particle heating, the Fusion Ignition Research Experiment (FIRE), was dropped in 2004 and the US involvement in magnetic fusion with conventional tokamaks was almost completely channelled into ITER. No large conventional tokamaks have since been built in the US. At Princeton a spherical tokamak, NSTX, was built instead (see Chap. 11).

The JET result of 1997 discussed above is essentially where we still stand today. In spite of the fact that JET failed to achieve reactor conditions for confinement and has remained far from breakeven, the scientists involved claim without blushing that JET has been an 'outstanding scientific success' (Keilhacker 2001). I always get a little suspicious when words like success have to be qualified by adjectives, like in this case scientific. Does that mean that in other respects it was not a success or a downright failure perhaps? Also, what does an outstanding success mean compared to a 'normal' success? There are hardly any 'normal' successes anymore these days, it seems.

Whether JET was a success or not, now close to 25 years later no further progress has been made; all trust in a future successful outcome of the fusion enterprise is still solely based on the JET result stated above. JT-60 and JT-60U, since they had no tritium operations, have not really added anything very significant, or it must be the development of steady-state scenarios.

The fact that JET did not reach breakeven was due to a variety of effects that had not been seen in previous machines operating at lower densities and pressures (making the scaling laws used rather suspect as they are based on the assumption that no unexpected things happen). Based on its results, and a number of advances in plasma shaping and divertor design, a new tokamak layout emerged, sometimes known as an 'advanced tokamak'. Basic features of an advanced tokamak design are a D-shaped plasma, superconducting magnets, operation in high-confinement mode, the presence of an internal divertor (that flings the heavier elements out of the fuel towards the bottom of the reactor, where a pool of liquid lithium is used as a sort of 'ash' tray), and maximising the non-inductive bootstrap current (which might do away with inducing a current through the plasma). As a reminder, this current through the plasma was always a defining core feature of the tokamak (providing the ohmic heating) and a fundamental difference with the stellarator which did without such a current from the very beginning. This advanced concept forms the basis of ITER.

An advanced tokamak capable of reaching breakeven would have to be very large and very expensive. ITER fits those requirements too, as we will see in a next chapter. No further progress, apart from undoubtedly very impressive technical advances in all kinds of fields and areas related to subsystems of the ITER tokamak, will be made for at least another 10 to possibly 20 years, and then, whatever ITER will bring, it will certainly be more than thirty years after the 1997 JET result before we know more, after which it will again take a similar amount of time before the next step may possibly be made. Mankind will need extraordinary stamina and patience to endure such a long and uncertain wait.

Part III
Intermezzo

Chapter 9
Summary of the World's Efforts so Far and Further Roadmap to Fusion

9.1 Synopsis

With the shutdown of JET an era closed. More than fifty years of trial and error ended with essentially no result. Not everybody feels the same about this first period as can be gathered from the ITER website which proudly calls this early period "60 years of progress". There undoubtedly was progress. When you start at zero, every step forward, however small, is progress. The question is not whether there is progress, but if progress was sufficient, in line with expectations and promises. That is not the case. The big tokamaks were the final step and they have dismally failed to fulfil their promises (and expectations), and they were not the first to do so. Many other promising devices had preceded them. At this point, i.e. the failure of the big tokamaks, it would have been sensible and wise to decide to close the fusion book and conclude that nuclear fusion on Earth, at any rate via the tokamak option, is a dead–end. Too many unknowns and too little knowledge left only two options in my view, either back to the drawing board or complete shutdown. These paths were however not followed. On the contrary, the failures were explained away and the time was considered ripe for a great leap forward: the solution lay in still bigger machines, so up towards ITER and beyond.

Before we continue our journey with a description of these further developments, let us pause for a little while, take stock of how the process has evolved so far and what the future has in store. In this intermezzo a few pivotal events and dates that determined the direction of the search for the 'ultimate energy source' will be reviewed.

The first important event is the 1958 "Atoms for Peace" conference in Geneva, which was held in an atmosphere of great optimism. Nuclear fusion was only a small part of the conference but took centre stage. The whole problem would be solved in a few decades, it was thought, and everybody was invited to take part in its realization. Global harmony and cooperation between otherwise implacable foes. That already should have aroused suspicion. Was it real cooperation or just a play for show, without any sincere intent on either side to be open and transparent? Whatever

© The Author(s), under exclusive license to Springer Nature Switzerland AG 2021
L. J. Reinders, *The Fairy Tale of Nuclear Fusion*,
https://doi.org/10.1007/978-3-030-64344-7_9

the case may have been, at that time nobody had a clear idea where the solution to the fusion problem would lie. In the years before, essentially just three countries, the US, the UK and the Soviet Union, as a by-product of their war-time research on the atomic bomb, had sizable and secret nuclear fusion research programmes. Progress was however very slow, much slower than initially thought. The secrecy did not serve any purpose, it was argued, as there were no direct military applications of nuclear fusion and so the above-mentioned conference, which for that matter was mainly concerned with nuclear fission, served as the vehicle for lifting this secrecy. The conference was accompanied by a large exhibition at which in particular the US and the Soviet Union tried to impress the general public with their achievements, including a glittering model of Spitzer's stellarator. The Soviets published all their research in four fat volumes, which were swiftly translated into English to make them accessible to the rest of the world. It resulted in a first flurry of activities in which, apart from the three original nuclear fusion pioneers, an increasing number of other countries started to take part.

After a decade of preliminary experiments with a bewildering array of devices, including various types of pinches, stellarators, mirror machines and tokamaks, the late 1960s saw the breakthrough of the tokamak when the Soviets, who had invented the tokamak concept in the early 1950s, presented their results with the T-3 tokamak at the 1968 Novosibirsk conference. This was the second pivotal date in the nuclear fusion story. Until then Bohm diffusion had been the great stumbling block and seemed to sound the death knell for the fusion dream. Bohm diffusion was an experimental finding which predicted that the diffusion rate of plasma across the magnetic field was much higher than follows from classical diffusion, namely that it would proceed linearly with temperature and inversely linearly with the strength of the confining magnetic field. The classical model predicted a dependence inversely with the square of the magnetic field and, if true, small increases in the field would result in much longer confinement times. If Bohm diffusion were true, magnetic fields would have to be far too high for confinement to be practical. The main problem at the time was that nobody really understood the behaviour of a plasma at the extreme conditions at which nuclear fusion would occur, a situation that continues for that matter to the present day. The Russians showed that the tokamak could provide a way out of the conundrum. The answer essentially was that the poloidal field, i.e. the field that goes around the torus the short way, should be smaller than the toroidal field, i.e. the field that goes around the torus the long way, or in other words that the current sent through the plasma should not be too large. In this connection the safety factor q was introduced, being the number of times a field line makes a toroidal turn for one poloidal turn. This safety factor must be larger than one, so more often around the torus the long way than the short way, for instabilities to be suppressed (the Kruskal-Shafranov limit).

These results unleashed the tokamak stampede and another tremendous surge of optimism that the dawn of fusion was near. The 1970s was truly a period of euphoria; the tokamak results combined with oil crises created a very favourable atmosphere for increasing public investment in nuclear fusion research, which was seen by many as a panacea for all prevailing energy problems. Support for fusion went hand in hand with energy crises due to hiccups in oil supply for political or other reasons. Not always for the proper reasons and even less backed up by any scientific evidence, scientific breakeven was predicted to be just a few years away. Even Lev Artsimovich, the usually very cautious Soviet chief nuclear scientist, could not resist the temptation to predict breakeven by 1978. A multitude of nations started to build tokamaks or acquire them by other means. The table of conventional tokamaks at www.tokamak. info gives an incomplete list of 185 conventional tokamaks (including spherical tokamaks the number grows to 226).[1] Most of them are no longer in operation of course, but perhaps 50 still are. This entailed an incredible amount of activity.

Optimism was again tempered in the late 1970s when seemingly insolvable problems with heating the plasma became apparent. Ohmic heating, i.e. heating the plasma by the current that is sent through it, has its limits since the resistivity of the plasma decreases when the temperature rises and hence the efficiency of ohmic heating decreases with increasing temperature. Other forms of heating must be found to reach temperatures at which fusion processes can take place. Neutral beam heating and ion or electron cyclotron resonance heating are possible candidates for this. When applying such methods and increasing the heating, however, the unsettling and essentially game-stopping phenomenon occurred that confinement times seemed to get worse. This was resolved in 1982 when the high-confinement mode (H-mode) was 'serendipitously' (i.e. accidentally) discovered on the German ASDEX tokamak: if the heating power applied by using neutral beams was increased beyond a certain critical value, the plasma spontaneously transitioned into a higher confinement state. The discovery of H-mode, which remains to be incompletely understood, has so far been the last pivotal moment in fusion research. It is now waiting for the first real breakeven event. If the machine is big enough, that event is bound to happen (unless nature has some nasty surprise in store). It will be a pivotal event in a completely different sense. It will more be a wake-up moment, the realisation that nuclear fusion will be so expensive that the construction of large numbers of power plants is out of the question.

As we have seen in Chaps. 6 and 7 numerous, even rather small countries ran at one time or another some sort of nuclear fusion programme with a tokamak. Some still do, but as fusion devices have become ever larger and consequently more expensive the tendency over the last 30 years or so has been a tremendous consolidation, while some large countries, notably the US and Russia, have substantially scaled down their domestic programmes. The first consolidation of fusion efforts took place in Europe when almost from the very beginning Euratom, the atomic energy arm of the European Communities, started to coordinate the nuclear research activities of the countries of the European Community. The Joint European Torus or JET, one of the big tokamaks whose construction was started in the 1970s, has so far been the culmination of this consolidated approach. Although JET failed to achieve breakeven, it can be considered a success, not only from a physics point of view, but also politically, as it has been the model for a still larger consolidation, the ITER project,

---

[1] Accessed on 11 September 2019; site last updated on 11 September 2018.

which is currently dominating the scene and will be dealt with in detail in the next chapter.

The American programme, which in the early days was by far the largest, is now mainly limited to its contribution to ITER, while the collapse of the Soviet Union has seen the Russian programme being dwindled down. Russia is still a partner in ITER though. The American magnetic fusion effort (in 2008 dollars) has dropped from a high of almost 900 million dollars per year in the late 1970s to early 1980 to just 300 million dollars in 2010, and this while the size and hence the cost of new devices, like ITER, has steeply increased. The big TFTR tokamak at the Princeton plasma physics laboratory has been the last large machine built in the US. As we will see in a future chapter, there is currently in the US more privately funded activity in fusion research than publicly funded. The decline in public funding for fusion research in the US, where it was always considered more a scientific research topic than a practical energy-generating option, is not that surprising considering that none of the predictions and promises of the fusion community have come true. No real progress was being made. The discovery of H-mode was a success but amounted to not much more than a patch-up. When TFTR failed to achieve breakeven, one would have expected all confidence to evaporate. But, no such thing. Fusion scientists are a cheerful lot and in spite of all the mishaps and lack of success, they still thought that there was not a cloud in the sky and that the future for fusion was bright and full of promise. Illuminating in this respect is Fig. 9.1 from a talk at the 1998 Royal Society Discussion Meeting on fusion power, called 'The approach to an ignited plasma'. It shows that roughly twenty years ago it was expected that ITER would start operations a mere seven years later in 2005 (the current expectation is 2025) and that a DEMO plant would start operating before 2025. As we will see this is no longer the roadmap to fusion. I do not know any field of (scientific) activity where predictions have been so blatantly wrong for so long. From the very first beginning

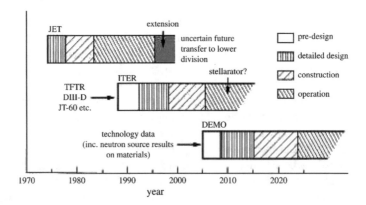

Fig. 9.1 From a talk at the 1998 royal Society discussion meeting, showing ITER operations to begin in 2005 (Bickerton 1999)

to the present day they have been off, not by a few months or years, but by decades, and each time again they manage to get away with it.

From the late 1970s the centre of gravity of nuclear fusion research has been shifting to Europe, and Europe's contribution to nuclear fusion, mainly through JET and ITER, has now soared to 1 billion euro annually, as can be seen from the approval by the European Parliament in January 2019 of 'the indicative Euratom contribution to the Joint Undertaking for the period 2021–2027 and the related supporting expenditure for the same period at EUR 6.07 billion (in current values)'.[2] It is rightly called an 'indicative contribution' as history has shown that estimates of costs in nuclear fusion are as unreliable as the predictions for scientific or any other form of breakeven in fusion. This EU contribution to ITER does not include the contributions of the individual member states to their own institutions within or outside the ITER framework.

Apart from Europe it is currently Asia that is surging ahead in the fusion exercise with China, Japan and South-Korea, all three participants in ITER, heavily banking on nuclear fusion to free their economies yet in this century from the dependence on foreign imports of fossil fuels. Japan has built a tremendous number of tokamaks in the 50 odd years of its research programme culminating in the large JT-60 tokamak, that is currently being upgraded to the first fully superconducting JT-60SA. A drawback of all of its devices is that none of them works with D-T plasmas.

Japan, China and South Korea are already busy designing projects as successors to ITER. So, after the shift from the US/Soviet Union to Europe, in the after ITER-era, or even before, we will in all likelihood see a shift from Europe to Asia. A demo-reactor, if it ever gets that far, will be built in Asia, and most probably in China. The Chinese, as we have noted in Chap. 7, even want to start the construction of their demo plant, the China Fusion Engineering Test Reactor, before ITER is ready, and will probably do so if there are further delays in the ITER construction. It will be interesting to see if after ITER has successfully run its course, the world will again succeed in setting up a collaboration of similar scope.

So, what progress has been made in these last 70 years (roughly from 1950 to 2020)? We have seen that the values of three quantities are essential in achieving fusion: the plasma density, the temperature and the energy confinement time. In Chap. 2 we have discussed the Lawson criterion which states that the triple product of these three quantities must be larger than 3×10^{21} m^{-3} keV s, while we need at the same time a temperature larger than 100 million degrees and a confinement time that is long enough for fusion reactions to take place. The best value for the triple product, while at the same time producing fusion energy, has been achieved at JET and is about one fifth of the value stated above.

Although ignition has not been reached with any of the big tokamaks, they have pushed magnetic confinement further to this ultimate gaol. Over time, results have steadily improved from the T-3 experiment in 1968 to the JET result of 1997. It is curious that 30 years have passed from the T-3 to JET, and that we will have to wait

---

[2] Amendment of Decision 2007/198/Euratom establishing the European Joint Undertaking for ITER and the Development of Fusion Energy and conferring advantages upon it, Com (2018) 445.

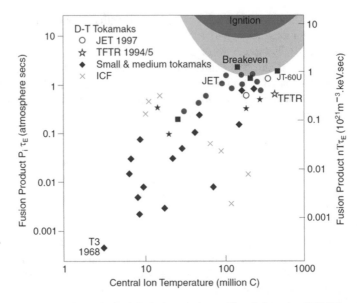

Fig. 9.2 Plot of the values of the triple fusion product achieved since the 1968 T-3 experiment. The ion temperature T is plotted along the horizontal axis and the $nT\tau_E$ fusion product (as well as pressure times confinement time) along the vertical axes (both on a logarithmic scale). The open symbols for JET and TFTR indicate the only results with D-T fuel; the JT-60 results are extrapolated from D-D plasmas. (*From* McCracken and Stott 2013, p. 116; ICF stands for inertial confinement fusion (Chap. 13))

another 30 years before the next result, if at all, will be forthcoming. This is now scheduled to be around 2035 when ITER will start its D-T operations. Progress is certainly not speeding up. Figure 9.2 summarises the progress made so far.

It plots the fusion triple product against the central ion temperature. In these first 30 years the Lawson triple product has increased by more than three orders of magnitude (a factor of 10,000) and is now within a factor of five of the value required for ignition. Temperature and density separately have actually already reached ignition values. One snag is that many of these values have been obtained in D-D plasmas and have been extrapolated to D-T, with only JET and TFTR having performed (rather limited) experiments with tritium. The increase in costs is however much more than a factor of 10,000 and this will eventually break the neck of nuclear fusion as a viable source of energy. To date JET remains the only tokamak in the world able to work with D-T mixtures, which, if you just care to contemplate this for a moment, is an astonishing fact given that, according to the Roadmaps discussed below, this source of energy is supposed to provide the world with 30% of its electricity before the century is out, another prediction that for sure will just be pie in the sky. The experience with D-T mixtures is virtually zero.

Figure 9.2 uses logarithmic scales. This tends to flatten things out and makes it hard to get a real grip on the numbers. The figure suggests in any case that $Q = 1$ (breakeven) has been achieved by both JET and JT-60U. As said before, the

Japanese devices have not operated with D-T plasmas and the results in the figure
are extrapolated values from pure deuterium plasmas. These results are questionable.
As we know, JET has failed to reach $Q = 1$ with real D-T plasma, so how can then
an extrapolated result from a pure deuterium plasma that gives $Q >1$ be trusted?
Only the full D-T results can be trusted, of which there are only three (two from JET
and one from TFTR). Figures like Fig. 9.2, a variant of which is still displayed by
EUROfusion, are perhaps not wrong, but at least slightly misleading and at any rate
confusing. There is no compelling reason, apart from propaganda, to include such
extrapolated results in the figure.

Whether the progress shown in the graph is impressive, depends on your point of
view. It is certainly not impressive when measured against the promises that were
made from the earliest days onwards. Scientific breakeven ($Q = 1$) should have been
achieved forty odd years ago (about 1980) and lies still at least fifteen years in the
future (as ITER will not start D-T operations before 2035).

To make the achievements (more) impressive fusion proponents like to show the
figure of Fig. 9.3, which compares progress in the value of the triple product with the
stupendous development of the number of transistors on a computer chip. The idea
here is that, when you are able to compare yourself favourably with a real success
story, some of that success will also rub off on you. In this respect Moore's law
(authored by Gordon Moore (b. 1929), a stupendously rich American businessman,
former chairman of Intel) is the observation that the number of transistors on an
integrated circuit chip doubles about every two years, and from 1970 up to about

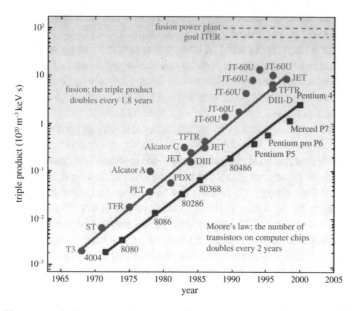

Fig. 9.3 The progress in the triple product towards fusion, compared with Moore's law for transistor
performance. Both showed a power law, doubling around every 2 years. (*From* http://www.fusenet.
eu/sites/default/files/moores_law.jpg)

2000 the increase in the Lawson triple product kept pace with this as can be seen in the figure. For chips this development did however not stop with the Pentium 4 as in Fig. 9.3, but continued unabated or at an even more dazzling pace during the first two decades of this century, reaching a current record of more than 200 times the number of transistors on the Pentium 4 chip just a dozen years later.[3] For fusion Stein's law (after the American economist Herbert Stein (1916–1999)) apparently set in: "Things that can't go on forever don't", something that also will happen in due course to the number of transistors on a chip. For fusion it happened to come earlier; it has already reached its zenith. Even if ITER achieves its expected result in 2035 or a little later, this plot will no longer be shown by fusion fans.

And if you then lay so much store on Gordon Moore, it is only fair to also quote his second law which states that the cost of a semiconductor chip manufacturing plant doubles every four years. This law in any case does not apply to nuclear fusion where new plants or experimental devices cannot even be built in four years.

Fusion scientists like to think that now, after the big tokamaks, they have arrived at a new stage, characterised by the transition from studying high-temperature plasma devices, with plasma parameters below scientific breakeven condition, to the design and manufacturing of technologies (such as superconducting magnets, vacuum vessel and in-vessel components, cryostats, heating and diagnostics, and remote handling and maintenance systems) essential for a fusion reactor like ITER, intended for steady-state operation (i.e. $Q > 5$). It is again a case of counting the chickens before they hatch. Breakeven has not even been achieved yet, let alone $Q > 5$, and steady-state operation is very far away indeed. The only thing that has been achieved is that at last, after a very long wait, some real fusion power has been produced with TFTR and JET, i.e. some power that originates from fusion reactions, but actually very little power and for a very short period of time, just a couple of seconds. The JET result was not stable in the sense that it could routinely be reproduced. The favourable H-mode conditions could not be maintained for long. Not only the temperature but also the density started to rise and became so high that the H-mode terminated (Braams and Stott 2002, p. 219).

And what is more, the energy balance for fusion is very negative, and will remain so for ITER, as set out in Fig. 9.4.

It is a figure that fusion aficionados do not like to show. The amount of 950 MWe, i.e. *electric* input power, for TFTR for instance is difficult to find; the website of the fusion critical *New Energy Times* says that they could only find it in the Princeton Alumni Weekly of 14 January 1980.[4] You certainly don't find the figure in the scientific papers of the TFTR experiments. In the end it is however the only figure that counts and gives an honest picture, whereby we still have to wait and see if ITER indeed will fulfil this promise of producing 166% of input *electric* power in *thermal* power, so not in electricity! It is gratifying to see though that in spite of its much

---

[3] *Wikipedia*, https://en.wikipedia.org/wiki/Transistor_count.

[4] http://news.newenergytimes.net/2018/03/28/experts-testify-before-congress-on-future-of-u-s-fusion-energy-research/.

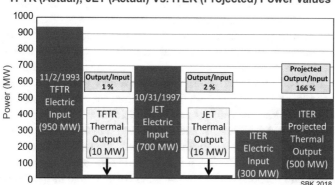

Fig. 9.4 Comparison of thermal output of various fusion devices compared to electric input (*from* https://news.newenergytimes.net/)

larger size, ITER will need much less electric input than TFTR or JET. That indeed is tangible progress!

9.2 Roadmap to Fusion

Before proceeding in the next chapter with a detailed description of ITER, the flagship of fusion research currently under construction, we will first describe the European roadmap to fusion, from the construction of ITER via a demonstration (DEMO) plant to the final step of a commercial fusion power plant. Most national fusion programmes were based on such a three-step strategy to commercialization, an ITER-like experiment, followed by a DEMO plant, that in turn was to be followed by a commercial reactor. The US for instance was designing an Engineering Test Facility (ETF). All this before the global ITER collaboration came off the ground.

As we have seen, past fusion research was fairly chaotic, with many countries running separate programmes, all in the impression that the realization of fusion as an energy source was just around the corner and in everyone's grasp. That view has now turned out to be completely mistaken. In the meantime, fusion devices have become so large, so complex and so expensive that large collaborations of many countries and careful planning and accounting practices are necessary. A detailed and substantiated story is required before being allowed to waste a few billion of public money.

The European fusion programme is currently the responsibility of two major organisations: Fusion for Energy (F4E), which is responsible for the European contribution to ITER construction and other major projects, and EUROfusion, which is responsible for the accompanying R&D programme. The EUROfusion consortium comprises 30 research institutes in 26 EU countries, plus Switzerland and Ukraine, and has come in the place of the bilateral agreements between Euratom and the various member countries, which were terminated at the end of 2013. Funding is now strictly aligned with the priorities of the Fusion Roadmap, with co-funding by the national programmes. EUROfusion is in charge of the fusion-related research carried out at JET, and its other fusion devices include the ASDEX Upgrade (Garching, Germany), the TCV Tokamak (Lausanne), the WEST tokamak (France), MAST Upgrade at Culham, the Wendelstein-7X stellarator, also in Germany, the Spanish TJ-II tokamak, the Plasma-Wall Interaction in Linear Plasma Devices (PSI-2) in Jülich (Germany), and the Pilot-PSI and MAGNUM-PSI devices in the Netherlands.

Fusion for Energy (F4E) is the European Union's Joint Undertaking for ITER and the Development of Fusion Energy. It is responsible for providing Europe's contribution to ITER and supports fusion research and development initiatives through the Broader Approach agreement concluded with Japan.

Since the alleged successes of JET as the largest tokamak in the world and the efforts to design and construct its successor ITER, the European Union has taken the lead in the global fusion effort, in which role it takes excessive pride, so it seems. In the 2018 *European Research Roadmap to the Realisation of Fusion Energy* (Donné et al. 2018) (in the following referred to as the Roadmap) it has been set out in detail how the fusion enterprise will have to be brought to fruition. This roadmap addresses the following three goals:

1. Technical demonstration of large-scale fusion power, which is the first goal of ITER (500 MW for 400 s);
2. Electricity delivered to the grid via a demonstration fusion power plant (DEMO) which would generate a few hundreds MW in electric power for several hours, operate with a closed fuel cycle (i.e. breed its own tritium) and allow extrapolation to early commercial fusion power plants;
3. In parallel, to establish a science, technology, innovation and industry basis to allow the transition from the demonstration fusion plant to affordable devices suitable for large-scale commercial deployment (stellarators might prove particularly attractive).

All three goals have to be realised in the context of the final goal of large-scale industrial production of fusion plants.

As can be seen from Fig. 9.5 the programme has been divided into short-term, medium-term and long-term prospects. In earlier documents a specific timescale in years was attached to such a programme. That is no longer done, as it entails the obvious danger of being pinned down on certain dates.

The short term encompasses the construction of ITER; R&D in support of ITER (which is ongoing all over the world); recommencement of the D-T operation of JET, which is designed to achieve and investigate ITER-like plasma regimes of operation (this operation will start in 2021, using tritium for the first time since 1997, showing how sensitive the use of the (radioactive) tritium is or is considered); conceptual design phase of DEMO (early conceptual designs should be ready by 2027 according to the Roadmap); R&D for DEMO; construction of a dedicated fusion materials testing facility (still certainly a decade ahead; see below and Chapter 16); scientific

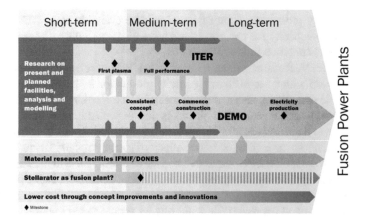

Fig. 9.5 The European Roadmap in a nutshell

and technological exploitation of the stellarator concept (well underway with the Wendelstein-7X (Chap. 13)).

The medium term must then see the first scientific and technological exploitation of ITER (so after December 2025 when first plasma is expected, but probably at least a year later due to the current Covid-19 pandemic (Clery 2020); the first exploitation of the fusion materials testing facility; engineering design phase of DEMO with industrial involvement; development of power plant materials and technologies; and possible further development of the stellarator concept (all still up in the air).

And the long term is reserved for high performance and advanced technology results from ITER; qualify long-life materials for DEMO and power plants; finalisation of the design of DEMO; construction of DEMO (to start a few years after high-performance D-T operation of ITER is achieved, which is now scheduled to start in 2035); demonstration of electricity generation; commercialisation of technologies and materials; and deployment of fusion together with industry (all this will probably take us into the next century).

Further down in the document the Roadmap speaks not of short, medium and long term, but of the first, second and third periods. In the first period ITER operation will be started and DEMO conceptual design(s) completed (< 2030); in the second period burning plasma on ITER and DEMO engineering design will be achieved (2030–2040); and the third period will see plasma and technology optimisation on ITER and the construction of DEMO (> 2040). As a precaution it is stated that the dates given here are target dates. They indeed will probably not all be met. These periods roughly coincide with the short, medium and long terms introduced above.

When consulting the 2012 document *EFDA Fusion Electricity: A roadmap to the realisation of fusion energy*, it can be found out that at any rate in 2012 short term meant up to and including 2020 (now probably a decade later), medium term the decade thereafter (quite a lot of work for just a decade, keeping in mind the slow and delay-prone practice in fusion research) and long term from roughly 2030 to 2050, so that electricity production from fusion should be available from 2050. DEMO, producing net electricity for the grid at the level of a few hundred megawatt, was foreseen to start operation in the early 2040s.

The 2012 EFDA Roadmap states explicitly that "fusion can start market penetration around 2050 with up to 30% of electricity production by 2100"! A stupendous prediction, and most likely also a stupendous folly (see also Chap. 19)! Since ITER is delayed by at least ten years (first plasma is now expected in December 2025 and D-T operation to begin only in 2035[5]), a few decades should at least be added to these forecasts, but even so the prediction is too audacious to contemplate. To predict that an unproven, far from realised technique will be able to capture 30% of a mature market within such a short time is just beyond comprehension. Global electricity production in 2040 will be around 40,000 TWh (terawatt hours), which is 40,000 million MWh, assuming that demand for electricity will grow by about 2% per annum.[6] It implies that global electricity generating capacity is about 7 TW.[7] In 2035, but probably still a considerable number of years later and, if all goes well, ITER will produce just 500 MW of *thermal* power for 400 s. If converted into electricity it is just a third of this, meaning that ITER will still consume more electricity than it produces. It will be a tremendous feat if the DEMO plant, planned after ITER, is capable of breaking even in that respect, i.e. produce as much electricity as it consumes. It will anyway not be producing more than a few hundred megawatts of electricity, as stated above. How then can we believe that around the year 2100 fusion will satisfy one third of global electricity demand, i.e. something like 10,000 million MWh of electric power or in other words have a steady-state capacity of more than 1 TW or 1 million MW?[8] Such predictions have no reasonable basis in fact; they are completely ludicrous! The follow-up to the 2012 *EFDA Roadmap* the 2018 *European Research Roadmap to the Realisation of Fusion Energy*, whose short version[9] rather cheekily starts with a section entitled "Combating climate change" as if fusion has anything to contribute to this (if anything, its heavy carbon footprint has just the opposite effect), repeats this folly: "The quest for fusion power is driven by the need for large-scale sustainable and predictable low-carbon electricity generation, in a likely future environment where the global electricity demand has greatly increased. This demand is expected to perhaps reach 10 TW [of global generating capacity (*LJR*)] in the second part of this

---

[5]https://www.iter.org/proj/inafewlines; the previous planning foresaw first plasma by 2020 and full fusion by 2023 (Reuters, Science News, 2 May 2016).

[6]Global electricity generation in 2018 amounted to 26,700 TWh, of which about 2/3 came from coal (38%), gas (23%) and oil (3%) (https://www.iea.org/geco/electricity/).

[7]By the end of 2014, the total installed electricity generating capacity worldwide was nearly 6.142 TW (million MW) which only includes generation connected to local electricity grids (https://en.wikipedia.org/wiki/World_energy_consumption).

[8]The capacity factor of a power plant ranges from 30% for wind/hydro plants to 90% for a nuclear power plant. A 1000 MW power plant with a 100% capacity factor would generate about $1000 \times 24 \times 365$ MWh $= 8.8$ TWh of electric power. If the capacity factor is less than 100%, the generated power is correspondingly lower.

[9]https://www.euro-fusion.org/fileadmin/user_upload/EUROfusion/Documents/TopLevelRoadmap.pdf.

century, by which time the vast majority of energy sources needs to be low-carbon. To make a relevant contribution worldwide, it is estimated that fusion must generate on average 1 TW of electricity in the long-term, i.e., at least several hundred fusion plants in the course of the twenty-second century." So, by that time fusion must have an *electricity* (not thermal) generating capacity of 1 TW, which means 200 5000 MW fusion plants running continuously throughout the year, so indeed several hundred fairly large plants, for a power station with a nameplate capacity of 5000 MW is not particularly a small one. The largest power stations are hydroelectric power stations with capacities of up to 22,500 MW (for the Three Gorges Dam in China), while the largest nuclear power stations (based on nuclear fission) have a capacity of about 7000 MW. I do not know what the capacity factor[10] of a fusion power plant will be, but assuming that it is 50% you will need double that number of fusion power plants, i.e. 400, for a generating capacity of 1 TW of energy. So, the long-term roadmap seems far off and ludicrously so, still apart from the cost of a nuclear fusion power station, which we will discuss in one of the final chapters dealing with the economic aspects of fusion.

Apart from the delays in the construction of ITER as stated above there are also substantial cost overruns. The project officially started in 2006 with an estimated cost of €5 billion and 2016 as the date for first plasma. Those figures quickly changed to €15 billion and 2019, and have now been further extended with an extra €5 billion added to the costs, bringing the total to €20 billion. The true cost of ITER is however almost impossible to determine. In the project agreement the manufacture of the necessary components was divided up among the partners: 45% for the account of the European Union (as host), and 9% for each of the other partners. How much each partner spends for manufacturing its components is the partner's individual concern and is not revealed (Clery 2015). The total cost figure quoted above adds another probably show-stopping complication to fusion as a viable source of electricity, namely that power plants will be much too expensive for most countries to construct. The ITER cost factor will also be discussed in detail in the next chapter.

So, let us leave aside the long-term predictions and cost issues of ITER and concentrate on the short-term and medium-term technical goals of ITER and a possible DEMO design. ITER is the key facility of the roadmap and indeed of nuclear fusion, for without ITER there will be no fusion. If ITER fails, there will probably never be a nuclear fusion power plant on Earth, at any rate not in another 100 years or so. So, there is an awful lot at stake here! Let us look in some more detail at the Roadmap[11] and quote from it.

ITER is the key facility and European laboratories should focus their effort on its exploitation. To make it a success JET and JT-60SA (construction completed early in

---

[10]The net **capacity factor** is the unitless ratio of actual electrical energy output over a given period of time to the maximum possible electrical energy output over that period. The maximum possible energy output of a given installation assumes its continuous operation at full nameplate capacity over the relevant period.

[11]The Roadmap is a living document, as they say, with updates and reviews to be performed at appropriate times. The latest version is from 2018.

2020) are being operated as main risk mitigation measures. Other small and medium-sized tokamaks, both in Europe and beyond, are used to study specific ITER-relevant topics (this is in full swing, to such an extent that hardly any independent research can be carried out with these devices and people move away from them to pursue their own goals (like spherical tokamaks; Chap. 11)).

Then the Roadmap continues by listing a few problem areas:

One of the main challenges is the development of a heat and power exhaust system able to withstand the large loads expected in the divertor of a fusion power plant. The risk exists that the baseline strategy pursued in ITER cannot be extrapolated to a fusion power plant. Hence, a programme on alternative solutions for the divertor is necessary. For this a dedicated facility, the Divertor Tokamak Test (DTT) facility, is envisaged in Italy, which will be constructed by the Italian ENEA at Frascati. Work on this facility has been delayed and will now probably start in 2020 (the agreement was signed in January 2020). It is supposed to be completed in 2027 at a cost of 600 million euro. Funding is both private and public and for the main part borne by Italian entities. EUROfusion is apparently not yet convinced of the necessity of such a facility and will only decide on the nature of its involvement after results from experiments on present devices are made available and after the construction of the Italian DTT will be close to completion, so somewhere towards the mid 2020s. The important point for Italy is probably not so much a divertor that is suitable for a real power plant, but that the project will have an impact of around 2 billion euro on national GDP and will create 1500 new jobs, including 500 for scientists and technical specialists. Activity for activity's sake, in the hope that someone will foot the bill later, one wonders?[12]

The second problem is a dedicated neutron source needed for material development. Studies of the irradiation of materials by a fusion neutron spectrum are needed before the DEMO design can be finalised. While a full performance International Fusion Materials Irradiation Facility (IFMIF) (to be discussed in Chap. 16) would provide the ideal fusion neutron source, this facility has been planned now for close to two decades and will probably not be available soon enough. The construction of an earlier DEMO Oriented Neutron Source (IFMIF-DONES, Europe) or the Advanced Fusion Neutron Source (A-FNS, Japan) must be speeded up to provide a source with a fusion-relevant neutron spectrum for materials testing. In addition, a comprehensive programme using materials test reactors (MTRs) is needed to establish data bases of neutron irradiated materials (a facility like the Fusion Nuclear Science Facility which is discussed in the US, but will most likely not be built in that country).

The next item to be addressed is that DEMO must be self-sufficient in tritium. Future power plants must not only be self-sufficient in tritium, but also breed enough surplus tritium to allow successor power plants to start up. Breeding occurs in a blanket surrounding most of the plasma, and the blanket is also the primary source for converting heat into electricity. The device should be structured such that the area available for breeding is maximised. The blanket must at the same time be able to

---

[12]https://www.enea.it/en/news-enea/news/energy-enea-and-eni-join-forces-for-international-dtt-project-worth-600-million-euros.

handle much of the exhaust heat from the plasma without excessive loss of neutrons before they reach the breeding material, and in addition must shield the vacuum vessel from neutrons. A test blanket programme on ITER will be indispensable (see Chap. 10). A sufficient range of blanket options for DEMO and commercial power plants should be explored to ensure a solution that meets the requirements as regards tritium breeding, materials and thermal efficiency.

DEMO design will of course benefit greatly from the experience gained with ITER construction. There will be a number of uncertainties for some time, e.g. lessons to be learned from ITER, but it is necessary to focus on a representative design point in order to uncover the key design integration issues, and steer the R&D. Alternative DEMO plant architectures must also be investigated in parallel to ensure that no opportunities are missed. Safety will be an all-encompassing element as well as environmental aspects such as waste minimisation and recycling strategies.

Figure 9.6 shows how the progress made with ITER will feed into DEMO with the corresponding periods this is supposed to happen.

A surprising point in the Roadmap is that the stellarator is seen as a possible long-term (whatever that may mean) alternative to a tokamak fusion power plant. It shows that confidence in the tokamak concept is no longer universal, unless there are political reasons for including the stellarator here. Germany currently operates the largest stellarator: Wendelstein 7-X. No other country involved in ITER or otherwise has made room in its research programme for stellarators as a viable alternative

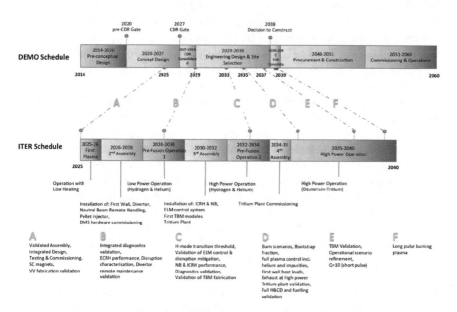

Fig. 9.6 Diagram depicting how information from ITER flows into the DEMO conceptual and engineering design activities. CDR = Concept Design Review. The dates are indicative (*from* European Roadmap 2018)

to the tokamak for a nuclear fusion plant. The Roadmap says that the EU stellarator programme should focus on the optimised Helias (Helical-Axis Advanced Stellarator) line, a stellarator optimisation approach based on modular field coils, i.e. on the most important stellarator currently in operation, the German Wendelstein 7-X, which will be discussed in Chap. 13. Work on other stellarator lines (e.g. Heliotrons) will continue as part of national programmes or within the framework of international collaborations. For the period 2014–2021, the main priority was the first-phase completion and commissioning of the Wendelstein 7-X machine (achieved in 2015). Further plans include the final completion of the device as well as its scientific exploitation. The most important goals are to validate the energy and particle confinement of optimised stellarators, and to demonstrate their special divertor for managing the heat and particle exhaust. Demonstration of high-performance plasma scenarios under steady-state conditions will be achieved beyond 2020. One of the main reasons for pursuing the stellarator line arises from the fact that they operate without a large plasma current, resulting in an inherent steady-state capability and the absence of plasma disruptions, which are both challenges for the tokamak. If Wendelstein 7-X confirms the good properties of optimised stellarators, a next-step HELIAS burning plasma experimental device may be required to address the specific dynamics of a stellarator burning plasma. The exact goal of such a device can be decided only after a proper assessment of the Wendelstein 7-X results. In the long run, it is expected that this strategy, together with the technology results from a tokamak DEMO and developments of plasma exhaust and plasma facing components, could allow a stellarator fusion power plant to be built. Stellarator development is however far behind the tokamak and, if it were adopted as the concept for a power plant, several decades will have to be added to the roadmap schedule based on the tokamak, even if tokamak results can be used as suggested above. A stellarator-based DEMO or power plant is out of the question in this century.

The Roadmap states that industry must be involved early in the DEMO definition and design. The evolution of the programme requires that industry progressively shifts its role from being a provider of high-tech components to a driver of fusion development. It is not made clear which industry is meant here. Surely not the industry that builds the components for ITER, DEMO or power plant, as they will not be interested in running a power station. Utility companies normally run power stations, but they do not know anything about nuclear fusion, neither are they at present involved in the construction of ITER, for instance. Fusion has so far been a fully government-sponsored research effort. The only industry one could remotely think of is the nuclear power industry, so the companies building nuclear fission power stations, but they too know nothing about fusion. Industry, the Roadmap continues, must be able to take full responsibility for commercial fusion power plants after successful DEMO operation. For this reason, DEMO cannot be defined and designed by research laboratories alone, but requires the full involvement of industry in all technological and systems aspects of the design. The Roadmap states: "Increased involvement of industry is especially required in the design and monitoring process from the early stage to ensure that early attention is given to industrial feasibility, manufacturability, costs, nuclear safety and licensing aspects. This is an evolution of the role of industry compared to that in ITER, and an early launch of the DEMO engineering design after the completion of ITER construction and beginning of operation would facilitate maintaining industrial competences and engagement. Industry involvement needs a policy to maintain industrial competence in fusion technology. It is also expected that industry will play a key role in developing effective, low cost and innovative manufacturing techniques, some of which may have applications outside fusion." I don't see ordinary utility companies contributing anything to the design of DEMO, nor do I see them running a nuclear fusion power station or taking full responsibility for it. As soon as DEMO or another power plant is connected to the grid, it will sell the generated electricity to a utility company, I suppose, which will then sell it on to the consumer, but will not itself run the nuclear fusion plant. A new industry will have to be built up for this with very different expertise than utility companies currently possess and, in view of the exorbitant cost and risk of a nuclear fusion power plant, with heavy involvement and far-reaching guarantees from governments.

Theory and modelling effort in plasma and material physics is crucial, the Roadmap further states. It is surprising that this has to be stated at all, as it seems obvious, but it reflects the fact that theory and modelling have been neglected in the past (the engineers have taken over and power generation is the goal; it doesn't really matter that we don't understand the physics behind it). It is true though that advances in computer science and technology and in big data can drastically transform modelling capabilities in the future, e.g. to multiscale modelling of the whole plasma or complete components (an example is the design of the exceedingly complex stellarator magnetic fields which has hugely profited from dramatic improvements in computer power and modelling capabilities, as will become clear in Chap. 13).

9.3 Summary

To summarise, the Roadmap has formulated the following, very ambitious eight missions:

M1. Demonstrate plasma scenarios (based on the tokamak configuration) that increase the success margin of ITER and satisfy the requirements of DEMO, i.e. stable burning plasmas with high fusion gain by minimizing the energy losses;

M2. Demonstrate an integrated approach that can handle the large power leaving ITER and DEMO plasmas, i.e. solve the exhaust problems;

M3. Develop materials that withstand the large 14 MeV neutron flux for long periods while retaining adequate physical properties;

M4. Find an effective technological solution for the breeding blanket to ensure tritium self-sufficiency;

M5. Ensure that safety is integral to the design of DEMO using the experience (to be) gained with ITER;

M6. Bring together the plasma and all the systems coherently, resolving issues by targeted R&D activities;

M7. Ensure the economic potential of fusion by minimising the DEMO capital and lifetime costs and developing long-term technologies to further reduce power plant costs (quite a task and probably impossible to achieve);

M8. Bring the stellarator line to maturity to determine the feasibility of a stellarator power plant.

Mission 1 should be fully accomplished by ITER and the accompanying research programme, together with an exhaust solution and the rest of the DEMO engineering design (Missions 2 and 6). Mission 2 can actually be considered the main challenge for ever realizing a fusion power plant. The exhaust problem and handling of the power output is already a demanding task for ITER, but it will be amplified for a power plant where the assumed linear dimensions are about 50% larger and the fusion power output at least 3 times higher. The various missions will be dealt with in the next chapters of this book.

Part IV
High Noon

Chapter 10
The International Thermonuclear Experimental Reactor

10.1 Introduction

In some sense this is the most important chapter of the book as it describes the culmination of more than half a century of efforts towards controlled nuclear fusion, the mammoth project of ITER, the International Thermonuclear Experimental Reactor. It is supposed to be the next, and probably the last step in the attempts to harness the inexhaustible source of nuclear fusion energy. It is undoubtedly a great scientific project, a jewel of technology, as the proponents like to call it, that has resulted in admirable, albeit not always smooth cooperation and collaboration between the leading nations on Earth, making it into a globe-spanning, transnational technological project. But there is a great chance that it will finally go down into the history books as one of mankind's greatest follies born out of sheer arrogance, a true case, if there ever was one, of misplaced confidence.

Not satisfied with the daily bath of energy coming from the Sun, man wants to bring the Sun to Earth and tap its source of energy at home and at will. A step further than Icarus who just wanted to go to the Sun but burned his wings and crashed. A post-Icarian case of pride and folly that is currently unfolding in the south of France.

Nations with varied traditions, cultures and attitudes, speaking a multitude of different languages are bundling their efforts to defy nature, reminiscent of the biblical story of the Tower of Babel. In that story humankind, which after the Flood spoke a single language, agreed to build a city and a tower tall enough to reach heaven, out of vanity and contempt for God. But God, observing their efforts, confounded their speech so that they could no longer understand each other, and scattered them around the world.

In more than one aspect the ITER tale is the opposite of the Tower of Babel story. Now, the various peoples scattered around the world, speaking different tongues and believing they can converse among each other in one of these languages, gather in the south of France, not to build a city and tower tall enough to reach heaven, but to create on Earth part of that heaven by taming the source of energy that has been reserved for the most exalted of celestial bodies, the stars.

Will the story have a similar ending? Let us tell it first before we make a reasoned judgement.

10.2   INTOR

In 1978, under the auspices of the IAEA, Japan, the US, USSR and the European Community (EC) joined in the International Tokamak Reactor (INTOR) Workshop. All four participants in the workshop were at the time constructing a big tokamak, which, apart from the Soviet T-15, which never worked to full capacity, were discussed in the previous chapter. The proposal for the world's fusion programmes to join together in designing and constructing a follow-up device at a time that was not very favourable for international cooperation, had surprisingly come from the Soviet side, delivered by the physicist Evgeny Velikhov, whom we already met in Chap. 5. His proposal was picked up by the IAEA. The broad objective formulated for the INTOR Workshop was: "To draw on the capability in all countries to prepare a report to be submitted to the International Fusion Research Council (IFRC)[1] describing the technical objectives and nature of the next large fusion device of the tokamak type that could be constructed internationally." INTOR, INternational TOkamak Reactor, would be the next step experiment in the progression from the generation of big tokamaks (JET, TFTR, JT-60) to DEMO (Braams and Stott 2002, p. 247). These big tokamaks were however not yet running. Far from it. The first one, TFTR, came online only in 1982. To start the design of a new, next generation machine before the previous one even has been built, is also a recurring feature in fusion land, while one would think that it would be sensible to let the experience gained from such preceding devices be guiding in any new design.

It is interesting to learn from Stacey's book (Stacey 2010) on the subject that in 1978 the EC, contrary to the other participants in the Workshop, was not yet ready for such an undertaking. In spite of the construction of JET that had gotten underway just a few years earlier as an EC collaboration project, the various national EC fusion research programmes were not yet subordinate to the common goal and had all kinds of separate interests they wanted to see protected. That was also the reason why Donato Palumbo (1921–2011), the head of the Euratom fusion programme, opposed the INTOR Workshop and had preferred if Europe had not taken part in it. The US had its own programme too and had been carrying out conceptual studies for a follow-up to TFTR, called the Engineering Test Facility (ETF), since 1978. ETF would provide a testbed for reactor components in a fusion environment (Steiner et al. 1981) and, once the INTOR Workshop got going, ways were sought for ETF and INTOR to work together, but to remain separate entities. It was decided that the materials developed for INTOR would also be used for the US ETF project and, conversely, that the work performed within the framework of the ETF project would serve as part of the US

---

[1] The IFRC was a permanent advisory body of the IAEA on plasma and nuclear fusion, made up of representatives of countries with major fusion programmes and meeting once annually.

input to the INTOR Workshop, but tension remained as in the US INTOR and ETF were competing for the same funds. The name ETF was soon changed into Fusion Engineering Device (FED) to agree with the 1980 US Magnetic Fusion Engineering Act which had called for the operation of a Fusion Engineering Device by 1990. The Act, sometimes called the 20-year, $20 billion plan for fusion, was overwhelmingly supported in Congress, and signed into law by President Carter in October 1980, one month before he was defeated for re-election by Ronald Reagan. The law was subsequently ignored by both Congress and Administration (Dean 1998a, p. 158).

The INTOR workshop was a very impressive affair. The required R&D was divided into 16 technical areas covering all topics of relevance for a future reactor and hundreds of fusion scientists and engineers in each participating country took part in a detailed technical assessment of the status of the tokamak confinement concept versus the requirements of an Experimental Power Reactor. At the end of Phase Zero, which was conducted during 1979, the workshop produced an impressive 650-page report, published by the IAEA in 1980.[2] It was the most comprehensive assessment of the status of fusion development ever undertaken. More than 500 physicists and engineers had contributed to it (Stacey 2010, p. 62; Stacey 1978), and the conclusion was that it was possible to undertake the design and construction of an experimental fusion power reactor based on the tokamak.

During the later phases of the Workshop it became increasingly clear that INTOR would not proceed to the design and construction stage, barring drastic changes in the international political situation. There was not enough support from the governments of the respective participants in the Workshop. The Soviet scientists however wanted at all cost to be part of such an international project, since around 1981–1982 the USSR government had already imposed constraints on fusion research within the Soviet system. The completion of the T-15 was its highest priority and prospects for a national INTOR-like project in the Soviet Union were considered small by knowledgeable scientists like Velikhov.[3] An international project was their best bet for remaining involved in cutting-edge fusion research. As said, independent national design efforts for INTOR-like devices had been initiated independently in the EC with the Next European Torus (NET) (Toschi 1986; 1989) and the US with the ETF/FED, leading to the design of TIBER (Tokamak Ignition/Burn Experimental Reactor) (see e.g. Henning and Logan 1987), while Japan wanted to get involved with the US and also had its own project, called the Fusion Energy Research (FER) Facility. However, none of these projects went well. In Japan first priority was the JT-60 and proposals for FER did not come off the ground, because of adverse economic circumstances in the early 1980s. The US maintained an independent effort throughout all phases of INTOR. The scientists involved in ETF were reluctant to support INTOR as it could bode ill for ETF. In the end however ETF/FED did not fare well either and, if

---

[2] A summary was published by the INTOR Group as International Tokamak Reactor (Executive Summary of the IAEA Workshop, 1979) in *Nuclear Fusion* 20 (1980) 349–388.

[3] Some design work was carried out in the Soviet Union on a hybrid fusion-fission tokamak reactor of ITER-size, called OTR (standing for Opytnyj termoyadernyj reaktor (experimental thermonuclear reactor)) (see Velikhov and Kartishev 1989).

the enthusiasm for INTOR among US scientific policy makers had been greater, it might actually have been beneficial for ETF. The Reagan administration, coming into power in 1980, was not supportive of any expensive national fusion initiative and fell back on the standard position of Republican administrations that the demonstration of fusion as an energy source was the role of industry.

The INTOR Workshop which started so promising and did impressive work in its Phase Zero never reached the design stage, let alone the construction stage. In the autumn of 1987, it was folded into the new ITER Project, publishing its final report in 1988. It had only partly achieved its objectives, but left an extensive database of the relevant scientific and engineering information to be used in future tokamak designs.

INTOR came too early. It should have waited for the results of the big tokamaks. In the early 1980s, as we have seen, tokamak physics had run into difficulties with additional heating, which were partly solved by the H-mode discovered on the ASDEX tokamak. In addition, extrapolation of scaling results to INTOR parameters failed to give assurance that the required plasma pressure and confinement time would be reached. It had assumed that ignition could be achieved with plasma currents in the range of 8–10 MA, but by that time JET had already operated at 7 MA and was clearly much too small to ignite (Braams and Stott 2002, p. 249). So, after some 10 years of extensive work, involving the best and brightest in nuclear fusion on the globe, it still had only produced a reactor design with parameter values that would certainly have failed. A sobering thought.

10.3   The Birth (Pangs) of the ITER Project

Evgeny Velikhov claims that it was he who made the proposal for collaboration on nuclear fusion between the West and the USSR to Mikhail Gorbachev, when the latter went on his first visit to France after having become General Secretary, in March 1985, of the Communist Party of the Soviet Union. It was a propitious time for formulating an idea for cooperation and he decided to propose cooperation in the area of nuclear fusion.[4] It should be realised though that the foremost political rationale of this proposal was scientific cooperation between the US and the Soviet Union. A 1990 CIA report[5] adds to this that one of the reasons for the Soviets to propose such collaboration is that they "must join an international collaboration if they are to have access to a fusion Engineering Test Reactor (ETR) (…) during the next 25 years. Because of economic and manufacturing constraints, they probably are unable to construct an ETR themselves" and "the advantages to the international fusion community of including the USSR in subsequent phases of the ITER programme are political, rather than technical." The cost estimates for the next generation of fusion machines ($1 billion or more) were beyond the means of

---

[4]https://www.iter.org/newsline/-/2326.

[5]"The Soviet Magnetic Confinement Fusion Program: An International Future" available from https://www.cia.gov/library/readingroom/print/175351.

Soviet national programmes and international collaboration presented a way out of this. Another issue was the inability of Soviet manufacturers to produce the necessary high-precision components for their domestic devices. The political rationale for collaboration and all further considerations of course ended when the Soviet Union collapsed in 1991. Russia took its place, but for considerable time it too was in equally dire straits.

Velikhov's initiative was taken up in 1985, at the Geneva summit meeting between Gorbachev and US President Ronald Reagan, where it was announced that the two countries would jointly undertake the construction of a tokamak Experimental Power Reactor as proposed by the INTOR Workshop. So, quite a few people around both Gorbachev and Reagan must have been convinced that such an undertaking could be successful, unless it were mainly political reasons that convinced Reagan that it was preferable to engage in collaboration in this field with the state he had called an 'evil empire' and the 'focus of evil in the modern world' just two years before. It must be added though that at the time of Reagan's 1983 speech Gorbachev had not yet assumed power and after he had done so in March 1985 the relations between East and West rapidly improved.

This proposal by Gorbachev is generally seen as the birth of ITER, while of course without the work carried out by the INTOR Workshop such a proposal would have been impossible. The INTOR Workshop is seldom mentioned though and does not get much credit. On the other hand, would there have been an ITER project without Gorbachev's proposal to Reagan? There probably would, but perhaps in a different form and later. The development of the various national programmes would have continued for some time, but the realisation would soon have dawned that without international collaboration they were bound to fail.

At Geneva it had apparently been agreed for the US and USSR to prepare for fusion cooperation specifically between their two nations. According to Stacey (2010, p. 144), the US jeopardised the project by unilaterally changing the approach and including, much to the chagrin of the Soviets, the Europeans and the Japanese, as well as involving the IAEA, under whose auspices the development of ITER was to take place. For Europe and Japan, the project entailed the danger that they were dragged into an international collaboration that was principally set up for political reasons as stated above, not because Gorbachev and/or Reagan were suddenly enamoured with nuclear fusion.

Not only because of this, but also for other reasons not everyone in Europe was convinced that such an international effort was in Europe's interest. European studies on NET were well underway and there was concern in some quarters that a renewed push for a major new international facility would prove a distraction. Throughout the 1980s, also during the INTOR Workshop, Europe's fusion research community maintained a cautious balance between building its next big fusion machine as a Europe-only effort and pursuing it via a broader international collaboration. A primary goal throughout this period was ensuring that European unity and cohesiveness remained intact. However, the appropriation of fusion energy research by political leaders like Reagan, Gorbachev and the French president Mitterrand to further their own political agenda was beyond the control of managers like Palumbo and Rebut. Top-down advocacy from heads of state proved in the end an essential ingredient in nudging the European fusion community towards international collaboration (McCray 2010, pp. 293–295).

One year after the Geneva summit an agreement was reached with the same former INTOR participants as parties and conceptual design work for an International Thermonuclear Experimental Reactor (ITER, which is also a Latin word for 'path' or 'journey', which is now the preferred meaning, as it avoids the apparently tainted word thermonuclear) started in 1988. In the agreement Euratom committed itself to a three-year conceptual design. The Americans too did not want to go further at this stage. Although Reagan thought he had agreed to the joint construction of ITER, objections by the US Defence Department, related to the possible transfer of sensitive US software and hardware capability, resulted in approval of Conceptual Design Activities (CDA) only (Dean 1998a, p. 159).

The goal of this first three-year period was to provide, by the end of 1990, all the information needed for one or several of the participants to build ITER in the mid 1990s. Between 1988 and 1990, the initial designs for ITER were drawn up with the aim of proving that fusion could produce useful energy. In 1992 the four parties (with the Soviet Union replaced by Russia) agreed to extend the collaboration with Engineering Design Activities (EDA), resulting, after some mishaps, in the final design for ITER being approved in 2001. The design studies (CDA and EDA) carried out between 1988 and 1998 cost more than US$1.1 billion, about a third of which came out of the European fusion budget.

10.4 The First ITER Design

The overall objective of ITER is to demonstrate the scientific and technological feasibility of fusion energy for peaceful purposes. This objective is to be accomplished by demonstrating controlled ignition and extended burn of a D-T plasma, with steady-state as an ultimate goal, by demonstrating technologies essential for a reactor in an integrated system, and by performing integrated testing of the high heat flux and nuclear components required in the practical utilization of fusion power.[6]

As a reminder, in magnetic confinement ignition is reached when the heating by the α-particles produced in the fusion reactions matches the energy lost from the plasma, meaning that once the fusion has been set in motion, enough energy is produced for keeping the plasma at the required temperature. The nuclear fusion reaction becomes self-sustaining and the fusion reactions heat the fuel mass more rapidly than various loss mechanisms cool it. The plasma is 'burning' by itself (for some short time) and releases enough energy to heat the plasma to the required temperature. If self-heating keeps the plasma hot enough so that no more external heating is needed, we will have ignition (Q is infinite) and the dream of fusion will

---

[6]ITER EDA Agreement, ITER-EDA Documentation Series No. 1 (IAEA 1992).

have come true. The condition for ignition has the same form as the Lawson criterion (McCracken and Stott 2013, p. 40).

ITER addresses the next logical challenge in tokamak design—to enter and explore the domain of burning plasma in which α-particles will be the main source of plasma heating and thus one of the major determinants of plasma behaviour. To achieve the above-stated objective, ITER would be operated in two phases: a physics phase, focused mainly on the achievement of the plasma objectives, and a technology phase, devoted mainly to engineering objectives and the testing programme (Gilleland et al. 1989).

The design activities started in 1988 with the Conceptual Design Activities (CDA), followed by the Engineering Design Activities (EDA) from 1992 to 1998. The objectives of the CDA were to define a set of technical characteristics and subsequently carry out the work necessary to establish ITER's conceptual design; to perform a safety and environmental analysis; to define the site requirements as well as the future research and development needs; to estimate the cost and manpower; and to prepare a schedule for detailed engineering design, construction and operation.

Following a two-year period of negotiations, the Engineering Design Activities (EDA) began in 1992 with the intent to decide, by the end of a six-year period, whether or not to proceed to construction jointly, separately, or not at all. The ITER process and its predecessor the INTOR Workshop had thus so far already spanned two decades (Dean 1998a, p. 159).

An ominous early warning sign of things to come was that it proved impossible to decide on a single site for the Joint Central Team for carrying out the EDA. The result was that the team was split between three sites, Garching in Germany, San Diego in the US and Naka in Japan, with an American director in Garching, a European overall director in San Diego with a Japanese and a Russian deputy director, and a European director in Naka. It must have been a sensitive balancing act to get this properly on the rails.

The technical objective of the EDA was to design a machine that would demonstrate controlled ignition and burn. The machine had to test and demonstrate technologies essential for a fusion reactor including superconducting magnets and tritium breeding, but did not have to be self-sufficient in tritium and of course would not generate electricity.

The design it came up with would be capable of full inductive operation at a plasma current of about 25 MA (meaning that the plasma current is generated completely by transformer action (induction) from the central solenoid) with a burn duration of about 1000 s, and a toroidal field of 6 T. A longer-term objective for ITER was to demonstrate steady-state operation, so using non-inductive current drive. The toroidal and poloidal fields would be produced by superconducting coils. Copper coils would require far too much power to be 'reactor relevant'. Superconducting magnets are capable of carrying higher currents and thus generate stronger magnetic fields, and consume much less electricity. The disadvantages are that they are much more expensive to produce and must be cooled to a very low temperature.

Plasma elongation is limited to 1.6 in order to avoid a complex control system. The device will have a single-null divertor (i.e. one X-point (see Chap. 4 and Fig. 4.10

in which two X-points have been drawn)) and a similar D-shaped plasma as JET (but more than twice the linear dimensions). The development of a fully coherent and robust divertor is one of the principal challenges to ITER. The maximum possible plasma has a major radius of about 8 m and a minor radius of about 3 m. Nominal fusion power is about 1.5 GW, which may be extended by a factor of about 2. The above-mentioned plasma current and size should provide the capability for ignition.

ITER differed from previous tokamaks in the exposure of machine components to radiation and heat fluxes. The superconducting coils must be shielded to reduce the heat flux from neutrons and to limit the radiation damage of the vacuum vessel to a repairable level.

Auxiliary heating power is required to make the transition into H-mode, to heat the plasma to ignition, to drive non-inductive currents, to control plasma equilibrium and fusion burn, and to suppress instabilities. The amount of heating power required for each of these tasks fell in the range of 100–200 MW, for which neutral beams, ion cyclotron and electron cyclotron heating were investigated (Shimomura 1994, p. 1613; Braams and Stott 2002, p. 250 ff).

The six-year period originally established for the Agreement on Cooperation in the Engineering Design Activities (EDA) for ITER ended in July 1998 and the results were laid down in the ITER Final Design Report. The report is the first comprehensive design of a fusion reactor based on well established physics and technology. With this design for ITER the bar is set extremely high. It will be a superconducting tokamak capable of controlled ignition and extended burn in inductive pulses with a duration of about 1000 s. It also aims to demonstrate steady-state operation using non-inductive current drive in reactor-relevant plasmas with α-particle heating power at least comparable to the externally applied power. It is designed to operate safely and to demonstrate the safety and environmental potential of fusion power. An important aspect of the technological mission for ITER is the demonstration of efficient remote handling capabilities for regular maintenance, re-fit and repair (Aymar 1999).

Construction was planned to take ten years and would be followed by a programme divided into a physics phase followed by a technology phase, each of about ten-year duration. The Final Design Report put the estimated construction costs at about $5.5 billion (1989 dollars, because of inflation 30% had to be added to this number to arrive at 1998 values); this number was within the anticipated target of $6 billion, set in 1989. Including all peripheral costs, construction would cost approximately 750 million dollars a year and annual operating costs would amount to 400 million. This must be compared with total global spending on fusion in 1998 of around 1400 million dollars ($500 million for Europe, about the same for Japan, the US $230 million[7] and the rest by other countries). So ITER was affordable, but a huge part

---

[7]The Department of Energy had consistently been asking for an increase of the fusion budget, but its attempts were frustrated by Congress. It had asked for an increase from about $370 million in 1995 to about $860 million in 2002 (an average of $645 million per year between 1995 and 2005), but Congress reduced it to a mere $230 million. In the original ITER agreement it was agreed that the EU and Japan would each bear one third of the cost, while the USSR and the US would share the other one third. With Russia bankrupt, the US would have to bear this one third on its own, which would consume its entire fusion budget and probably more.

of the total fusion budget would have to be spent on it. All seemed very positive and everybody seemed enthusiastic, but the design nevertheless failed to receive approval, as the US government judged the cost of the construction project to be unaffordable within its funding priorities.

By 1998, the US research community had raised a multitude of concerns, including ITER's costs, claims that it would not achieve its technical goals, frustrations with the lengthy design process, and a political climate reluctant to allocate large amounts of money to support an international energy-related project for which the original political rationale (East-West collaboration in the climate of an easing Cold War) had disappeared (Dean 1998a, p. 156).

It should also be realised though that the final ITER design came just after the failure of TFTR. While the fusion community considered the TFTR results as achievements, the reaction of the bureaucrats of the Department of Energy was different: the attempt to reach breakeven had been unsuccessful and a new approach was clearly necessary. The very credibility of the tokamak concept was undermined. Even the use of the word "tokamak" was discouraged (Zakharov 2019).

The Americans had lost faith in themselves and in ITER. ITER was thought to be too expensive and too big, which was not all that surprising in view of the increasingly dwindling US budget for fusion. Congress, helped by negative media coverage undermining ITER's credibility, instructed that the US withdraw completely from the venture. As usual, all kinds of political reasons also played a role here. The Soviet Union had collapsed and the collaboration could no longer be advertised as a means to improve East-West relations, now that the East had shrivelled to nothing with a bankrupt Russia and China not yet in the picture. Long-term concerns about energy resources and climate change still existed, but there was no sense of crisis yet in this respect (Braams and Stott 2002, p. 256).

10.5   The Revised ITER Design

After the withdrawal of the US a new course had to be set out. The three remaining parties remained positive, especially Japan, but were reluctant to increase spending. Critics would repeat long-standing and quite valid objections that so much still remained to be understood about the behaviour of plasmas that it would be foolhardy to proceed with such an expensive step as ITER, an argument that had been made many times before, but had consistently fallen on barren ground. The proponents argued that the scientific basis was much more reliable and secure than 25 years before, which was not such a strong argument, of course, for if the scientific basis had not improved one could rightly ask what they had been doing all that time. Without the US and with Russia bankrupt, Europe and Japan[8] would essentially have to

---

[8]Russia remained a party but due to its difficult economic situation limited its participation. Canada had joined the project in the late 1980s, but as an associate member participating as a member of the European team.

shoulder the greater part of the burden. The withdrawal forced them to aim for a 50% reduction in cost and a redesign, although a further reason for the redesign was that Japanese scientists too had raised doubts about the burgeoning cost of the project which now stood already at $10 billion (Glanz and Lawler 1998; Glanz 1998). The reduction in size and cost was achieved largely by compromising on the machine's main scientific objective, ignition in a burning plasma. This hugely important objective was abandoned and ignition was no longer aimed for. Instead, the new design will aim for a burning plasma in which α-particles provide at least 50% of plasma heating. It will produce at least 10 times as much energy as consumed for heating the plasma ($Q > 10$) (meaning just the energy that actually goes into the plasma, not including losses and any other energy needed, e.g. for cooling the magnets and suchlike), generating 400 MW of power in bursts of 400 s, instead of the original 1.5 GW in 1000-second bursts (Redfearn 1999). The three remaining parties soon agreed to go back to the drawing board and design a less ambitious reduced-size version of ITER that would still satisfy the overall objective of demonstrating scientific and technical feasibility of fusion power, but at 50% of the cost. The result was a considerably smaller device initially called the ITER Fusion Energy Advanced Tokamak (ITER-FEAT) and expected to cost $3–4 billion.

The new design was published in 2001 and proposed a tokamak device with major and minor radii as well as plasma current roughly three-quarters of the earlier 1998 design. Changes to the design were made until 2014, when the construction of ITER was already in full swing. In that year the design of the reactor was 'frozen', and no further changes accepted. One of the consequences is that ITER will be out of date before it is actually running. No use can be made of any new technological developments (e.g. high-temperature superconductors) between 2014 and 2035 when D-T experiments will hopefully start.

The parameter values of the various designs have been summarised in Table 10.1.

From the table we see that several designs were meant to reach ignition. Ignition is of course more desirable than just $Q > 10$, which is ITER's goal. The curious thing about Table 10.1 is that the Next European Torus (NET), which was overtaken by INTOR and is considerably smaller than ITER 2001, would actually be able (at any rate according to its design with the usual caveats for extrapolation, scaling and suchlike) to reach ignition and have a burning pulse length of several hundred seconds, and so would be closer to DEMO than ITER (Toschi 1986, p. 325). So, why has the NET design, which was ready in 1985, been abandoned or why was it not revived when ITER 1998 had to be scaled down?

Figure 10.1 shows a cutaway view of the ITER tokamak next to JET on the same scale.

It shows how vastly larger the ITER device is going to be. Its plasma volume will be 840 m^3 (to JET's 100 m^3) inside a vacuum vessel of 1400 m^3. It will weigh 23,000 tonnes, as heavy as three Eifel towers, and will be housed in a building, measuring 73 m high (60 m above ground and 13 m below). All these and other figures are proudly displayed on the ITER website.[9] They are indeed very impressive, but also

---

[9]www.iter.org.

Table 10.1 Comparison of the parameters of INTOR, the designs of various national test reactors and the two ITER designs

Name	INTOR 1986	NET	FER	TBER	OTR	ITER 1998	ITER 2001
Origin		EU	Japan	US	USSR		
Plasma current, I (MA)	8	10.8	8.74	10	8.0	21	15
Major radius, R (m)	5.0	5.18	4.42	3.0	6.3	8.14	6.2
Minor radius, a (m)	1.2	1.35	1.25	0.83	1.5	2.8	2.0
Elongation, κ	1.6	2.05	1.7	2.4	1.5	~1.6	1.7–1.85
Toroidal field, B_T (tesla)	5.5	5.0	4.61	5.55	5.8	5.68	5.3
Safety factor, q	1.8	2.1	1.8	2.2	2.1	3.0	3.0
Average n_e (10^{20} m^{-3})	1.6	1.7	1.14	1.06	1.7	0.98	1.0
Average T_i (keV)	10	15				12.9	8.1
Average β (%)	4.9	5.6	5.3	6.0	3.2	2.2–3.0	2.5
Confinement time, τ_E (s)	1.4	1.9	1.7	0.44	1.7	5.9	3.7
Energy multiplication, Q	Ignition	Ignition	>20	>5	>5	Ignition	10
Nominal fusion power (MW)	600	650	406	314	500	1500	500
Neutron wall load (MWm^{-2})	1.3	1	1	1	0.8	1.0	0.57

Adapted from Braams and Stott 2002, p. 251

deeply disturbing, since they do not bode well for a future DEMO or real power reactor. If they still have to be proportionally bigger, they will become far too big and, what is more important, far too expensive for a realistic power plant option (we will discuss these and other aspects of the economics of power generation by nuclear fusion in a later chapter). The expected growth can be seen from Fig. 10.2, which illustrates the increase in plasma volume and major radius of successive tokamaks, and foresees a plasma volume of 2200 m^3 for DEMO[10]. The vacuum vessel will probably be about double that. Compare this to the volume of a 1GWe Pressurized Water Reactor (PWR), the most common type of nuclear fission power station, which is typically something like 35 m^3, a factor 60 smaller! (Tucker 2019, p. 35). For a

---

[10]This plasma volume, for that matter, is only slightly bigger than the volume of 2000 m^3 foreseen in the ITER 1998 proposal.

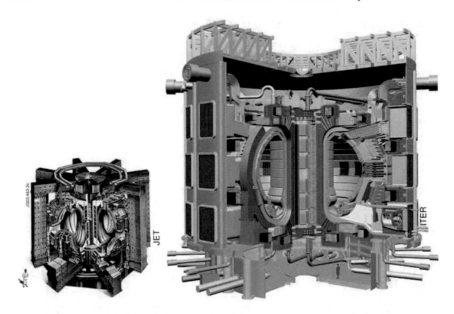

Fig. 10.1 Cutaway view of the ITER tokamak next to JET on the same scale

Fig. 10.2 Evolution of the
dimensions of some relevant
tokamaks compared with
ITER and DEMO
(*from* Lampasi et al. 2017)

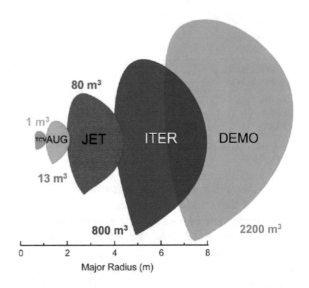

PWR the Pressure Vessel, which contains the nuclear reactor coolant, the surrounding stainless steel cylinder (core shroud) and the reactor core, is roughly 200 m³.

After the approval of the final ITER design and motivated by the renewed prospect of a positive next step in magnetic fusion research, the US signalled in 2002 its willingness for a renewal of US participation in ITER, resulting in President George

W. Bush announcing in 2003 that the United States would re-join the collaboration. The decision was undoubtedly influenced by the fact that China and South Korea were also negotiating their participation. The US did not want to lose out, now that other countries were joining in. For what it is worth, as US policy regarding fusion is rather fickle, in November 2003, US Secretary of Energy Spencer Abraham announced that ITER would be the top priority in the 20-year facility development plan of the DOE Office of Science, but its support has remained lukewarm and already in 2008 it failed to pay its financial share (Chen 2011, p. 303). The twenty years Abraham was talking about are almost over now and will be over before construction of ITER has finished.

China and South Korea joined the project in 2003, followed by India in 2005, indicating the broad international appeal of and support for the project.

With these additions ITER has become by far the greatest consolidation of nuclear fusion efforts ever, on a truly global scale, with all major economies (except Brazil) taking part. By 2005, the megaproject represented over half of the world's population. Total global GDP in 2019 is estimated to be about 87 thousand billion dollars according to the International Monetary Fund. The countries participating in ITER account for about 66 thousand billion of this, or about 75%, and on that scale 20, 30 or even 40 billion for a project such as ITER is indeed just a very, very tiny amount.

The ITER Members currently taking part in the design, construction and operation of this experimental device (see Fig. 10.3) are the EU (actually the European Atomic Energy Community (EURATOM) which comprises the 28 member states of the EU (including Britain), plus Switzerland and Ukraine), Russia, India, China, South Korea, Japan and the US, while technical cooperation agreements have been concluded with Australia (in 2016) and Kazakhstan (in 2017). A Memorandum of Understanding has been agreed with Canada about exploring the possibility of

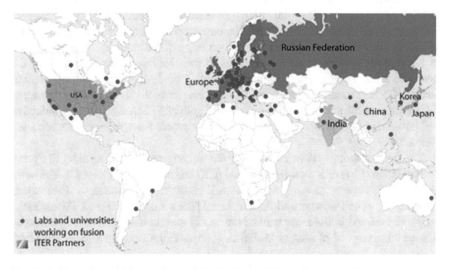

Fig. 10.3   Countries and laboratories participating in the ITER project (*from* F4E)

future cooperation and a Cooperation Agreement has been concluded with Thailand (2018).[11] In accordance with the ITER Agreement each of the seven ITER Members has established a separate legal entity (the Domestic Agency) through which the contributions to ITER will be made. For the EU this Domestic Agency is Fusion for Energy (F4E). The official name of this organization is European Joint Undertaking for ITER and the Development of Fusion Energy and its seat is in Barcelona.

10.6   Details of the New ITER Design[12]

To a great extent the newly designed ITER is an enlarged version of JET. It is (or better will be) a long-pulse tokamak with elongated (D-shaped) plasma and a poloidal divertor (i.e. in which the weaker poloidal field lines (going the short way around the torus) are diverted, which are weaker and therefore easier to divert than toroidal field lines). It is expected to produce 500 MW of fusion power in a D-T plasma with a burn length of 400 s (so the plasma will burn for 400 s, which is extremely long for a tokamak), with the injection of 50 MW of auxiliary heating (the 500 MW of fusion power divided by this 50 MW of auxiliary heating gives the scheduled Q value of 10).

The major components of the ITER magnet system are the 18 superconducting toroidal field (TF) coils, and the 6 superconducting poloidal field (PF) coils, whose combined magnetic field will confine, shape and control the plasma inside a toroidal vacuum vessel. The coils are extremely heavy, in line with everything else on ITER. The toroidal field coils produce a maximum magnetic field of 11.8 T around the torus. Its primary function is to confine the plasma particles. The poloidal field magnets pinch the plasma away from the walls and contribute to maintaining the shape and stability of the plasma. The poloidal field is variable and induced both by the magnets and by the main plasma current. This latter current is induced by the changing current in the central solenoid, which is essentially a large transformer, and the 'backbone' of the magnet system. It also contributes to the shaping of the field lines in the divertor region, and to vertical stability control. The design of the central solenoid enables ITER to access a wide window of plasma parameters, enabling the testing of different operating scenarios up to a plasma current of 17 MA and covering inductive and non-inductive operation. Figure 10.4 shows where the various components are situated.

The total magnet system consists of the already mentioned toroidal (TF) and poloidal field (PF) coils, a central solenoid (CS) and correction coils (CC). The latter consist of 18 superconducting coils that will be distributed around the tokamak at three levels. Much thinner and lighter than ITER's massive TF and PF magnets, they will be fixed to the inner wall of the vessel and used to control the plasma so that certain types of plasma instabilities, such as Edge Localized Modes (ELMs)

---

[11] https://www.iter.org/proj/inafewlines#5.

[12] Taken from the ITER website at www.iter.org.

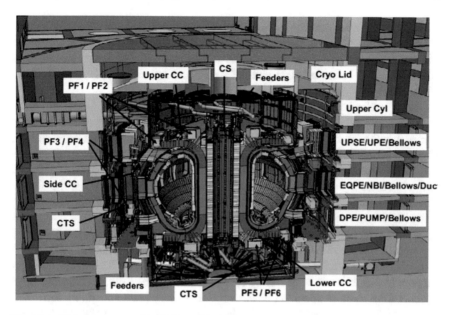

Fig. 10.4 Major components of ITER outside the vacuum vessel

are mitigated. ITER's success will partly depend on its capability to control these instabilities. The positions of these magnets are indicated in Fig. 10.4.

The six ITER poloidal field coils are wound in double pancakes. All pancakes are cooled in parallel with liquid helium. The helium inlets are located at the innermost turns whereas the helium outlets are located at the outer radius. The pancake model of these coils is shown in Fig. 10.5.

The ITER magnet system will be the largest and most complex ever built. The superconducting material for both the central solenoid and the toroidal field coils is an alloy of niobium and tin (Nb_3Sn). The poloidal field coils and the correction coils

Fig. 10.5 The poloidal field coils completely encircle the ITER vacuum vessel and toroidal field magnet system (*from* ITER Organization)

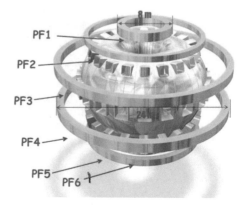

use a more standard and cheaper niobium-titanium (NbTi) alloy. These alloys are ordinary superconductors, i.e. not high-temperature superconductors, which makes it necessary to cool the magnets to −269 °C, just four degrees above absolute zero. ITER will use an estimated 500 tonnes of Nb_3Sn and 250 tonnes of NbTi, in the form of strands tightly wound into cables.

The central solenoid consists of six separate superconducting niobium-tin coils. With its height of eighteen metres, width of four metres and weight of one thousand tonnes it is the largest solenoid ever built for a fusion device. A maximum field of 13 T will be reached in the centre of the central solenoid, the strongest of all ITER's magnet systems. Its main function is to induce the current in the plasma, which in turn will create a poloidal magnetic field that helps to confine and heat the plasma (to an insufficient 20 million degrees) (Libeyre et al. 2019). The complete magnet system is shown in Fig. 10.6.

All these superconducting coils have to be connected to power supplies. Normally copper cables would be used for this. For ITER this is no good as the heat would leak along the copper from room temperature to the coils at −269 °C. Therefore, high-temperature superconducting (HTS) current leads will be used, transferring large currents from room-temperature power supplies to very low-temperature superconducting coils at a minimal heat load to the cryogenic system. The HTS leads would need only 0.63 MW of refrigeration power, while copper leads would require 2.2 MW.

The coils of the magnet system are manufactured all over the world in Europe, Japan, Russia and China to very detailed technical specifications in order to make sure that they will be as much as possible identical. More than 1000 people worldwide are involved in the production of ITER's magnets. When in the end all these magnets turn out to be compatible and to fit smoothly into their envisaged place in the system, it must be considered a great feat of engineering and management.

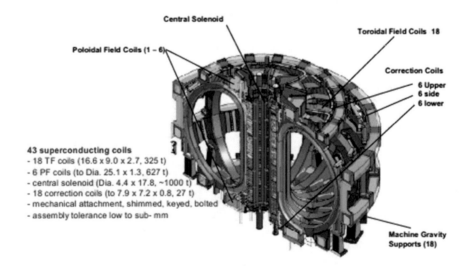

Fig. 10.6 Further details of the magnet system of the ITER tokamak

The practical impossibility of the entire project, and of future reactors of even larger proportions, is nicely illustrated by the effort needed to transport such a coil (one of in total 19) to the ITER site. An ITER news item of 23 March 2020 says about the transport of a coil made in Europe: "It's big, it's heavy, it's precious and it's highly symbolic: the toroidal field coil that was unloaded at Marseille industrial harbour on 17 March is the most massive, most sophisticated ITER component to arrive in France, and the first to belong to the very core of the ITER Tokamak. (...) It is a unique piece of high technology: twelve years of work, involving more than 700 people and 40 different companies, went into its making. Ensuring safe delivery to the ITER site has come with many daunting challenges. Global logistics provider DAHER, the ITER Organization, and the European Domestic Agency Fusion for Energy collaborated closely for three years to establish procedures and to design and manufacture lifting tools and a transport frame. The resulting transport and storage frame, a 100-tonne structure that brings the total load to 420 tonnes, is in itself a remarkable achievement." This procedure has to be repeated for each of the other 18 coils. Imagine having to do this for even larger coils for dozens or hundreds of future power stations all over the world. And this just concerns the coils! It certainly can be done but requires an immense effort and huge expense.

The entire tokamak is enclosed in a cryostat, a sort of giant thermos flask, as a thermal shield between the hot components and the cooled magnets. The toroidal and poloidal field coils lie between the vacuum vessel and this cryostat, where they are cooled and shielded from the neutrons of the fusion reactions. Imagine the enormous temperature gradient in the machine over a few metres from the 150 million degrees in the plasma vessel to the -269 °C of the liquid helium for cooling the magnets; from the highest temperature in the universe to almost the lowest possible. The cryostat, whose parts are manufactured in India and then welded together on site, is a large (29 × 29 m), stainless steel structure surrounding the vacuum vessel and magnets, providing a super-cool, vacuum environment, as well as structural support to the tokamak. It is made up of two concentric walls with the space between the walls filled with helium gas that acts as thermal barrier. It will encase the entire reactor including all the magnets. The cryostat has many openings, some as large as four metres in diameter, to provide access to the vacuum vessel for cooling systems, magnet feeders, auxiliary heating, diagnostics, and the removal of various parts. During operation the structures of cryostat and vacuum vessel will experience thermal contraction and expansion by as much as 5 cm under the influence of the magnetic fields and temperature changes from room temperature to -269 °C. The huge forces exerted on the machine will be transferred to the ground by eighteen chrome-plated spherical bearings, each weighing approximately 5 tonnes.

The vacuum vessel has various internal, replaceable components, such as blanket modules, divertor cassettes, limiter, heating antennae, test blanket modules, and diagnostics modules, which absorb the radiated heat as well as most of the neutrons from the plasma and protect the vessel and magnet coils from excessive nuclear radiation. Many of these components are being tested and their design adjusted and refined for ITER in smaller tokamaks around the world, like JET and JT-60. The heat deposited in the internal components and in the vessel is exhausted to the outside by means of the cooling water system.

Remote handling will have an important role to play in the ITER tokamak. When the machine is in operation, it will no longer be possible to make changes, conduct inspections, or repair any of the components in the activated (radioactive) areas other than by remote handling. These handling techniques must be very reliable and robust to manipulate and exchange components weighing up to 50 tonnes. Their reliability will also impact the length of the machine's shut-down phases.[13]

The vacuum vessel is a hermetically-sealed steel container inside the cryostat, which both are sucked vacuum to a pressure of one millionth of normal atmospheric pressure. The magnet system is then switched on and the low-density gaseous fuel fed in. The vessel acts as a first safety containment barrier. The central solenoid will induce a current in the gas, which will be maintained during each plasma pulse, ionise the gas and transform it into a plasma. The plasma particles continuously spiral around in the vessel's doughnut-shaped chamber without touching the walls. The size of the vacuum vessel dictates the volume of the fusion plasma; the larger the vessel, the greater the amount of power that can be produced. The ITER vacuum vessel will be twice as large and sixteen times as heavy as the JET vacuum vessel, which has been the biggest so far. It has an internal diameter of 6 m and will measure a little over 19 m across (outer diameter) by 11 m high; the empty vessel weighs in excess of 5000 tonnes (and 8500 tonnes when fully equipped). The vessel will be built partly in Europe and partly in South Korea.

The inner surface of the vacuum vessel is covered by the blanket, a thick complex structure consisting of 440 modules, also called bricks. Each brick, of which there are about 100 different types, will be about 2 m high and 1 m wide and will weigh up to 5000 kg. In ITER it will serve two major purposes: (1) to capture the neutrons produced by the fusion reactions and convert their energy into heat; and (2) to provide shielding of the superconducting magnets from these high-energy neutrons. This blanket is designed to withstand a thermal load of 700 MW (i.e. 700 MW in heat, the output of a small power station. In a real power reactor this heat will be converted into steam and subsequently into electricity). In future power reactors part of the blanket will also be used to breed tritium and ITER will be a testing ground for this. ITER will obtain the tritium it needs (probably the entire global supply) from the market, but tritium is expensive and the global supply, which is about 25 kg and increases by about half a kilogram per year, is insufficient to cover the needs of future power plants. There may even not be enough tritium available to start up a DEMO plant after ITER (Ni et al. 2013). The biggest sources of tritium are Canada Deuterium Uranium (CANDU) nuclear fission reactors that use heavy water (D_2O) as moderator (i.e. to reduce the speed of the fast neutrons produced in the fission reactions). Today there are 31 CANDU reactors in use around the world. Apart from Canada, which is the world's largest producer of tritium with 19 of such reactors, they are operational in Argentina, India, Pakistan, China, South Korea and Romania. A typical CANDU reactor produces about 130 g of tritium per year. Tritium can only be extracted from

---

[13] https://www.iter.org/mach/RemoteHandling.

Fig. 10.7 Main layers of the tokamak vessel wall

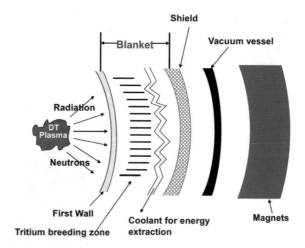

the heavy-water moderator by means of a Tritium Removal Facility (TRF), of which only two are currently in operation, one in Canada and one in South Korea, with plans for a third one in Romania.[14] For DEMO about 300 g of tritium will be needed per day to produce 800 MW of electric power.[15] Hence, it is essential that future devices will be able to breed their own tritium. This will be further discussed in Chap. 19, but it is clear that commercial development of fusion energy will be out of the question, if self-sufficiency in tritium cannot be achieved. ITER will test tritium-breeding concepts by testing breeding blanket models, called Test Blanket Modules or TBMs, which contain lithium and can produce tritium through the reaction

$$^{6}_{3}\text{Li} + n \rightarrow {}^{4}_{2}\text{He}(2.05\text{MeV}) + {}^{3}_{1}\text{T}(2.75\text{MeV}).$$

Figure 10.7 shows how the wall of the vacuum vessel will look like: the neutrons will pass through the first wall into the tritium-breeding zone and from there into a coolant to extract their energy (which can be converted into electricity).

A cross section through the torus of the tokamak in its cryostat with the various components is shown in Fig. 10.8.

Forty-four ports (openings in the vessel wall) will provide access to the vacuum vessel for remote handling operations, diagnostic systems, heating, and vacuum systems.

The temperatures inside the ITER tokamak must reach 150 million degrees for the gas in the vacuum chamber to reach the plasma state and for fusion reactions to occur. The hot plasma must then be sustained at these extreme temperatures in a controlled way in order to extract energy. Three sources of external heating will be used to provide the input heating power of 50 MW (comparable to the amount in

---

[14]This is a sensitive matter as CANDU power plants emit tritium, which is radioactive, into the environment.

[15]https://www.iter.org/mach/TritiumBreeding.

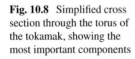

Fig. 10.8 Simplified cross section through the torus of the tokamak, showing the most important components

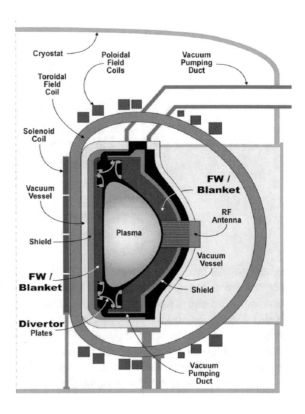

JET) required to bring the plasma to the temperature necessary for fusion. These are neutral beam injection and ion and electron cyclotron heating (ICRH and ECRH). These methods have been described in Chap. 2.

The large plasma volume in ITER will impose new requirements on the neutral beam heating: the particles will have to move three to four times faster than in previous systems in order to penetrate far enough into the plasma, and at these higher velocities the positively charged ions become difficult to neutralize. At ITER this problem will be circumvented by working with a negatively-charged ion source (N-NBI). Such a source has already been used in JT-60U and will be perfected in JT-60SA, especially for ITER (see Chap. 8). Although the negative ions will be easier to neutralize, they will be more challenging to create and to handle than positive ions. The additional electron that gives the ion its negative charge is only loosely bound, and consequently readily lost.

Two neutral beam injectors—each one delivering a deuterium beam of 16.5 MW with particle energies of 1 MeV—are currently foreseen for ITER. A third neutral beam will be used for diagnostic purposes.

ECRH is also used to deposit heat in very specific places in the plasma, as a mechanism to minimize the build-up of certain instabilities that might cool the plasma. In comparison to ICRH, ECRH has the advantage that the beam can be transmitted through air which simplifies the design and allows the source to be far from the plasma, simplifying maintenance.

The 600 m^2 interior surface of the vacuum vessel is covered by the blanket, which is one of the most critical and technically challenging components in ITER: together with the divertor it directly faces the hot plasma. Because of its unique physical properties (low plasma contamination, low fuel retention), beryllium has been chosen as the element to cover the first wall, the part of the blanket facing the plasma. The rest of the blanket modules will be made of high-strength copper and stainless steel. During later stages of the ITER operation, some of the blanket modules will be replaced by specialized modules to test materials for tritium breeding. As said, a future fusion power plant producing large amounts of power will be required to breed all of its own tritium. ITER will test this essential concept of tritium self-sufficiency.

The divertor is positioned at the bottom of the vessel (see Fig. 4.11 for a sketch of the divertor position in ITER) at a place where the magnetic field strength is almost zero. Particles will leave the plasma flowing along magnetic field lines and then naturally fall into the 'ashtray'. In this way the divertor extracts heat (sixty per cent of the plasma exhaust is designed to go into the divertor) and ash (helium) produced in the fusion reactions, minimizes plasma contamination by removing impurities, and protects the surrounding walls from thermal loads. It is one of the key components of the device and is in effect a giant nuclear ashtray. The ITER divertor is made up of 54 remotely-removable cassettes, each holding three plasma-facing components, or targets. The targets are situated at the intersection of magnetic field lines where the high-energy plasma particles strike the components. Their kinetic energy is transformed into heat. The heat flux received by these components is extremely intense (10–20 MW/m^2, ten times higher than the heat load a spaceship experiences when entering the Earth's atmosphere and for much longer periods) and requires active water cooling. The choice of the surface material for the divertor is an important issue. Only very few materials are able to withstand temperatures of up to $3000\,^\circ\text{C}$ for the projected 20-year lifetime of the ITER machine. For this reason, the divertor will be made of tungsten, which has the advantage of a low rate of erosion and thus a longer lifetime, meaning that the divertor needs to be replaced only once during the lifetime of the tokamak. To demonstrate that the tungsten components can withstand the demanding thermal conditions of the ITER machine, full-scale prototype sections of the plasma-facing components undergo testing in a high-heat flux test facility specially conceived and built for ITER (the Divertor Test Facility in St. Petersburg, Russia). The components are manufactured in Japan and shipped directly to St. Petersburg for testing at the Russian facility. Another special divertor test facility, the ITER Divertor Test Platform Facility, is operating since early 2009 in Tampere, Finland. This facility develops and tests the maintenance robot and remote handling operations necessary for replacing parts of the ITER divertor. Experiments to test the melting behaviour of tungsten, which has the highest melting point ($3422\,^\circ\text{C}$) of all elements, have been run on JET, while the WEST project in France (see Chap. 6)

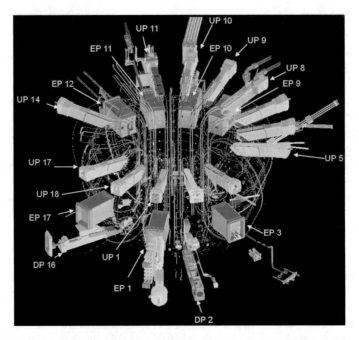

Fig. 10.9 The ITER device is surrounded by 60 instruments to measure 101 parameters. (*from* https://www.iter.org/newsline/-/3360)

will contribute key insight into the steady state operation of a tungsten divertor and its impact on plasma performance.[16]

Diagnostic systems play an essential role for ITER, to make sure that the reactor will operate as efficiently as possible. The device will be surrounded by about 60 measuring systems. Figure 10.9 gives an impression of this.

Magnetic diagnostics will measure currents in the plasma or in structures; the energy stored in the plasma; and control plasma shape and position.

Neutron diagnostics such as neutron cameras, neutron spectrometers and neutron flux monitors will measure fusion power from the neutrons released in the fusion processes.

Optical systems will measure temperature and density profiles at the core or the edge of the plasma.

Bolometric systems situated all around the vacuum vessel will provide information on the spatial distribution of radiated power in the main plasma and divertor region.

Spectroscopic instruments and neutral particle analysers will deliver information on plasma parameters such as impurity species and density, input particle flux, ion temperature, helium density, fuelling ratio, plasma rotation, and current density.

---

[16]https://www.iter.org/mach/Divertor.

Microwave diagnostics will probe the main plasma and the plasma in the divertor region in order to measure plasma position.

A further large array of diagnostic instruments will provide the measurements necessary to control, evaluate and optimize plasma performance and to further the understanding of plasma physics.

Plasma-facing and operational diagnostics will aid in the protection and operation of the machine. Several visible and infrared viewing systems will monitor the conditions in the main chamber and the divertor.

All this shows that ITER will be an extremely complex machine. In total it will have 1 million components comprising 10 million pieces (Claessens 2020, p. 79) that all have to work in concert to guarantee proper operation. Operators in the future ITER control room will have to play with and adjust the magnets, the external heating, the density of the gas and other parameters to stabilise the plasma against turbulence and instabilities. It may take several months, if not years to sufficiently master and control the plasma and be able to run an experiment.

10.7 Site Selection and Construction

The New Final ITER Design Report was presented in 2001 and after its approval by the ITER Council, the supervising body of the ITER Organization, construction of ITER could start, if a site had been selected. This issue turned out to be almost as difficult as designing the machine itself, in spite of the fact that there were essentially only two candidates (Japan and the EU) for siting the device, since by 1999 Japan and the EU had remained as the primary parties still interested in building ITER. With the earlier difficulties in mind in selecting a site for JET and for the EDA Joint Central Team, we should perhaps not be too surprised about this.

Site selection can indeed often be a thorny matter, even for less costly scientific projects. Scientists might choose several sites that meet their scientific and technical criteria. Other factors that may be relevant include accessibility to scientists from around the world—possible visa restrictions; licensing for nuclear materials; the desire for a green-field site versus the benefits of using existing roads and other infrastructure; incentives offered by potential hosts; proximity to a university; the reputation of local schools; and jobs for spouses. The interplay of these and other factors is specific to each project, but if the price tag is high enough, politics inevitably plays a role (Feder 2004).

In 2001 Canada, which was not a full partner to ITER, but just participated as an associate member of the European team, offered a site next to the Darlington Nuclear Power complex on the shore of Lake Ontario. The site appealed to some who believed that building the facility there might entice the US to re-join the ITER project. One of the other attractions of the offer was that ITER could use the tritium produced as a by-product of Canada's nuclear fission plants. Moreover, Canada's cheaper labour and electricity would save an estimated 15–20% in ITER operating costs compared to Japan, for instance. When some time later the Canadian offer was rejected, the country withdrew from the project altogether in 2004. Only recently in April 2018 a Memorandum of Understanding was signed between the ITER Organization and the Canadian government agreeing to explore the possibility of future cooperation.

In the meantime, Europe and Japan had also come up with site proposals. Japan offered to host the device at Rokkasho, Aomori, in the north of its main island Honshu, which is home to multiple facilities of the Japan Atomic Energy Agency. Like many other regions in Japan, it is prone to earthquakes, reason why China (after it joined ITER) was vehemently against this choice as the site for ITER. Two sites in Europe were proposed, Spain offering a site at Vandellos, about 140 kms from Barcelona, for which it was prepared to even double its contribution—upwards to $1 billion—to the project, and France the Cadarache site near Aix-en-Provence in southern France. The Spanish site had the advantage that it was much closer to the sea than Cadarache, which would make transport of components from overseas easier, but it had no research facility nearby as Cadarache, which already for a long time had been the home of the French CEA. With the US, China and South Korea also having joined the ITER project, there were now six parties of which Europe, China, and Russia insisted on Cadarache, while Japan, South Korea, and the US voted for Rokkasho. The US vote for Japan was seen as a revenge action for France's opposition to the US-led invasion of Iraq early in 2003, while the support from China and Russia for the French site had clearly to do with their opposition to the Iraq war. The US even for some time backed the Spanish site to reward Spain for its support over the Iraq conflict and to spite the French. This attitude of a party that was going to enjoy a ride on the ITER carrousel while contributing a mere 10% towards its costs and had re-joined just recently was not appreciated in Europe. There was an impasse because everything had already been agreed, and there were not very many things left to negotiate.

It took the EU (consisting at the time of 15 member states) to late 2003 to decide to throw its support behind the Cadarache site, yielding to pressure from the eternal French-German axis in European politics, as some Spanish politicians grumbled. From this point on, the choice was between France and Japan and a new stalemate ensued. In the competition to host the project, the main issues were global prestige and regional economic benefits. As both Cadarache and Rokkasho had been deemed suitable by the ITER site evaluation team, the eventual choice had little to do with engineering and instead involved 'financial, political, and social' issues. Also here, the US let politics play a role by supporting the Japanese site as a reward for Japan's contribution to the US's Iraq adventure.[17] Japan was so keen to host ITER that it offered to pay a substantial fraction of the project's cost for the privilege, to which Europe responded that it would do the same. Some horse trading was going on, whereby the party that does not get ITER will host a €1 billion support facility—the International Fusion Materials Irradiation Facility (IFMIF). At some point the EU, led by France, threatened to build ITER as a Europe-only project, for which France, Italy, Spain, and Switzerland volunteered to increase their contributions (Feder 2005).

---

[17]See https://fire.pppl.gov/abraham_japan_010904.pdf (accessed 2 October 2019) for the US Secretary of State's grovelling remarks to his Japanese hosts.

Throughout 2004 negotiations continued in countless meetings, until finally, in May 2005, French President Chirac, back from a visit to Japan, declared that he had reached agreement with the Japanese.[18] ITER would be built at Cadarache in the small commune of Saint-Paul-lez-Durance. When the details of the agreement emerged, it became clear that it involved major concessions to Japan. The EU and France would contribute half of the then estimated €12.8 billion total cost, with the other partners—Japan, China, South Korea, US and Russia—just putting in 10% each. Japan will get 20% of the industrial contracts, host the IFMIF and have the right to host a subsequent demonstration fusion reactor. Finally, 20% of the project's scientists would come from Japan and it would get to choose ITER's first director-general.

Further negotiations established the ITER Agreement to detail the construction, exploitation and decommissioning phases, as well as the financing, organization and staffing of the ITER Organisation. In November 2006, the seven members, with India also having joined, signed the ITER implementing agreement. The total cost of ITER comprises about half for the ten-year construction and half for 20 years of operation (McCray 2010, p. 296ff).

With this problem out of the way, construction could begin and site preparation works at Cadarache commenced in January 2007, in spite of all the bickering, less than a year later than originally planned. The project almost immediately ran into delays and cost increases, resulting in a management shake-up in 2010 when the first Japanese Director-General, a diplomat with a nuclear engineering background, was replaced by the second Japanese Director-General, this time a physicist by training. It did not help. Poor management, especially human resources management, is partly to blame. First concrete for the buildings was poured in December 2013, but delays and cost overruns kept occurring. It is hard to keep track of the real and rumoured versions of them. In 2015 the Frenchman Bernard Bigot, former head of CEA, was appointed as ITER's Director-General and he has since managed to keep the project on track. Fact is that, according to the original planning, experiments (first plasma) were due to begin in 2016 (Chen 2011, p. 305), but first plasma is now expected at the end of 2025. [19] This will mark the end of the construction phase and the beginning of operations. But assembly will actually continue through 2035, the planned date for the first ignition experiments using D-T plasma. The schedule up to 2035 is shown in Fig. 10.10.

Most large technological projects, and certainly any that have to do with anything nuclear, suffer delays and some extra delays were to be expected, not only due to the complexity of the ITER device itself, but also due to the extra complexity built in by the fact that it is built by a global cooperation effort unprecedented in scale and involving half of the world. Some delays were unavoidable and could not be foreseen, such as the year-long delay in the Japanese contribution, and consequently in the entire project, caused by the 2011 Fukushima earthquake and tsunami. Most delays, certainly the ones given as examples by Claessens in his book (Claessens

---

[18]For an extensive account of the whole unseemly proceedings, see Claessens 2020, pp. 48–54.

[19]In the original planning ITER would be decommissioned in 2027, which is now roughly the date at which construction is expected to be finished.

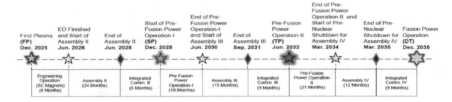

Fig. 10.10 ITER experimental schedule from first plasma (*from* Bigot 2019)

2020, p. 109ff. See also Chap. 9 in his book, which is an eye opener as regards the ITER management), are rather mundane and due to poor planning and inadequate understanding.

As mentioned above the ITER device has in total 1 million components, comprising 10 million pieces. Specifications of all these components have been stored in an electronic package of over 2 terabyte in size. The package contains detailed three-dimensional models of all components, which took 20 years to finalize and are constantly being improved and updated. The value of the construction and manufacturing of each component was estimated, and their manufacture divided among the ITER Members in accordance with the contribution allocation key for the ITER project (45.46% for the EU, 9.09% for the other Members). This was finalised in 2007 and the ITER Organisation started to place procurement contracts with the various Domestic Agencies, which in turn have sourced out the work to thousands of companies in their respective countries. Work was started, but modifications were soon proposed which led to difficulties as components became more expensive due to the proposed changes. In some cases, such changes also had consequences for other components or other parts of the project. A multitude of Project Change Requests was submitted and had to be decided upon, resulting in delays and discord as Domestic Agencies refused to take responsibility for these changes or to bear the extra costs. This explains a good part of the delays mentioned above. A further complication arose as the manufacture of key systems was allocated to more than one member. For instance, Europe, Russia and Japan are collaborating on the divertor, while it would obviously have been easier if one of them would have taken sole responsibility for the complete divertor. Even strictly identical components are sometimes built in different countries, and once they arrive in France, they must be compatible and in full conformity with the specifications and comply with the necessary standards and requirements. All this is mentioned to illustrate how herculean the task of constructing and coordinating the construction of ITER actually is. Nevertheless, in December 2017 50% of the construction work to first plasma was announced completed. In July 2020 in a special ceremony it was announced that assembly could begin, although not all parts had arrived at the site yet. The gigantic jigsaw puzzle must now be put together, which is expected to take about five years by a 3000-strong assembly team. Then it will become clear whether they all indeed fit as intended and ITER is not a modern Tower of Babel.

ITER is a machine of such daunting complexity that nobody has or can have a full grasp of it. This does not only apply to the machine, but also to the equally complex organization spanning half the globe. In addition, ITER is as much a political as a technological project, with all the associated political sensitivities. From the very beginning it has been a political project, with the most crucial decisions taken by politicians at the highest level, starting with Reagan and Gorbachev in 1985 and ending with Chirac in 2005. At first it was portrayed as a means to ease the tensions between East and West, and when it no longer could fulfil that function with the collapse of the Soviet Union, the US took a back seat leaving the EU and Japan to fight for the glory. For reasons of politics and prestige France, supported by the EU, went to extraordinary length to make sure that Cadarache was chosen as the ITER site. The ITER Council, which oversees the project, is essentially a political body, comprised of mostly ministerial level officials; governments not only pull the purse strings, but through the Council also have influence on the decision-making, e.g. in respect of contracts. Negotiations taking place on the Council are completely different from discussions on the board of a company; the representatives will first and foremost have the interests of their respective country in mind when they commit themselves to one thing or the other. In the end their governments must be satisfied with the outcome.[20]

We now just have to wait and see whether the latest revised schedule will be met. The assembly of the tokamak and other facilities will severely test the acumen of the ITER Organisation. The ITER Council is convinced that the 2025 deadline will be met but will probably be forced to revise the schedule. The Covid-19 pandemic that broke out early in 2020 and stalled work in various countries for several months slowed work down and will probably result either in a further delay of one year or in extra costs (Clery 2020). Progress can be tracked on the ITER website where the entire construction and assembly process through to first plasma (2025) has been divided up into milestones.[21]

When construction is finished, a relatively quiet period of ten years of continued machine assembly and periodic plasma operations with hydrogen and helium will follow. These gases do not produce any fusion neutrons, and permit the resolution of problems and the optimization of plasma performance with minimal radiation hazards. Plasma instabilities must be kept at bay to ensure adequate energy confinement, so that the reacting plasma can be heated and maintained at high temperature.

---

[20]Claessens 2020, p. 80ff, p. 120ff, who stresses the complexity and the engineering and logistical challenges of the project, as if he wants to hedge against any major mishaps with the excuse that it was obviously too great and too complex a task.

[21]https://www.iter.org/proj/itermilestones#150.

10.8   Cost

A big issue that tends to recur time and again in the media and is the cause of many delays is the cost of ITER. The report published after the 1988–1990 Conceptual Design Stage (Tomabechi et al. 1991, p. 1212) gave the first indications of the construction costs. They were then estimated at about $5 billion. The reduced design of 2001, made necessary after the withdrawal of the US, had a price tag of less than $4 billion. Since then costs have exploded and currently a figure of more than $20 billion is quoted by the ITER Organisation for the construction costs alone. Below it will be shown that total costs will be at least a factor of two larger before the game is over.

Recently the US Department of Energy nearly tripled its cost estimate for ITER to $65 billion. The ITER Organisation however stuck by its figure of $22 billion, which it claims is sufficient to bring ITER to 2035. Although DOE has maintained in the past that the US contribution could balloon, this is the first time that the agency has publicly challenged the ITER Organisation's overall cost assessment.[22] The fact is though that the ultimate cost of ITER may never be known. A large part of the project is managed directly by individual member states and the central organization has no way of knowing how much is actually being spent. In this connection the eventual total US contribution, which includes an enormous central electromagnet capable, it is said, of lifting an aircraft carrier, has been estimated at about $4 billion (*New York Times* 2017).

Let us try to bring some order in the ITER cost picture and get some hard figures from the ITER agreement documents and from EU documents that lay down the spending by the EU. If we know how much the EU spends, we can multiply by two to get a rough idea of the total amount spent, since as host, the EU is paying about 45% of ITER's construction cost, five times the share of each of the other six partners: China, India, Japan, Russia, South Korea, and the US.

The ITER Agreement itself mentions three types of contributions: contributions in kind (consisting of specific components, equipment, materials and other goods and services in accordance with the agreed technical specifications, and staff seconded by the members), contributions in cash (i.e. financial contributions by the members to the budget of the ITER Organisation) and additional resources received either in cash or in kind (which are not further specified; they just have to be within limits and under terms approved by the ITER Council). The first type of contributions is by far the largest and each member country has taken responsibility for a package of component parts. A nominal value has been assigned to these components and the total nominal value of the package corresponds to that member's contribution to ITER. The member must supply ITER with the agreed package of components and must do so at whatever the actual costs turn out to be. The components are manufactured in the member's own country and it is perfectly possible of course that for one member country the costs will balloon far beyond the nominal value, while others manage within the original budget. In many cases they will even not be prepared to disclose how much

---

[22] *Physics Today*, Politics and Policy, 16 Apr 2018.

they have spent in manufacturing a certain component. The fact that each of the seven participants has its own currency complicates matters even further. It is clear that such a system is very opaque and can cause problems, because in the end the real costs have to be met and, if they deviate much from the nominal value assigned to them, it could impact fusion budgets in some countries. It would have been much simpler and more cost-effective to have funded the ITER construction centrally, but that was at the time politically unacceptable to some countries (McCracken and Stott 2013, p. 146).

The ITER Agreement was signed on 21 November 2006 and concluded for a total period of 35 years; ten years for construction, 20 years for running experiments and five years for decommissioning. The agreement itself does not mention any numbers; these are contained in the documents "Value Estimates for ITER Phases of Construction, Operation, Deactivation, Decommissioning and Form of Party Contributions" and "Cost Sharing for all Phases of the ITER Project". The first of these documents states that total construction costs for ITER will be €3.6 billion, plus an extra €400 million if required, so in total roughly €4 billion (in agreement with the figure originally quoted for the reduced 2001 design). The document also states that the figures are at January 2001 values. The EU will contribute 45.46% towards the construction costs and the other six members/parties 9.09% each. This only covers the construction phase, which was to take 10 years, so until 2017. For running the device €200 million per year is envisaged and finally €800 million for deactivation and decommissioning. For these latter two phases another allocation key is used with the EU contributing 34% and the other parties 13% (US and Japan) or 10% (China, India, South Korea, Russia). The total amount envisaged for construction, operation and decommissioning is therefore €8.8 billion (2001 values).

The next document we will look at is a European Union document: EU Council Decision 2007/198/Euratom[23] which establishes the Euratom Domestic Agency as a Joint Undertaking[24], it carries the date 27 March 2007 and states that "the indicative total resources deemed necessary for the Joint Undertaking (…) shall be EUR 9653 million" (in current i.e. 2007 values) of which €4.1 billion would be spent between 2007 and 2016 (so for construction according to the original schedule) and €5.5 billion between 2017 and 2041 (for operation, deactivation and decommissioning). The Euratom contribution will be €7.6 billion of this and the rest has to come from "contributions from the ITER Host State (France)[25], the annual membership contributions and voluntary contributions from members of the Joint Undertaking other

---

[23] In full: Council Decision of 27 March 2007 establishing the European Joint Undertaking for ITER and the Development of Fusion Energy and conferring advantages upon it (2007/198/Euratom).

[24] The Joint Undertaking is called "Fusion for Energy" (F4E). It has an extensive website https://fusionforenergy.europa.eu/aboutfusion/ with links to the ITER organizations of the other participants.

[25] France contributes to ITER as a member of the EU, but as the host country for the project it has undertaken to provide the site for ITER (180 ha), carry out all preparatory work, install networks for electricity and water, build roads (it has spent an estimated €110 million adapting roads to accommodate transport of ITER components), establish an international school for employees' children, and construct the ITER headquarters buildings. It is not clear if and where the costs for these facilities are included in the ITER accounts.

than Euratom". So here we see that compared to the ITER Agreement and accompanying documents the cost for ITER construction alone has more than doubled (by using 2007 values compared to 2001 values, but inflation alone cannot account for such a dramatic increase in six years). Total construction costs will now amount to €4.1 billion/0.4546 = €9 billion, and total costs to about €25 billion (in the post-construction phases the EU (or better its Domestic Agency) will pay 34% of the costs, which means that its contribution of €9.6 billion minus €4.1 billion = €5.5 billion must be tripled to obtain the total costs for those phases).

So, the conclusion is that in 2007 total ITER costs stood at €25 billion, with construction costs at €9 billion; these numbers are hard figures that follow directly from figures mentioned in EU documents.

We now skip a few years and go to EU Council Decision 2013/791/Euratom of 13 December 2013 amending the Decision 2007/198/Euratom discussed above. It states that the resources deemed necessary for the Joint Undertaking during the ITER construction phase for the 2007–2020 period amounted to €7.2 billion (in 2008 values). So, the EU spent €7.2 billion in this first 13-year period (up from the €4.1 billion mentioned above) towards the construction of ITER. This is not everything spent by the EU on ITER or in support of ITER, since it does not include the contributions from the individual EU member states towards programmes in their own countries for testing certain components or aspects of ITER. In this connection it is for instance not clear to what extent the costs of JET, when carrying out programmes in support of ITER, have been included in the EU figure stated above. From the €7.2 billion EU contribution it follows that total construction costs in that period must have been about €16 billion. This is another hard figure: €16 billion construction costs paid for from 2007 to 2020, and the construction is not yet finished in spite of the earlier construction schedule of 10 years.

Another illuminating document is Com (2018) 445 containing an amendment to EU Council Decision 2007/198/Euratom, approved early in 2019, in which spending for ITER for the six-year period from 2021–2027 is set at about €6 billion, so €1 billion per year for the EU alone for that period just for ITER construction. This again will have to be met by a contribution of about €7 billion from the other ITER parties, bringing the total for ITER construction up to 2027 to €29 billion (in 2008 values) (the €16 billion already spent to 2020 plus the €13 billion to be spent up to 2027). The construction phase is scheduled to end one or two years before 2027, but I doubt that ITER construction costs will be lower than the figure just mentioned. In 2019 euro values it will be close to €36 billion (assuming an inflation rate of just 2%) or about $40 billion.[26]

This €13 billion (in 2008 values) spent or to be spent by the EU in the period from 2007 to 2027 (the €7.2 billion up to 2017 plus the €6 billion from 2020 to 2027) is one of the few rare hard figures around; they all come from official documents and

---

[26]This figure agrees fairly well with the value given by Claessens 2020, p. 115, but he then apparently is shocked by his own calculation as it is much higher than the official figure, and via a contorted reasoning he arrives at the conclusion that the actual figure must be somewhere between €13 and €41 billion, while he must have known that the EU alone has already spent more than his lower limit.

state exactly was has been spent or will be spent. If deviations occur, they have to be laid down in amended documents. The only assumption in arriving at the figure of €29 billion for ITER construction costs in 2008 values is that the contribution of the EU will be met by contributions from the other ITER parties in accordance with the ITER agreement. Apart from the US, they have so far all done so. It concerns just the construction phase of ITER and the total value of €29 billion (in 2008 values) of total ITER construction costs spent or to be spent from 2007 to 2027 must be considered a hard figure.

Extrapolating to 2035, while assuming that from 2027 annually also €1 billion from the EU will be needed (it probably will be more), the EU alone will spent an extra €8 billion bringing its total contribution to €21 billion (in 2008 values), about the same as the total budget the ITER Organisation talks about. The period after 2027 must be considered as the post-construction phase, so if the EU spends €1 billion per year, the other parties will have to spend about €2 billion, making the total costs for that period €24 billion and total ITER costs up to 2035 €53 billion, which is not far from the US DOE figure mentioned above. And 2035 is not the end date since experiments are scheduled to continue until 2047 after which deactivation and decommissioning will still follow.

It may be that the figure of €3 billion per year for ITER's operating costs as assumed above is somewhat high. However, spending on ITER is currently about €2 billion per year, assuming for simplicity that the EU is paying half. Normally when such projects get to the major operational phases the budget actually goes up. It would therefore not be unlikely that after 2025 the budget will start going up to €3 billion or €4 billion per year. Even if we assume that annual spending from 2027 will not increase and remain at €2 billion per year, the total cost amount up to 2035 will be €45 billion (in 2008 values) or $50 billion, twice the amount the ITER Organisation still assumes. It gets a little speculative, but if experiments continue until 2047 another €24 billion will have to be added to this, after which deactivation and decommissioning are still to follow.

All figures quoted at a certain time in the media or anywhere else are greatly distorted by inflation, by the use of different currencies and changing exchange rates. For a proper comparison it would be necessary to calculate them with respect to a reference year and in a single reference currency. This is virtually impossible and makes it very difficult to get a real grip on the figures. There can be no doubt though that the costs are soaring in a fabulous, unprecedented manner, which is a great worry, not so much as regards ITER, but for the prospects of any commercial electricity generation from fusion. If it costs so much to just show that power generation from nuclear fusion is in principle possible (not net power generation, mind you, but just that it is possible to have a burning plasma that can almost sustain itself for some rather short time), what will then be the cost of a real nuclear fusion power station? And who will be able to afford it? We will discuss these issues in a later chapter. I end here with the remark that it is of course an utter disgrace that the citizens of Europe, who have to pay for this extravagance, have no way of knowing in any precise way how much is being spent on their behalf. The Chinese and Russians may be clubbed into trusting their government and be prevented from asking any awkward questions, but the Americans, Japanese, Indians and South Koreans have likewise no way of knowing how their tax money is being wasted. One may accept such lack of transparency for a highly secret defence project, but it is unbecoming for a scientific project, especially when it has overrun its initial budget by such a huge margin.

10.9   Conclusion

What has ITER brought so far and what can we still expect? Technical delays, labyrinthine decision-making (true to its nature as a modern Tower of Babel) and opaque cost estimates that have soared from five to perhaps 45 billion euro (with some of the additional costs hidden in contributions in-kind, whose price is difficult to assess, and in contributions from domestic programmes formally outside the ITER project[27]) have saddled the ITER project with the reputation of being a money pit.

In fact, ITER is not really a scientific project. The underlying science has not yet come of age and is actually not yet ready for such a huge technological enterprise, embedded in global politics with the EU at the helm. Politics demands it to be a success; at stake are the prestige and credibility of the EU that has fought so hard to get the project on its soil, for essentially no other reason than the vanity of a French president. Any of the other members can procrastinate, complain about the soaring cost or other aspects, withdraw or suspend (part of) their contribution (as the US and India have done), the EU must and will continue to foot the bill. In addition, it is an almost impossible project to manage logistically and organizationally with so many countries involved.

The blaming game that is always going on around such delays and cost overruns is almost as old as the history of fusion itself. It has become abundantly clear by now that creating an experimental nuclear fusion reactor is an undertaking that faces both technical/scientific and political challenges. One of the general complaints of fusion scientists and administrators is that the political challenges are greater than the technical and scientific ones they are facing and that politicians are "more uncontrollable" as they "belong to the political world".[28] That would suggest that the scientists actually think themselves capable and able to meet the scientific and technical challenges posed. I am sure though that, if they indeed had done so, it would just have been plain sailing as far as the political challenges are concerned. The politicians would be fighting to support them. The ITER project is not the first project that promises to

---

[27]Fusion machines all over the world have re-oriented their scientific programs or modified their technical characteristics to act either partially or totally in support of ITER. These machines are conducting R&D on advanced modes of plasma operation, plasma-wall interactions, materials testing, and optimum power extraction methods, contributing to the success of ITER and the design of the next-phase device (https://www.iter.org/sci/BeyondITER).

[28]https://www.euractiv.com/section/energy/news/nuclear-fusion-project-leader-laments-uncontrollable-political-forces/; see also Dean (2013), a book that consistently lays the blame with the politicians, especially their reluctance to put up the money asked for.

show that nuclear fusion can be carried out on a large scale, and apparently scientists often forget and don't understand that their confidence in getting this done is not matched by the impression outsiders get when they look in on their projects. From the early days of nuclear fusion, they have greatly underestimated the scientific and technical problems, and then tend to later blame the politicians (or in other words society) when they understandably become reluctant to foot the ever larger bills. For more than half a century they have been making promises and spending lots of money that politicians must in the end account for to the general public. It is the first and foremost task of the scientists to convince these same uncontrollable politicians, and through them the general public, that the money will be and has been spent well. They have notably failed to do so. Where they were saying that the problems would be solved in a couple of years or in at most two decades, one now hears them say that there still is a long way ahead given "the complexity of the necessary technologies involved". Has that only dawned on them in the last few years or so, while Lawrence Lidsky for instance was saying this already in 1983?

That does not alter the fact that ITER is a beautiful and impressive machine as can be seen in Fig. 10.11. It is a wonder of technical and engineering skills, with unfortunately not the scientific (plasma physics) knowledge to match. Where the engineers have invented and developed the most amazing devices to create the

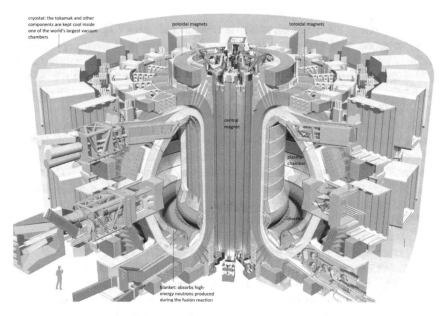

Fig. 10.11 Cutaway view of the ITER tokamak showing some of its vital components (*Source*: ITER Organization)

circumstances in which fusion reactions can take place, basic knowledge of the behaviour of a plasma at very high temperatures in a magnetic field is still wanting.[29]

In December 2017 ITER proudly announced[30] that "50% of the total construction work scope through first plasma is now complete", so in ten years half of what was promised has actually been accomplished. True to form, the terms used here do not have the meaning most people would assign to them. Total construction work scope is not just total construction work, but, I quote, "the performance metrics used in ITER assign a relative weight to every activity category within the project. Design, for instance, accounts for 24%; buildings construction and manufacturing for 48%; assembly and installation for 20%. After having compounded the percentage of completion of each category, the metrics produce a figure for the totality of the work scope through the launch of operations ("first plasma"). Design, which accounts for approximately one-fourth of the scope, is now close to 95% complete; manufacturing and building, which represents almost half of the total activities is close to 53% complete. Do a little math and the result is clear: in terms of activities that need to be completed, ITER is now halfway to its first operational event." Here the quote ends and I will now do the math, but I first note that the three categories mentioned do not add up to 100%, but just to 92%. Is there still a fourth category? And doing the math as suggested (on the figures of 24% (design) and 48% (buildings construction and manufacturing) mentioned above) I arrive at 48%, which is still 2% away from 50. Is it really that hard to be precise?

The first real ITER milestone will be reached by the end of 2025 when the project hopes to produce its first plasma and "will be followed by a staged approach of additional assembly and operation in increasingly complex modes, culminating in D-T plasma in 2035." This period has been defined in more detail as follows: four years of hydrogen-hydrogen and hydrogen-helium plasma pulses, a relatively short period of D-D plasma operation and a three-year period of D-T operation.

There is still a long way to go and all kinds of mishaps can and will occur. ITER may be hit by a violent disruption, damaging vital components of the machine (this is indeed considered a serious threat to ITER's mission and has received special attention; in 2018 the disruption mitigation system was estimated to cost €175 million, more than twice the initial estimate (Claessens 2020, p. 102)), old and new kinds of instabilities may play havoc with the plasma, etc.

Sometime after 2035, if all goes well and the machine behaves according to expectations, ITER will produce 500 MW of fusion power (in heat produced by fusion reactions) during pulses of 400 s and longer. This is presented as a ten-fold return on energy, which to a certain extent is true. Once the nuclear fusion reactions start, i.e. when the plasma starts to burn, the dominant mechanism for keeping the plasma at the required temperature will be heating by the α-particles produced in the fusion reactions. This is however not sufficient; an extra 50 MW of additional heating is needed to keep the reactions going. That is the origin of the factor 10, Q

---

[29] In a recent seminar M. Abdou of UCLA lamented that very important R&D identified in the 1970s and 1980s has not been done yet!! (Abdou 2019).

[30] https://www.iter.org/newsline/-/2877.

$= 10$. If this 50 MW were not needed Q would be infinite. Total plasma core heating needed is 150 MW, of which 100 MW must come from α-particles and 50 MW from additional heating. This 50 MW is the net heat that goes into the plasma, and a true statement would be that 'during the pulse the ITER plasma will create more energy than it consumes'. In order to get at the stage where the additional heating is fed into the plasma, energy is needed. This energy is not included here. For the additional heating it would have been more honest to quote the 'energy at power source', so the total energy needed to get this 50 MW into the plasma. Neutral beam heating is very efficient, but other types of heating much less. Also completely ignored is the energy needed to bring the plasma at the required temperature in the first place, the losses encountered in the (inductive) heating process, the power needed to cool the superconducting magnets etc. We have seen in an earlier chapter that for JET 700 MW (in electric power) was needed to produce in the end 16 MW (in heat) of nuclear fusion power. ITER is supposed to do better, in spite of its larger size, since its superconducting magnets require much less energy than JET's copper coils. The ITER website states that electricity requirements for the ITER plant and facilities will range from 110 MW up to 620 MW for peak periods of 30 s during plasma operation.[31] Can we conclude from this that 620 MW is the equivalent of JET's 700 MW? Or is it 110 MW?[32] For your information, 620 MW is the output of a small power station and enough to supply a medium-sized city. Of course, ITER does not continuously draw such amounts of electric power from the grid, only during plasma operation, but also, when the device is not operating, the continuous electric power drain for its auxiliary systems varies between 75 and 110 MW (Jassby 2018).

Contrary to the ITER website, the JT-60SA website is brutally honest about what ITER will achieve:[33] "The efficiency of the heating systems is ~40%. Other site power requirements lead to a total steady power consumption of about 200 MW during the pulse (*this does not seem to agree with the ITER website quoted above, which talks about 110 MW up to 620 MW (LJR)*). Now the fusion power of ITER is enhanced by about 20% due to exothermic nuclear reactions in the surrounding materials (*hence increasing the 500 MW to 600 MW (LJR)*). If this total thermal power were then converted to electricity at 33% (well within reach of commercial steam turbines), about 200 MW of electric power would be generated. Thus, ITER is about equivalent to a zero (net) power reactor, when the plasma is burning. Not very useful, but the minimum required for a convincing proof of principle (*and much better than JET (LJR)*). In ITER the conversion to electricity will not be made: the production of fusion power by the ITER experiment is too spasmodic for commercial use." Here the quote from the website ends and now we know, thanks to the Japanese,

---

[31] https://www.iter.org/mach/powersupply, accessed 6 June 2020.

[32] Claessens 2020, p. 74, states that the plant requires 110 MW, of which 40% is consumed by the cooling-water system, 30% by the cryoplant and 10% each by the tritium plant and building services.

[33] https://www.jt60sa.org/b/FAQ/EE2.htm, accessed 6 June 2020.

who are probably still taught to be honest. It can be summarised as: ITER will not establish much.

And that while ITER's website greets you with the proclamation "UNLIMITED ENERGY" in huge, boldly printed letters, against a star-studded background. Indeed, a bold statement, but also a big lie, as ITER will not produce any energy at all; it is not even intended to produce energy! Its carbon footprint will be unfathomably large. The energy used in its construction and operation will partly come from France's nuclear power stations and partly from fossil fuels, enormous amounts of fossil fuel, and will never be paid back! In this respect it is stupefying to learn that the European Commission has found another (deceptive) use for ITER. In 2019 the German journal *Der Spiegel* (Becker 2019) reported that the European Commission intends to book the expenditure for ITER for 100% as climate protection measure. After all, fusion energy is free of carbon emission. The fact that ITER has not and will not produce any carbon-free energy is apparently not relevant. Such a bookkeeping trick, which no ordinary company would be permitted, will help achieve the goal of spending 25% of the EU budget on climate protection in the years to 2027.

Apart from the production of zero energy, ITER's goals also include: [34]

"Demonstrate the safety characteristics of a fusion device." In 2012 the ITER Organisation was licensed as a nuclear operator in France, a bit prematurely as construction was far from complete and such license will not be needed until tritium will actually be processed (at the earliest after 2035). At that time the basic premises on which the license has been granted may have changed. It is not clear either what this safety demonstration would entail. ITER is supposed to burn deuterium and tritium, and especially this tritium is a hazard, both by itself, as it is radioactive, and by the neutrons produced in the fusion reactions. When it is working as designed, ITER is nothing more than the producer of a constant stream of very high-energetic neutrons that produce large quantities of radioactive material by bombarding the vessel and its associated components.

Figure 10.11 only shows the actual fusion power machine. It will be surrounded by a multitude of other buildings and auxiliary systems as can be seen in the final picture of Fig. 10.12.

We finish the discussion on ITER by summarising its problems or actually the reasons why it is a fatally flawed project:

1. **ITER is too big**. It would imply that an ITER based power plant would be 60 times more massive than a conventional fission core. That is just its core. In addition, the plant would need a vacuum, cryogenic, handling and environmental recovery system.
2. **ITER is too complex**. The machine has roughly one million components. Imagine the cost of doing maintenance and repair on such a machine.
3. **ITER is too expensive**. More than €53 billion. And any fusion power plant may well cost considerably more.

---

[34]https://www.iter.org/sci/Goals, accessed 6 June 2020.

Fig. 10.12 Artist's view of the completed ITER site. The machine of Fig. 10.11 will be housed in the purple building (*from* ITER Organisation)

4. **ITER will not be finished in time**. Delays are slowly running into decades now, and whatever fusion may do in the future, it will not be able to contribute to combating climate change. It just comes too late.
5. **ITER is not safe**. ITER creates two safety issues: plasma disruptions and quenching, apart from radioactive waste and radioactive fuel. If disruptions accidently happen, it would be expensive and dangerous. The heat in a disrupted plasma can be ten times higher than the melting point of the first wall and the divertor. The second problem is quenching, when a superconducting magnet suddenly become a normal electromagnet—and releases its energy. ITER's coils contain the same energy as 10 tonnes of TNT. Such quenching has already happened 17 times in tokamaks. It causes overheating and melting of components.
6. **ITER needs new technologies**. No solid material can reasonably handle ITER's steady state. For instance, the hot helium produced buries itself in the metal walls causing blistering.[35]

The criticism of ITER and the entire fusion effort will be dealt with in detail in Chap. 18.

---

[35] http://www.industrytap.com/iter-will-never-lead-commercial-viability/32484.

Chapter 11
Spherical Tokamaks

11.1 Introduction

In this chapter we will discuss a fairly recent development within the field of fusion research with tokamaks: so-called spherical tokamaks. The attention for this design is on the upsurge and most new tokamaks built in this century are actually spherical tokamaks. The reasons for this are that they can still be built rather cheaply and, more importantly, bring some new insights. Most public money made available nowadays for fusion research is drained away by the insatiable demands of ITER or by devices that are mobilised to test parts of the ITER design. It is therefore mostly left to private companies to pursue the route to a possible power station based on alternative designs. However, this seems to be changing, now that the UK government has recently allocated a sizable amount of money for research towards a power generation plant based on a spherical tokamak design with its STEP programme (see below).

It is all still very young and so far no fusion power whatsoever has been produced in spherical tokamaks. Their development is still some years behind conventional tokamaks, as devices like ITER are called, but since spherical tokamaks are much smaller (at least so far) and cheaper, they can be built quickly and are catching up fast. No easily accessible exposés of the development and working of spherical tokamaks are currently available in the literature, so a fairly extensive description of what is and has been going on in this field is given in this chapter.

The claims made about spherical tokamaks are in any case much bigger and grander than their differences with conventional tokamaks, which differences are actually fairly small. Essentially the only difference is that, by discarding all non-essential components from the inner side of the plasma, the central hole for the central column can be made much smaller. This seemingly simple change has profound consequences, which, so the advocates want us to believe, could open up a faster way to a fusion power plant that would match ITER's promise without the massive scale and cost. As a reminder, a large solenoid normally runs through the hole down the centre of a tokamak. By varying the electrical current in the solenoid, a current is induced in the plasma. This current heats the plasma and contributes to the poloidal

L. J. Reinders, *The Fairy Tale of Nuclear Fusion*,
https://doi.org/10.1007/978-3-030-64344-7_11

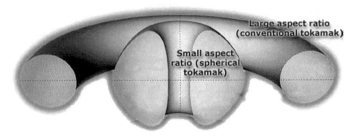

Fig. 11.1 Plasma shape of spherical (small aspect ratio) tokamak compared to conventional (large aspect ratio) tokamak

magnetic field that together with the toroidal (and other poloidal) fields created by external coils completes the forces that hold the hot, charged gas together.

In a spherical tokamak (ST) the size of the doughnut hole is as much as possible reduced, resulting in a plasma shape that is almost spherical, hence spherical tokamaks look very much like the cored apple of Fig. 4.7, where obviously little room is left for a central column and, more importantly, neither for the central part of the toroidal field coils. See Fig. 4.4 to remind you of the basic setup of a tokamak. In terms of the parameters we have used before, an ST has a small aspect ratio A, the ratio of major radius R to minor radius a, $A = R/a$. The ultimate is of course when there would be no hole at all, $R = a$ and consequently $A = 1$, and we would have a sphere. Aspect ratios of the early tokamaks in the 1960s were rather large, the famous Russian T-3 for instance had an aspect ratio of 8.3. When tokamaks became bigger, aspect ratios also went down and for the big tokamaks like ITER, JET and JT-60 they are a little larger than 3, although the original JET design (without divertor) actually had an aspect ratio of 2.37 and came very close to an ST. For STs aspect ratios are smaller still, equal to or smaller than 2, with extreme ones as small as 1.1. The number 2 does not seem to have any special value and is quite arbitrary in this respect. Figure 11.1 shows the difference of the plasma shapes for an ST and a conventional tokamak with large aspect ratio.

In the early days of the spherical tokamak it was emphasized that the ST can be operated in regimes where the physics processes are expected to be broadly similar to larger (i.e. conventional) aspect-ratio tokamaks, but also in regimes where the novel features of the ST come to the fore. Nowadays ST advocates go much further and see them as a viable power generating device and as an alternative to the increasingly cumbersome and costly conventional tokamaks. Since STs exhibit most of the characteristics of conventional tokamaks, they are also able to exploit the extensive knowledge base built up in relation to conventional tokamaks, leading to very rapid progress in ST physics development and understanding.

11.2 Alleged Advantages of the Spherical Tokamak

So why would a compact or spherical tokamak be advantageous or preferable?

As we will see in the following, in STs plasmas are confined at higher pressures for a given magnetic field and the toroidal fields can support a larger plasma current than in conventional tokamaks, or stated differently the toroidal field needed to maintain the same current is lower. The greater the pressure, the higher the power output, and the more cost-effective the fusion device. Research in fusion has so far mainly focused on increasing the fusion power, i.e. the power produced in the fusion reactions. This is not unreasonable as fusion power was and still is the great unknown. After all, the task was to show that any net fusion power can be created. But, once that question has been settled, it is evidently as important to try to diminish the various losses and especially to increase the efficiency (lower the amount of energy needed to run the reactor). We have seen that for the celebrated JET event, which produced just 16 MW of fusion power, in total 700 MW was needed to run the reactor, most of which was lost through radiation and heat losses in the magnets. Much could be gained if the efficiency were improved. STs work with much lower toroidal magnetic fields (at the same current) and hence the magnets consume much less energy, significantly improving efficiency and reducing cost.

As Fig. 11.1 shows, the cross section of the plasma in an ST is elongated in the vertical direction, while its magnetic topology remains virtually the same as that of a conventional tokamak, with a toroidal field generated mainly by toroidal field coils and a poloidal field generated by the plasma current and poloidal field coils. Its compact geometry makes the ST different from the standard tokamak.

Potential advantages of the spherical tokamak (ST) shape were first suggested in the second half of the 1970s. Daniel Jassby, Alan Sykes and Martin Peng (Sykes et al. 1979; Peng and Dory 1978) are most commonly mentioned as having been the first scientists to show the theoretical advantages of the spherical shape, which were put on a firmer footing in the 1980s. Jassby proposed a spherical tokamak with the SMARTOR design (Small Aspect Ratio Torus) in 1978; a design that never left the drawing board (Jassby 1978).

Based on a large number of calculations of MHD instabilities of tokamak plasmas with aspect ratios ranging from 2.5–5, it was shown that the maximum stable *beta* value (the ratio of the thermal plasma pressure to the magnetic pressure) increases with decreasing aspect ratio and with increasing elongation. Just decreasing the aspect ratio by a factor of 2 would double the value of *beta*! As we have seen in Chap. 4 a high *beta* is a desirable feature and with other factors being constant such an increase would be a considerable and important improvement for magnetic confinement fusion. Conventional tokamaks operate at relatively low *beta*, just 4 or 5%, with the record being just over 12%, but calculations show that practical designs would need *beta* values as high as 20%, which would be easily achievable in an ST. The higher *beta* the higher the fusion power, since fusion power rises like *beta* squared. At the same time STs need less toroidal field, the plasma volume can be reduced and the fusion power can be produced in a smaller device. So, an ST would mean high power density in a small device.

These considerations were worked out further in 1986 by Peng and Strickler (Peng and Strickler 1986; see also Peng 2000) who pointed out a number of further theoretical advantages of tokamaks with low aspect ratio. One of the important things they noted was that the vertical (poloidal) field needed in tokamaks has the natural tendency to elongate the plasma and that this elongation increases as the aspect ratio goes down, so reinforcing the increase of *beta*. This elongation can clearly be seen in Fig. 11.1. If the aspect ratio goes down from 2.5 to 1.2, the elongation, indicated by κ, increases from 1.1 to 2, and the toroidal magnetic field required to give the same quality or safety factor q (see Chap. 4) for a given plasma current falls by a factor of 20 (Sykes 2008)! This implies that the value of *beta* would also increase by a factor of 20. Their calculations also showed that if the aspect ratio is bigger than 2.5, the natural plasma elongation is still present, but less than 1.4, and shaping-coil currents are needed to force the elongation up to two or more, but will eventually lead to instability and require active feedback control. Thus, in a compact configuration, such shaping coils would be unnecessary and some further savings in the cost of the reactor magnet systems could be made.

So, in summary the low aspect ratio and large elongation imply that the *beta* values that can be reached by STs are very high (about 40%). Improving *beta* means that less energy is needed to generate the magnetic fields for any given plasma pressure (or density), and reactors operating at higher *beta* are less expensive for any given level of confinement. The magnetic fields are used more efficiently.

Another important difference is that, while in a conventional tokamak the toroidal field is much stronger than the poloidal field, the poloidal field in the outer region of an ST can be comparable to and actually be larger than the toroidal field, while the fields are comparable in the inner region.

The force exerted by a current in a magnetic field is perpendicular to the magnetic field. These forces will move the plasma around until a certain balance has been reached, which is the case when the current is everywhere parallel to the magnetic field, so that there is no perpendicular force.

Hence, the plasma in an ST is more stable and especially near the tokamak's central hole particles enjoy unusual stability (Fig. 11.2). In a conventional tokamak the particles spiral along the magnetic field lines going around in the tube, while in an ST the magnetic field lines wind tightly around the central column, where they hold the particles for extended periods (in a region of high field (i.e., with good curvature, meaning that instabilities are suppressed)) before they return to the outside surface (where the field is much lower and instabilities are more prominent). Let us consider what this means for the safety factor q, which was defined in Chap. 4 as the ratio of the number of times a toroidal field line goes around the torus the long way (i.e. around the hole in the torus) to the number of times a poloidal field line goes around the short way. According to the Kruskal-Shafranov limit q should be greater than 1 in order to suppress instabilities. From Fig. 11.2 it is now clear that "going around the torus the long way" for an ST is actually the short way, and the field lines in an ST go around the torus hole much more often, implying that q will be large and the

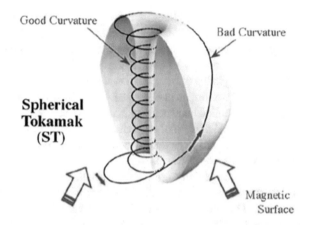

Fig. 11.2 The magnetic configuration in a spherical tokamak (from Peng 2000, p. 1683)

plasma in an ST will be more stable. In addition, the D-shaped cross section of the plasma also helps suppress turbulence and improve energy confinement.

The fact that the particles are held for extended periods by the tightly-wound magnetic fields also implies that their drift (due to the non-uniformity of the magnetic field), which played an important role in early tokamaks and stellarators (see e.g. Chaps. 3 and 4), is almost absent. There are large so-called near-omnigeneous regions which are free from such particle drifts and where in the absence of turbulence and collisions the particles are perfectly confined.

11.3 Alternative Current Drive

With the central hole of the torus squeezed as much as possible, there is little room left to fit in all the necessary equipment for a central solenoid, needed if a current is to be inductively driven, the shielding it requires, and the inner limbs of the toroidal magnetic field coils, which if superconducting material is used for them likewise require extensive shielding. This implies that in large devices based on the spherical tokamak configuration other current drive methods must be applied. The current must be driven by something other than through induction from a central solenoid. Important for the viability of the spherical torus concept is that progress has been made in advanced alternative (non-inductive and other) current drive schemes, also inspired by the fact that an eventual reactor must be able to operate in a steady state, and not as a pulsed device. Several forms of non-inductive start-up are under active investigation. Some of these are very inefficient, especially neutral beams and radiofrequency waves have efficiencies as low as 0.1%, so other means must be found.

One method of non-inductive current drive is by so-called helicity injection. (Magnetic) helicity describes the degree of linkage between magnetic field lines within a volume. Combining the toroidal and poloidal magnetic fields gives a twist, i.e. helicity, to the field lines that run around the torus. It is a measure of how the field lines wrap and coil around one another. The helicity in a tokamak is proportional to the plasma current, and sustaining the tokamak poloidal field against ohmic dissipation (loss of energy by conversion into heat due to the resistance the current feels when flowing in the plasma) can be considered as a form of (constant) helicity injection, since if nothing is done the helicity will slowly vanish. Other forms of helicity injection are called local helicity injection (LHI) and coaxial helicity injection (CHI). The latter is a favourite of the Princeton Plasma Physics Laboratory.[1] Its efficiency has been predicted to be tens of percent, which is considerable if compared to using neutral beams, for instance.

Another especially promising method is called **merging compression** (Gryaznevich and Sykes (2017)) which is based on **magnetic (field line) reconnection**. The method goes back more than 25 years, but has more recently been pioneered for heating plasmas (Ono et al. 2015). In tokamaks the main task of the magnetic fields is to confine the plasma. These fields contain energy and it would be a considerable step forward if the energy contained in the fields could also be used for heating and starting up the plasma. If you manage to create the confining field configuration by magnetic reconnection, the magnetic energy freed in the reconnection can actually be a very efficient means of heating the plasma. This has been demonstrated now on several devices (e.g. START, MAST, ST40). In particular, in experiments on Tokamak Energy's ST40 (see below) it has been shown that up to 90% of the magnetic energy can be converted into thermal energy, with 90% of this thermal energy going into accelerating the ions in the plasma. Remember that it is the ions that have to fuse, so it is important that they become hotter.

Magnetic reconnection is a process that converts the magnetic energy of reconnecting magnetic fields into kinetic and thermal energy of the plasma through the breaking and topological rearrangement of magnetic field lines. The magnetic field lines in a plasma snap apart (like the release of tension when you let go a stretched rubber band) and violently reconnect, whereby the released magnetic energy is converted into thermal energy that will accelerate (heat) the particles in the plasma. This phenomenon is most prominent in the form of solar flares on the surface of the Sun, and it also powers the northern lights, gamma-ray bursts, and other violent natural phenomena. The geomagnetic storms on earth that follow solar flares demonstrate how much energy can be released by magnetic reconnection

On a much smaller scale than in astrophysics, this phenomenon can also (be made to) occur in magnetically confined plasmas. It rearranges the magnetic topology, i.e. the structure and linkage of magnetic field lines, and converts magnetic energy into kinetic energy, thermal energy and particle acceleration. Magnetic field lines embedded in the plasma come together, break apart, and reconnect.

The success of such methods would remove the necessity of having a (full) solenoid to induce the plasma current. An additional disadvantage of such a solenoid is that the device necessarily has to operate in pulses, while in a real power plant

---

[1]See e.g. https://www.pppl.gov/news/2016/01/pppl-physicists-simulate-innovative-method-starting-tokamaks-without-using-solenoid.

steady-state operation must be possible. In this respect STs might kill two birds with one stone, although such current drive methods can of course also be applied in conventional tokamaks.

In magnetic compression this is realised, the energy embedded in magnetic fields is released, the plasma is started up and heated. This extraordinary process is illustrated in Fig. 11.3, both schematically and in the form of camera images. The starting point is the vacuum vessel with the toroidal fields switched on and filled with hydrogen gas. Inside the vessel magnetic field coils are located (in Fig. 11.4 these induction coils are indicated for the START device) and when running high currents through these coils two rings of plasma are created around them, by the magnetic field produced by the current. As the current in the coils is reduced to zero, the plasma current is going up (induced by the current reduction in the coils). The plasma current will in principle, if there are no losses, reach the value of the current that was initially in the coils. The current in the coils will become zero and the two plasma rings detach, attract each other and start to combine. In this combining process the magnetic fields reconfigure by magnetic reconnection. Stretched field lines break and release huge amounts of energy, in the same way as a stretched catapult releases enough energy to

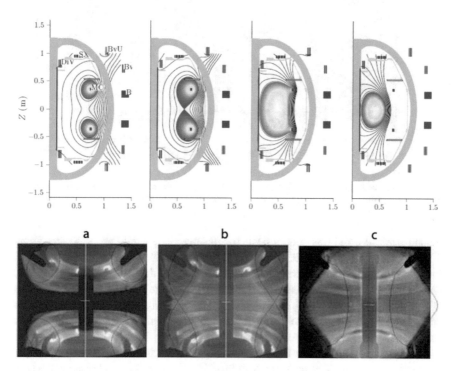

Fig. 11.3 Progression of merging compression seen in one half of the ST (top) (*from* Tokamak Energy), and visible images (bottom) of the merging compression plasma start-up on MAST (*from* Kirk et al. 2017)

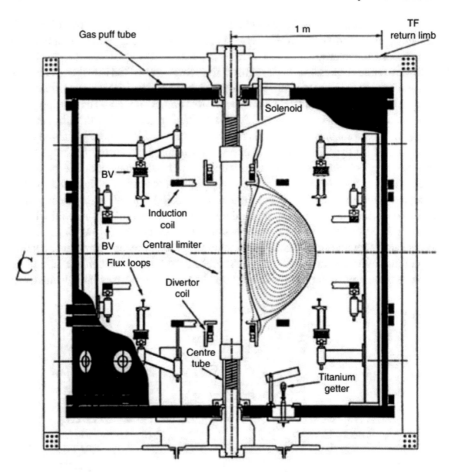

Fig. 11.4 Poloidal cross-section through the START vessel showing the solenoid, poloidal field coils and an outline of a typical START plasma (*from* Akers et al. 2002)

launch a missile, and this energy heats the plasma.[2] The whole process takes about 6 ms. Application of a vertical field from other coils then compresses the plasma into the required ST configuration (the fourth image in the top picture of Fig. 11.3) and the plasma current increases even more. The programme can also be conducted with coils outside the vessel, in which case the field rings are formed between the coils; this is called double-null merging compression, which has been pioneered on some Japanese STs as we will see later and is illustrated in Fig. 11.9. ST-plasmas with currents up to 450 kA are routinely obtained by this merging–compression technique and temperature increases from 100 to 1000 eV, so from a rather cool plasma to fusion relevant temperatures (Cox et al. 1999; Sykes et al. 2001; Kirk et al. 2017).

---

[2]http://www.100milliondegrees.com/merging-compression/, accessed 10 June 2020.

11.4 The START, MAST and NSTX Spherical Tokamaks

The Culham Centre for Fusion Energy (CCFE) at the JET site in Culham near Oxford in the UK has pioneered the spherical tokamak fusion concept and still plays an important role in its current development. JET had already reduced the aspect ratio to around 2.4, but it was the START (Small Tight Aspect Ratio Tokamak) device of the Culham Centre that revolutionized the tokamak by changing the previous toroidal shape into the tighter, almost spherical, doughnut shape and in the subsequent experiments verified the aforementioned advantages of STs predicted by Peng and Strickler in 1986 (high elongation, high *beta* and drift-free (omnigeneous) regions). As already indicated above, the new shape increased efficiency by reducing cost over the conventional design, while the toroidal field required to maintain a stable plasma was a factor of 10 less. START and its successor MAST (Mega Ampere Spherical Tokamak), plus their American equivalent NSTX, have provided the most information on STs and by doing so they were the instigators of the current, still modest, upsurge in spherical tokamak research.

- **START**

START was a relatively low-cost device; it was small, had a tight aspect ratio and was mainly constructed from existing equipment. No flashy pictures seem to be available of this piece of equipment, which broke new ground and made the case for the spherical tokamak. First plasma was obtained in 1991, at the time when JET still got most of the attention and spherical tokamaks were not yet taken that seriously, and it was shut down on 31 March 1998 when the whole team moved to MAST.

The geometry of the START device is shown in Fig. 11.4. It employed a large cylindrical vacuum vessel 2 m diameter $\times$ 2 m high, and hot, high current plasmas were obtained at aspect ratios as low as 1.25 (major radius $R = 0.36$ m and minor radius $a = 0.25$ m).

Tight aspect ratio plasmas are obtained in START by the method of merging compression explained above, without using the flux of a central solenoid. In the first year START had no central solenoid at all and plasma start-up was achieved by merging compression alone. Then a solenoid was introduced to further ramp up the plasma current. Peak plasma currents of up to 350 kA after compression and aspect ratios as low as 1.3 were achieved in START (Sykes et al. 1992).

A compact central solenoid can induce additional current and maintain the discharge for up to 40 ms, or alternatively induce ST plasmas directly at low major radius. START was the first spherical tokamak to provide experimental results on hot spherical tokamak plasmas ($T_e \approx 500$ eV (6 million degrees), still far from the temperature needed for fusion; in most discharges temperatures for both electrons and ions remained around 100–150 eV).

The main achievements of START include attainment of record tokamak *beta* values by successful NBI heating (Akers et al. 2002) and a demonstration of NBI-heated H-mode operation. A number of experiments in START reached *betas* of 40% (Gryaznevich et al. 1998), where the previous world record for *beta* in a tokamak was 12.6% achieved on the DIII-D tokamak, which has an aspect ratio of 2.5 (see Chap. 5). A feature of the ST, as discussed above, is the increasingly efficient use of the toroidal field due to the large increase in safety factor produced by toroidal effects (Sykes et al. 1999).

At the end of March 1998, the START experiment finished. It has since been disassembled and transferred to the ENEA research laboratory at Frascati in Italy where its vacuum vessel has been used for the construction of the even more innovative Proto-Sphera ST. It aims to form an ST confinement plasma not around a metal central column (as in a tokamak), but around a central plasma column. Magnetic plasma instabilities will 'kink' the central plasma column, such that it produces a spherical torus around it.[3]

• MAST

In 1995 the START team began designing MAST, with the design approach closely following START with a cylindrical vessel providing both the vacuum boundary and the mechanical support structure for the poloidal and toroidal field coils. So, the geometry is very similar to the one shown in Fig. 11.4. Construction started in 1997, first plasma was obtained in 1998 and the MAST experiment was operated at the Culham Centre until 2013. Building on START's success, MAST is a larger, more sophisticated device with extremely advanced diagnostics for analysing the plasma. Its aspect ratio is 1.3 (major radius $R = 0.85$ m and minor radius $a = 0.65$ m).

Its stainless-steel vacuum vessel has a height of 4.4 m and a width of 4 m, making the volume almost nine times larger than the START vessel volume. Although still small compared to the big conventional tokamaks or to ITER, the sizes of STs are also already considerable and still growing.

MAST's objectives were to provide improved understanding of tokamaks with a view to the ITER design, testing ITER physics in new regimes (e.g. effects of plasma shaping), and to investigate the potential of the spherical tokamak route to fusion power.

MAST is nominally a 1-MA machine (i.e. a plasma current of 1 MA), but designed to withstand the forces associated with 2-MA plasmas. For plasma start-up MAST also used the merging–compression technique, without the need of any input from the central solenoid. Figure 11.5 shows a beautiful picture of an almost spherical plasma in MAST.

MAST produced hot ($T_i \leq 3$ keV, $T_e \leq 2$ keV), dense ($n_e = (0.1 - 1) \times 10^{20}$ m^{-3}) and highly shaped plasmas (triangularity $\delta \leq 0.5$, $1.6 \leq \kappa \leq 2.5$) at moderate toroidal field $B_T \leq 0.62$ T. Apart from this parameter regime, which is typical for spherical tokamaks, MAST plasmas also show many similarities to conventional aspect-ratio tokamaks, and there is considerable synergy between the development of the ST and the development of conventional medium and high aspect-ratio tokamaks.

As regards investigating the potential of the spherical tokamak route to fusion power, which was one of MAST's other objectives, it does not seem that MAST has made considerable progress in bringing this goal any further. MAST has essentially

---

[3]For more details about this project see: https://www.afs.enea.it/project/protosphera/.

Fig. 11.5 Camera image of
plasma in the MAST reactor.
Note the almost spherical
shape of the outside edge of
the plasma. The high
elongation is also evident.
The central post is part of the
toroidal field coil

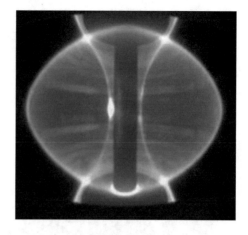

only shown that the regime (parameters and physics) obtained on START could be
obtained on a bigger machine.

- **MAST-Upgrade**

Almost as soon as MAST had produced its last plasma in 2013, the engineers moved
in to start preparing the facility for a £30 million upgrade, the MAST-Upgrade or
MAST-U. Building on the mission of MAST, the specific goals of MAST-U are to
test a range of novel exhaust concepts including the Super-X divertor; contribute
to the ITER/DEMO physics base to improve predictive capability, and explore the
feasibility of the spherical tokamak as a future fusion device (Lloyd 2019). This last
goal of further development towards a fusion power plant based on the ST concept was
not among the original missions of the upgrade, its main function being to contribute
to ITER and DEMO. This has only recently changed with the announcement of
STEP, Britain's programme for a fusion power reactor based on the ST design (see
below). The second goal of MAST-U, i.e. to contribute to ITER, will now probably
get less attention, while the third goal will get prime place.

  MAST-U will make the case for a fusion Component Test Facility (CTF)[4], which
would test reactor systems, components and materials for DEMO. Such a CTF would
be a post-ITER device. A spherical tokamak is seen as an ideal design for such a
facility (See e.g. Peng et al. 2005; Morris et al. 2005.) The innovative Super-X divertor
is a high-power exhaust system that reduces power loads from particles leaving the
plasma. Its advantages and disadvantages will be studied in MAST-U, as well as in
other tokamaks. In this divertor the plasma escaping from the core will be guided
along a longer exhaust path, cooling the particles and spreading them over a larger
area so that the impact on the wall of the machine is reduced. The aim of the divertor is

---

[4]Such component test facilities are not uncommon for power plants. For fission power plants
they were widely used during the 1950s–1960s, but also for fossil fuel power plants they are
not uncommon. For instance, in 2013 the EU has published a CTF design for a 700 °C coal-based
power plant to test high temperature durable new materials needed to realise such a power plant
with efficiencies above 50%.

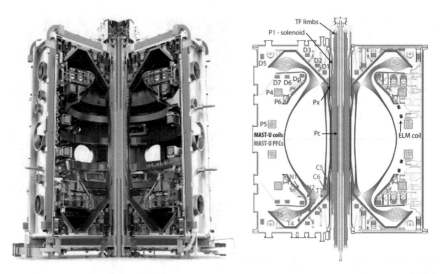

Fig. 11.6 Cut-open view of the MAST upgrade next to cross section of the vessel (*from* http://www.ccfe.ac.uk/mast_upgrade_project.aspx and Kirk et al. 2017)

to take a 50 MW/m^2 heat load and reduce it to just 5 MW/m^2. If successful, Super-X could be used in DEMO and any other future fusion device.[5]

A cut-open view of the MAST-Upgrade is shown in Fig. 11.6 next to a cross section of the vessel.

The upgrade will be implemented in three stages. Funding has been agreed for stage 1a, which will be ready for plasma operations in late 2019 or early 2020, an amazing 6 years after starting the upgrade. No proper reason has been given why it had to take so long to get the upgrade on the rails. A new centre column and the Super-X divertor have been installed and the neutral beam heating system has been improved. Two additional phases (stage 1b and stage 2) will follow later subject to funding, which now seems to be secured after STEP and further additional funding was announced.

MAST-U is being supported by EUROfusion, the European consortium of national fusion research institutes, within its Medium Sized Tokamak (MST) programme, complementing two other tokamaks, namely ASDEX Upgrade (Germany) and TCV (Switzerland). This will probably end with Brexit. EUROfusion formulates an annual work plan for exploitation of these devices to be supplemented by UK-funded operations. MAST-U has chosen for support by EUROfusion because there are many aspects which require research and development that are either complementary to ITER, or go beyond ITER parameters and the spherical tokamak line may help with this research, so it seems that as far as EUROfusion is concerned MAST-U will just be a tool to assist in the development of ITER and

---

[5]http://www.ccfe.ac.uk/mast_upgrade_project.aspx.

to prepare the ground for a future DEMO plant (MAST Upgrade Research Plan, November 2016).

- **STEP**

A little premature in my view, since MAST-U has not yet been shown to work, but in line with the trend in fusion research, a follow-up of MAST-U is already being planned at the Culham Centre in the form of STEP (Spherical Tokamak for Energy Production), which as its name indicates is rather ambitious, going even beyond the future DEMO for conventional tokamaks. Its website (http://www.ccfe.ac.uk/ste p.aspx) states that the STEP programme aims to deliver an integrated concept design for a fusion power plant based on the spherical tokamak, and develop and identify solutions to the challenges of delivering fusion energy. Its equally ambitious technical objectives include the construction of a fusion power plant by 2040 (invoking so it seems the long-established magical two-decade rule for fusion power to be realised) based on a design that should be ready in 2024. Its aims are to "deliver predictable net electricity greater than 100 MW; to innovate to exploit fusion energy beyond electricity production (*whatever that may mean; one would think that electricity production was the end goal (LJR)*); to ensure tritium self-sufficiency; materials and components qualification under appropriate fusion conditions; and a viable path to affordable lifecycle costs." In addition, it is said that UKAEA will continue to play a full role in the international fusion programme, e.g. JET, ITER, EU-DEMO (Lloyd 2019).

The design for STEP will take account of the results from MAST-U, which has now changed its Research Programme to include research focusing specifically on the ST route towards fusion power.

The UK government has announced £220 million of funding over four years for the conceptual design of STEP. The programme is still in a very early stage with £20 million having been made available for the first year in September 2019. While the ink on this announcement was barely dry, in November 2019 the government announced further extra funding of £184 million for the next five years for MAST on the condition that it will contribute to STEP (it can now leave behind the pretext of working for ITER). This amount comes on top of the £220 million mentioned above. A further £86 million was promised for a Nuclear Fusion Test Facility (NFTF). So, in total the UK government has now committed £490 million for STEP studies, material studies etc., bravely starting on the march towards power generation by nuclear fusion via spherical tokamaks.

- **NSTX**

In the United States, after the shutdown of TFTR, the Princeton Plasma Physics Laboratory shifted its attention to spherical tokamaks. With a plasma current of 1 MA, NSTX (National Spherical Torus Experiment) and its upgrade belong to the most powerful STs in the world. The experiments on NSTX are a gigantic international effort with participation from physicists and engineers from 30 US laboratories and universities and 28 international institutions from 11 countries.

According to a document published on its website (https://nstx-u.pppl.gov/ove rview), the mission of the NSTX programme is to establish the scientific potential of the ST configuration as a means of achieving practical fusion energy and to contribute unique scientific understanding of magnetic confinement in research areas such as electron energy transport, liquid metal plasma-material interfaces, and energetic particle confinement for burning plasmas. I have not seen any claims that it has fulfilled its mission and that the ST configuration has indeed been shown to have "scientific potential as a means of achieving practical fusion energy", nor has it been explained what is necessary for such potential to be apparent.

First plasma was obtained on NSTX in February 1999, so roughly a year later than on MAST. Its parameters were very similar to those of MAST, with aspect ratio 1.5, $R = 0.85$ m and $a = 0.68$ m. While the two devices have similar device and plasma parameters, they also have important complementary features. NSTX has a near-spherical vacuum vessel with a set of stabilizing plates near the plasma to offer effective wall stabilization at high *beta*. See Fig. 11.7 for schematic views of both the NSTX and its upgrade the NSTX-U. In a somewhat contrasting design (Fig. 11.6), MAST has a large cylindrical vacuum vessel with internal poloidal field (PF) coil sets which provide flexibility for plasma shaping and divertor configuration. The internal PF coils, being closer to the plasma, generally require less power for plasma control, and can also be used, as we have seen, for PF coil-based plasma start-up by merging compression. MAST also has a large number of internal control coils. Both have demountable toroidal field coils. And both are heated by NBI and various radiofrequency heating systems. With this the NSTX managed to increase the electron temperature from about 5 million to above 50 million degrees (Ono and Kaita 2015).

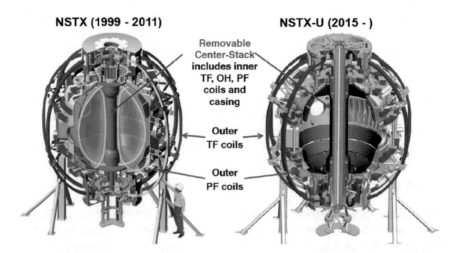

NSTX (1999 - 2011)

NSTX-U (2015 -)

Removable
Center-Stack
includes inner
TF, OH, PF
coils and
casing

Outer
TF coils

Outer
PF coils

Fig. 11.7 Schematic view of the NSTX and the NSTX-U (*from* Ono et al. 2015) (OH coils are ohmic heating coils)

The results obtained with NSTX include the following: [6]

- In 2002 about 60% of the total plasma current was obtained from a combination of neutral-beam driven current and self-generated bootstrap current (see Chap. 4), relaxing the need for induction (from a central solenoid) to sustain the current. We have already seen in Chap. 8 that in TFTR researchers had managed to drive the entire plasma current by means of the bootstrap effect and neutral-beam injection, so this result of NSTX, although important, does not make the ST configuration stand out.
- A few years later NSTX also copied START's record of a toroidal *beta* of about 40%.
- The energy confinement time in NSTX has been consistently 1.5–2.5 times larger than expected from the results accumulated from conventional tokamaks that were run in their basic mode of operation. This is a very favourable result for the ST's potential in fusion energy development.
- The high power levels anticipated in any tokamak fusion reactor will create large heat loads on the plasma facing components (PFCs). It is therefore important to develop and test both advanced PFCs capable of withstanding high heat fluxes and methods to spread the heat over the PFCs. In NSTX the evaporation of lithium coatings on PFCs has been shown to suppress the return of particles that escape from the plasma and to be effective in improving the plasma confinement and preventing ELM instabilities.

In 2008 a new NSTX Five-Year Research Plan for the years 2009–2013 was formulated, in which several upgrades to the capabilities of NSTX were proposed. In 2012 the NSTX was shut down as part of an upgrade programme and became NSTX-U.

- **NSTX-U**

The upgrade has boosted the principal capabilities of the NSTX reactor and made it the most powerful spherical tokamak in the world. It has doubled the field strength to one tesla, and the electric current flowing in the plasma has been increased to 2 MA. For this the central column had to be widened, decreasing the minor radius and hence increasing the aspect ratio which however stayed below 2 (at 1.7). New heating capability comes from the installation of a neutral beam injector. For further details on the first results obtained with NSTX-U see Menard et al. (2017) and on the capability and goals of NSTX-U see Ono et al. (2015).

The major questions NSTX-U will try to answer include:

- Can the device continue to effectively contain plasma when the temperature rises, as it could make confinement more difficult? If the upgrade can effectively control the hotter plasma, it means high fusion power could be achieved in a fairly compact machine.

---

[6]https://nstx.pppl.gov/nstxweb_2009/info/NSTX_information_bulletin_2009.pdf.

- Can new ways be found to start and sustain the electric current that creates the plasma? Future compact reactors will operate under conditions that would damage the spherical tokamak reactor's solenoid and the more compact the reactor the less space there is for such a solenoid. If the solenoid can be eliminated, this would give a boost to the ST concept for the next-step machine. These questions were also studied on START and MAST.
- Can the hot plasma particles that escape the confinement and reach the reactor walls be tamed? This heat flux can damage interior surfaces, drive impurities back into the plasma and shut down the reaction. As noted above, parts of the present plasma facing components of the NSTX torus were coated with lithium, a metal that turns liquid when struck by stray particles and sponges up the impurities. Liquid metal PFCs will be investigated in NSTX-U operations. The power flux expected in the upgrade will be much increased and how the increased flux will be handled in NSTX-U could serve as a model for DEMO.

NSTX-U was completed in the summer of 2015. First plasma (Fig. 11.8) was subsequently achieved and H-mode very quickly accessed. NSTX-U surpassed NSTX-record pulse-durations and toroidal fields.

Just after its first 10 weeks of running time, in the summer of 2016, the NSTX-U had a mishap ('significant technical issue') when one of the machine's 14 magnets, a poloidal field coil, shorted out. Replacing it, and a second identical coil on the opposite side of the tokamak chamber, took about a year, which implied that the machine would be out of commission for a year. The mishap cost the PPPL director his job and the expected one-year delay has now stretched to more than three and the upgrade is not expected to run before 2022.

As also apparent from some of the adjusted goals of NSTX-U, the US now also seems to have fully changed direction in the sense that fusion research should be oriented at the generation of power, and no longer be merely considered as scientific research. The Final Report of the Committee on a Strategic Plan for US Burning Plasma Research recommends that "Along with participation in international fusion

Fig. 11.8 Camera image of an NSTX-U plasma in H-mode. Note the similarity with the MAST plasma shown in Fig. 11.5 (from Menard et al. 2017)

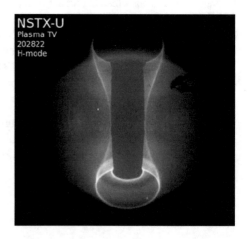

research, including the ITER partnership, the US (…) should start a national program of accompanying research and technology leading to the construction of a compact pilot plant, which produces electricity from fusion at the lowest possible capital cost." (National Academy of Sciences 2018, p. 89.)

- **NSST and SHPD**

NSTX-U can also help determine the path to a possible next-generation spherical torus that would produce a burning plasma to complement the output of ITER. Such a spherical torus would be roughly twice as powerful as NSTX-U and could be used to test components for a commercial fusion reactor by around mid-century, it is thought. A Next Step Spherical Torus (NSST) experiment has been proposed and is to take advantage of the tritium experience with the former TFTR facility at PPPL in order to test the ability of the ST to confine D-T fusion-producing plasmas. It is envisaged to play a role for STs similar to that of JET, JT-60, and TFTR for conventional tokamaks. The subsequent DEMO device could then either be a conventional tokamak or an ST (Ono et al. 2004). Although studies are underway for the design of an NSST experiment, there are currently no plans in the US to provide public funding for a spherical tokamak beyond NSTX-U.

But plans are changing, which in principle is not such a bad thing. It seems to me, however, that the fusion community is unable to make up its mind on what is the best way (in spite of '*iter*') to proceed to a power generating facility. And when then after much toing and froing a decision is taken on how to proceed, not much is needed to throw things off balance again. Typically, the sign of an immature field and that after seventy years of activity.

The newest development is that a recent National Academy study considered a Sustained High Power Density (SHPD) Facility as the next step for the US, to be designed and constructed between 2026 and 2035, followed by half a decade of operation. This facility is described as a possible bridge to a Compact Fusion Pilot Plant and would either be a new facility or an upgrade to an existing facility. The proposed SHPD mission would be to close the gap between confinement and sustainment, in practical terms to increase both the Lawson triple product and pulse duration by 2–3 orders of magnitude. A range of SHPD configurations are currently being investigated (Menard et al. 2019).

Apart from PPPL several other institutions in the US were and still are involved in STs. Their devices will be discussed in the next section.

11.5 Overview of Other Spherical Tokamaks in the World and Their Research

The construction of STs really took off from the early 1990s after the START device had shown that the theoretical advantages of STs as explained above held up in practice. A few STs date from before that time, starting with the Japanese Asperator

T-3, which goes back to 1974 (see Nagao et al. 1974). It was a small device with major radius 27 cm and aspect ratio 1.6. It was followed by the Erasmus tokamak in Belgium, dating from 1976 and vying for second place with the South African Tokoloshe tokamak (De Villiers et al. 1979). In southern African folklore, a tokoloshe is a mischievous and lascivious hairy manlike creature of short stature. So, it seems an apt name for a tokamak, which always has some surprises up its sleeve (Speake and LaFlaur 2002). The project started in 1975 and centred on an aspect ratio of 2. It had a major radius of 0.52 m and a minor radius of 0.26 m. Judging from the paper that the South African scientists wrote in the *South African Journal of Science* in 1979, they were already aware of most of the potential benefits of a small aspect-ratio tokamak, including high *beta* and lower toroidal magnetic field, well before the 1986 Peng and Strickler paper.

A large number of spherical tokamak devices have since been built and are or have been operational in more than a dozen countries (Australia, Brazil, China, Costa Rica, Egypt, Italy, Japan, Kazakhstan, Pakistan, South Korea, Russia, Turkey, UK and USA). The website www.tokamak.info gives a list of about 40 STs. Some of these will be discussed in greater detail below.

- **Japan**

An extremely highly elongated ($\kappa = 10$) low aspect-ratio tokamak has been produced by a theta-pinch device at Nihon University. A conducting rod is installed into a vacuum vessel along the central axis of a theta-pinch coil to produce a toroidal configuration. The device was built in 1998 and called NUCTE-ST, where NUCTE probably stands for Nihon University Controlled Thermonuclear Experiment. It has a major radius of 4.9 cm, minor radius 4.4 cm and aspect ratio 1.1. It was later modified to improve the plasma parameters (Narushima et al. 1999).

At Tokyo University a number of STs (TS-3, TS-4, TST-M, TST-2 (TST standing for Tokyo Spherical Tokamak) and UTST (standing for University of Tokyo Spherical Tokamak)) have been operating from 1986 onwards. On TST-M (aspect ratio 1.3) experiments on helicity injection and turbulence-induced transport were carried out with plasmas of relatively short pulse length (several milliseconds) and low plasma current.

UTST is a new addition to the impressive series of STs at Tokyo University. It has major radius 0.4 m and aspect ratio >1.2. It has been constructed for the purpose of exploring the formation of ultrahigh-*beta* ST plasmas using the plasma merging compression method. Start-up by merging plasma has been demonstrated in the TS-3 and TS-4 devices using coils inside the vacuum vessel, like the procedures in START and MAST discussed above (Fig. 11.3), and the TS-3 plasma obtained a 50% *beta*. In order to demonstrate such start-up in a more reactor-relevant situation, UTST has placed all poloidal field coils outside the vacuum vessel. The procedure here with coils outside the vessel (a setup with one coil outside and one inside the vessel can also be imagined) is the so-called double-null merging compression, mentioned above. The point is that the plasma rings are not formed around the coils, but inside the vessel in the 'null' between the coils. When they merge to form a single plasma,

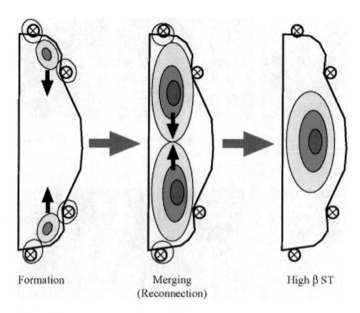

Formation   Merging   High β ST
                              (Reconnection)

Fig. 11.9 Plasma formation in a spherical tokamak by double-null (between coils) plasma merging (from http://fusion.k.u-tokyo.ac.ip/research/UTST/UTST.html)

magnetic field lines reconnect, and magnetic field energy is converted into plasma kinetic energy, increasing the plasma *beta*. Its working is illustrated in Fig. 11.9 (Imazawa et al. 2012).

QUEST (Q-shu University Experiment with steady-state Spherical Tokamak) is the newest and largest spherical tokamak at Kyushu University. Its mission is to study issues related to steady-state operation in particular, to develop a fully non-inductive current drive scheme (i.e. working without a central solenoid) that is effective in ST plasmas with high *beta* and to develop a divertor in an ST magnetic configuration which can handle power and particle fluxes in steady state. It came into operation in 2008 and device parameters are major radius 0.7 m, making it one of the biggest STs around, minor radius 0.48 m, hence aspect ratio 1.47 (Ono and Kaita 2015).

- **Russia**

Among the few early tokamaks of the 1980s is the first Russian ST, called GUTTA, built in 1980 at the Ioffe Institute in at that time Leningrad. A drop of glycerine floating in a glass of water, will become a D-shaped torus, if you put a glass rod through the drop (Fig. 11.10). This inspired the construction of GUTTA (which means 'drop' in Latin).

Figure 11.11 shows a picture of the vacuum vessel of this device, which illustrates how far removed such devices were from the big tokamaks like TFTR and JET that were then in operation or under construction. It was a very small table-top device, very much reminiscent of the early 1960s tokamaks, with a major radius of just 16 cm

Fig. 11.10   Picture of a drop of glycerine becoming a D-shaped torus when a glass rod is put through it (*from* M. Gryaznevich, private communication 2019)

Fig. 11.11   Picture of the vacuum vessel of the low-aspect-ratio desk-top ST 'Gutta' (*from* Gusev et al. 2003)

and aspect ratio 2, but with a plasma current of 150 kA making it a high-performance machine. In 2004 it was moved from the Ioffe Institute, where it had operated until 1985, to St. Petersburg University, where it is involved in start-up and ECRH studies.

From 2002 to 2017 the Ioffe Institute ran the Globus-M (Globus-Modified) project, dedicated to basic spherical tokamak physics research up to the 0.5 MA plasma current level, with emphasis on auxiliary heating and current drive using various radio frequency methods: ion cyclotron heating, high harmonic fast waves, and lower hybrid schemes and neutral beam heating at high magnetic field. The aspect ratio of this device is 1.5 with a major radius of 0.36 m. In 2017 it was replaced by an upgraded version, the Globus-M2, of the same dimensions, but a higher toroidal magnetic field and higher plasma current. The device is used for ITER-related diagnostics development and material testing and the development of a novel programme of compact hybrid neutron sources.

- **United States**

In 1994 studies with spherical tokamaks started at the University of Wisconsin in Madison with the very small MEDUSA (Madison EDUcation Small Aspect ratio tokamak) experiment, a tokamak with a major radius of just 12 cm and aspect ratio of 1.5. It has since been moved to the Costa Rica Institute of Technology. Experiments at Madison have continued with the Pegasus Toroidal Experiment, which was constructed just two years later in 1996. It is a medium size ST with an ultra-low aspect ratio of 1.15-1.3 and has been focusing on plasma start-up and MHD research. It is a university scale experiment (major radius $R \sim 0.45$ m, aspect ratio 1.2, plasma current 0.3 MA and $B_T = 0.15$ T) in the US ST-research programme. The device has a highly engineered ohmic heating solenoid, with very high magnetic field capability within very tight spatial constraints. With a very slender central post, the device can access a regime of ultra-low aspect ratio with ohmic current drive. Because of the high ohmic resistivity, Pegasus has achieved *beta* of 100% and H-mode by ohmic heating alone (Bongard et al. 2019). In 2019 Pegasus is being upgraded (e.g. by removal of the central solenoid) to build the Unified Reduced A Non-Inductive Assessment (URANIA) experiment or Pegasus-III. This enhancement to the Pegasus facility will remove the central solenoid entirely and investigate a variety of non-inductive start-up techniques (For a recent update see Raman 2019; https://news.wisc.edu/federally-funded-upgrade-reenergizes-fusion-experiment/).

At PPPL the Current Drive Experiment-Upgrade (CDX-U) was the first spherical tokamak operating in the US, beating the MEDUSA experiment by one year. Its name suggests that it is an upgrade of CDX. CDX was however a completely different device and not an ST. It operated at Princeton in the late 1980s to early 1990s. CDX-U operated from 1999 to 2005 and was the first fusion device to study liquid lithium as a plasma facing component (PFC). The choice of material for the plasma-facing portions of the reactor vessel, also known as the first wall, is an ongoing issue in fusion research. One major problem is that escaping fusion fuel when hitting the first wall cools and returns to the fuel mass at a lower temperature, cooling the fuel as a whole. This is known as "recycling". Another problem is that these reactions can also eject metal atoms from the first wall (sputtering), which due to the high atomic mass (high Z) of these atoms give off copious amounts of X-rays, which also cools the plasma fuel. One of the attractive features of liquid lithium as a plasma facing material is that it virtually eliminates recycling, since it easily binds with atomic hydrogen, which is then retained in the plasma facing material. Lithium also has a low atomic number Z. This gives the lowest possible energy loss by radiation from PFC material that may end up in the plasma, because radiation strongly increases with increasing Z.

As the first test of large-area liquid-lithium PFC, CDX-U had a circular tray at the bottom of the vacuum vessel to contain the lithium. Even with this partial non-recycling PFC, major improvements in plasma performance were obtained. Impurities were reduced, and a dramatic improvement (times 6) in energy confinement was observed in 2005.

In 2009 the CDX-U facility was significantly upgraded to become the Lithium Tokamak Experiment (LTX). It is the first device with a full, liquid-lithium wall and began operations with lithium wall coatings in 2010. The main objective of the LTX experiment is to investigate the tokamak plasma performance enhancement under extremely low wall recycling.

LTX is designed to operate with a pool of liquid lithium, similar to the lithium-filled tray used in CDX-U. In LTX lithium evaporates onto the inner liner surface, making a full "wall" of liquid lithium. The temperature is maintained just above the melting point. Lithium is so light—it is the only metal that floats on water—that simple surface tension is enough to pin down a film of molten metal film on the liner wall. In October 2010, technicians coated the LTX liner with lithium for the first time (See the Lithium Tokamak Experiment (LTX) Fact Sheet; https://www.pppl.gov/sites/pppl/files/LTX.pdf). Although liquid-lithium systems were expected to provide a further reduction in recycling, as was observed on CDX-U, the thin hot films employed so far in LTX are rapidly made inactive (Schmitt et al. 2013).

Recently the Lithium Tokamak eXperiment has undergone an upgrade to LTX-β, a major part of which consists of the addition of neutral beam injection. It will carry out further research on these issues.

11.6   Private Companies Pursuing Spherical Tokamak Research

Since a decade or so private companies, mainly start-ups, have been involved in (spherical) tokamak research. In Chap. 15 we will discuss the research and prospects of private companies that bet on other concepts than ST and have burst onto the scene in recent years funded by venture capital. Here we will only deal with private companies that specifically take the ST concept as their starting point, of which there are surprisingly few. These are the British firms Applied Fusion Systems (founded in 2013) and the by now fairly well-known company Tokamak Energy (founded in 2009), which were recently joined by an effort launched in China by the billion-dollar energy company ENN.

Applied Fusion Systems started life as Bionica Systems Ltd., and early in 2019 it was still boasting to be in the process of privately financing (seeking £200 million) the construction of its own British made tokamak reactor, called STAR (Small Toroidal Atomic Reactor), but the company apparently is no more. The reactor was to generate 100 MW and be slightly larger than MAST, but the STAR has faded. It seems that the company has re-emerged as Pulsar Fusion, which has an impressive website (https://www.pulsarfusion.com/), but the company's statements filed with Companies House do not reveal much activity and we will pay no further attention to it.

ENN Energy Holdings (ENN Energy, formerly known as "XinAo Gas Holdings Limited") is a company listed on the Hong Kong Stock Exchange. It belongs to the ENN Group, one of China's largest private energy groups. The subsidiary ENN

Energy is one of the largest clean energy distributors in China. It started to explore fusion technology in 2017, and its main research direction is compact, pollution-free and low-cost energy technology. ENN spent about $10 million over 2 years to construct the EFRC-0 (ENN Field Reversed Configuration), based on an American Axisymmetric Tandem Mirror device and has now built a medium-sized spherical tokamak fusion experimental device, the EXL-50, in Hebei province in north China, the first provincial-level compact fusion lab in China. XL stands for Xuanlong, which apparently means "Artificial Sun" in Chinese. It is China's first medium-sized spherical tokamak fusion experimental device (in addition to the small SUNIST device operating since 2003 at Beijing University). Construction started in October 2018 and first plasma was already achieved in August 2019. Its basic parameters are $R = 58$ cm, aspect ratio $A \geq 1.45$. Its aim is to extend the physics potential for commercial fusion power (Peng et al. 2019).

- **Tokamak Energy**

The most serious competitor and undoubtedly the leader in the private ST field is Tokamak Energy (TE), formerly known as Tokamak Solutions UK Ltd. and based on an industrial park in Oxfordshire, not too far from the Culham CCFE lab. It has so far been a fairly small player on the fusion playing field of private capital, but is slowly building up its position through a well-thought-out research and construction plan. The company grew out of the Culham laboratory. Some of its leading scientists were heavily involved first with START and MAST and later with JET, experience they now use to build their own fusion devices.

In 2019 they were still claiming to produce net fusion power just after 2025, while the latest version of the website[7] makes the heavenly promise of clean and abundant energy by 2030. Another page of the same website just speaks rather modestly of demonstrating "the feasibility of fusion as an energy source by 2030". Such partly deceptive statements seem to be part of the industry's characteristics. It is hoped that the tough-minded venture capitalists also take it lightly. Fortunately, early in 2020 Nature sent an easy excuse for a few years delay in the form of the Covid-19 virus. TE has so far raised about $150 million from private investors, but it needs far more to realise its plans. The company has also received grants and R&D tax credits from the UK government.

It has set out building a series of spherical tokamaks and started in 2012 with the first version of its ST25 (Sykes et al. 2018). ST stands, quite unimaginatively, for Spherical Tokamak. Its major radius was, again not surprisingly, 25 cm. The denotation ST for the machine is marginal—as it has aspect ratio 2, which is just a borderline case for being called a spherical tokamak. Unusually, the whole table-top device was contained in a small room and powered from only a 32A 415 V supply, i.e. ordinary three-phase electric power. Its achievements include tests of electron Bernstein wave heating and plasma pulses of 27 s. As noted in Chap. 6, the ST25 is now on permanent loan to the Technical University of Denmark and renamed NORTH.

---

[7]Its website is https://www.tokamakenergy.co.uk/, accessed 12 June 2020.

The second version of the machine, ST25-HTS, which is now in the Science Museum in London, used high-temperature superconducting magnets and provided the world's first demonstration of a tokamak magnet system with all the magnets made from HTS. All coils (toroidal and poloidal) were wound from YBCO HTS tape, a type of ReBCO (rare-earth (either yttrium or gadolinium) barium copper oxide). Coils made with this material will generate at least twice the field strength of ITER's magnets for confining the plasma, which are made from conventional, low-temperature superconductors.

The advantage of HTS is that in addition to operating at relatively high temperatures, they can also produce and withstand relatively high magnetic fields. The coils are cooled by helium gas to 20–50°K (−253–223 °C). Using conventional low temperature superconductors (LTS), meaning superconductors made from material that becomes superconducting just a few degrees above absolute zero (−273 °C) and has to be cooled to that temperature, would need thick shielding (of about 1 m) to prevent neutrons from heating the superconductors. This shielding would have to be installed in the central column and would make the device extremely large. HTS promises a solution to this problem, although the necessary shielding will still increase the aspect ratio.

Tokamak Energy also runs an extensive programme for developing tokamak magnets from high temperature superconducting (HTS) materials. In September 2019 they announced to have succeeded in producing a magnetic field of 24.4 T in magnets made from ReBCO tape. For these magnets cooling to 'only' 21°K (−252 °C) would be needed, which in superconducting terms is a relatively high temperature. These values seem similar to those achieved by Commonwealth Fusion Systems, which uses a conventional tokamak design with HTS to speed up fusion (Kramer 2015. See also Chap. 15.) Additionally, the magnets are extremely robust, reliable and easy to manufacture. It is claimed that this achievement is an important milestone on the route to commercial fusion energy, but such claims are rather commonplace in the fusion business. It is undoubtedly true though that high-temperature superconducting materials will facilitate the higher magnetic fields needed for efficient fusion reactors. The next step is to scale up these magnets into the configuration required for tokamaks.

HTS is not all roses though. In the first place there is the increased pressures the coils are producing on themselves, because the fields are much higher. HTS magnets, which are completely differently constructed and much more compact, are however much stronger than LTS magnets. A second disadvantage is the vastly increased neutron wall load. The ST devices constructed by TE are many times smaller than ITER, but the power density and neutron load will be similar, e.g. the divertor loads are equal.

STs have so far operated at toroidal fields of less than or just equal to 1 T. For high fusion performance, devices operating at 3 T or above are needed. This will require innovative engineering solutions especially for the central column. To develop and demonstrate such solutions, TE has recently constructed the ST40 (Fig. 11.12 shows a cutaway view and Fig. 11.13 the full device), which has copper magnets and is intended to operate at magnetic fields up to 3 T. It will not be equipped with HTS

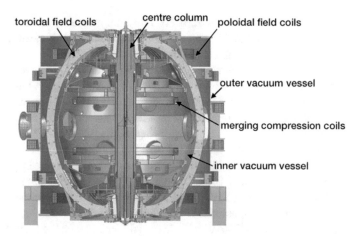

Fig. 11.12 Engineering drawing of TE's ST40, showing the steel support rings above and below the vessel, and the merging-compression coils which provide an initial high current, hot plasma without need of the central solenoid (*from* Sykes et al. 2018)

Fig. 11.13 The vacuum chamber surrounded by copper toroidal field coils on Tokamak Energy's ST40 spherical tokamak. The device is a predecessor to the tokamak the company plans to build with high-temperature superconducting coils

magnets, but its successor devices will. If HTS toroidal magnets were installed the inner legs would have to be fitted into the central column and would need shielding. For copper magnets such shielding is not needed.

ST40's main objective is to demonstrate the feasibility of the ST concept with such high fields and the advantages of high fields in ST, aiming at plasma temperatures in the 10 keV (100 million degrees) range (Sykes et al. 2018). The experiments on this machine will start with parameters somewhere between START and MAST, and will end at parameters of up to a factor of ten higher than MAST, i.e. burning plasma conditions, not enough to obtain fusion of the actually used D-D plasma, but sufficient when extrapolated to D-T. If this is successful, it will have been shown that fusion conditions can be achieved on a comparatively small machine, much smaller than the big tokamaks like JET and TFTR, let alone ITER. The intention is to also use D-T plasmas in the ST40, but if it ever gets that far is not clear.

The centre column of the ST40 contains two vital components: the central parts of the toroidal field magnets, and a solenoid. The solenoid maintains a current flowing through the plasma, which is important for plasma stability, and the toroidal field magnets generate the magnetic field that keeps the plasma confined.

ST40 has a design toroidal field of $B_T = 3$ T at a major radius of 40 cm. Use of copper for the toroidal field coil has the advantage of combining structural strength with good conductivity (especially when cooled to liquid nitrogen temperature). Whereas existing STs have operated typically at 0.3–0.5 T, with the recent MAST, Globus-M and NSTX upgrades striving for 1 T, innovative design features are employed to enable ST40 to operate at up to 3 T; for comparison the much larger ITER has a design vale of 5.3 T for B_T. Stresses are high at such high fields and an external support structure (the steel support rings in Figs. 11.12 and 11.13) is needed to accommodate the resulting forces.

The device will be equipped with two neutral beam injectors, which will inject heat, fuel (deuterium), momentum and current into the tokamak plasmas.

In summary the ST40 parameters are $R = 0.4$ m, $a = 0.22$, $A = 1.8$ (with the copper magnets, when HTS magnets are installed A will go up), $I_p = 2$ MA, $B_T = 3$ T, elongation $\kappa \approx 2.5$, and triangularity $\delta \approx 0.3$. As of November 2019, $B_T = 2$ T and $I_p > 0.4$ MA were achieved (Gryaznevich 2019). Compared with the large STs discussed above its physical dimensions are considerably smaller than MAST and NSTX, but its toroidal field is much higher. With this the ST40 is the first high-field spherical tokamak and a necessary step towards a steady-state ST reactor. The plasma inside the tokamak will reach more than 100 million degrees, and this while just 40 cm away the copper coils are cooled with liquid nitrogen to -196 °C, a temperature gradient that is unparalleled in the entire universe. With HTS magnets the gradient will be larger still.

In 2018 in the first experimental campaign with the ST40, not yet at design parameters, the company achieved start-up using the compression technique illustrated in Fig. 11.3. Two toroidal plasma rings were inductively formed around two high voltage poloidal field coils located inside the primary vacuum vessel. The position of these coils can be seen in Fig. 11.12. When the current in each plasma ring is sufficiently high, the two plasma rings are attracted towards each other and merge to form a single plasma. To develop such amazing start-up scenario was one of the main objectives of the campaign. The plasma was maintained with an approximately constant plasma current for about 0.2 s after merging. After merging no additional external heating was applied. Ion temperatures up to 1.5 keV (about 18 million degrees) were obtained.

The experience obtained with the low-field HTS device ST25-HTS and the high-field ST40 will be used to design and construct a high-field ST with HTS magnets. The main objectives of the successor devices, named ST-F1 and ST-E1, are the demonstration of efficient production of neutrons and the demonstration of Q ~ 1 (scientific breakeven) with an upgraded NBI system in D-D plasmas and in D-T plasmas subject to site availability and a licence to work with (radioactive) tritium. Tokamak Energy intends to move very fast and have at least the ST-F1 running before 2025.

A series of upgrades of the ST40 are foreseen in the near future until it reaches its design parameters of 2 MA for the plasma current, 3 T for the toroidal field (2 T has already been achieved) and up to 4 MW of auxiliary heating, supplied by two 1 MW NBI systems and 2 MW from Electron Bernstein Wave (EBW) heating/ECRH. The programme of experiments on the ST40 will last till after 2022 (Buxton et al. 2019; McNamara et al. 2019).

Tokamak Energy's proposals for its next devices are the ST-F1 (demo; major radius probably 1.4 m) and the ST-E1 (power plant model; major radius probably 2 m). It will be clear that these devices will be vastly smaller than ITER. The parameters for them were found from analysing a system code based on an established physics model to explore possible steady state, high-gain fusion devices. A wide parameter scan was undertaken to establish possible regions of parameter space that could potentially offer high fusion power with acceptable engineering parameters. Apart from the high aspect ratio, large tokamak solution, i.e. the ITER solution, a region of parameter space at low aspect ratio and relatively small major radius, and so small plasma volume, was identified. Tokamak Energy's proposals are still on the drawing boards and the feasibility of the devices depend critically on obtaining satisfactory solutions for some critical engineering problems, but if such relatively small fusion module is proved feasible, an alternative supply of fusion power based on a modular concept may be possible. The energy confinement in STs at high field, and the thickness of shielding needed to protect the HTS, especially on the central column, will however have a strong impact on the minimum size (Sykes et al. 2018).

11.7 Some Considerations of Power Plant Design Based on the ST Concept

The hope is of course that all the effort described above will eventually result in a real power plant. In many respects the ST is now equal to existing medium-sized conventional aspect-ratio tokamaks, which together with the large devices JET, JT-60U and TFTR formed the basis for the design of ITER. The ST equivalent of these large tokamaks is still lacking, reason why it may be somewhat premature to talk about a power-plant design based on the ST concept. There have however been developments, in particular high-temperature superconductors, which according to some have also greatly changed the playing field in respect of power plant designs based on conventional large-aspect ratio tokamaks. An example is the above-mentioned Commonwealth Fusion Systems, which has announced that it will shortly construct the world's first fusion power plant, based on the ARC tokamak concept. ARC has a compact, but conventional advanced tokamak layout, with as basic novelty the use of ReBCO HTS magnets. It will be slightly larger in size than JET and operate at a considerably higher field. A compact conventional aspect-ratio tokamak requires a very high field of about 12 T to achieve high fusion gain, while the physics advantages (such as high *beta*) of low aspect ratio potentially enable a compact ST to achieve a high fusion gain at a comparatively modest toroidal field of around 4 T. The elongated shape of the plasma in an ST results in significant reductions in major radius and/or field for comparable fusion performance. So, the device can be considerably smaller than mammoth constructions like ITER.

But size is not the only thing that plays a role. Because the energy confinement time of tokamak plasmas scales positively with plasma size, it is generally expected that Lawson's fusion triple product (of ion density, ion temperature and energy confinement time) also increases with size. This has been one of the reasons that tokamak devices have become increasingly bigger. Tokamak plasmas are however also subject to operational limits and two important ones are a density limit and a *beta* limit, discussed in Chap. 8. When these limits are taken into account, the triple product becomes almost independent of size; and depends mainly on the fusion power. In consequence, the fusion power gain, which is closely linked to the triple product is also independent of size. Further, it has been found that the triple product is inversely dependent on *beta*, a result that tends to favour lower power reactors. It implies that the minimum power to achieve fusion reactor conditions is driven mainly by physics considerations, especially energy confinement, while the minimum device size is driven by technology and engineering considerations, such as wall and divertor loads. These latter aspects are evolving in a direction to make smaller devices feasible (Costly 2016).

So, when the density limit is taken into account, the relationship between fusion power and fusion gain is almost independent of size, implying that relatively small, high performance reactors should be possible, perhaps with a major radius of 1.5–2.0 m (where ITER, not even a reactor, has a major radius of 6 m), volume of 50–100 m^3 (ITER's plasma volume being bigger than 800 m^3) and operating at relatively low power levels, 100–200 MW. The lower power requirement is especially advantageous. Further beneficial properties of STs from a reactor standpoint are operation at high plasma pressure relative to the pressure of the confining magnetic field (high *beta*), and the generation of higher levels of self-driven current within the plasma. Smaller devices would make a modular approach to fusion power generation possible, whereby single or multiple, relatively small, low-power devices could be used together to generate the required power (Costley 2019).

There are however a number of critical engineering problems for which solutions must be found before an ST power plant will be feasible. Many of these challenges equally apply to conventional large devices. They include handling the stress in the central column while simultaneously accommodating the HTS toroidal field magnet; handling the plasma exhaust in the divertor region, where power loads will be at the limit of available materials; the inboard shielding needed to protect the HTS tape from the intense neutron and gamma radiation, for it to have an acceptable lifetime; and reducing the neutron heating to a level that can be handled with a reasonable cryogenic system.

Various proposals for power plant designs based on the ST configuration have appeared in the literature (e.g. Morris et al. 2005; Ono et al. 2004; Wilson et al. 2004; Menard et al. 2016; Voss et al. 2002), some going back more than twenty years. Especially Stambaugh et al. (1998) seriously presented the spherical tokamak path to fusion power. Their research into a proper power plant, not just an experimental reactor like ITER was carried out in 1996, and one wonders why their findings were not incorporated into the modified ITER-FEAT design, made necessary after the withdrawal of the US in 1998 and published in 2001. Because of the extremely high-power densities that STs can deliver, the neutron wall load will be immense and the limiting factor in performance will be a neutron wall load constraint. Taking this constraint to be 8 MW/m^2 (about ten times the average wall load in ITER and not far from the divertor neutron load expected in ITER), Stambaugh et al. proposed a nuclear fusion power plant based on the ST design with major radius 5.25 m, minor radius 3.75 m, central column radius 1.5 m, allowing for shielding of 1 m for the low-temperature superconducting magnets installed in the central column. Their proposals were not picked up at the time; the reason for this is not clear. The other parameters and envisaged output of their plant are not of interest here, neither does this specific design, as the higher fields and higher operating temperature of recently developed high-temperature superconductors may open up new opportunities, and can do this equally for conventional tokamaks of course.

11.8 Conclusion and Prospects

The advocates of spherical tokamaks claim that STs could result in more economical and efficient fusion power and that designs demonstrating the feasibility of ST power plants have already been developed. The latter is a bit far-fetched as so far no fusion power at all has been produced by an ST device; no breakeven of any form, not even 'scientific breakeven' has been shown. Nor has any route towards a power plant based on the ST configuration been laid out in any clarity, in spite of the inclusion of this task in mission statements of several of the big STs discussed in this chapter. In this respect it is surprising that the UK on its own has launched a £200 million design effort for a fusion power plant based on the spherical tokamak (the STEP programme discussed above), while at the same time still being involved in ITER or ITER-related programmes (at least until Brexit has been clinched). But the doubt in ITER is rising. Does anyone really have confidence that it will work and that it is a viable route towards *affordable* fusion power? Affordable is the important word in this respect. Even people who have worked all their active life in fusion-related research, like Daniel Jassby, one of the pioneers of the spherical tokamak and referred to above, have lost all faith in fusion reactors (Jassby 2018). And when doubt creeps in, the alternative suddenly looks much better, even if it has nothing to show for yet. It offers an open field for a good salesman.

STs have however indeed unmistakable advantages:

1. A compact device with plasmas confined at higher pressures for a given magnetic field. The greater the pressure, the higher the power output, and the more cost-effective the fusion device.
2. The magnetic field needed to keep the plasma stable can be up to ten times less than in a conventional tokamak, also allowing for more efficient plasmas.
3. STs will cost less to build as they do not need to be as large as conventional machines for the same performance. This still has to be shown; current devices are still fairly small but growing in size.

The disadvantage of the spherical tokamak is also clear. The need for a slender centre column imposes severe engineering issues and space constraints. There is hardly any space for the inner limbs of the toroidal magnets, and because of problems with the shielding of sensitive components against damaging neutrons produced by fusion events, many STs are forced to decrease the minor radius and hence make the aspect ratio of the device bigger. Using low-temperature superconductors (like in ITER) would require a neutron shield thicker than a metre and make a spherical tokamak unfeasible (Windsor 2019). However, this could be remedied by making the device bigger. ITER has a major radius of about 6 m and a minor radius of 2 m, resulting in an aspect ratio of about 3. Making the minor radius 4 m would still leave space for shielding and decrease the aspect ratio to about 1.5.

The CCFE website also states that, as STs are in a relatively early stage of development, they will probably not be used in the first generation of fusion power plants. This statement is undoubtedly true in the sense that any statement about an empty set is true. There are no fusion power plants and envisaging generations while the first working plant must still see the light is just preposterous, even for predictions in fusion.

In the nearer term, STs could serve as component test facilities to ensure that the systems and materials used in power plants can withstand the bombardment by neutrons from the fusion reactions. It seems that at present the publicly funded STs are mainly geared to a future as such test facility, having all but abandoned their own route towards a working power plant, which has mainly been left to private companies. If STs indeed show better prospects for fusion power to be realised, one would expect more public funds to be channelled into their development.

Additionally, plasma physics studies in STs are feeding into the development of ITER. As their geometry differs from conventional tokamaks, they are providing insight into how changes in the characteristics of the magnetic field affect plasma behaviour, revealing trends that otherwise would be difficult to spot. This is true, but if the conventional tokamak is a dead-end street it would make more sense to make a U-turn now and stop wasting more money on it. You either believe in the ST concept for a power plant or you don't. Vacillating between the two design options will jeopardize them both (http://www.ccfe.ac.uk/ST.aspx).

A note of caution is however appropriate. When reading through papers reporting the results of research on spherical tokamaks, you do not get the feeling that any real progress is being made. They grapple with the same problems as conventional tokamaks did before or still do, e.g. instabilities, only still at a lower parameter regime. Temperatures are still low and there is hardly any mention of the Lawson condition for instance, which has to be fulfilled for STs too before any fusion can become a reality. All the research reported so far has not really brought 'power on the grid' from a spherical tokamak much closer.

Chapter 12
Inertial Confinement Fusion

12.1 Basics

Inertial confinement fusion (ICF) has always been a sideshow to the main fusion efforts, which have mainly concentrated on magnetic confinement. It has also always been part of weapons research and consequently most of it was classified. Some of it still is, although in December 1993 much of the research was declassified. (Dean 2013, p. 121.) The basic idea goes back to the hydrogen bomb which had shown that fusion can be initiated by a sufficiently strong compressive force on a small amount of fusion fuel. The first step in this respect was taken in 1951 by the second atom bomb test in the Operation Greenhouse series, called George, which was the world's first thermonuclear burn. The George device was shaped as a torus and, apart from a nuclear fission device, it had a small amount of liquid deuterium and tritium placed at its centre. The vast majority of its yield derived from the fission device, the energy output from fusion being insignificant in comparison, but nevertheless in this test the small amount of deuterium and tritium fused, demonstrating this possibility. Its role in George was to generate a strong flux of fast neutrons to spark more fissions in the uranium nuclei present.

George was a development stage for thermonuclear weapons and, as recalled in Chap. 1, these were eventually realized by using a nuclear fission bomb (an 'ordinary atomic bomb') as a 'matchstick' for the fusion part of it. In the Teller-Ulam configuration[1] of the thermonuclear or hydrogen bomb the primary stage consists of a fission bomb that provides the necessary temperature and pressure for the second stage, the fusion bomb, to go off (Fig. 12.1).

Each stage would undergo fission or fusion (or both) and release energy, much of which would be transferred to another stage to trigger it. The actual workings of the device are still classified, so how the energy is transferred from the *primary* to

---

[1] Named after Edward Teller and Stanislav Ulam (1909–1984). Ulam was a Polish-American mathematician and physicist. The Teller-Ulam configuration originated from an idea Ulam had in 1951. It was first tested in 1952 in the Ivy Mike, the first test of a full-scale thermonuclear device.

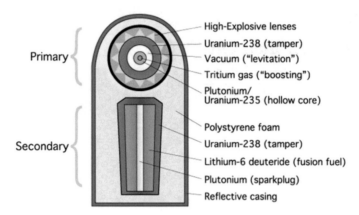

Primary:
- High-Explosive lenses
- Uranium-238 (tamper)
- Vacuum ("levitation")
- Tritium gas ("boosting")
- Plutonium/Uranium-235 (hollow core)

Secondary:
- Polystyrene foam
- Uranium-238 (tamper)
- Lithium-6 deuteride (fusion fuel)
- Plutonium (sparkplug)
- Reflective casing

Fig. 12.1 A possible Teller-Ulam configuration of a thermonuclear bomb (*Wikipedia*)

the *secondary* has been the subject of some disagreement in the open press, but it is thought to be transmitted through the X-rays and gamma rays emitted from the *primary* fission device. This energy is then used to compress the *secondary*. The crucial details of how this pressure is created are still disputed in the unclassified press but are of no concern to us.

As can be imagined it is impractical to use a fission bomb to get a fusion process going in a power plant. The idea behind ICF is however similar to that of the hydrogen bomb set out above: fuel is compressed and heated so quickly that conditions for fusion are reached before the fuel has time to escape. The fuel's inertia (the tendency of a physical object to resist any change in its motion, either the magnitude or direction of such motion) will delay its escape and cause the fuel to burn significantly before the explosion takes place that blows the target apart–hence the name inertial confinement fusion. Thus, the challenge of ICF as an energy source is to scale down the yield of the thermonuclear explosions to values small enough for the energy of such explosions to be captured and converted into electricity.

It was the American physicist John Nuckolls (b. 1930) who in the late 1950s investigated what would happen if the thermonuclear secondary in the hydrogen bomb was scaled down to a very small size.[2] Nuckolls completed computer calculations of the radiation implosion, ignition, and burn of 1 mg of D-T fuel, and calculated that the gain in energy would be a factor of 10, so the energy output would be 10 times the energy input. A fusion power plant in his vision would then be a device in which very many small droplets of fusion fuel would be exploded in rapid succession in order to produce a steady output of energy. One such explosion would release something like 50 megajoules in energy, roughly the equivalent of 1 kg of gasoline. His idealised calculations suggested that at very small sizes, on the order of milligrams, very little energy would be needed to ignite such a device, much less than a fission "primary", although at the time he did not yet know of a power source that could provide the

---

[2]Nuckolls has written a report of his early work on ICF in a 1998 Lawrence Livermore National Laboratory report "*Early Steps Toward Inertial Fusion Energy (IFE) (1952–1962)*", UCRL-ID-131075.

Fig. 12.2 Schematic picture
of target capsule
(*from* Pfalzner 2006, p. 14)

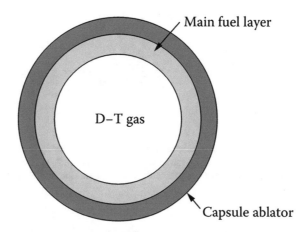

necessary energy. As with all energy generating devices, it all hinges of course on the energy balance. Is it possible to get more energy out than has to be put in? Although Nuckolls had computed a theoretical gain of a factor 10, the problem still waits to be resolved in practice. (See e.g. Seife 2008, p. 159 ff. about this.)

In the early 1960s, after the invention of the laser[3], scientists inside and outside weapons laboratories began to speculate as to whether a fusion reaction of practical interest could be initiated by using a laser as driver, i.e. focusing a high-power laser on a small millimetre-sized capsule or pellet containing fusion fuel. Lasers are unique tools for transporting extremely high powers over large distances, but the transfer of such power from the photons of the laser light to matter in small volumes is a very complicated problem. First of all, the interaction proceeds very far from equilibrium. The photons have energies of a few electron volts and must heat plasma to temperatures thousand times higher. Second, these processes are strongly nonlinear, as they involve the transfer of the energy of a large number of photons to a much smaller number of charged particles in extremely small volumes and at very short time scales (Tikhonchuk 2019).

In 1972 Nuckolls and collaborators (Nuckolls et al. 1972) were the first to publish the idea of compressing a tiny target with high-power lasers to bring thermonuclear fuel to ignition conditions. Figure 12.2 shows a schematic picture of how such fuel capsules would look like. They typically consist of a small (3–5 mm diameter) plastic or metal sphere filled with tritium and deuterium. The outer shell (called the capsule ablator in Fig. 12.2) is made of a high-Z material, a chemical element with a large atomic number, while the inner region contains deuterium and tritium, first a dense layer of frozen deuterium and tritium and at the core a D-T gas. More intricate designs use multiple layers of different materials, e.g. a layer of cryogenic D-T ice inside a polystyrene (CH) shell. The ice layer thickness typically ranges from 0.16 to 0.6 mm (about the diameter of a human hair).

---

[3]See the Glossary for a brief explanation of a laser.

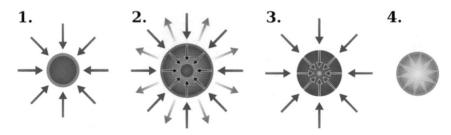

Fig. 12.3 Schematic picture of the stages of ICF, using lasers. The blue arrows in 1. represent the laser radiation delivered to the target; orange in 2. is blow-off, the ablation of the outer shell that pushes the shell inwards to the centre; light blue in 3. is inwardly transported thermal energy (*Wikipedia*)

Figure 12.3 schematically shows the various stages of the process envisaged in ICF.

In stage 1 laser beams or laser-produced X-rays rapidly strike and heat the outer layer of the fusion target, so that it evaporates almost immediately (called *ablation*, hence the term 'capsule ablator' in Fig. 12.2), forming a surrounding plasma envelope. Above a certain *critical* density, the laser light can no longer penetrate this plasma envelope and will no longer be able to reach the fusion target. The laser energy is than deposited at the critical surface (where the density of the plasma envelope is equal to the critical density; Fig. 12.4) and carried further by heat conduction. The location of the critical surface depends strongly on the wavelength, intensity, and pulse length of the laser beam used and the choice of these parameters is therefore essential for an efficient absorption of the laser energy by the target.

When in stage 1 the outer shell of the capsule burned into plasma blasts off, it results in a force in the opposite direction (like a rocket is accelerated upwards by blowing off fuel), causing the main fuel layer of the pellet to be forced inwards by an inward-propagating spherical shock wave through the pellet to the middle (stage 2 in Fig. 12.3). Subsequently, during the final part of the capsule implosion (stage 3),

Fig. 12.4 Schematic picture of the critical density surface formed when a laser interacts with the target (*from* Pfalzner 2006, p. 18)

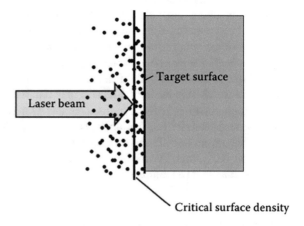

the fuel core reaches 20 times the density of lead and ignites at 100 million degrees. This causes fusion reactions to start. If the rate of such reactions is high enough, the heat generated will cause the thermonuclear burn to spread rapidly through the compressed fuel, yielding many times the input energy (stage 4). The power required to drive this process is very high: to heat a 1 mm fuel capsule to 10 keV (about 120 million degrees) requires 10^5 J. This latter amount is actually not so much, only 0.03 kWh, but it must be delivered in a few picoseconds (a picosecond is 10^{-12} or one trillionth of a second) to the outer part of the target shell, which requires an enormous power. This burst of energy will immediately heat up, ionise and evaporate the shell.

12.2 The Lawson Criterion

Just as for any other fusion device, the Lawson criterion sets the requirements that must be satisfied for ICF to be possible (see Chap. 2). ICF follows a different route than magnetic confinement fusion in fulfilling the Lawson criterion. For magnetic confinement, as we have seen, the plasma is confined at very low densities for several seconds or so. For inertial confinement, on the contrary, confinement times are as minimal as can be; very small, just the small, but finite time (due to the inertia of the fuel) it takes for an assembled fusing fuel to disassemble when driven to do so by its own high pressure. So, it may not come as a surprise that the confinement time and capsule size are closely connected, and can be estimated as roughly equal to the capsule radius divided by the speed of the shockwave travelling through the pellet (in stage 2 of the process in Fig. 12.3). This shockwave speed is close to the speed of sound (about 10^6 m/s at 100 million degrees) (Nuckolls et al. 1972, p. 139) and a typical capsule radius measures one to two millimetres. This very rough estimate results in a confinement time of 10^{-9} s (one nanosecond or one billionth of a second). More detailed calculations give 10–20 ns (Pfalzner 2006, p. 15). For such small confinement times the Lawson condition requires the density to be very high (10^{31} nuclei per cubic metre, a factor of about one million times higher than the density of air, and about 100,000 million times higher than the plasma density in magnetic confinement fusion). Hence, the fusion conditions for the two approaches (magnetic or inertial confinement) are indeed vastly different, making them two completely different attack routes at the same problem.

12.3 Direct and Indirect Drive

At present lasers are the most common drivers in ICF-experiments. In this respect two main general approaches can be distinguished: *direct drive* and *indirect drive*. In the case of direct drive, the laser beam is focused directly at the fuel capsule, while for indirect drive the target is heated indirectly.

Fig. 12.5 Mockup of a
gold-plated hohlraum
designed for use in the NIF
(*Wikipedia*)

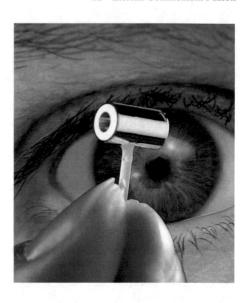

The indirect-drive method was developed at the Lawrence Livermore National Laboratory (LLNL) in 1975 with the aim of increasing the degree of uniformity of heating the capsule and through this to avoid instabilities (see below). The fusion capsule (with a diameter of a few millimetres) is held at the centre of an enclosure in the form of a cylindrical chamber, typically a centimetre across, 5–6 mm in diameter and with walls of a heavy metal, such as gold e.g. 30 μm thick, so equal to the breadth of a rather thin hair (Fig. 12.5). The cylindrical chamber is known as a *hohlraum*, a German word meaning cavity, or radiation case. Its actual size is not so critical, the important parameter is the wall to hole surface area (the hole through which the lasers must enter the hohlraum) which in current designs is approximately 2:1. The holes should not be too small as in that case the laser light can be refracted and absorbed at the hole edges before entering the hohlraum. The laser beams are not focused directly at the fuel capsule, but onto the interior walls of the hohlraum, which absorb and reemit the energy in the form of X-rays. The X-rays bounce around inside the hohlraum, are absorbed and reemitted many times, rather like light in a room with walls completely covered with mirrors. (McCracken and Stott 2013, p. 76.) Of course, energy is lost in this process, but current designs achieve a 70–80% conversion of laser energy into X-rays. A schematic working of the hohlraum is shown in Fig. 12.6. The hohlraum serves the same purpose as the bomb casing in a hydrogen bomb (Fig. 12.1), trapping X-rays inside, acting as an oven and irradiating the fuel. The main difference is that the X-rays are not supplied by a primary within the shell, but by an external driver, like a laser. The X-rays strike the fuel capsule many times and smooth out any irregularities in the original laser beams. Such indirect heating achieves highly uniform compression, as well as highly uniform heating of the outer layer of the fuel pellet, to start the process of Fig. 12.3, without the need for precise positioning of incoming energy beams as is required for direct-drive

Fig. 12.6 Schematic picture of indirect drive (*from* Pfalzner 2006, p. 19)

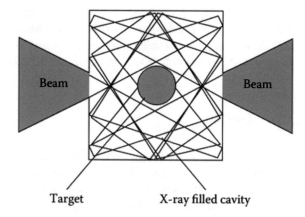

Target   X-ray filled cavity

heating. Consequently, beam uniformity, beam energy balance, and beam alignment requirements are less stringent than for direct drive. For example, for direct drive, a typical beam alignment tolerance might be 20 μm. The baseline indirect-drive target (Fig. 12.14) of the National Ignition Facility, to be discussed below, can tolerate a beam misalignment of about 80 μm.

A further considerable advantage of this approach is that much of the capsule physics of indirect drive is almost independent of the driver. After all, the X-rays generated from the hohlraum wall impinge on the target, not the laser light itself. This implies that large amounts of information learned from indirect-drive experiments with lasers are directly useful for indirect-drive with other drivers. Because of the appalling energy inefficiency of laser drivers, such alternatives will be absolutely necessary.

The high efficiency of coupling the laser energy to the imploding fuel is usually considered the most important advantage of direct drive. In the case of indirect drive, a substantial fraction (at least 20–30%) of the laser energy is used to heat the hohlraum wall.

Another potential advantage of direct drive is the chemical simplicity of the target. Laser direct-drive targets usually contain little material with a high atomic number (high-Z material). In contrast, indirect-drive targets require a hohlraum made of some high-Z material such as lead or gold. For this reason, the indirect-drive waste stream (from target debris) contains more mass and is chemically more complex than the direct-drive waste stream.

Hence, direct drive and indirect drive each have their advantages and disadvantages and there is still no consensus which scheme would be better in terms of civil energy applications of ICF. The indirect-drive method is favoured in military applications, both in the US and in France, and may indeed be a good setup if a single explosion in a matter of days is what you are after. For a power plant the process of igniting a capsule has to be repeated at a rate of seconds rather than days which is easier in the direct-drive approach and would make this scheme more favourable.

Recently a third method, called *hybrid drive*, which is a mixture of direct and indirect drive, is being developed and tested in China. In this method a layered fuel capsule inside a spherical hohlraum with an octahedral symmetry is compressed first by indirect-drive soft-X rays and then by direct-drive lasers.

12.4   Ignition

The term ignition is also used in ICF, but with a slightly different meaning. In magnetic confinement fusion the aim is to achieve steady-state operation and ignition occurs when the heating by the α-particles produced in the fusion reactions is sufficient to maintain the plasma at the required temperature for fusion to continue. ICF is inherently pulsed (a rapid succession of mini-explosions). In the final stage of the process, fusion neutrons and X-rays are produced together with charged particles (α-particles in the case of D-T fuel). They deposit their energy in the D-T fuel, bringing more compressed fuel to fusion temperatures and leading to a propagating burn in a so-called *hot spot*—a high-temperature, low-density region surrounded by a lower-temperature, higher-density D-T shell (stage 4 in Fig. 12.3). This is commonly referred to as ignition in this approach. As ignition is reached, the fusion energy produces an outward directed pressure, which soon overcomes the imploding wave and the capsule blows apart. The resulting energy is used to generate electricity and the process must then be repeated all over again with the next fuel capsule.

The first experiments carried out in the late 1960s showed that inertial fusion was indeed possible, at least in principle. In these early days the expectation was that fusion could be achieved relatively quickly (a common thought it seems for any fusion enterprise), as the required energy did not seem that demanding. It was thought that all the fuel in the capsule should be compressed as a whole to fusion conditions at the end of the compression stage. This concept is called *volume ignition*, but turned out to require an unrealistically high driver energy, as not all the energy in the driver can actually be used for ignition. A lot of energy is lost through various conversion processes from the laser to the final burn. Two key points are at issue here: it takes more energy to heat fuel than to compress it and the compression of hot material is more energy-consuming than of colder material. More powerful lasers were needed, and were indeed built in the 1970s and 1980s, culminating in the National Ignition Facility (NIF) at LLNL, the grandest attempt at ICF so far, and to be discussed in some detail later in this chapter.

Another development aimed at using the available energy more efficiently led to the *hot-spot* concept mentioned above. The concept involves the creation of a small central mass of fuel heated to temperatures that are sufficient to start efficient thermonuclear burn (about 10 keV or 100 million degrees). This concept is considered more likely to achieve the fusion goal. The aim is to make the process of ablation and compression (the first two stages in the process of Fig. 12.3) as efficient as possible. The heating of the target is quite a balancing act, as it must be avoided to heat the core of the capsule too strongly during the early part of the compression process (stage 2 in Fig. 12.3) because much more energy is then needed to compress the hot plasma. Ideally the fusion material starts to burn in a small central area (*hot spot*) (approximately 1 μm in size and with a lifetime of 100–200 picoseconds) at the time of maximum compression. Once this core starts to burn, the energy produced by the fusion reactions (α-particles) will heat up this central area very quickly. The neutrons resulting from the fusion reactions and fast electrons will rapidly transport the energy from the hot spot to the outer fuel area and raise the temperature of this outer region, so that the fusion reaction will rapidly spread outwards into the main fuel region. This entire process takes about 10 ps. During this time a very high pressure builds up that will eventually blow apart the remaining fuel. The ICF-cycle ends here. In a reactor the next target has to be injected and the whole process will start all over again. (Craxton et al. 2015, p. 4; McCracken and Stott 2013, p. 78; Pfalzner 2006, pp. 22–23.)

Because less material needs to be heated in the hot-spot scheme, it is more energy efficient than volume ignition and has the advantage that the external dense fuel layer provides better confinement. If the target is constructed in such a way that the central hot spot contains just 2% of the total fuel mass, heating the hot-spot mass and compressing the remaining fuel will need comparable energy. One important issue is that any premature heating of the material has to be avoided, because this would completely jeopardize the compression. This is done by introducing doping agents (dopants), e.g. germanium, that will effectively absorb the X-rays and prevent them from preheating the fuel. The hot-spot method is currently the prevailing scheme for inertial confinement facilities under construction or in operation.

The ignition scenario discussed above relies on simultaneous compression and ignition of a spherical fuel capsule in an implosion. An alternative route to ignition, termed *fast ignition*, separates these two stages. It uses the same hardware as the hot-spot approach, but adds a high-intensity, ultrashort-pulse laser to provide the "spark" that initiates ignition. A D-T target is first compressed to high density, in either direct or indirect drive, by a traditional driver, so far mostly lasers but alternatives are also possible. Then the short-pulse laser beam delivers energy to ignite the compressed core, analogous to a sparkplug in an internal combustion engine.

An advantage of this approach is that density and pressure requirements are less strict than in central hot-spot ignition, so in principle the need to maintain precise, spherical symmetry of the imploding fuel capsule may be relaxed, which might reduce target production costs. In addition, the fast-ignition concept is attractive as it requires less total energy input to achieve ignition, yet provides an improved energy gain estimated to be as much as a factor of 10–20 over the central hot-spot approach. With reduced laser-driver energy, substantially increased fusion energy gain—as much as 300 times the energy input—and lower capsule symmetry requirements, the fast-ignition approach could provide an easier development path towards an eventual inertial fusion energy power plant. The first compression stage is identical to the ordinary ignition discussed above. X-rays generated by laser irradiation of the hohlraum wall deposit their energy directly on the outside of the ablator shell, which rapidly heats and expands outward. This action drives the remaining shell inwards, compressing the fuel to form a uniform dense assembly. Then a short-pulse laser

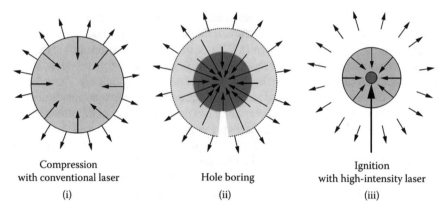

| Compression
with conventional laser | Hole boring | Ignition
with high-intensity laser |

Compression
with conventional laser
(i)

Hole boring
(ii)

Ignition
with high-intensity laser
(iii)

Fig. 12.7 The three stages in the fast-ignition concept, **i** compression, **ii** hole drilling and **iii** ignition (*from* Pfalzner 2006, p. 194)

beam drills a hole in the fuel capsule, after which a high-energy, high-intensity laser deposits a large amount of energy in a few picoseconds (trillionths of a second) in a 35-micrometre spot to ignite the fuel assembly. Figure 12.7 schematically shows the three stages in this process. The term "fast ignition" was coined to reflect that the dense core is ignited by particles (electrons, protons, or other ions) that are generated by the short, ultra-intense, high-energy laser pulse. The method is still in development and the physics basis is currently not yet as mature as that of the central hot-spot approach. (Craxton et al. 2015, p. 105.) Experimental demonstration of ignition via the fast-ignition scheme will require a high intensity multi-petawatt (10^{15} W) laser coupled to high energy multi-beam systems; a combination that will not be available in the near future.

Two fast-ignition approaches are being studied: (i) the "hole-boring" method, discussed above, and (ii) the "cone-in-shell" method. The first approach has been successfully tested at the OMEGA device of the University of Rochester's Laboratory for Laser Energetics. The second approach, tested successfully on the GEKKO XII laser in Japan, uses a small gold cone that cuts through a small area of the capsule shell; the fuel implosion produces dense plasma at the tip of the cone, while the hollow cone makes it possible for the short-pulse-ignition laser to be transported inside the cone without having to propagate through the coronal plasma and enables the generation of hot electrons at its tip, very close to the dense plasma. A variant cone concept uses a thin foil to generate a proton plasma jet with multi-MeV proton energies. The protons deliver the energy to the ignition hot spot. (Onega and Ogawa 2016, p. 773; Key 2000.)

Experimental investigations of the fast-ignition concept are challenging and involve extremely high-energy-density physics: ultra-intense lasers; very high pressures in excess of 1 gigabar; very high magnetic and electric fields.

A third alternative ignition scheme, called *shock ignition*, also separates the compression and heating stages, but uses a single laser to first compress the fuel capsule without heating it. Then the same laser launches a strong, spherically converging shockwave into the compressed fuel. This ignition shock is timed to collide with the return shock near the inner shell surface. This collision results in a third inward shock that further compresses the hot spot. If timed correctly, the shock-induced pressure enhancement triggers the ignition of the central hot spot. Shock ignition can be understood as shock-assisted central ignition. The fuel mass is typically greater for shock ignition than for hot-spot ignition. The large mass of fuel leads to high fusion-energy yields and large burn-up fractions, so to high predicted gain. The ignition shock is required because at low velocities the central hot spot is too cold to reach the ignition condition with the conventional ICF-approach.

Shock ignition has been experimentally studied at the Rochester OMEGA laser facility, demonstrating that a properly timed final shock has the potential to significantly enhance the neutron yield. The advantage of shock ignition for laser-based inertial fusion energy is a reduction of the required energy (potentially by a factor of five), partly also because it uses a single laser, while fast ignition would need two laser systems. (Tabak et al. 2014, p. 4; McCracken and Stott 2013, pp. 161–162; Schmitt et al. 2010.)

12.5 Instabilities

One of the beauties of ICF was supposed to be the absence of instabilities. No such luck. New instabilities, as well as familiar ones showed up. Experiments from the mid 1970s to mid 1980s revealed the importance of instabilities, caused among other things by nonuniformities in the laser beam. This made the uniformity of the laser beam or laser beams (plural as in present-day experiments it is common to work with multiple lasers), which reach the target from different directions, one of the most important issues in the field. It is paramount that the fuel capsule is heated uniformly from all sides, so that the compression on the fuel capsule surface is highly homogeneous and results in a perfectly spherical implosion of the fuel pellet. In reality this ideal can never be achieved, as laser energy is necessarily deposited somewhat nonuniformly according to the intensity distribution of each beam, and the number of laser beams is limited. If the nonuniformities are too large, instabilities can arise and lead to loss of energy. Such nonuniformities can roughly be distinguished into macroscopic and microscopic nonuniformities. An obvious example of the former is the geometrical effect caused by the limited number of beams. In the direct-drive scheme such macroscopic nonuniformities can be reduced by taking a larger number of beams, but this makes such systems very expensive, as well as technically challenging. Other examples are the existence of power imbalances between the beams, or a less than perfect timing of the beams.

It is also clear that the quality of the fuel capsule, in particular its sphericity, plays an important role. It should be as close as possible to a perfect sphere. The quality of the finish of the surface is crucial, because even small machining marks from the production process or the crystalline structure of the material can become the seed of disturbances. The same applies to slight variations in the thickness of the various target layers. Microscopic nonuniformities can be due to spatial fluctuations within a beam. Whatever their cause the important thing is that both types of nonuniformities can lead to instabilities in the compression stage.

The dominant instability is the familiar Rayleigh-Taylor instability (RT-instability), whose devastating effect is due to the fact that it initially grows exponentially, so that even very small, seemingly insignificant disturbances can reach a size that can threaten the whole compression. As the gas heats up in an ICF-capsule, it begins to push back on the shell and slows the compression down. Rayleigh-Taylor instability occurs, and elements of the gas are pushed outwards, while elements of the shell are pushed inwards. The RT-instability has already been discussed in Chap. 2, where it is stated that it can occur at an interface between two media, normally two fluids with different densities (like oil and water). Here, in inertial confinement fusion, there is essentially only one "fluid". Instead of two fluids we have a lower density (hot) plasma on top of a higher density (cold) plasma, while the ablation pressure, the pressure on the outer surface from the laser light, which evaporates (ablates) the fuel capsule surface, acts as the transmitting force (instead of gravity in the case discussed in Chap. 2). When the target is compressed, hot plasma pushes onto colder plasma. The higher density plasma tries by all means to escape to the outside. If it succeeds, hot plasma will mix with cold plasma, leading to an undesired cooling of the hot plasma. To prevent such escape the ablation pressure should be perfectly homogeneous on the entire spherical surface of the fuel capsule. A small perturbation can grow out exponentially fast and cause the target shell to break up. In ICF RT-instabilities play a role in the initial compression stage, where the ablation process forces the fuel towards the centre (stage 2 in Fig. 12.3), and also in the deceleration stage where the fuel reaches its final stages of compression. (Pfalzner 2006, p. 129.)

Since it is impossible to create a perfectly uniform contact between laser or laser produced X-rays and the capsule surface, RT-instabilities are unavoidable in ICF, whatever the scheme. One has to live with them and design the targets in such a way that they are minimised as far as possible. Research has shown that the ratio of the fuel capsule radius to the thickness of the ablator layer (the capsule ablator in Fig. 12.2) is crucial in this respect. This ratio is called the *in-flight aspect ratio* and has to be of the order of 25–40, a firm requirement for fuel capsule design.

Other instabilities related to the Rayleigh-Taylor instability are the Richtmyer-Meshkov instability and the Kelvin-Helmholtz instability. The latter instability goes back to publications by Lord Kelvin (1824–1907) and Hermann von Helmholtz (1821–1894) in the late 1800s and concerns instability that occurs due to turbulent flows in fluids or other media of different densities moving at various speeds. So, two fluids are in motion and encounter a change in the velocity component parallel to the interface between the two fluids. It can also be observed, quite spectacularly, in clouds (Fig. 12.8) and deep in the oceans. In ICF they can develop once RT-instabilities have occurred, so the suppression of the latter automatically suppresses Kelvin-Helmholtz instabilities.

Fig. 12.8 Kelvin-Helmholtz instability in clouds, known as *fluctus,* a relatively short-lived wave formation, usually on the top surface of the cloud in the form of curls or breaking waves (*Wikipedia*)

The same is true for the Richtmyer-Meshkov instability, which is essentially the RT-instability for the case that two fluids of different density are accelerated, e.g. by the passage of a shock wave as happens in stage 2 of Fig. 12.3. The instability is caused by laser-beam nonuniformities.

Other notable instabilities are due to nonlinear plasma interactions with the incident laser wave in the plasma envelope formed in the ablation process (see Fig. 12.4). In the case of the two-plasmon–decay (TPD) instability the incident wave excites two waves of plasma electrons (such waves are called plasmons). Other such instabilities are called the stimulated Brillouin scattering (SBS) instability and the stimulated Raman scattering (SRS) instability. The latter comes in two varieties, termed the backward SRS instability and the forward SRS instability, which are both also due to the laser ray generating waves in the blown-off plasma. Incoming laser light reflects off these waves. The reflected wave then interferes destructively with the incoming laser light, strengthening the plasma waves etc. The net result is that much of the laser light is reflected back towards the laser, and less light will reach the target. The different names (Brillouin and Raman) are related to the fact that two kinds of plasma waves can be generated, as the plasma consists of ions and electrons. In the case of stimulated Brillouin scattering it concerns an ion acoustic wave (an oscillation in the plasma that behaves like a sound wave), and in the case of stimulated Raman scattering it is an electron plasma wave. SRS and TPD are especially dangerous as they can generate superhot electrons exceeding 100 keV, which can preheat the fuel. (Craxton et al. 2015, p. 4, 62; Chen 2011, pp. 398–399.) As mentioned above, preheating must be avoided as it makes the compression of the fuel more difficult. SBS instability can occur anywhere below the critical density of the plasma envelope (the density below which the laser light can still penetrate the plasma envelope) and can lead to a loss of energy. In one form of SBS, known as cross-beam energy transfer, laser energy can be lost by being scattered from incoming rays of one laser beam into outgoing rays of another.

These instabilities illustrate how complicated the situation is and how many unexpected events can happen. When experiments continue and come closer to ignition, a further set of instabilities, either already known or unknown, will undoubtedly emerge and have to be suppressed. Their suppression will be challenging as many laser plasma instabilities are sensitive to plasma densities and temperatures, which are very complex and uncontrollable in experiments.

12.6   Fuel Capsule Design[4]

Figure 12.2 shows a schematic picture of what a fuel capsule or target typically looks like. The outside (ablator) shell is exposed to the enormous burst of energy from the laser causing it to blow off and forcing the second layer, consisting of frozen D-T fuel, to move inwards. This layer forms the main fuel, while the rest of the capsule is filled with a low-density D-T gas. When the ablator layer blows off, momentum conservation requires that the D-T fuel layer implodes towards the centre of the capsule. In this process it is compressed to high densities and thermonuclear temperatures. In the final compressed state pressures up to 200 gigabars (one gigabar being 10^9 bars) exist. It ideally consists of two regions, the central hot spot containing between 2% and 5% of the fuel, and a cooler main fuel region containing the remaining mass. Ignition occurs in the central region and the thermonuclear burn propagates outward.

The laser drives a target capsule inwards at nearly a million miles per hour. Because the targets are subject to extreme temperatures and pressures during experiments, the targets must be designed, fabricated, and assembled with extreme precision. For example, components must be machined to within an accuracy of 1 micron (1 millionth of a metre). Many material structures and features can be no larger than 100 nanometres, which is just 1/1000th of the width of a human hair. And a capsule must have a smoothness tolerance approaching 1 nanometre—equivalent to removing all features on the Earth's surface taller than 60 m.[5]

Ignition, let alone fusion, has however not yet been achieved in the inertial confinement approach and a single optimal target does not yet exist. Targets differ depending on the purpose of the research the target is used for, and targets for future power plants will probably be vastly different from the targets currently used in trying to achieve ignition in e.g. the National Ignition Facility. Also, the indirect-drive and direct-drive approaches require different targets that also depend on the individual properties of the laser(s) used. They must fit in with the characteristics of the laser, and there are certain performance characteristics that all targets must fulfil.

---

[4]Pfalzner 2006, p. 155 ff.; McCracken and Stott 2013, p. 78.

[5]NIF website accessed 18 June 2020 (https://lasers.llnl.gov/about/how-nif-works/seven-wonders/target-fabrication).

The main tool in designing fuel capsules is computer simulation, which models the dynamics of a target in the ICF-process. These can be hydrodynamic codes, i.e. codes using fluid models for the fusion plasma, that simulate the whole ICF-process, or specialized codes that analyse certain aspects relevant to the target design, such as RT-instabilities, transport of superhot particles, deposition of driver energy, or energy transport. The main design parameters are: the amount of deuterium-tritium (the so-called fuel loading) in the capsule, the shell structure of the target and the areal fuel density, expressed in grams/cm^2 and defined as the product of the density of D-T fuel in the spherical hot-spot area, created by the D-T shell imploding inwards, times the radius of that sphere. The required fuel loading depends on the efficiency of the burn, which itself depends on the areal fuel density. The larger this density, the higher the expected burn fraction of the fuel; the burn fraction being the ratio of the burned fuel to the total fuel. The fraction is always smaller than one, as it will never be possible to burn all of the fuel. The areal fuel density is one of the most important parameters in ICF. The requirement that the target must release fusion energy before it blows apart, results in the requirement, following from the Lawson criterion, that this density must be larger than 0.3 grams/cm^2.

As discussed above, the primary aim of the fusion experiment is to compress the fuel to high densities and temperatures with the minimum possible energy input. In other words, the goal is to push the D-T shell when moving inwards to the highest possible velocity in the most efficient way. For this it would be ideal to have a long distance over which the shell could be accelerated towards the centre. Therefore, a large capsule with a very thin layer would be favoured, as a larger, thinner shell that encloses a greater volume can be accelerated to a higher velocity than a thicker shell of the same mass. However, above we also mentioned that the *in-flight aspect ratio* (the ratio of capsule radius to shell thickness) is limited by RT-instabilities, which will amplify any nonuniformity, either from initial surface imperfections or in the heating. The longer the acceleration process the more time there is for such instabilities to grow, and in extreme cases the shell might even be destroyed, certainly when it concerns a thin shell. This limits the in-flight aspect ratio to 25–40 and requires extreme precision in target manufacturing to make sure that the capsule surface is as smooth and uniform as possible.

In the hot-spot scenario described above, the layer of dense D-T fuel in the shell compresses a small mass of D-T gas in the centre. Here too, RT-instabilities play an important role. In forming the hot-spot area, the low-density 'fluid' of the D-T gas pushes against the high-density 'fluid' originally comprising the solid D-T ice layer. A mixing between the two regions sets in and the hot spot is cooled. To avoid the hot spot from being cooled below the threshold temperature required for ignition, it is essential that the D-T shell is initially very smooth and highly uniform. Hollow thin-walled spherical capsules of plastic filled with D-T gas are currently the most favoured approach to fulfil the above requirements.

In current designs of direct-drive and indirect-drive targets, the layer of D-T ice is essential. These so-called *cryotargets*, which have been developed since the late 1970s, operate at a temperature of $\sim$18 °K (-255 °C) and it is important to keep this relatively constant to avoid instabilities. There are alternatives to cryogenic targets, but the designs considered so far have a much lower performance. The task for the target designers is to find a configuration that helps to achieve fusion conditions, i.e. the required peak temperature and peak density in the hot spot area, but also the total areal fuel density, the implosion velocity, and others.

The designs of fuel capsules in use are very diverse and mainly adjusted to the particular experiments that are carried out. In the next section when discussing the National Ignition Facility some specific capsule designs will be discussed.

12.7   National Ignition Facility

In the United States three major facilities presently conduct research in ICF: the National Ignition Facility at LLNL in California, the OMEGA Laser Facility at the Laboratory for Laser Energetics (LLE) of the University of Rochester, and the Z facility at Sandia National Laboratories.

LLNL, whose self-proclaimed mission (https://www.llnl.gov/missions) is strengthening the United States' security by developing and applying world-class science, technology and engineering, has a great tradition in laser physics and the use of lasers in ICF-experiments, not so much for creating a future source of energy, but for developing new nuclear weapons. The first laser, built for this purpose just a couple of years after the pioneering paper of Nuckolls et al. in 1972, was the two-beam high-power infrared neodymium[6] doped glass laser, called Argus. It advanced the study of laser-target interactions and paved the way for its successor the 20-beam Shiva laser, named after the multi-armed Hindu god Shiva because of the laser's multi-beam structure, although the number of Shiva's limbs were limited to just six. Shiva, which came online in 1977, was instrumental in demonstrating the compression of targets with lasers, leading to a major new device being constructed to address this problem, the Nova laser. The Nova laser was in operation from 1984 to 1999, and was the first laser built with the intention to reach ignition. It failed, but the data it obtained defined the ignition problem as being mostly a matter of magnetohydrodynamic instability, which led to the design of its successor the National Ignition Facility. In the spring of 1986, the Nova laser came within a factor of ten of fulfilling the Lawson criterion, the main thing lacking was a high enough temperature. Nova also generated considerable amounts of data on high-density matter physics, which is useful both in fusion and nuclear weapons research. This usefulness for weapons research was one of the main reasons that this research continued to be funded by Congress even when various administrations wanted to put the axe to it and the funding for other nuclear fusion programmes in the 1990s was actually severely reduced.[7]

---

[6]Neodymium is a rare earth metal with atomic number 60 belonging to the lanthanides. Neodymium-doped crystals can generate high-powered infrared laser beams.

[7]Dean 2013, p. 91ff, p. 130, p. 135 and other places. For instance, because of cuts by Congress the Tokamak Fusion Test Reactor at the Princeton Plasma Physics Laboratory was shut down in 1997.

This current facility, the National Ignition Facility (NIF), is the largest and most energetic ICF-device built to date (Fig. 12.9). Its construction was proposed by the Clinton administration in the budget for fiscal year 1996 and was justified as a vehicle to study weapons physics in the absence of the possibility of underground nuclear weapons testing. In those years the US nuclear weapons programme shifted emphasis from developing new designs (no new weapons have been developed by the US since 1992) to dismantling thousands of existing weapons and maintaining a much smaller enduring stockpile. Underground nuclear testing ceased, and the Department of Energy created the science-based Stockpile Stewardship and Management Program to maintain the safety, security, and reliability of the US nuclear deterrent without full-scale testing. The mini-explosions that lasers create are powerful enough to mimic the effects of hydrogen bombs. The data they produce are needed to ensure that the nuclear capabilities of the United States are not eroded as nuclear weapons age, and can be used to develop new weapons without having to resort to underground or other testing. Because NIF is the only facility that can create the conditions that

Fig. 12.9 An inside picture of the National Ignition Facility. The blue object in the background is the vacuum chamber (see Fig. 12.11) (*from* NIF website)

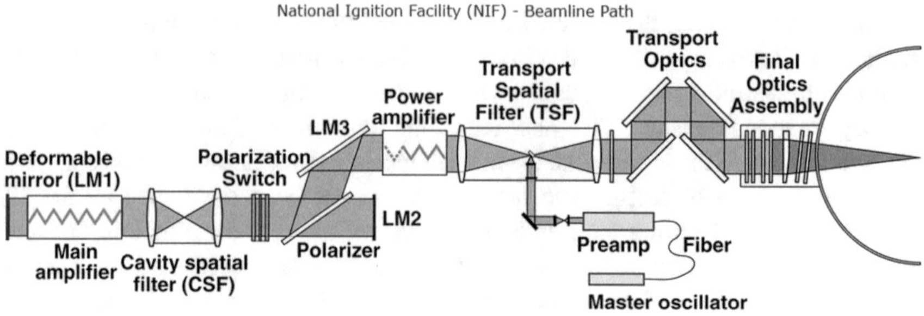

National Ignition Facility (NIF) - Beamline Path

Fig. 12.10 The beamline path of the National Ignition Facility

are relevant to understanding the operation of modern nuclear weapons, it is a crucial element of this stockpile stewardship.[8] Its task within this programme is to provide experimental insight and data. The experiment represents an important milestone in the continuing demonstration that the stockpile can be kept safe, secure and reliable without a return to nuclear testing.

Construction started in 1997 and it was completed on schedule in 2009. In the spring of that year NIF fired its first full-power laser beam shots into a test cell. During the construction period the costs for the machine had escalated to 4 billion dollars from an original estimate of 1 billion dollars.

NIF is also part of the National Ignition Campaign, a multi-institutional effort to achieve fusion ignition by ICF established by the US Department of Energy in 2005. It started in 2006 and ended in September 2012. The goal of this campaign was to develop and integrate all the capabilities required for a precision ignition campaign and, if possible, to demonstrate ignition and gain by the end of September 2012. Partners in the campaign were LLNL, Los Alamos National Laboratory, Sandia National Laboratories, the University of Rochester's Laboratory for Laser Energetics, and General Atomics.

NIF operates in an indirect-drive configuration but can also be configured in a direct-drive arrangement. This is important, although too far in the future to merit serious consideration at this point in time, as direct drive is supposed to be favoured for an inertial fusion energy plant (simpler target design and significantly more energy into the compressed core than with indirect drive) (Dean 2013, p. 221). NIF's principal goal is to achieve ignition of a deuterium–tritium fuel capsule by creating a single 500 terawatt (5×10^{14} W) peak flash of light that reaches the target simultaneously within a few picoseconds from numerous directions. For this it operates a 192-beam laser system, which uses neodymium glass lasers to produce light in the infrared region of the spectrum; this light is converted into ultraviolet light (with a wavelength of one third of the wavelength of the original light).

The starting point is the generation of a very low energy pulse of high-quality infrared light, generated by a master oscillator (Fig. 12.10 presents a sketch of the beamline path). This light is split into 48 beams which pass through 48 identical

---

[8]https://lasers.llnl.gov/science/stockpile-stewardship.

Fig. 12.11 The target chamber, 10 m in diameter and weighing 130 ton, being lifted into place at the NIF

preamplifiers for initial amplification and beam conditioning. The energy of each beam is increased a billion times in these preamplifiers. Then follows a further split of each beam into four, making a total of 192 beams, which pass back and forth several times through a series of amplifiers. At the end the total energy is one million billion (10^{15}) times the energy of the pulse that started the operation. The beams then pass through an array of mirrors, are converted into the desired ultraviolet light and positioned in groups of four, symmetrically around the target chamber (Fig. 12.11) to precisely line them up on the target. The final energy of the beams is surprisingly small, in view of all the effort to produce them, and approximately equal to the energy released from burning 135 grams of coal or 50 ml of gasoline, but the instantaneous power of 500 terawatt is enormous and equivalent to the combined output of 250,000 large (2 gigawatt) power plants. (McCracken and Stott 2013, p. 150ff.)

The 192 high-energy laser beams enter a hohlraum through entrance holes at its ends. The hohlraum is about 9 mm high and 5 mm across. It has been filled with He gas, which is prevented from escaping by windows over the entrance holes made of the polymer polyimide. The beams are arrayed in 8 cones at various specific angles to the hohlraum axis, containing 4, 4, 8 and 8 quadruplets of beams on each side. Beams coming in at different angles have a slightly different wavelength (or colour), which makes it possible to transfer energy from one set of beams to another and control in this way the symmetry of the radiation flux. As discussed above, the X-rays that result from heating the walls of the hohlraum ablate material from the spherical shell surrounding the fuel, which is mounted in the centre of the hohlraum and held in place by thin polymer diaphragms (called the tent; Figs. 12.12 and 12.13). The resulting implosion compresses and heats the central fuel to fusion conditions.

Fig. 12.12 All the energy of
NIF's 192 beams is directed
inside the gold cylindrical
hohlraum, which is about
5 mm across. The fuel
capsule inside the hohlraum
is suspended by ultra-thin
polymer tents (*from* NIF
website)

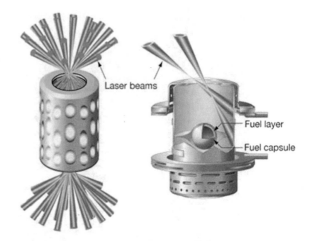

Fig. 12.13 Cutaway model
of a NIF hohlraum showing
the fuel capsule suspended
by the ultra-thin polymer
tents (*from* NIF website)

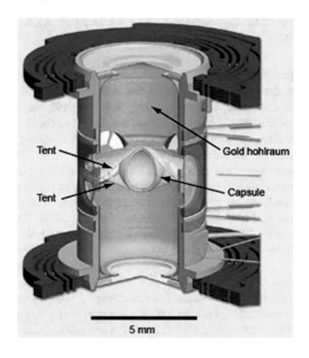

The central fuel capsule is made up of thin concentric spherical shells (for an example of a possible capsule structure, see Fig. 12.14). The outer ablator shell can be made of plastic (CH) or beryllium, high-density carbon or another material with low atomic number. For some time, beryllium was preferred. (See e.g. Hurricane et al. 2014, Wilson et al. 2015.) Such a capsule is made by depositing beryllium on a smooth, perfectly spherical plastic shell. As the shell is rotated, a 150 micrometre (μm) thick layer of beryllium slowly builds up on its surface. After a capsule is

Fig. 12.14 Layered
structure of the NIF baseline
capsule with dimensions
(*from* Basko 2005)

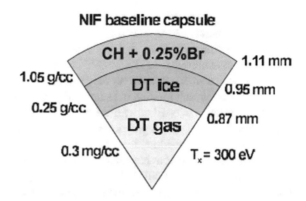

polished, a laser is used to drill a 5 μm fill hole, and a 10 μm tube is attached to the
capsule so that it can be filled with D-T gas.

The layers of the shell must be very smooth to minimize seeds of hydrodynamic
instabilities. Therefore, manufacturing requirements for all NIF-targets are extremely
rigid. Components must be machined to within an accuracy of 1 micrometre and the
capsule's surface must be smooth to within 1 nanometre (10^{-9} m; a sheet of paper
or a human hair is about 100,000 nm thick). Many material structures and features
of a capsule can be no larger than 100 nanometres (1/1000th of a human hair). In
order to minimize instability growth at the interface between the ablator and the
fuel layer, the ablator includes concentric layers of doping agents (either silicon or
germanium), which absorb preheat X-rays. The ablator encloses a spherical shell of
D-T fusion fuel, kept solid by keeping the entire assembly at cryogenic temperatures.
The interior of the shell contains D-T vapour in equilibrium with the solid fuel layer.
(Moses 2009, pp. 4–5; Lindl et al. 2014, p. 3.)

In September 2010 NIF-scientists performed the first integrated ignition experiments firing all 192 laser beams on a cryogenically layered capsule containing
tritium, deuterium and hydrogen. Precision experiments devoted to ignition began
in May 2011 and researchers were confident that ignition would be achieved by
2012. Advances were made, but ignition has remained illusive, mainly because of
hydrodynamic instabilities and laser-plasma interactions that were not foreseen in
computer simulations. (Kramer June 2018.) The advances made include demonstrating the self-heating process for the first time in the laboratory. In this process,
alpha particles released by fusion reactions in the hot spot at the centre of the target
capsule deposit their energy in the cooler fusion fuel outside the hot spot, leading
to additional heating of the fuel. This is a critical step on the path to ignition and
has now become routine. Subsequent experiments, however, have stopped short of
ignition, due largely to two major factors: implosion asymmetry, and perturbations
(instabilities) caused by engineering features in the fusion targets.

An overall performance parameter used by the LLNL group is the Experimental
Ignition Threshold Factor (ITFx) (Glenzer et al. 2012). The ITFx has been derived
from the results of hundreds of computer simulations of ignition targets to find

a measurable parameter which gives an indication of the performance as regards ignition. The parameter is related to the Lawson criterion and is normalized such that an implosion with ITFx $= 1$ has a 50% probability of ignition. To date, the highest value achieved for ITFx in D-T layered implosion experiments on NIF is about 0.1, which shows how far off they still are.

The failure to reach ignition does not really come as a surprise. Already in 1995 before construction of NIF had started Stephen Bodner, in a very explicit and detailed paper, "predicted that the highly intense NIF laser would create instabilities in the plasma. That, plus the formation of unpredictable magnetic fields, would prevent the symmetrical implosions required for ignition." (Bodner 1995, Kramer 2016.) He blamed the design of the NIF hohlraum for the impending failure; the walls of the hohlraum were too close to the fuel pellet, but increasing the hohlraum size would cause the ignition energy to rise. His comments were ignored as is clear from a detailed paper on NIF-fuel capsule design from 2011, just before the actual ignition experiments started, which makes no mention of Bodner or his paper (Haan et al. 2011). It now seems that the Department of Energy is no longer sure either that the NIF can ever reach ignition. A 2016 review report[9] states: "Barring an unforeseen technical breakthrough and given today's configuration of the NIF laser, achieving ignition on the NIF in the near term (one to two years) is unlikely and uncertain in the mid-term (five years). The focus of the Laser-driven Indirect Drive Program over the next five years should be on the efficacy of NIF for ignition. The question is *if* the NIF will be able to reach ignition in its current configuration and not *when* it will occur. The focus of integrated experiments in the Laser-driven Indirect Drive Program should not be on high-gain capsules simply because codes and models predict they will perform well. The codes and models themselves are not capturing the necessary physics to make such predictions with confidence. A lack of appreciation for this, combined with a failed approach to scientific program management, led to the failures in the National Ignition Campaign". In other words, the physics is not understood, and the models used are no good (as had also been pointed out by Bodner in the paper referred to above). It means that ICF is probably not going to work as a power plant approach. It may not be possible to get more energy out of the machine than has to be put in.

In the most recent experiments (LePape et al. 2018, Berzak Hopkins et al. 2019) the ablator shell consisted of a 20 μm thick layer of high density carbon (diamond doped with a thin layer of tungsten) and the hohlraum was made of depleted uranium, instead of gold. The imploding shell, composed of the remaining nonablated high-density carbon and the D-T cryogenic layer, was driven to velocities of the order of 380 km/s. These experiments achieved, for the first time, a fusion yield which was twice the kinetic energy of the imploding shell. The density, temperature and pressure of the hot spot are the closest on earth to conditions in the Sun. Again, as so often in the past and in spite of the pronouncement in the review report cited above, the chief scientist could not resist the temptation to declare that NIF is "at the threshold of

[9]Office of Defense Programs, 2015 Review of the Inertial Confinement Fusion and High Energy Density Science Portfolio, May 2016.

achieving a burning plasma state". A less attentive person might think that a burning plasma sounds very much like ignition, but it is not. A burning plasma is a state where alpha particle heat deposition from deuterium–tritium fusion reactions is the leading source of energy input to the D-T plasma. If achieved, it would be a step forward, but is not ignition, and others note that still 10 times as much alpha heating is needed and the pressure at 360 gigabars is still 30% short of what is needed for ignition. Just call it something else, and another breakthrough can be claimed.

The goal of current NIF-experiments is to increase the density of the hot spot by a factor of three at about the same temperature as already achieved. Under those conditions, the website states, the fusion reaction rate would be sufficient to generate ignition.[10] So, apparently the scientists and engineers working at NIF have not given up hope or may just continue to fool themselves, but may be saved by the budgetary axe of the Trump administration. However, their repeated failures in achieving what was promised, in combination with using confusing language and not completely honest reporting of 'breakthroughs', which later turn out not to be breakthroughs at all,[11] undermine confidence in their judgement and may revive funding arguments that have been going on since the beginning of the fusion project. There can be no doubt though that the science and engineering of the project, although chasing a chimera, are terrific, a quality the world has never seen before and as such it may be a worthwhile activity to pursue, but please leave out the hyperbole.

The National Ignition Campaign ended in September 2012 without the NIF having achieved ignition. Articles in the press suggested that funding arguments would start again and that the NIF would shift its focus back towards materials research. It has indeed done so, but still has its eyes on ignition too.

---

[10]NIF website (https://lasers.llnl.gov/science/ignition/ignition-experiments), accessed 18 June 2020.

[11] An example is the reporting in 2013 of a fusion shot that produced many more neutrons than ever before, and also noted that the reaction released more energy than the "energy being absorbed by the fuel", described as "scientific breakeven" and referred to as a "milestone". A number of researchers pointed out that the experiment was far below ignition and did not represent a breakthrough as reported. Others noted that the definition of breakeven was when the fusion output was equal to the energy of the laser used. This definition of breakeven was not repeated in the *Nature* publication of 20.02.2014. In the report, the term was changed to refer to the energy deposited in the fuel, not the energy of the laser. Moreover, the method used to reach these levels was claimed not to be suitable for general ignition. (*Wikipedia*)

Intermezzo
Different hohlraum designs

The hohlraum is a critical component in ICF. Hohlraums shaped like cylinders have been the workhorse of ICF-research for three decades. The cylinder geometry has however its limits in accommodating a larger capsule—which makes one wonder why the cylinder has been the dominant design for hohlraums all this time. The goal of testing new hohlraum shapes is to increase the level of energy absorption by the capsule (the energy coupling efficiency or the hohlraum-to-capsule efficiency) using the same amount of energy produced by laser beams.

Research using a hohlraum shaped like a rugby ball increased the level of laser-induced energy absorbed by a single-shell fuel capsule to about 30%. That is about double the level of energy absorption of 10–15% with a standard cylindrical hohlraum used at the NIF. Figure 12.15 illustrates different hohlraum designs.

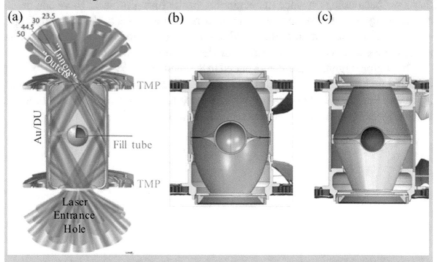

Fig. 12.15 Illustration showing different hohlraums: **a** cylinder, with inner and outer beams entering through laser entrance holes and reflecting off the hohlraum walls; **b** rugby; and **c** Frustraum

The rugby hohlraum (Fig. 12.15b), like the ball used in its namesake sport, has a wide centre that can accommodate a target capsule with a radius about 50% larger than used in a cylindrical hohlraum, and obviously the larger target can absorb more of the laser energy. The rugby experiments (https://lasers.llnl.gov/news/rugby-hohlraum-kicks-up-nif-energy-efficiency) inspired development of the Frustraum, which is constructed from two truncated conical halves (or frusta) joined at the waist. It has been developed for the next stages of ICF research at NIF. (Amendt et al. 2019.) The name Frustraum combines part of the familiar word hohlraum with the Latin word *frustum*, for "piece broken off." It could significantly boost the amount of laser-induced energy absorbed by the fuel capsule.

As with other hohlraums, each end is open to allow laser beams to enter. An associated larger waist volume above the capsule allows fielding ~50% larger capsules than the nominal 1 mm (radius) scale.

A complimentary design to the Frustraum is the I-Raum (Fig. 12.16), a cylindrical hohlraum with recessed pockets in the wall near the laser entrance holes. (https://lasers.llnl.gov/news/papers-presentations/2018/january)

Fig. 12.16 Maintaining implosion symmetry in indirectly driven inertial confinement fusion relies heavily on the ability to control the laser energy reaching various portions of the hohlraum wall at all times. The I-Raum geometry adds a small pocket at the location where the outer cones hit the hohlraum wall

Computer simulations for the Frustraum design showed it had promise. Implosion symmetry is affected by imperfections of the target shell. With a larger target capsule, implosion symmetry becomes less susceptible to those imperfections. By having larger capsule and surface area, the energy absorbed by the capsule almost triples compared to what is common for a cylinder.

Both the Frustraum and its more egg-shaped rugby cousin can handle a larger target. But the flatter sides tapered towards the top and bottom openings reduce the wall surface area, which results in energy savings while still allowing a larger middle bulge to accommodate a bigger fuel capsule.

The next step is a full-scale campaign for testing the Frustraum which will still take some time.

12.8   Funding of ICF in the US

In the US funding for ICF was always more generous than for magnetic confinement fusion. The ICF research came under the Department of Energy's weapons programme,[12] since the US interest in this concept primarily lay in the construction of small nuclear bombs and not in energy generation in a power plant. This always caused friction. From 1998 the US Congress wanted the Department of Energy to set up a civilian inertial fusion programme.

For a fusion power plant to provide a steady output of energy a device would be needed that could ignite at least ten capsules every second, while for weapons development one or a few capsule explosions per day are sufficient. Ten explosions per second sounds a lot, but in a car a few thousand explosions per minute is nothing special. Since the fuel capsule in an ICF power plant would be destroyed in the explosion, a new one would have to be inserted at least ten times per second. This is probably hard, but it is possible to imagine a sequence of devices that fire consecutively (like the cylinders of a car). The weapons developers of the Department of Energy apparently were reluctant to do research into such a power plant, although Congress provided funds for this on its own initiative, without the Department of Energy having requested such funds. For a few years a civilian oriented programme called the HAPL (High-Average Power Laser) program, was implemented, but soon terminated. (Dean 2013, pp. 155–156; 235.)

In general, the US fusion landscape is littered with abandoned, revived and again abandoned programmes. Since the early 1980s US federal energy policies in respect of fusion have been far from consistent, partly due to partisan politics which caused succeeding administrations to drastically shift emphasis and budget allocation. It gives the proponents of fusion the excuse that the budget collapse in the mid-1990s is the main cause for the goal of commercial fusion energy to have receded to far in the future.

There are or have been however quite a number of US facilities, supported by the ICF-programme of the US Department of Energy: the Argus, the 20-beam Shiva, the 10-beam Nova, and the National Ignition Facility at LLNL; the 4-beam DELTA, the 6-beam ZETA, the 24-beam OMEGA, the 60-beam OMEGA and the OMEGA EP[13] facilities at LLE of the University of Rochester; and the 8-beam Helios laser at Los Alamos National Laboratory (LANL). The OMEGA project is in operation since 1985, but scheduled to be shut down during a three-year period from October 2018 in the Trump administration's budget for the Department of Energy (Kramer Feb. 2018). Originally operating with 24 beams, upgraded in 1995 to 60 beams, it has been successfully used for testing the physics of direct-drive implosions. A further upgrade was completed in 2008 and a further four laser beams were added, two of which can produce the ultrashort picosecond pulses required for fast-ignition experiments. The OMEGA EP laser system is eventually, if it ever gets the chance,

---

[12]Other fusion research comes under the Office of Fusion Energy Sciences of the Department of Energy.

[13]EP stands for extended performance.

expected to reach an intensity of 10^{21} W/cm^2, inducing an electric field so large that the electrons for the fast ignition will be accelerated to a velocity close to the speed of light. It complements the work at the much larger NIF.

The US was the dominant innovator and user of high-intensity laser technology in the 1990s, but now Europe and Asia have taken the lead. Currently, 80–90% of the world's high-intensity ultrafast laser systems are outside the US, and the highest-power research lasers currently in construction or already built are also abroad. To counter this development the press reported on 1 November 2018 that a new US-wide national research network called LaserNetUS was set up, involving nine institutions across the country. Together these facilities will provide US scientists with improved access to high-intensity, ultrafast lasers. The project is funded by the US Department of Energy's Office of Fusion Energy Sciences and will receive (the pittance of) $6.8 million over the next two years. Such a small amount will not achieve much.

12.9 Activities in Other Countries

The United States has always been the dominant player in ICF-experiments, for many years virtually the only one. In the Soviet Union some research on laser construction was carried out and several lasers (Iskra-4 and Iskra-5) were built with a view to the possibility of achieving conditions for fusion by imploding spherically symmetrical fuel capsules. This was done on the initiative of Andrei Sakharov, who suggested such research in 1960–1961 soon after the construction of the first laser in the US (Sakharov 1990, p. 149). Others (McCracken and Stott 2013, p. 70) give this honour to the Soviet laser pioneers Nicolai Basov (1922–2001) and Alexander Prokhorov (1916–2002), who were co-recipients of the 1964 Nobel Prize in physics. However, no large-scale experiments like in the US were set up in the USSR.

The dominance of the US is slowly but surely diminishing and R&D on ICF, at any rate on laser facilities that are in principle capable of imploding fusion fuel capsules, currently also continues in the European Union, Russia, Japan and China. Consequently, there now exists in the world a remarkably large number of multibeam laser facilities. They have the same problems as NIF and none of them has achieved ignition.

Russia has the 9-beam Delfin laser at the Lebedev Physics Institute in Moscow, and the 12-beam Iskra-5 and 128-beam Iskra-6 lasers at the Russian Federal Nuclear Centre in Sarov,[14] which is part of Rosatom and builds on the experience gained with earlier Iskra lasers. The Iskra-6 laser is under investigation and, if built, its power would be close to NIF and the French Laser Mégajoule. In China the Shen Guang lasers SG-III (48-beam) and SG-IV lasers are in operation at the Research Centre of Laser Fusion. (Craxton et al. 2015, p. 3.) The Chinese programme is focused on the

---

[14]Because of the presence of the Federal Nuclear Centre Sarov is a closed town, which can only be visited with special permission. From 1946 to 1995 it was known as Arzamas-16, where the Soviet atomic bomb project was carried out.

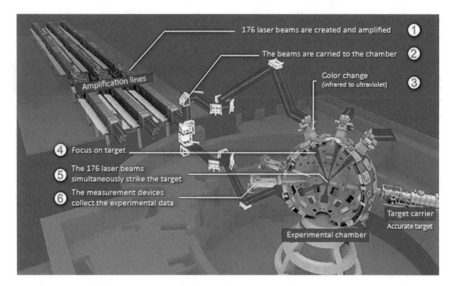

Fig. 12.17   Schematic working of the Laser Mégajoule (*from its website*)

development of new laser drivers and fast ignition. The near-term goal is to achieve
ignition and plasma burning around 2020.

In the European Union activities are mainly concentrated in France and the UK.
France had the 8-beam OCTAL laser at the Commissariat a l'énergie atomique et
aux énergies alternatives (CEA) at Limeil near Paris, which has been dismantled by
the CEA to concentrate its efforts on the Laser Mégajoule currently being built near
Bordeaux. The Laser Mégajoule is the largest inertial confinement fusion experiment
outside the US and about as energetic as its US counterpart, NIF. It is designed to
deliver, in a few billionths of a second, more than one million joules of light energy
to targets measuring a few millimetres in size. It was officially commissioned on
23 October 2014, with the performance of a first series of experiments.[15] As in the
US, one of its primary tasks will be refining fusion calculations for France's nuclear
weapons, for which the so-called Simulation Program has been set up. Since France
indefinitely halted nuclear testing in 1996, this programme has been, according to
the Laser Mégajoule website, the key element that allows physicists to guarantee the
reliability, safety and performance of the country's nuclear weapons. Its purpose is
to use calculations to reproduce the various phases of operation of nuclear weapons
to ensure their performance without having to resort to further nuclear testing. Any
ICF-experiments are just an extra to this.

The facility uses a series of 176 neodymium glass laser beams (compared to
the 192 of NIF), divided into 22 bundles of 8 beams and installed in the four laser
halls located on either side of the experimental chamber. Like NIF, it employs the
indirect-drive approach with the laser light heating a hohlraum. Figure 12.17 presents
a schematic rendering of how the Laser Mégajoule works.

---

[15]Laser Mégajoule website, http://www-lmj.cea.fr/, accessed 9 November 2018.

The facility is not in full operation yet and is gradually being built up. In addition, associated with the Laser Mégajoule, the PETAL (PETawatt Aquitaine Laser) has been developed. It adds one short-pulse ultra-high-power, high-energy beam to the Laser Mégajoule facility, which could among others be used for fast-ignition experiments.

In the UK the 8-beam Vulcan laser operates at the Rutherford Appleton Laboratory and the 12-beam Orion laser at the Atomic Weapons Establishment. Vulcan is a Petawatt laser facility available to the UK and the international research community. The facility delivers a focused beam which for 1 picosecond is 10,000 times more powerful than the entire UK National Grid. The Orion laser is used to study nuclear warhead science, again showing the close connection of inertial confinement science and laser physics with weapons research. In this respect it fulfils the same role for the UK as the Laser Mégajoule does for France. Orion is a large neodymium glass laser facility and designed to study high energy density physics, generating conditions several times denser than solid, with temperatures up to 10 million degrees (hence far below ignition temperatures). The high temperatures, pressures and compressions achieved during Orion's 'shots' can also aid research in understanding the conditions relevant to inertial fusion energy, planetary and solar physics, high energy particle acceleration and even black holes. Its 12 beams are divided into 10 long-pulse (nanosecond) and two short-pulse (sub-picosecond) beamlines. (Orion Factsheet.)

HiPER, the High Power laser Energy Research facility,[16] at the Central Laser Facility of the UK Rutherford Appleton Laboratory is a proposed experimental fusion device undergoing preliminary design for possible construction in the European Union to continue the development of the laser-driven inertial confinement approach. Its website states that it will provide a 'state of the art' laser facility for Europe, enhancing existing high-power single-shot laser experimental capabilities provided by Orion and Vulcan (UK), Laser Mégajoule, LULI and PETAL and Jena (Germany). As usual in this kind of business, the ambitions are sky-high, so high that they will probably never be fulfilled. According to its website, which for that matter seems to have been dormant since 2015, "Demonstration of the scientific proof of principle (scientific breakeven or ignition) for laser fusion is expected in the next couple of years at the National Ignition Facility in the USA. The HiPER Project will drive the transition from scientific proof of principle to a demonstration power plant, capable of delivering electricity to the grid". As we have seen, it is questionable whether NIF will ever deliver on that point, but even if that were to be the case the road from there to a (demonstration) power plant will still be very long and bumpy. HiPER is the first experiment designed specifically to study the fast-ignition approach to generating nuclear fusion. Using much smaller lasers than conventional designs, it yet produces fusion power outputs of about the same magnitude with a reduction in construction costs of about ten times, so one wonders why it has not been tried in the first place. Theoretical research since the design of HiPER in the early 2000s has cast doubt on fast ignition as a route to practical inertial confinement fusion, but a new approach

---

[16]http://www.hiper-laser.org/.

known as *shock ignition* (see above) has been proposed to address some of these problems. It is currently unclear how the HiPER Project will be affected by Brexit, the UK leaving the EU. The latest EU note on the matter in 2017 stated, among a lot of the usual hullabaloo, that construction is not foreseen for the next few years. The start of reactor design is planned for 2026 and operation for 2036.[17]

Germany, Europe's largest economy, does not itself run any large laser facility for the simple (fortunate) reason that it does not have a nuclear stockpile to maintain. German scientists participate of course in several of the international, especially EU, cooperative efforts.

Japan has the 4-beam GEKKO IV and the 12-beam GEKKO XII glass lasers (the GEKKO XII is neodymium-doped) operating at the Institute for Laser Engineering at Osaka University. These lasers and their predecessors have been in operation since the early 1980s. Japan also developed the KOYO-F fusion reactor design for testing the fast-ignition concept and the laser inertial fusion test (LIFT) experimental reactor.

The ultra-sharp laser beams at these facilities are, for that matter, not only useful for inertial confinement implosion experiments, but can be and have been used everywhere where extremely precise cuts must be made, from laser machining to corrective eye surgery.

12.10   Direct Drive

In the earlier part of this chapter we have mostly discussed indirect-drive experiments, since the two largest devices that are currently in operation, NIF and Laser Mégajoule, are based on the indirect-drive approach. Direct drive is the main mission of the OMEGA lasers at the Laboratory for Laser Energetics at Rochester University, the Japanese GEKKO lasers and the Chinese Shen Guang lasers.

Let us first review the four main stages in a typical direct-drive implosion with somewhat more detail than described for the process depicted in Fig. 12.3. The following is adapted from Craxton et al. 2015, p. 4ff. The target capsule normally consists of a layer of cryogenic ice of deuterium and tritium, surrounded by a plastic shell, but could also include a foam layer filled with deuterium and tritium. Typical target capsules have diameters from 3 to 5 mm and ice layer thicknesses from 0.16 to 0.6 mm, so roughly one tenth of the capsule diameter. The interior of the capsule contains low-density D-T gas in thermal equilibrium with the ice. At first (Fig. 12.18a), similar to stage 1 of Fig. 12.3, laser light is absorbed by the target, resulting in the ablation of target material to form a hot plasma. The laser irradiation typically starts with a sequence of one to three low-intensity pulses, resulting in a sequence of shock waves that propagate into the target, compressing it. The laser pulse intensity is then rapidly increased, launching a strong shock wave that merges with the earlier shock waves around the time that the shock waves break through the inner surface of the ice layer. During this transit stage of the shock waves, target

---

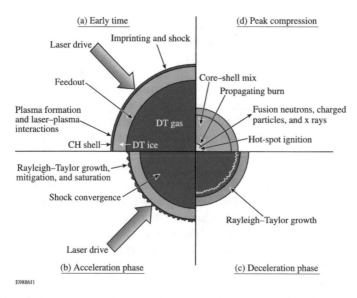

(a) Early time

Laser drive

Imprinting and shock

Feedout

Plasma formation
and laser–plasma
interactions

CH shell → ← DT ice

DT gas

Rayleigh–Taylor growth,
mitigation, and saturation →

Shock convergence

Laser drive

(b) Acceleration phase

(d) Peak compression

Core–shell mix

Propagating burn

Fusion neutrons, charged
particles, and x rays

Hot-spot ignition

Rayleigh–Taylor growth

(c) Deceleration phase

E9886J1

Fig. 12.18 Schematic representation of the four main stages in a direct-drive experiment (*from* Craxton et al. 2015, p. 5)

nonuniformities (imperfections in fabrication or modulations imprinted (induced) by laser-beam nonuniformities) evolve as a result of a Richtmyer–Meshkov-like instability. When the shock wave reaches the inner surface, a rarefaction wave (the opposite of compression) starts to move outwards towards the ablation surface, and the target shell and ice layer (collectively known as the shell) begin to accelerate inwards towards the target capsule centre (stage 2 of Fig. 12.3). All this takes place in the outer layers of the target capsule in a few billionths of a second. Precise shock timing is crucial to achieving high compression of the shell while it is in flight. A highly compressed shell acts like a rigid piston, efficiently transferring its kinetic energy to the central plasma contained within. Optimum shell compression is achieved when all shocks break out of the shell nearly simultaneously, minimizing the decompression resulting from rarefaction waves launched at each shock breakout. Laser-plasma interactions can have undesirable effects including the production of high-energy electrons, leading to fuel preheat. Calculations have shown that ignition can fail if 1–2% of the shell kinetic energy is deposited in the main fuel due to electron preheat. X-rays from the hot plasma surrounding the target can also lead to preheat.

During the acceleration phase (Fig. 12.18b) the laser intensity increases. Ablation-surface modulations grow exponentially because of the Rayleigh–Taylor instability, while the main shock wave within the D-T gas converges towards the target centre. During this phase the greatest concern is the integrity of the shell, which is at peril from exponentially growing RT-instabilities.

The main shock wave moves ahead of the shell accelerating towards the centre. Soon after it reflects from the target centre and reaches the converging shell, the deceleration phase begins (Fig. 12.18c). As the shell decelerates, its kinetic energy is converted into thermal energy and the fuel is compressed and heated. The compression that can be achieved depends on the temperature of the fuel at the start of the deceleration phase, and the maximum temperature depends on the kinetic energy of the shell. The greatest issue during the deceleration phase is the hydrodynamic instability (the Rayleigh-Taylor growth) of the inner surface of the shell, seeded by ablation-surface modulations.

Peak compression occurs in the final stage (Fig. 12.18d). Fusion neutrons and X-rays are produced together with charged particles (α-particles for D-T fuel) that deposit their energy in the D-T fuel, bringing more compressed fuel to fusion temperatures and leading to a propagating burn. In these targets, the first fusion reactions occur in a central "hot spot"—a high-temperature, low-density region surrounded by a lower-temperature, higher-density D-T shell. The hot spot results from the compression of hot fuel in the deceleration phase and typically accounts for about 10% of the compressed fuel mass. It is critical that the hot spot has sufficient energy production and areal fuel density for significant α-particle energy deposition to occur. This typically requires a hot-spot temperature of 10 keV and, as discussed before, areal fuel density of around 0.3 g/cm^2. It will result in hot-spot self-heating if the rate of α-particle heating exceeds the rate of hot-spot energy loss.

Another concern at this stage is mixing between the hot fuel in the core and the cooler shell material, which would reduce the temperature of the hot fuel.

As mentioned several times, one of the main issues for all schemes, both direct and indirect drive, is to realize high irradiation uniformity of the target capsule, in order to get an effective implosion and symmetric compression of the capsule. In general, the indirect-drive scheme provides better uniformity because of the quasi-isotropic X-rays generated by the laser beams that irradiate the interior of a hohlraum. By comparison, the direct-drive scheme, in which the target capsule is directly irradiated with intense lasers, can provide higher energy efficiency, but laser imprinting may occur on the surface of the ablator material due to irradiation non-uniformities. Spatial perturbations of the laser imprinting are then amplified by the Rayleigh-Taylor instability during the shell acceleration, as indicated above. So, high illumination uniformity is especially necessary in direct-drive schemes, although in the indirect-drive NIF experiments it were also Rayleigh-Taylor instabilities that turned out to be impossible to control. The required high illumination places heavy demands on the ablator material used, which should be as stiff as possible to reduce laser imprinting. This favours ablators made of high-density carbon (diamond). It is estimated that the illumination nonuniformity must be less than 1% to avoid Rayleigh-Taylor and Richtmyer-Meshkov instabilities, which may result in failure of the compression.

The most powerful laser systems for direct-drive experiments currently in use are the 60-beam OMEGA Upgrade (Fig. 12.19) at the University of Rochester and the 56-beam Nike laser (a Krypton fluoride laser) at the Naval Research Laboratory in Washington DC. The OMEGA is a frequency-tripled (i.e. the resulting frequency is three times that of the input laser, a technique also used in the National Ignition

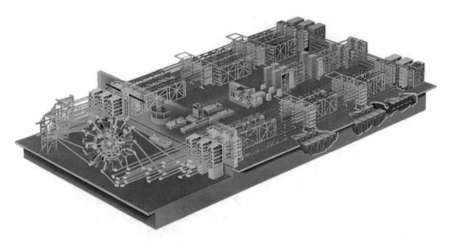

Fig. 12.19 Cutaway illustration of the OMEGA laser facility at LLE of the University of Rochester

Facility) neodymium glass laser system. Because of the energy limitations of the OMEGA laser, direct-drive targets for OMEGA are much smaller than targets on NIF, containing about ten times less D-T fuel. The use of a large number of beams improves the spatial uniformity of the illumination of the target. The beams overlap each other, so that the energy falling on any point is averaged over several beams. In 1996 the OMEGA project at Rochester achieved a fusion gain of about 1 per cent.

Outside the US, the most important research work on direct-drive is carried out in Japan with the GEKKO XII laser at Osaka University in Japan, and the Shen Guang lasers at the Research Centre of Laser Fusion in China. None of these installations are even close to ignition. The GEKKO XII system consists of 12 beams and has been redesigned for fast ignition research by adding an ultrashort-pulse laser (the Fast Ignition Realization EXperiment (FIREX)). The GEKKO XII setup is special in the sense that it converts the infrared light into visible light in the green part of the spectrum with a wavelength of one half of the infrared light (frequency doubled), unlike other ICF lasers which convert it, as we have seen above, into ultraviolet light with a wavelength of one third of the original light (frequency tripled). In the conversion into visible light less energy is lost, but it still remains to be seen whether visible light is as effective as ultraviolet light.

Recently direct-drive experiments have also been carried out on NIF, in spite of being optimized for indirect drive, to study laser-to-target coupling and various other problems.

12.11 Alternative Drivers and Methods

It has been noted above that for a power plant based on inertial confinement with a steady output of power, be it direct or indirect drive, at least ten target capsule explosions per second will be needed. A problem is not only that in the explosion the target capsule is destroyed and will have to be replaced, but also that it is not possible for a glass laser to pulse that rapidly. When in operation, the glass in the laser will heat up to the point of cracking and will take hours to cool down. With earlier lasers two shots a day was about the maximum that could be achieved.

One further restriction is that the wavelength of the laser light should not be too large, as certain instabilities cannot be controlled at large wavelengths, of let's say μm (10^{-6} m) scale. Wavelengths should preferably be a couple of hundred nm (10^{-9} m).

A further problem with glass lasers is their inefficiency, less than 1% of electrical energy is converted into ultraviolet light (i.e. less than 1% of the light from the flashlamps used to pump the gain medium is eventually converted into laser light), and in the indirect-drive approach there is further loss in generating X-rays in the hohlraum. For a commercial reactor such a waste of energy would be out of the question. Vastly greater efficiency is required, in order for a fusion power plant to be able to yield a surplus of energy. Fusion power plants, if they ever become reality, will require vast amounts of energy to build and run, and all this spent energy must eventually be earned back by running the plant.

Alternative drivers have therefore to be considered. Four main systems are currently being studied as potential drivers of inertial fusion plants: diode-pumped solid state lasers (such as the neodymium glass laser mentioned before), krypton fluoride (KrF) gas lasers, heavy-ion or other particle beams from accelerators, and pulsed (electric) power drivers.

Using diodes[18] instead of flashlamps to pump a solid-state laser could allow a much higher repetition rate plus a better efficiency. Of particular interest are the projected lifetimes of large diode laser arrays for pumping a laser driver for inertial fusion energy. Based on recent measurements, the operational lifetimes are projected to be greater than 13.5 billion shots, or greater than 100,000 h at a 37 Hz repetition rate. For diode-pumped solid-state lasers (lasers that pump a solid gain medium (crystal) to give the emitted light another, mostly shorter wavelength) an efficiency of around 10% is expected. Apart from their higher (still rather poor, but steadily improving) efficiency and higher repetition rate, the ultra-high brightness of the beams would make them attractive as drivers for ICF-reactors. The aim is now to build high-power specimen of such lasers and test which irradiation uniformities can be obtained in inertial confinement experiments. The laser driver for inertial fusion energy is a significant component (~25%) of the capital cost of inertial fusion energy and is therefore the subject of research and development aimed at maximizing the performance, availability, and reliability of diode-pumped solid-state laser drivers for inertial fusion energy in Europe, Japan, China, and the United States. (Committee on the Prospects for ICF 2013, p. 50; Pfalzner 2006, p. 46.)

Krypton fluoride (KrF) gas lasers, so-called excimer lasers discovered in 1970 by Nikolai Basov, V. A. Danilychev and Yu. M. Popov, at the Lebedev Physical Institute

---

[18] A diode is a sort of electronic one-way valve as it conducts current primarily in one direction and not in the other. It has low (ideally zero) resistance in one direction, and high (ideally infinite) resistance in the other.

in Moscow and developed since the 1980s, use no glass at all and can pulse more rapidly (current level is at five pulses per second, but at low power). KrF lasers have a high beam uniformity, short wavelength, and an adjustable spot size (essentially the radius of the beam itself), as well as a much higher efficiency than glass lasers (the theoretical maximum is at around 20%, still not impressive, and practical values are much lower). The Nike laser at the US Naval Research Laboratory has demonstrated that a large KrF laser can be built and achieve highly uniform target illumination, but its efficiency was only around 5%. There is currently no ignition-level facility available at the KrF wavelength of 248 nm, but calculations have shown that cost-effective power generation could be possible with KrF-driven inertial fusion energy. (Committee on the Prospects for ICF 2013, p. 224.) There are or were several KrF laser facilities in the world, e.g. in Britain, Japan and Russia, but not necessarily used (solely) for ICF-experiments.

Apart from lasers as drivers, a consortium of laboratories in the United States led by the Lawrence Berkeley National Laboratory explores particle beams as drivers and the Z-machine at Sandia National Laboratories uses pulsed magnetic fields.

Beams of ions are under study for use as drivers but will not be realised any time soon. The advantage of heavy-ion beams would be a high repetition rate for the driver delivering its energy into the target, and a much better efficiency (2–4 times higher than laser beams). The problem is that they are hard to focus at a small target and also have to be pulsed in order to be at use in inertial confinement setups. It is, however, usually easier to focus ions at higher kinetic energy and higher mass, so most of the emphasis is currently on heavy-ion fusion as opposed to light-ion fusion. A drawback in this respect is that heavy ions, such as xenon, caesium and bismuth, need to be accelerated to much higher energies than light ions, such as lithium. Energies required would be in the order of 10 GeV, for which a large accelerator would be needed.

The initial interaction process also differs significantly when particle beams are used as drivers instead of lasers. Basically, laser light interacts only with the surface of the matter it encounters, whereas particle beams penetrate a certain distance into the material. Such penetration has been studied for over a century, but only for low-intensity beams in cold matter. How they behave in hot matter is not (yet) known.

For indirect drive with ion beams, the fuel capsule (ablator and fuel) is essentially the same as for laser indirect drive. The primary difference lies in the physics of the beam–target interaction and conversion of beam energy into the radiation that must strike the capsule. Thus, experience with laser indirect drive on NIF will provide much of the physics basis for heavy ion-driven targets and the same computer codes are used in target simulations of both driver options.

Heavy-ion beam facilities are however still far from delivering sufficient energy onto a target. Laser systems are much more advanced in this respect and it is likely that the first inertial fusion reactor, if it ever gets that far, will still use a laser driver.

The final approach that we will mention briefly is pioneered with the Z pulse machine at Sandia National Laboratories. The Z Pulsed Power Facility (Fig. 12.20) at this laboratory, informally known as the Z-machine, is the largest high-frequency electromagnetic-wave generator in the world and is designed to test materials in

Fig. 12.20   The Z Pulsed Power Facility at Sandia National Laboratories

conditions of extreme temperature and pressure. The machine is in use since 1996 as an ICF-facility. Its origin, like most of ICF-research, can be traced back to the need to replicate the fusion reactions of a thermonuclear bomb in a lab environment to better understand the physics involved. The idea of the Z-machine is to generate sufficiently high magnetic field pressures to compress and heat magnetized, pre-ionised fusion fuel contained in a cylindrical target to ignition conditions, so the implosion is here driven by electromagnetic fields.

The Z-machine uses the well-known principle of Z-pinch (see Chap. 3) where the fast discharge of capacitors through a tube of plasma causes it to be compressed towards its centre line by the resulting Lorentz forces. The Z-machine layout is cylindrical. On the outside it houses huge capacitors (see Glossary) discharging through high-voltage pulse generators which generate a one microsecond high-voltage pulse. This time is then divided by a factor of 10 to enable the creation of 100 ns discharges. Initially they were discharged into a tube, resulting in the effect of the crushed lightning rod illustrated in Fig. 3.1. This did not work as the current flow was highly unstable and rotated along the cylinder which twisted the imploding tube and decreased the quality of the compression. A Russian scientist, Valentin Smirnov (b. 1937) then had the idea of replacing the tube (called the "liner") with a wire array consisting of hundreds of tungsten wires, each thinner than a human hair, enclosed in a small metal container. This container serves to maintain a uniform temperature. The flow of energy through the tungsten wires dissolves them into plasma and creates a strong magnetic field that forces the exploded particles inward. The speed at which the particles move is equivalent to traveling from Los Angeles to New York—about 3000 miles—in slightly less than one second. The particles then collide with one another along the z axis (hence the name Z-machine), and the collisions produce intense radiation that heats the walls of the container to approximately 1.8 million degrees Celsius.

In every shot the machine consumes only about as much energy as it would take to light 100 homes for a few minutes. Each shot from Z carries more than 1000 times the electricity of a lightning bolt and is 20,000 times faster. [19]

A special version of this approach is Magnetized Liner Inertial Fusion which uses a combination of lasers for heating and Z pinch for compression. The targets being considered for Magnetized Liner Inertial Fusion are at present beryllium (conducting) cylinders that contain the fusion fuel at high pressure. As the magnetically driven implosion of the cylinder is initiated, a laser pre-ionises and preheats the gaseous fuel, which is then compressed and heated to ignition by the imploding metal cylinder in less than 100 ns. Magnetic implosion offers the possibility of significantly higher implosion efficiency than the other approaches.

Both indirect and direct drive have been studied for pulsed-power inertial fusion. Many of the considerations for laser and heavy-ion targets also apply to pulsed-power targets.

12.12 Conclusion

The main goal of the National Ignition Facility to achieve ignition by September 2012 was not achieved, and now more than seven years later it still hasn't. It took about ten years to put a man on the moon and, not counting all the preparatory work from 1970 to the end of the twentieth century, roughly 20 years have now passed since the start of the construction of NIF and still ignition has been illusive. So, what has all the effort in ICF led to? It will be clear from the above that power generation by this approach is very complicated and an astounding feat if ever realised. The sheer idea of having a plant that generates power from rapidly repeated explosions of tiny thermonuclear bombs blows the mind and sounds very much like science fiction, or rather engineering fiction, with the emphasis on fiction. Still, scientists and engineers are pursuing this goal now for close to fifty years, and although progress is slow, they are nudging forward, albeit at a snail's pace and at enormous expense. The ultimate prize is however not in sight and seems to be fading ever further into the future.

In view of the fact that suitable drivers that can keep input energy costs within reasonable margins currently do not even exist, it is without doubt fair to say that, unless a real breakthrough of one kind or another is achieved, it will never get to the stage that it makes sense to even start thinking about designing a real power plant. And even if it does, the costs will most probably be incalculable.

Recently some privately funded activity in ICF has started up. First Light Fusion, a British company based in Oxford, and the American companies Proton Scientific and Fusion Power Corporation make big, but probably empty claims about ICF, as will be detailed in Chap. 15.

---

[19] See the very clear and instructive webpage of Sandia for further explanation: https://www.sandia.gov/z-machine/about_z/how-z-works.html, accessed 20 June 2020.

Chapter 13
The Revival of Obsolete Practices: Stellarators, Magnetic Mirrors and Pinches

13.1 Return of the Stellarator

In Chap. 3 we discussed the early stellarator designs studied at PPPL by Spitzer and his collaborators. After Spitzer left the field in 1966, experiments continued for a couple of years with the Model-C stellarator, but without any major success. Major problems were a low power density (i.e. low *beta*, see the Glossary or Chap. 2 to remind you of the meaning of this fundamental plasma parameter) and the fact that confinement remained consistent with Bohm scaling, i.e. a confinement time inversely proportional to the electron temperature and linearly proportional to the magnetic field strength (instead of the field strength squared).

Stellarator development at PPPL ended, when in 1970 as part of the tokamak stampede the Model-C stellarator was quickly, in four months time, converted into a tokamak and used to reproduce the promising Soviet results with the T-3 tokamak. During the rest of the century research concentrated on tokamaks, a trend that is still continuing today, with comparatively little attention to other designs (with the exception of inertial confinement fusion, see previous chapter). The stellarator idea did not lay completely dormant though. Several groups in various parts of the world maintained faith in this type of machines and continued their experiments. Their efforts will be discussed in this chapter, and it is perhaps a little unkind to rank them among obsolete practices, although it is unlikely that they will play a significant role in power generation by nuclear fusion.

When tokamaks started to develop problems of their own, which for that matter were very similar to the problems encountered earlier with stellarators, the interest in stellarators experienced a revival and currently several stellarators are in operation. Since over the years much more effort and money has been invested in tokamak research, stellarators are still lagging behind, at any rate in size, but also as regards stage of development. Nobody engaged in stellarator research talks as yet about "power on the grid in 20 years" or so. They are already happy when they see some plasma being confined in their intricate vessels.

© The Author(s), under exclusive license to Springer Nature Switzerland AG 2021   349
L. J. Reinders, *The Fairy Tale of Nuclear Fusion*,
https://doi.org/10.1007/978-3-030-64344-7_13

Stellarators and tokamaks have in common that they both use magnetic fields to confine the plasma in a torus-like device. We have seen in earlier chapters that the resulting nonuniformity of the magnetic fields causes the particles to drift. For a plasma to be confined this drift has to be cancelled. The basic difference between stellarators and tokamaks is how they cancel this drift, to make sure that the magnetic field lines form a closed surface inside the tube and do not end up in the wall of the tube, taking the plasma particles with them. To achieve this the magnetic field is helically twisted. When discussing the early tokamaks in Chap. 3 we introduced the *safety factor* q, which is the ratio of the number of times a toroidal field line goes around the torus the long way to the number of times a poloidal field line goes around the torus the short way. We also saw that the Kruskal-Shafranov limit implies that q must be bigger than one for the kink instability to be suppressed. By convention in stellarators the term *rotational transform* is used which is just the inverse of q. They both are a measure of the degree in which the magnetic field is twisted.

In tokamaks a poloidal field is usually created by a toroidal electric current (a current travelling in the plasma around the torus, which also takes care of (part of) the heating of the plasma), in addition to magnetic coils for generating the toroidal field. In stellarators only coils are used. There is no need for a current flowing in the plasma. This is *the* fundamental difference between the tokamak and stellarator approach to nuclear fusion. The absence of this current is a great advantage as it is a source of perturbations in the plasma and gives rise to instabilities.[1]

Since the plasma current in the tokamak also produces part of the heating, it is impossible in a tokamak to study confinement in isolation from the heating process and from any other plasma current effects, while the stellarator allows for such experiments. The absence of the current also excludes the major concern of a massive energy dump into the walls of the vessel, as happened both with the JET and PPPL tokamaks in the past (see Chap. 2) and remains a hazard that must be closely watched.

In addition to poloidal coils, producing the toroidal field, the stellarator has (1, 2, 3 or more) pairs of helical windings that complicate the design, but make it possible to create the required closed magnetic surfaces in vacuum without the plasma playing a role in this. Indicating the number of pairs by l, an $l = 2$ stellarator has 4 helical windings, like the one shown in Fig. 13.1.

A second very important advantage of the stellarator over the tokamak is that it can operate in a steady-state regime, while tokamaks are almost by definition pulsed devices. For a nuclear fusion power reactor steady-state operation is obviously crucial. But, an advantage normally comes with a disadvantage, which for power reactors operating in steady state is that the plasma needs to be refuelled continually and that impurity control and 'ash' removal (the residue from the burning plasma) are needed. For this they are equipped with a divertor: a device enabling the removal of waste material from the plasma via open magnetic field lines while the reactor is operating. Divertors are also used in tokamaks and their functioning has been

---

[1]Since this current is also used to heat the plasma (ohmic heating) it implies that in stellarators other ways must be found to heat the plasma. Current-carrying stellarators experienced anomalous electron heat loss. This problem was only solved by 1980.

Fig. 13.1 Principle setup of
a classical stellarator with
helical coils. Note that the
current flows in opposite
directions in adjacent
windings of these coils
(*from* Wagner 2013, p. 14)

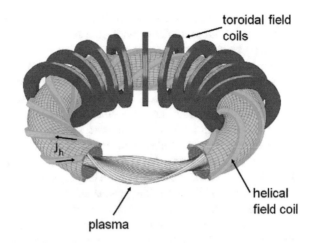

toroidal field
coils

J_h

helical
field coil

plasma

discussed in Chap. 4. They allow control over the build-up of fusion products in the
fuel, and remove impurities in the plasma that have entered from the vessel lining and
might cool the plasma. A divertor design has been suggested as early as the 1950s
by Spitzer (see Spitzer 1958, p. 263, which refers to a 1951 AEC report by Spitzer).
He proposed that the magnetic field lines at the edge of the plasma be deliberately
diverted or led away into a separate chamber where the particles carried with them
can interact with the wall and do no harm (McCracken and Stott 2013, p. 100).

Finally, the geometric parameters are also generally very different between tokamaks and stellarators. For tokamaks the aspect ratio (the ratio R/a of the major radius
R and the minor radius a of the torus, see Fig. 3.21) is usually in the range from 2.5 to
4, and even smaller for spherical tokamaks. To avoid certain resonances stellarators
are designed to have larger aspect ratios (5–12), which allows for easier access for
maintenance and suchlike. A consequence is that the effective plasma volume in
tokamaks is much larger than in stellarators. In line with the development of spherical tokamaks, there also have been attempts to develop stellarators with smaller
aspect ratios of 3–5 instead of larger than 10. One of these attempts was the National
Compact Stellarator Experiment (NCSX) at PPPL, which had an aspect ratio of 4.4.
Funding had been obtained and construction had started when it went under in the
2008 worldwide financial crisis. Since it proved difficult to build, repeatedly ran over
its timelines and budget (in total more than $70 million was spent), it probably would
also have been cancelled without an economic crisis.

13.1.1 Spatial and Classical Stellarators and Torsatrons

Stellarators have great flexibility and different stellarator configurations (and combinations of them) have been developed. They go under various names, but no accepted

usage of names has been established. The original figure-8 design is sometimes indicated as a *spatial stellarator*, while the name *classical stellarator* (Fig. 13.1) is used for a toroidal or racetrack-shaped design with separate helical coils to create the closed magnetic surfaces. Both were designed by Spitzer. The set of toroidally continuous helical windings with current flowing in opposite directions in adjacent windings provides the poloidal field. Because of the oppositely directed currents in adjacent windings there is no net toroidal field in such a configuration. A toroidal field must then be obtained from a separate set of planar (flat) poloidal coils (in Fig. 13.1 indicated as the toroidal-field coils). One of the advantages of such a setup is flexibility as the helical and toroidal-field components can be varied independently, allowing for a wide range of configurations from large or small toroidal currents (tokamak-like configuration) to no net toroidal plasma current (pure stellarator) in a single experimental device. A disadvantage is that the interaction between the toroidal-field coils and the helical windings may cause large radial forces, which alternate in direction from one helix to the next, posing serious problems for supporting the helical windings.

These designs were followed up by a design called *torsatron* (known in Japan as *heliotron*), which was proposed at the Novosibirsk conference in 1968 (Gourdon et al. 1968), and independently already in 1961 at the Kharkov Institute for Physics and Technology (KIPT) (Tolok 2001, p. 127). In a torsatron the currents in the helical field coils flow in the same direction (unidirectional). The 'basic' torsatron retains poloidal-field coils to cancel the vertical field produced by the helical coils, while the 'ultimate' torsatron design ('ultimate' meaning that a special winding law for the helical field coils is selected) dispenses with the toroidal-field coils altogether, but requires the windings of the helical coils to be carefully specified (Braams and Stott 2002, p. 146). So, this would essentially be the setup of Fig. 13.1 without the toroidal-field coils and with unidirectional helical field coils. A simple design is shown in Fig. 13.2. An advantage is that the complication of the classical stellarator with two sets of interwoven windings is eliminated. Early torsatrons were constructed in Kharkov, Culham (UK), and Nagoya and Kyoto (Japan). In Japan the National

Fig. 13.2 Simple torsatron design with a single helical coil (yellow) and six turns of the coil around the vacuum chamber with the plasma (purple). The design shown here is the basic design of the TJ-IU, an upgrade of the TJ-I, constructed at CIEMAT (Madrid)

Institute for Fusion Science was set up in 1989 to especially build and operate such heliotrons.

13.1.2 Torsatrons at Kharkov (Ukraine)

After having seen the American stellarator at the 1958 Geneva exhibition (Fig. 3.18), Kurchatov also wanted to have one in the Soviet Union and the first stellarator, the L-1, was duly built in 1962 at the Lebedev Physics Institute in Moscow, which would remain active in stellarator research until the early 1970s with the successor stellarators L-2 and TOR-1. Promising results with the L-1, notably in respect of overcoming the Bohm diffusion restriction, were reported at the Culham nuclear fusion conference in 1965. Unlike American and German stellarators, the Soviet stellarator was perfectly round without any straight sections viz. racetracks. Soviet researchers decided from the very outset to do without the racetracks because of magnetic field distortions that could considerably deteriorate plasma confinement and are inevitable in the transition region from the rounded sections to the racetracks (Voronov 1988, p. 138).

Now that stellarator research had been started at the Lebedev Institute, this institute would have been the obvious choice for expanding the programme, but Kurchatov chose the Kharkov Institute for Physics and Technology (KIPT) in Ukraine for a concentrated effort in such research and for the construction of a device that would be even bigger than Spitzer's Model C. There was considerable opposition to his plans, e.g. from Artsimovich, who vigorously promoted the tokamak, but Kurchatov's standing in Soviet scientific life and his clout with the authorities were such that nothing could stop him. After his premature death in 1960 at age 57 the opposition became stronger, which caused things to be put on hold for some time. After some bickering Efim Pavlovich Slavsky (1898–1991), the minister of Medium Machine Building, which also included the Soviet nuclear industry, and a friend of Kurchatov[2], cut the knot by just brushing aside the objections of other scientists and decided that the programme should go ahead in honour of Igor Kurchatov. And so it did, but at a much slower pace; construction took 7 years instead of the 2 years planned by Kurchatov and the eventual design was considerably smaller, being scaled down repeatedly in the innermost recesses of the Ministry at the urging of the opponents to the project. The end result was that in 1967, just a few years before the Americans abandoned stellarator research because of the success with the Soviet T-3 tokamak, the Soviets built a stellarator facility called "Uragan" (Hurricane) in Kharkov, which was roughly of the same scale as Princeton's Model C. Model C had not been able to overcome the problem of Bohm diffusion, but the Uragan-1, still an ordinary stellarator, actually did (by a factor of 30!). It encouraged the construction of further stellarators and especially with a torsatron configuration. At KIPT the first stellarator, called "Sirius", a very small device, serving as a test reactor, had already

---

[2]They shared the dubious fame of being the creators of the "nuclear shield" for the Soviet Union.

Fig. 13.3 Layout of the flexible heliac stellarator TJ-II operating in Spain, in which the twisted magnetic axis can clearly be seen (http://fusionwiki.ciemat.es/wiki/TJ-II)

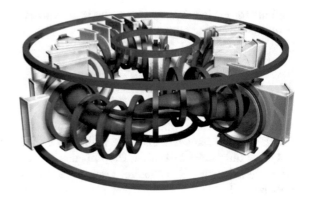

been constructed in 1964, followed by the above-mentioned Uragan-1 and a whole family of torsatrons (e.g. Saturn, Vint-20, Uragan-2, Uragan-3, and its modification Uragan-3 M) in later years. Saturn was the world's first torsatron (it could actually operate both as a torsatron and a stellarator) and started experiments in March 1970. In the 1980s larger torsatrons (Uragan-3 and Uragan-3 M) were built, which are still in use but have suffered considerably from the collapse of the Soviet Union (Tolok 2001). There are currently two torsatron-type devices operating at KIPT. The bigger one is the Uragan-2 M facility (with major radius of 1.7 m and minor radius 0.22 m, which came into operation in 2006, and the smaller Uragan-3 M (major radius 1.0 m, minor radius 0.125 m), which is an upgrade of the Uragan-3 and dates from 1990 (Moiseenko et al. 2016). So, these machines are tiny compared to tokamaks in operation and under construction, and smaller, but not that much, than the Wendelstein 7-X, to be discussed below. Useful experiments are currently still being carried out with these facilities, mostly as supporting experiments for the German Wendelstein 7-X.[3]

13.1.3 Heliacs

A further stellarator-type design is the *heliac* (helical axis stellarator), in which the magnetic axis (and the plasma) follow a twisted (helical) path to form a toroidal helix rather than a simple ring shape. The twisted plasma induces twist in the magnetic field lines to effect drift cancellation. It typically can provide more twist than the torsatron or heliotron, especially near the centre of the plasma (magnetic axis). The original heliac consists only of circular coils, and the *flexible heliac* (Fig. 13.3) adds a small helical coil to allow the twist to be varied. A stellarator with a flexible heliac design (H1-NF) was built and operated for some time in Canberra (Australia), before being shipped lock, stock and barrel to China in 2017, to add to their collection of fusion facilities, which already included four tokamaks, and to hedge their bets just

---

[3]V.S. Voitsenya (KIPT), private communication.

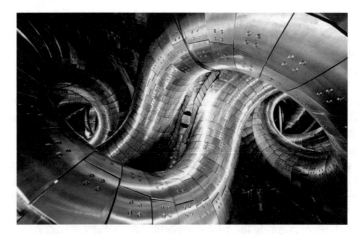

Fig. 13.4 The twisted vacuum chamber of the Large Helical Device (Japan)

in case the stellarator might turn out to be the holy grail of fusion. Flexible heliac facilities also operate in Spain (TJ-II), designed in collaboration with Oak Ridge National Laboratory and the Max Planck Institute for Plasma Physics in Garching (Germany), and in Japan (Tohoku University (TU) Heliac) at the National Institute for Fusion Science in Toki.

13.1.4 Large Helical Device (LHD)

The Large Helical Device is a heliotron/torsatron configuration operating since 1998 at the National Institute for Fusion Science in Toki in Japan, It is equipped with an additional helical coil to confine the plasma, together with a pair of poloidal field coils to provide a vertical field. Toroidal field coils can also be used to control the magnetic surface characteristics. Its vacuum chamber has a weird twisted form (see Fig. 13.4). The machine demonstrated that large superconducting coils can be manufactured and reliably operated for years.

The LHD has outperformed tokamaks in plasma density, as well as in a *beta* value of 5%. It is the second largest superconducting stellarator (i.e. using superconducting magnets) in the world after the Wendelstein 7-X, to be discussed below.

13.1.5 Wendelstein 7-X

The most important of the newer stellarator designs is however the *Helias* (Helical-Axis Advanced Stellarator) configuration, which in the European Roadmap is seen as a possible long-term alternative to a tokamak fusion power plant. It is the design of the most important stellarator currently in operation: Wendelstein 7-X. Wendel is the German word for helix or spiral, which explains the name. It is also a reference to the Wendelstein which is the highest peak (1838 m) of the Wendelstein massif in the Bavarian Alps.[4]

The Wendelstein 7-X device is located at the Max Planck Institute for Plasma Physics in Greifswald (IPP Greifswald) in the northeast of Germany, a branch institute of the Max Planck Institute for Plasma Physics in Garching (IPP Garching). IPP Greifswald was founded in 1994, while the one at Garching dates back to 1960 (see Chap. 6). The construction of Wendelstein 7-X took from 1991 to 2014 at a cost of slightly more than 1 billion euro and became operational at the end of 2015.

It is the successor of Wendelstein 7-AS (*Advanced Stellarator*), which in turn was the successor of Wendelstein 7-A. The latter device went into operation in 1975 after a series of smaller devices were explored from 1960 onwards, which makes Germany the most persistent player in the stellarator field. The first stellarator, Wendelstein 1-A, a very small machine compared to current-day devices, made its appearance in the Institute's foundation year 1960. In Chap. 3 we have seen that Spitzer and collaborators failed to beat Bohm diffusion with their stellarators. At the 2nd IAEA Conference in Culham in 1965 German researchers from IPP however already showed that their Wendelstein stellarators could overcome Bohm diffusion with relatively cold and low-density plasmas and that the good plasma confinement as predicted by theory was actually possible with a stellarator (Eckhartt et al. 1965). They were met more with mistrust than admiration. After all, the Model-C stellarator at Princeton, the world's biggest machine of that type with the largest team of physicists and engineers, was still unable to do the same. The result was at first ridiculed as the 'Munich mystery'. In order to be believed it had to be reproduced and confirmed somewhere else, which only could be done in Princeton, the only other place that performed research with stellarators. Günter Grieger (1931–2012) from IPP was even offered a position at PPPL to solve their problems with Model C. He however preferred his job at IPP and declined the offer as he considered Model C a poorly constructed device which according to him was beyond improvement. Research continued on further devices in the Wendelstein series, in particular Wendelstein 2-A, and at the 1968 Novosibirsk conference a confirmation of the earlier results was presented. It was made clear that the so-called pump-out in Model C (the phenomenon that particles drift to the edge of the confinement vessel faster than predicted by theory and by doing so take heat out of the plasma) and its failure to overcome Bohm diffusion should rather be designated the 'Princeton mystery' instead of speaking of the positive Munich results as the 'Munich mystery' (Eckert 1989, pp. 133–134).

The Novosibirsk conference was also the one at which Artsimovich triumphed with the T-3 tokamak results, which unleashed the tokamak fever. IPP also caught it and in 1968 embarked on tokamak research, but the stellarator was not abandoned, and since then IPP is the only institute in the world that studies both types of devices, tokamak as well as stellarator.

---

[4]The Wendelstein is not the highest mountain in the German Alps as is sometimes erroneously claimed; the highest is the Zugspitze with 2962 m.

Wendelstein 7-A, mentioned above, gave a major boost to the stellarator line, as in 1980 it solved the heating problem caused by the absence of a plasma current. It was able to demonstrate the "pure" stellarator principle with a hot plasma for the first time in the world, i.e. confinement without plasma current, showing that stellarators were back in business.[5]

On the basis of these successes IPP Garching operated the 7-AS device from 1988 to 2002. The major radius of its torus was 2 m and the minor radius 0.2 m, creating a plasma volume of 1 cubic metre. Previous stellarators had an almost circularly symmetric magnetic field, like tokamaks, in which the circular current in the plasma necessitates circular symmetry of the whole configuration. But stellarators by nature also have to use fields that do not possess circular symmetry. The 7-AS was called an "advanced stellarator" as the fast computers that had meanwhile become available made it possible to select the best, i.e. the most stable and most thermally insulating fields. It was the first of a new class of stellarators with modular coils that achieve the required magnetic fields through their twisted form.[6]

Sophisticated computational techniques were developed making it possible to first specify the field shape optimised for its plasma properties and then to design the coils that would produce the fields. The large number of degrees of freedom was used to approximate the generated magnetic field to the theoretical optimum. The experiment demonstrated that the innovative coil system can be manufactured to exact specifications. This approach revolutionised stellarator design (Braams and Stott 2002, p. 17). With plasma temperatures of 70 million degrees for the electrons and 16 million degrees for the ions, energy confinement times of up to 60 ms, and reactor-grade plasma densities, Wendelstein 7-AS broke all stellarator records in its size group.

Its successor, Wendelstein 7-X, has a *helias* configuration, using a single optimized modular coil set (i.e. a set made up of separate units consisting of a number of the strangely shaped coils depicted in Fig. 13.5 and some planar (flat) coils) designed to simultaneously achieve high plasma currents and good confinement of energetic particles. The planar coils are used to make fine adjustments to the magnetic field. The helias is currently thought to be the most promising stellarator concept for a power plant. Figure 13.5 shows its schematic design and Fig. 13.6 shows a non-planar coil as installed in the device. Such a coil is about 3.5 m high.

Wendelstein 7-X comprises five large, almost identical modules. Each module consists of a section of the plasma vessel, its thermal insulation, ten superconducting non-planar coils and four planar coils together with the interlinking connections, the piping for the cooling of the coils, and also a section of the support ring. The modules were separately constructed and set up in a circle along the plasma ring in the experimentation hall. Assembly began in April 2005 with the installation of the first half module. Seven superconducting magnet coils were strung along one tenth of the steel plasma vessel. Two half modules were then combined into a whole

---

[5] https://www.ipp.mpg.de/3951949/wendelstein7a.

[6] Max Planck Institute for Plasma Physics, *50 years of research for the energy of the future*, https://www.ipp.mpg.de/17194/geschichte.

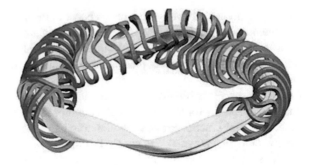

Fig. 13.5 Stellarator design with a single set of twisted external coils (like the one in Fig. 13.6), as used in the Wendelstein 7-X experiment, in addition to other coils. A series of non-planar magnet coils (blue) surrounds the plasma (yellow). A magnetic field line is highlighted in green on the yellow plasma surface (*from* Wendelstein website)

Fig. 13.6 Drawing of one of the non-planar superconducting coils as installed in the Wendelstein 7-X device (*from* Thomas Klinger, The Construction of Wendelstein 7-X, ITER seminar)

module, after which it was furnished with cooling pipes and other peripherals. In the experimentation hall the five modules were one by one equipped with their cryostat sheath and joined together. The base machine was completed in 2011.[7]

The principal objective of Wendelstein 7-X is to investigate the suitability of this type of device for a power plant. It will test an optimised magnetic field for confining

---

[7]https://www.ipp.mpg.de/1727365/zeitraffer_w7x.

the plasma, with temperatures up to 100 million degrees and confinement times up to 30 min. The field will be produced by the system of superconducting magnet coils. This is the technical core piece of the device.

It will be clear from Figs. 13.5 and 13.6 that the fields produced by these awkwardly twisted coils are very complex. The complexity of the stellarator magnetic fields makes them much more difficult to analyse than the fields in a tokamak. In this respect, however, stellarator design has hugely profited from dramatic improvements in computer power and 3-dimensional modelling capabilities in the last few decades, making it possible to accurately model such fields, as well as plasma behaviour and various other components of the device. This more than cancels out the complexity disadvantage of stellarators compared to tokamaks.

Wendelstein 7-X is the first large-scale, fully optimised stellarator and the world's largest device of its kind. The machine weighs about 750 tonnes; its major radius is 5.5 m, minor radius 0.52 m, and plasma volume 30 cubic metres. The fact that it uses superconducting magnets implies that it also is a cryogenic device and has to be kept at the very low temperature of -270 °C (425 tonnes of the total 750 are cooled with liquid helium to this temperature). Figure 13.7 shows the layout of the intricate coil system of the device, consisting of 50 non-planar coils, 20 planar coils and five so-called trim coils. These external trim coils, of which four are shown in the figure, are used to fine-tune the shape of the plasma. The trim coils have the size of a barn door and were designed and manufactured for Wendelstein 7-X by PPPL in collaboration with local industry in the US. PPPL also provided other components of the device.

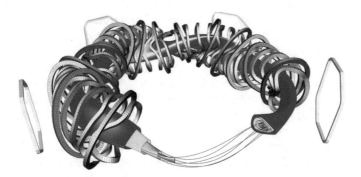

Fig. 13.7 Layout of the coil system of Wendelstein 7-X. Some nested magnetic surfaces are shown in different colours in this computer generated figure, together with a magnetic field line that lies on the green surface. The coil sets that create the magnetic surfaces are also shown, planar coils in brown, non-planar coils in grey. Some coils are left out of, allowing for a view of the nested surfaces. Four out of the five external trim coils are shown in yellow. The fifth coil, which is not shown, would appear at the front of the figure (*from* Nature)

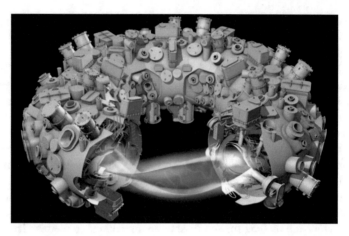

Fig. 13.8 The Wendelstein 7-X device with all its fittings and trimmings. A 16-m wide container enclosing all the magnetic coils and their helium cooling liquid, with 250 access ports

Around the coils we have the cabling, piping and central support ring, then comes the outer vessel with its diagnostic ports resulting in the bizarre structure shown in Fig. 13.8, which shows little resemblance to any of Spitzer's original designs.[8]

It is expected that plasma equilibrium and confinement in Wendelstein 7-X will be of a quality comparable to that of a tokamak of the same size. But it will avoid the disadvantages of a large current flowing in a tokamak plasma. With plasma discharges lasting up to 30 min, Wendelstein 7-X is to demonstrate the essential stellarator property, viz. continuous operation (Wendelstein website https://www.ipp.mpg.de/w7x). Its primary purpose is to investigate whether it is suitable for extrapolation to a fusion power plant design. The future will tell whether that is the case. First plasma was produced at the end of 2015 and ignition conditions have not yet been reached. Its latest campaign ended in late 2018—during which operators were able to achieve discharge times of up to 100 s with 2 MW of input heating power or 30 s at 6 MW. It is now being prepared for the next round of experiments that is expected to generate plasma pulses of up to 30 min.[9]

The successor to Wendelstein 7-X, providing that the design shows to be suitable for a power plant, will be much larger and consequently much more expensive (comparable to the cost of ITER perhaps), most probably too expensive to be financed by a single country. In anticipation of the success of Wendelstein 7-X, the Helias reactor, an upgraded version and straightforward extrapolation of Wendelstein 7-X, is the first attempt at such a reactor (Beidler et al. 2001). The problem in this respect is that, if ITER fails, there will be little appetite for another global venture of similar scale and, if ITER succeeds, international fusion effort is likely to continue along the tokamak road, so it is hard to see any future for the stellarator as a power plant.

---

[8]Various beautiful videos showing the construction and layers of the Wendelstein stellarator can be viewed on YouTube (an example is https://www.youtube.com/watch?v=u-fbBRAxJNk).

[9]https://www.ipp.mpg.de/4828222/01_20, dated 16 March 2020; accessed 21 June 2020.

13.1.6 Other Stellarators and Hybrid Devices

Over the years various institutions in the US have started and subsequently abandoned the construction of stellarators of various types. Funding turned out to be erratic and changed with the political winds (See Dean 2013 about this). Many small stellarator-type devices were constructed for research purposes, not with the aim to actually build a fusion power reactor.

The construction of the **Advanced Toroidal Facility** (ATF) at Oak Ridge National Laboratory started in 1985. It was the first large stellarator to be built in the US for 20 years. It was a torsatron-type stellarator and for some time the world's largest. The primary goals of ATF were to study high *beta* and transport, and to demonstrate steady-state operation. It theoretically had the capability to operate at *beta* $\geq$ 8% in steady state. The achievement of high *beta* was made possible by a carefully chosen combination of helical and poloidal magnetic field coils. During its final operating period ATF achieved pulse lengths of over one hour (4667 s). After a few years of operation, it was terminated in 1992 within the framework of a policy of the US Department of Energy to eliminate non-tokamak devices in order to focus on tokamak research (Dean 2013, pp. 110–111).

We also mention the **CIRCUlar coil Stellarator** (CIRCUS) and the **Columbia Non-neutral Torus** (CNT) that were constructed in the early 2000s. CIRCUS is a remarkably simple, table-top tokamak-torsatron hybrid, with circular coils, vertically tilted compared with a tokamak. Like a stellarator or torsatron, CIRCUS is expected to generate helical field lines and confine a current-free or nearly current-free plasma. At the same time, the plasma is expected to be more uniform in the toroidal direction (in some sense, less three-dimensional), similar to a tokamak. CNT was built to study non-fusion plasmas that were mostly or exclusively made of electrons. It is now operating as an electron-cyclotron-heated neutral plasma for fusion studies relevant to bigger stellarators (https://pl.apam.columbia.edu/stellarator-research).

Auburn University in Auburn (Alabama) operates the **Compact Toroidal Hybrid** (CTH) device, an experiment for magnetically confining high temperature plasmas. CTH can operate as a pure torsatron, but has an ohmic heating system to drive plasma current allowing the researchers to study disruptions and magnetohydrodynamic phenomena in current-carrying stellarator plasmas.

The **Helical Symmetric Experiment** (HSX) is a stellarator with modular coils optimized for quasi-helical symmetry at the University of Wisconsin-Madison. Plasma physics research goals include investigation of transport, turbulence, and confinement in a quasi-helically symmetric magnetic field. HSX is the only device in the world that has a magnetic field structure termed Quasi-Helically Symmetric (QHS). Quasi-helically symmetric stellarators possess, to a good approximation, a direction of symmetry and are therefore topologically equivalent to a tokamak without plasma current. Its research is continued by the private company Type One Energy (see Chap. 15).

The **Hybrid Illinois Device for Research and Applications** (HIDRA) is a toroidal plasma device at the University of Illinois at Urbana-Champaign, formerly known as WEGA when it operated in Greifswald. The WEGA fusion device, a classical stellarator, is a member of the Wendelstein family. It started operational life as a joint Belgian-French-German project under the name "Wendelstein Experiment in Grenoble for the Application of Radio-frequency Heating". Scientists from IPP Garching and from Centre d' Etudes Nucléaires at Grenoble had jointly planned, built and operated WEGA. Then, after a ten-year interlude at the University of Stuttgart, it was started up again in 2001 at IPP Greifswald. Now renamed "Wendelstein Experiment in Greifswald für die Ausbildung" (Wendelstein Experiment at Greifswald for Training), it continued until 2013 and was used for training young scientists in basic research, for investigating magnetically confined low-temperature plasmas and for testing diagnostics and experiment control for Wendelstein 7-X. It was *the* plasma experiment at IPP Greifswald. New heating antennas, measuring facilities and control systems for Wendelstein 7-X were tested in the readily adaptable WEGA device. At the age of almost 40 years WEGA is, if not the oldest, certainly one of the long-lived fusion experiments. After its shutdown at IPP, it was passed on in 2014 to the University of Illinois in the US, where it has since been used for plasma physics and fusion research (https://www.ipp.mpg.de/1055873/wega). The HIDRA vacuum vessel has a circular cross section and a steady state toroidal magnetic field. Since HIDRA has the ability for long pulse steady-state operation via the classical stellarator configuration, HIDRA has an actual toroidal magnetic field, just like a tokamak. It also has the capability to operate as a tokamak, hence a pulsing capability. The steady-state and pulsed capabilities of HIDRA make it an ideal test bed for liquid-lithium science and technology, where flow, ejection and recycling can be assessed. HIDRA will study plasma material interactions and develop the technology needed for new plasma facing components (https://cpmi.illinois.edu/2016/04/26/hidra-hybrid-illinois-device-for-research-and-applications/).

Even small countries like Costa Rica have joined the playing field. The **Stellarator of Costa Rica 1** (SCR-1) is a research project that seeks to implement a small-scale modular stellarator magnetic confinement device for research in high temperature and low-density plasmas.

The Institute of Plasma Physics in Stuttgart (Germany) operates the **TJ-K** experiment, a fusion device of the torsatron/heliotron type. Toroidally closed magnetic flux surfaces are generated by three coil sets: a helical one, which winds six times around the torus-shaped vessel and two vertical-field coil systems. TJ-K was constructed at CIEMAT (Centro para Investigaciones Energéticas, Medioambientales y Tecnológicas) in Madrid. Its name at the time was TJ-IU, where TJ stands for 'Tokamak de la Junta de Energía Nuclear', and it is one of CIEMAT's series of TJ devices. In 1999 it came to Kiel in Germany, and was renamed TJ-K, with the letter K standing for Kiel. In 2005 it moved to Stuttgart.

From this example and from others in the text it can be seen that there is a lively movement of fusion devices in the world.

13.2 Revival of Magnetic Mirrors

As related in Chap. 3, in the early period of fusion research the mirror device was seen as an attractive method for confining hot plasmas, and experiments were carried out in the US, France, UK and Russia. The technology (diagnostics, real-time controls, material radiation limits) is simpler in mirror confinement systems than in toroidal systems, and mirror systems also present lower technological challenges. In the US the mirror programme centred at Lawrence Livermore National Laboratory and such research almost completely stopped in 1986 when the MFTF (Mirror Fusion Test Facility), constructed at a cost of 372 million dollars, was promptly shut down after its completion, without ever having operated. The support for mirror research was terminated as the Department of Energy gave its ear to the tall stories of fusion power generation within 20–30 years by the tokamak.

Before discussing some developments after 1986 that have revived the interest in mirror devices, although it has never become an extensive industry, let us first recap briefly what magnetic mirrors are all about.

The simplest geometry for plasma confinement one can think of is that of a straight cylinder with a magnetic field to keep the plasma confined. The problem of this geometry is, however, that plasma particles can escape at both ends (Fig. 3.9). Such leaks can be greatly reduced by forming two 'magnetic mirrors', which simply means that additional magnetic coils are used to increase the magnetic field strength at the ends, to pinch the endpoints of the cylinder. As we have seen, a charged particle describes a helical (corkscrew) motion around a guiding centre in the direction of the external magnetic field, hence it can be seen as a tiny current loop (i.e. a mini magnet) drifting along its guiding centre. The magnetic field of this little magnet is always opposite to the externally applied field (the particle is said to behave diamagnetically). The plasma particles are therefore repelled by the stronger end magnetic fields (the mirror fields), containing them within a 'magnetic bottle'. Machines based on this concept are called mirror machines (Ongena et al. 2016). Good confinement could however never be achieved in such devices, mainly because of their inherent instability due to the 'wrong' curvature of the magnetic fields towards the plasma (as pointed out by Teller in his famous 1954 talk) and the instabilities generated by end losses. Indeed, the end mirrors do not reflect all particles, only those that have a sufficiently large perpendicular velocity component (i.e. are sufficiently strong mini magnets). Particles with a velocity directed mainly along the magnetic field line are not stopped by the end mirrors, and promptly escape. Therefore, the particle population confined in the bottle is depleted of particles with small perpendicular velocity. This implies that the velocity distribution of the particles is non-Maxwellian, which is a source of instabilities.

Moreover, since there is no such thing as an absolutely collisionless plasma, collisions take place at a low rate even in very hot plasmas, which causes diffusion of particles. In a mirror machine collisions continuously scatter trapped particles changing their velocities whereby particles that end up with a small perpendicular velocity component slowly leak out of the device. For these reasons, magnetic mirror machines are not particularly successful plasma confinement devices, and attempts to achieve nuclear fusion using this type of device have been mostly abandoned

The obvious solution to prevent end losses is to wind the cylinder onto itself, i.e. to have a toroidal shape configuration, like a tokamak, and this has been the path that most research in nuclear fusion has followed.

However, in spite of the difficulties some people did not want to give up and continued working on magnetic mirrors, especially in Russia and Japan, and many pioneering developments in mirror research are indeed due to Russian scientists. Two developments since the 1980s are the Gas Dynamic Trap (GDT), which operates at the Budker Institute of Nuclear Physics (BINP) in Akademgorodok, near Novosibirsk in Russia, and the Gamma-10 experiment in Tsukuba, Japan.

Current interest in such systems has also been raised as the open magnetic field line configuration of mirror fusion devices is particularly well suited for propulsion system applications, since they allow for the easy ejection of plasma that is used to produce thrust. An example of such application is the Variable Specific Impulse Magnetoplasma Rocket (VASIMR), an electrothermal (magneto)plasma thruster under development at the Ad Astra Rocket Company in the US for possible use in spacecraft propulsion. In these engines, a neutral, inert propellant is ionized and heated using radio waves. The resulting plasma is then accelerated with magnetic fields to generate thrust (Romanelli et al. 2006; *Wikipedia*).

13.2.1   *The Gas Dynamic Trap*[10]

A gas dynamic trap is a magnetic mirror device with axial symmetry, using only simple circular coils, a large mirror ratio, i.e. the ratio of the minimum and maximum magnetic field in the device (the fields at the centre of the trap and at the location of the mirror), and a long mirror-to-mirror distance, larger than the effective mean free path of the ions, so that the ions in the plasma collide frequently. Due to these frequent collisions, the plasma confined in the trap is very close to being Maxwellian, and, therefore, many instabilities, potentially dangerous for classic magnetic mirrors with a collisionless plasma, generally do not arise. The rate at which plasma is lost at the ends is governed by a set of simple gas-dynamic equations, from which the device derives its name.

The Gas Dynamic Trap (Fig. 13.9) has been running at BINP for several decades, and has been continuously modernized, reconfigured and upgraded to investigate a wide range of magnetic mirror scientific topics. It fills a cylinder with a length of 7 m and a diameter of 28 cm. The magnetic field strongly varies along the tube. At the centre the field is low; reaching (at most) 0.35 T. The field rises to as high as 15 T at the ends. This change in field strength, implying a large mirror ratio, is needed to reflect the particles and get them internally trapped. The plasma is heated

---

[10]Ivanov and Prikhodko 2013, 2017.

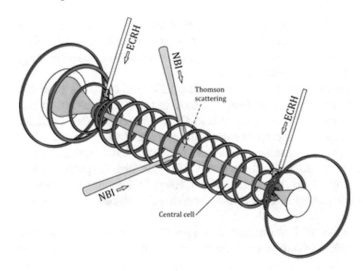

Fig. 13.9 The structure of the Gas Dynamic Trap at BINP, showing the magnets (in red) and two methods of heating the plasma (Neutral beam injection) and (Electron Cyclotron Resonance Heating). Also shown is the magnetic field profile across the machine

by three methods (neutral beam injection, ion cyclotron resonance heating (ICRH) and electron cyclotron resonance heating (ECRH)).

Figure 13.9 shows the general structure of the GDT and Fig. 13.10 presents a schematic view which gives a clear picture of the in essence simple setup of the device.

The mirror concept was thought to have three unattractive features. The magnets are complex, the plasma is plagued with micro-instabilities and the electron temperature would never approach required keV levels. Persistent research with the GDT in Russia has overcome a number of these deficiencies. As of 2016, the machine has

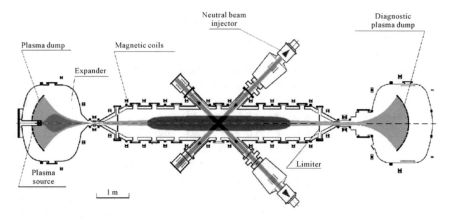

Fig. 13.10 Schematic drawing of the Gas Dynamic Trap at BINP (*from* Horton et al. 2014)

achieved MHD stability with a *beta* of 60% for 5 ms in a simple circular axisymmetric magnetic mirror coil configuration, without complex magnets. Micro-stability of energetic mirror-trapped ions has been achieved with densities reaching 10^{20} m^{-3} by skew injection of neutral beams.

Many previous experiments were plagued by ion-cyclotron instabilities which led to rapid loss of energetic ions. The GDT has reached an electron temperature of 1 keV (about 12 million degrees) using ECRH. This temperature would be sufficient for a GDT type D-T neutron source for the purpose of materials and sub-component testing. Plasma escapes at the ends of the mirror, but is replenished at a rate to maintain the density inside the machine. It is argued that the progress made provides strong motivation to re-examine the potential of linear mirror systems given their inherent advantages in construction, operation, maintenance and repair (Simonen 2016).

Initially the GDT was proposed as a possible concept for an open-ended fusion reactor, which would produce power in a long, axially symmetric, high-*beta*, magnetic solenoid. However, a more recent potential application of the GDT concept is as a 14 MeV neutron source for fusion materials development. For such applications, requirements to provide reactor-grade plasma lifetime or net energy production become less important. Generally, mirrors offer many potential advantages: high *beta* values, topological simplicity, the absence of axial plasma current and disruption-like events, intrinsically steady-state operation, natural He-ash removal out of the ends and low heat loads on plasma-absorbing surfaces in the end tanks. Although mirror devices very much remain a sideshow to the dominant toroidal devices, a new device, a gas dynamic multiple-mirror trap (GDMT) is being developed at BINP, also with the prospect of using it as a neutron source (Molvik et al. 2010).

13.2.2   Gamma-10 Experiment

Gamma-10 is the successor of Gamma-6, which in the mid 1970s was the first so-called tandem mirror, a device we also encountered in Chap. 3, whereby several magnetic mirrors are placed in tandem to prevent leakage. The endpoints are plugged by special coils to generate a plasma of high density at the end mirror traps. Gamma-10 is the world's largest mirror device and operates at the University of Tsukuba in Japan. The strength of its magnetic field is 0.5 T at the central mirror and 3.2 T at both ends. The vacuum vessel for confining the plasma is 27 m long, with 6 m for the central cell, and has a volume of 180 m^3. Plasma experiments on Gamma-10 commenced in 1983 and it has been operating ever since. The central cell is heated by NBI and ICRH and the end cells by NBI and ECRH.

When the problem of edge loss from the mirror magnetic field is solved, the problem of plasma loss in the radial direction is still a problem. This loss (called neoclassical transport) is caused by the non-axial symmetry of the minimum magnetic field (minimum-B) used to secure the stability of the plasma. Such loss can be kept small. Gamma-10 is designed such that the ions in the central mirror can be confined

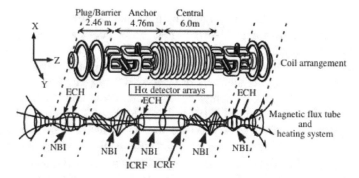

Fig. 13.11 Schematic view of the Gamma-10 tandem mirror; showing the magnetic coil set (top) and the magnetic-flux tube with heating systems

with an axially symmetric mirror magnetic field, while ensuring stability with a minimal magnetic field. So, Gamma-10 is an effectively axisymmetric device and in fact, it demonstrated little neoclassical transport. Figure 13.11 presents a schematic view of the magnetic coil arrangement and the flux tube and heating system. Gamma-10's end region consists of a minimum-B anchor between the central cell and the axisymmetric end plug.

Both the GDP and GAMMA-10 experiments have successfully demonstrated the effectiveness of the electrostatic plug in suppressing ion end losses (ion confinement times up to 1 s have been achieved in GAMMA-10); the ability to maintain MHD stability by using minimum-B anchor cells; and the possibility of suppressing high-frequency micro-instabilities.

13.2.3 Other Devices

At the 12th International Conference on Open Magnetic Systems for Plasma Confinement, held in Tsukuba in 2018, a new mirror device, the Wisconsin Axisymmetric Mirror (WAM), was announced. It is a GDT-like device with superconducting mirror coils producing a field of up to 6 T. It will serve as a prototype for the above-mentioned GDT based neutron source for materials testing. The 3-meter device has a central solenoidal field up to 0.3 T with mirror coils producing up to 6 T. The plan is to use the latest generation of ReBCO high temperature superconducting tape for the mirror coils. A second mission is using lithium-coated plasma facing materials for particle handling and pumping. Third, this device will serve as a prototype for the design and construction of a higher-field, longer-pulse device that may serve to advance the GDT fusion-neutron source concept for a materials and components test facility.

In January 2020 this was followed up by the announcement that the Wisconsin HTS Axisymmetric Mirror (WHAM) is now under construction at the University of Wisconsin-Madison. So, its name has apparently been changed to explicitly include HTS (High-Temperature Superconductors). A public-private partnership (UW Madison, MIT and Commonwealth Fusion Systems (see Chap. 15)) has been formed to build and operate a compact, high-field simple mirror that will aim at sustaining good confinement and MHD stability. MIT and Commonwealth Fusion Systems have been involved as they can provide the key technology for building the high field superconducting magnets needed. If successful, the design would qualify for a low-cost BEAT (Break-Even Axisymmetric Tandem) experiment and offer a simpler and more economical approach to fusion energy than offered in conventional toroidal schemes. This development shows that also in the US the interest in mirror systems is again on the rise and that high-temperature superconducting magnets are apparently very hot in fusion research. Many groups, also in tokamak research (see Chaps. 11 and 15) are planning to install or are actually installing such magnets in their devices.

13.2.4   Conclusion

Linear mirror systems have a number of advantages, including:

(1)   The mirror plasma has a natural magnetic divertor that sends the radially-escaping plasma particles with their high kinetic energy out at the ends through an expanding magnetic nozzle as shown in Fig. 13.10. The end nozzle means that the divertor is external to the plasma confinement region in contrast to tokamaks where the divertor is part of the confinement vessel. The external divertor promises the possibility of direct conversion of the energy of the (charged) particles escaping from the plasma into electric power. Such direct conversion of plasma energy into electricity—thus avoiding the conversion of neutron *thermal* energy into electricity—has considerable advantages. The efficiency of thermal energy conversion is below 50%. GAMMA-10 has already demonstrated such direct conversion into electric power on a small scale from the exhaust plasma in the exterior divertor chamber. In spite of this it is very unlikely that such mirror systems will ever be made into power plants.

(2)   The mirror plasma is an ideal neutron source. As mentioned above, the mirror group at the Budker Institute is currently designing a mirror machine to produce a 14 MeV neutron source for investigating wall materials exposed to high neutron and plasma fluxes.

(3)   Because of its linear geometry the device is simpler to build and maintain in a radioactive environment. The architecture of the diagnostics is simpler in the linear cylindrical geometry. Mirror traps have excellent single-particle confinement in contrast to both tokamak and helical/stellarator magnetic traps.

Thus, the role of a new, well-diagnosed mirror plasma experiment has the potential of advancing fusion science and building a broader engineering and science base for future magnetic fusion plasma reactors (Horton et al. 2014). So, it could well be that once it has dawned on the fusion community and the wider scientific world that nuclear fusion reactors for power generation are not feasible, the revival of mirror research will gain in strength, as it promises better prospects for some of the useful spin-offs of fusion research.

13.3 Revival of Research in Pinches

Pinches were the very first devices tried in nuclear fusion research as related in Chap. 3. Their main feature is that, contrary to the tokamak, the poloidal magnetic field (going around the torus the short way) produced by the current in the plasma is much larger (and pinches the plasma) than the toroidal field produced by external coils.

Recently (Zhang et al. 2019; see also https://newatlas.com/nuclear-fusion-bre akthrough-z-pinch/59257/), at the University of Washington, a breakthrough was announced in Z-pinch research. Although breakthroughs in nuclear fusion research often have to be taken with a pinch of salt, this one has to be mentioned in this book, if only to show how open the situation in nuclear fusion actually still is that such an old chestnut can still be reeled out.

The Z-pinch as a means of compressing and confining plasma with magnetic fields is prone to instabilities, like the kink instability of Fig. 3.3, since the plasma can escape between the parallel magnetic field lines. The magnetic field forms circular loops around the plasma column and confines the plasma radially, but the plasma can form bulges, like an aneurysm, which locally weaken the magnetic field allowing the bulge to grow. Such problems have plagued the Z-pinch approach since its inception in the 1950s and effectively resulted in it being abandoned.

Through making slight adjustments to the behaviour of the plasma by inducing what is known in fluid dynamics as shear(ed) axial flow, it was possible to break new ground in a 50 cm long Z-pinch plasma column. In a shear flow (or shearing flow), adjacent layers of fluid move parallel to each other with different speeds. The shear flow stabilizes the plasma by constantly smoothing the plasma surface and preventing the bulges from developing.

Shear flows are nothing new and have been explored in Z-pinches for years, but now for the first time "evidence of fusion neutron generation from a sheared-flow stabilized Z-pinch" has been produced. More specifically, a flowing plasma was held in place 5000 times longer than a static plasma, and energetic neutrons that are the tell-tale signs of nuclear fusion were observed.

Such shearing of the magnetic field as having a stabilizing effect on almost all instabilities has actually also been observed in tokamaks, so it is not such a surprise that a similar effect is present in a Z-pinch.

Chapter 14
Non-mainstream Approaches to Fusion

14.1 Introduction

A bewildering multitude of fusion approaches has been tried in the past and/or is still being explored. All these devices have an equally bewildering number of inter-connections and are in one way or another focused on nuclear fusion, but very few actually have the pretension to be useful as a power generating device. The majority is used for other purposes or just as a research tool to study problems in plasma physics.

Because of their interconnections it is not easy to classify them in distinct groups. A first division that can be made is into thermal and non-thermal approaches, meaning that in thermal approaches the plasma is (almost) in thermal equilibrium. The ion and electron populations in these plasmas have energy distributions close to Maxwellian (see Chap. 2). In the preceding chapters we have so far only discussed plasma confinement systems that are essentially in thermodynamic equilibrium and, as we have seen, in principle capable of reaching fusion conditions when a mixture of deuterium and tritium is used as fuel. They all belong to the two main approaches to fusion: magnetic confinement fusion and inertial confinement fusion. Magnetic confinement fusion approaches can be divided into two categories: toroidal devices, which all employ a torus-like construction in which the plasma is trapped by toroidal and poloidal magnetic fields [(spherical) tokamaks, pinches, stellarators and variants] and linear mirror devices (discussed in Chap. 3). Inertial confinement fusion attempts consist of laser-driven (direct drive, indirect drive and fast ignition) and ion-beam driven contraptions and have been described in Chap. 12.

In Table 14.1 an attempt has been made to divide all fusion approaches into six groups with in the first and last column the two main approaches of magnetic and inertial confinement fusion. The middle four columns contain the non-mainstream approaches, of which the first concerns approaches that are purely magnetic confinement, while the approaches in the other three columns all borrow elements from the main approaches combined with unique features of their own. Of these, cusp confinement is most closely related to magnetic confinement (in particular mirrors)

L. J. Reinders, *The Fairy Tale of Nuclear Fusion*,
https://doi.org/10.1007/978-3-030-64344-7_14

Table 14.1 Table of Fusion approaches

Magnetic confinement Fusion	Non-mainstream fusion approaches				Inertial confinement fusion
	Magnetic fusion	Cusp confinement	Magneto-Inertial fusion	Inertial electrostatic fusion	

1. Toroidal devices
a. Tokamak
  – Spherical tokamak
b. Stellarator
  – Torsatron
  – Heliac
  – Heliotron
  – Helias
c. Pinches
  – Z Pinch
  – Theta Pinch
  – Screw Pinch
2. Mirrors
	a. Compact toroids/self-organised plasmas
	  – Spheromak
	  – Rotamak
	  – Reversed field pinch
	  – Dynomak
	b. Levitated dipole
	c. Pinches
	  – Dense Plasma Focus
		Lockheed Martin
		Biconic Cusp
		Picket Fence
		Tormac
		Surmac
		(Polywell)
			Plasma Liner Experiment
			Magnetized Liner Inertial
			Magnetized Target Fusion
				Periodically Oscillating Plasma Sphere
				Penning Trap
				Fusors
				Insulated Fusors
				Polywell
					1. Laser Fusion
					  – Direct Drive
					  – Indirect Drive
					  – Fast Ignition
					2. Ion Beam Fusion

and inertial electrostatic fusion to inertial confinement, with magneto-inertial fusion borrowing from both categories. Inertial electrostatic fusion is non-thermal, while the others are all thermal. The Polywell has close relations with both approaches and could be placed in either category, or in both. There are innumerable other interconnections between the various approaches which cannot be shown in such a simple table.

Apart from such alternative devices, there are also alternative fuels, different from the dominant D–D and D–T plasmas used in all the devices we have spoken of in earlier chapter. We will deal with these alternative fuels first before turning to non-mainstream devices.

14.2 Alternative Fuels

Another approach, and perhaps even more challenging than trying to construct alternative devices, consists in looking for clean and still efficient fusion power generation using other fuels. The D–T reaction is the least difficult fusion reaction as it can be achieved at the relatively low ion temperature of around 100 million degrees, all other combinations requiring much higher temperatures. In spite of this, D–D plasmas, instead of D–T, are used in test devices as they do not require the use of radioactive tritium, but all devices discussed in earlier chapters eventually intend to use D–T and the only devices ever having produced some fusion power, JET and TFTR, used D–T in these power producing operations. The D–T fusion reaction is however a terrible fusion reaction, apart from complications arising from the radioactivity of tritium, such as remote handling and other precautions, the D–T cycle has two further principal disadvantages: (i) certain sensitive parts of the reactor require extensive shielding against the very high-energy neutrons released in the fusion processes, while other parts are damaged and activated by these neutrons, severely shortening their lifetime and needing frequent replacement during which time the reactor will necessarily be shut down; (ii) the need to breed tritium in the reactor itself, which results in considerable extra complexity and cost, and requires space for a lithium blanket. The in situ breeding of tritium can also result in large onsite stocks of tritium (in the lithium blanket and tritium recovery system) raising both safety and nuclear proliferation concerns.

Let's go back for a moment to deuterium-deuterium. The fusion reactions are:

$$D + D \rightarrow {}^3He \ (0.82 \ MeV) + n \ (2.45 \ MeV) \ (half \ of \ the \ time)$$

$$\rightarrow T \ (1.01 \ MeV) + p \ (3.03 \ MeV) \ (half \ of \ the \ time),$$

As can be seen, the fusion products in these reactions (^{3}He and neutrons or tritium and protons p) are 5–6 times less energetic than the products (^{4}He and neutrons) of D–T fusion. The energy of the neutrons is 2.45 MeV compared to 14 MeV for the neutrons released in D–T fusion.

It has been proposed to recycle the tritium and ^{3}He produced in the D–D reactions (so-called catalysed D–D fusion). The tritium produced when the D–D reaction goes the second way will quickly undergo a secondary reaction with deuterium and produce a 14 MeV neutron, but there will be much fewer of them, so less radiation damage, than when D–T is used as fuel. And secondly the ^{3}He produced in the first primary reaction will also fuse with deuterium to produce a proton and ^{4}He. There would then be no need for tritium breeding, as the process is started with D–D, and there would be much fewer 14 MeV neutrons to harm the reactor, although the neutron flux would be augmented by the 2.45 MeV neutrons from one of the possible primary D–D reactions. It has been estimated that the first wall of the reactor, i.e. the shield positioned between the plasma and the magnets to protect outer vessel components from radiation damage, would only need replacement once in a 30-year lifetime. Forty per cent of the energy will come out as charged particles (p, tritium, ^{3}He and ^{4}He), which can keep the plasma hot and/or give up their energy electrically instead of through a thermal cycle (direct conversion). These advantages are however dwarfed by the fact that the D–D fusion rate is much lower than the D–T rate. Nonetheless, the private company Type One Energy wants to take this course as will be related in the next chapter.

Other light elements are also candidates for fusion and in principle there are some 100 possible fusion reactions. The most promising would be reactions that would not produce any (high-energy) neutrons (so called aneutronic reactions, as activation of the reactor structure would be avoided) or reactions that would only produce charged particles (which can be kept contained by the magnetic field). Some possibilities are:

Deuterium – Helium-3	$D + {}^3He \rightarrow {}^4He$ (3.6 MeV) + p (14.7 MeV)
Deuterium – Lithium-6	$D + {}^6Li \rightarrow 2\,{}^4He$ (22.4 MeV)
Proton – Lithium-6	$p + {}^6Li \rightarrow {}^4He$ (1.7 MeV) + ^3He (2.21 MeV)
Helium-3 – Lithium-6	$^3He + {}^6Li \rightarrow 2\,{}^4He + p + 16.9$ MeV
Helium-3 – Helium-3	$^3He + {}^3He \rightarrow {}^4He + 2p + 12.86$ MeV
Proton – Lithium-7	$p + {}^7Li \rightarrow 2\,{}^4He + 17.2$ MeV
Proton – Boron-11	$p + {}^{11}B \rightarrow 3\,{}^4He + 8.7$ MeV
Proton – Nitrogen	$p + {}^{15}N \rightarrow {}^{12}C + {}^4He + 5.0$ MeV
Tritium – Helium-3	$T + {}^3He \;\rightarrow {}^4He + p + n + 12.1$ MeV (51%)
	$\rightarrow {}^4He$(4.8 MeV) + D(9.5 MeV) (43%)
	$\rightarrow {}^5He$(2.4 MeV) + p(11.9 MeV) (6%)

Of these the most commonly considered are the D–^{3}He, T–^{3}He, p–^{6}Li and p–^{11}B reactions.

The helium-helium reaction is part of the proton-proton cycle taking place in the Sun, as can be seen in Fig. 1.7. Since ^{3}He has two charged particles (protons) in its nucleus, helium–helium fusion is not a practical possibility on Earth. The other reactions too involve particles with several protons in the nucleus (Li and B even have 4, respectively 5 protons), which makes fusion virtually impossible due to the much stronger Coulomb repulsion.

Hence, the greatest difficulty with these alternative reactions is that fusion rates are much lower than for D–D and especially for D–T, as can be seen for some reactions in Fig. 14.1, which shows that D–T fusion is by far the most favourable reaction. This is the principal reason why these reactions have so far not played any role in practice, with the exception of proton-boron fusion, which is still actively pursued, although with little success, as we will see later in this chapter. It indeed has several advantages, but the prohibitive disadvantage seems to be that the temperature for fusion to occur must be much higher (roughly 1 billion degrees, a factor ten higher than for the fusion of hydrogen and its isotopes). Also, when considering reactions with elements with Z above 2, such as lithium and boron, the number of protons and neutrons in the fusing nuclei is rather large and after fusion with a proton they can recombine in various ways, in other words there are competing reactions apart from the ones given above, or the main fuel can react with products of the fusion reaction itself, e.g. ^{6}Li and a proton fuse into ^{4}He and ^{3}He, after which the newly produced ^{3}He can react with still remaining ^{6}Li, regenerate the proton and leave in the end only ^{4}He particles. This possibility of multiple outcomes applies to most of the reactions given above. The more neutrons and protons there are available in the fusing nuclei, the more possible combinations in which they can recombine after the fusion process and the more possible outcomes.

So, unfortunately, for all these fuels the temperatures, the confinement properties and the general plasma conditions required to obtain breakeven (net fusion power exceeding plasma losses) are very severe and most likely prohibitive compared to a D–T burning plasma. The conclusion must be that fusion generators based on devices,

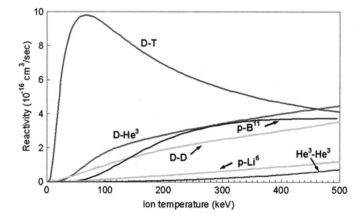

Fig. 14.1 Reactivity or reaction cross section (the average of the probability of a fusion reaction) versus ion temperature for some possible fusion reactions discussed in the text (*from* Chen 2011, p. 366; Dean 2013, p. 7)

such as tokamaks and stellarators, with plasmas in almost thermal equilibrium and using such alternative fuels, are probably not feasible.

Let's consider a little more closely the aneutronic D–^{3}He reaction. We have already encountered it above as a secondary reaction in catalysed D–D fusion. It is the first reaction in the above list and is often held to be the answer to the disadvantages of D–T fuels mentioned above. When using a deuterium-helium mixture as fuel, some neutrons will still be produced by parasitic D–D reactions, i.e. deuterium fusing with other deuterium ions instead of with helium, but the number will be small and could be minimized by a careful choice of fuel mixture—in particular a so-called 'D-lean' fuel mix (with more than 75% ^{3}He). As can be seen from the D–^{3}He reaction above, the fusion energy is released in the form of charged particles (the protons having as much energy as the neutrons in the D–T reaction, but being charged are kept under control by the magnetic fields). This would make it possible to use some form of 'direct conversion' of fusion energy into electricity (as with the charged particles in the catalysed D–D fusion discussed above) at higher efficiency than in a D–T reactor, where neutrons carry away 80% of the energy, which is deposited as heat. In the subsequent conversion of this heat into electricity 60% is typically lost due to the constraints imposed by thermodynamics. That is why it has been suggested that a commercial tokamak reactor using D–^{3}He might be competitive with one using D–T. The physics conditions for D–^{3}He are however much tougher (in order to overcome the Coulomb repulsion a plasma temperature of around 70 keV (a staggering 840 million degrees) is needed for this fuel mixture) and the reaction rate is much lower than for D–T fusion (Fig. 14.1). These drawbacks might be compensated by a reduction of the severe engineering difficulties for D–T reactors, as have become apparent e.g. for ITER. The lower reaction rate makes the confinement requirement for ignition in the D–^{3}He fuel cycle much higher (requiring a confinement triple product of $nT\,\tau_E = 2.4 \times 10^{23}$ m^{-3} keV s (about 100 times higher than the triple product for D–T fusion)). The fusion power density at fixed plasma pressure is also much lower (Nevins 1998).

In addition, an obvious and very serious limitation to the use of D–^{3}He is the lack of an adequate terrestrial source of ^{3}He. Lunar mining of ^{3}He or its manufacture in a D–D fusion reactor, which on itself is already an almost impossible undertaking, has been proposed as a lunatic solution to this problem. Lunar mining would require the development of an enormous industrial base on the Moon, which does not seem realistic at present or in the near future. A second possibility to produce helium would require not only D–^{3}He reactors but also D–D reactors, doubling the problems and losing the advantages as tritium (from which the ^{3}He would be extracted) and neutrons would also be produced. Careful investigations by several scientists into the use of D–^{3}He fuel have led to the carefully worded conclusion that "it seems unjustified to claim that aneutronic fuels and direct conversion offer comparable promise to conventional D–T fuelled magnetic confinement system" (Stott 2005; Rosenbluth and Hinton 1994). Taking in mind that the promise of D–T fuelled confinement systems is also very slim indeed, this assessment seems deadly.

Of the other aneutronic reactions the p–^{11}B reaction is the most attractive, although it has been concluded more than once in the literature that the fusion rate (see

Fig. 14.1) for p–^{11}B fusion is not large enough to achieve ignition because of electron bremsstrahlung losses (X-rays resulting from electrons scattering on ions) (Nevins 1998; Nevins and Swain 2000). When struck by a proton with energy of about 500 keV, the reaction produces three alpha particles with an energy of 8.7 MeV. A major advantage is that the reaction products are not reactive and only helium is produced. Since there are no neutrons, no shielding and no blankets would be required; no tritium breeding and/or recovery; no remote handling, while the reaction rate would be even better than for D–D, but still vastly lower than D–T, for ion temperatures above 150 keV. Below 100 keV the reaction rate is almost negligible even compared to D–D. In the process only hydrogen and boron are used, which are both plentiful on Earth, with ^{11}B being the main isotope of boron. All the energy comes out as charged helium particles which together carry away 8.7 MeV in energy. They can be slowed down with electric fields and produce electricity directly (Chen 2011, p. 367). Quite a number of advantages, so no wonder that it has always drawn a fair amount of attention.

This has recently been revived by the private company LPPFusion, where LPP stands for Lawrenceville Plasma Physics. Another company studying the possibilities of proton-boron fusion is TAE Technologies. We will discuss the efforts of these companies in Chap. 15, which is devoted to privately funded fusion research.

14.3 Non-mainstream Magnetic Confinement Devices

In any quest, even for fusion energy, Mother Nature sometimes seems to help. Certain plasma configurations exhibit properties known as "self-organization" where the plasma tends to converge to a certain optimum state; the plasma alters an externally applied magnetic field in a way that improves the confinement properties needed for fusion. Examples include the reversed-field pinch, the spheromak, the field-reversed configuration, included in the first column of non-mainstream approaches in Table 14.1, and magnetized target fusion. Tokamaks too show some features of plasma self-organisation, but still require external control.

Self-organisation is a generic process in nature, which describes the spontaneous formation of ordered structures. In the universe it is a common process for plasma and magnetic fields to evolve together in a turbulent way, but then to rapidly relax to simple, self-organised structures. The magnetic reconnection phenomena happening on the Sun and discussed in Chap. 11 are an example of this. Solar flares erupt from the photosphere tangled and chaotic but relax and straighten via magnetic reconnection with the release of huge quantities of energy. Several fusion devices are based on this self-organisation principle.

14.3.1 Reversed-Field Pinch (RFP)

A reversed-field pinch (RFP) is a toroidal pinch, like the ones used in the early days (see Chap. 3); a torus with a strong poloidal field that can pinch the plasma. It has comparable field strengths for the fields in the toroidal and poloidal directions, unlike the tokamak, which has a much larger magnetic field in the toroidal direction than the poloidal direction. The main idea of RFP is to have the toroidal magnetic field, so the magnetic field lines going around the torus the long way, run in one direction inside the plasma and in the opposite direction outside the plasma, giving rise to the term reversed field. This field is then complemented by the poloidal field produced by a toroidal plasma current. Such a configuration can be sustained with comparatively lower fields than a tokamak configuration of similar power density. A typical RFP has a field strength approximately one half to one tenth of that of a comparable tokamak. One of the disadvantages is that it tends to be more susceptible to non-linear effects and turbulence. The main problem is that plasma confinement in the best RFPs is only about 1% as good as in the best tokamaks. This makes them lousy candidates for fusion reactors, but useful systems for studying non-ideal magnetohydrodynamics (MHD).

The largest Reversed-Field Pinch device presently in operation is the RFX (Reversed Field Experiment) (aspect ratio $R/a = 2/0.46$) in Padua, Italy. Others include the MST (Madison Symmetric Torus) $(R/a = 1.5/0.5)$ in the United States, EXTRAP (External Ring Trap) T2R $(R/a = 1.24/0.18)$ in Sweden, RELAX (REversed field pinch of Low Aspect ratio eXperiment) $(R/a = 0.51/0.25)$ in Japan, and KTX (Keda Torus eXperiment) $(R/a = 1.4/0.4)$ in China.

14.3.2 Field-Reversed Configuration (FRC)

An FRC is an ultra compact axisymmetric toroidal configuration in which a plasma is confined on closed magnetic field lines, like tokamaks and stellarators, but without a central hole. There is no ohmic transformer passing through the centre of the device. FRCs have zero toroidal magnetic field everywhere and hence have no toroidal field coils. The plasma is confined solely by a poloidal field. The lack of a toroidal field means that the FRC has no magnetic helicity (no twisted field lines) and that it has a high *beta*. This makes the FRC in principle attractive as a fusion reactor and well-suited to aneutronic fuels because of the low required magnetic field. The basic configuration is shown in Fig. 14.2. If the plasma current is strong enough, the magnetic field direction along the axis will be reversed. A compact toroidal plasma configuration may result, but field reversal alone is not enough to ensure that the field lines are closed.

The FRC is an inherently pulsed device; one that is quite simple from the point of view of its technological structure. They were a major area of research in the 1960s and into the 1970s. The experiments so far have been small and had problems scaling

Fig. 14.2 Field-reversed configuration: a toroidal electric current is induced inside a cylindrical plasma, making a poloidal magnetic field, reversed in respect to the direction of an externally applied magnetic field (*from* Post 1976)

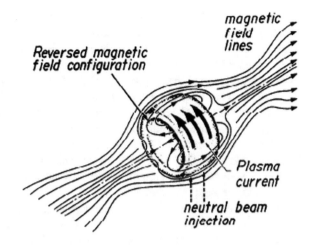

up to practical fusion triple products. Interest returned in the 1990s and as of 2019 FRC is an active research area. A large number of experiments with FRCs have been and are still being carried out. (See the lists in https://en.wikipedia.org/wiki/Field-reversed_configuration.) The configuration is strongly MHD unstable and must be carefully shaped to avoid instabilities. A number of such experiments have shown that FRC plasmas appear to be substantially more stable than predicted by ideal MHD, which was one of the reasons that the interest in FRCs remained large. In general, performance of FRCs was better than expected at small size and prognoses were worrisome for large size FRCs. This led to the suggestion that an FRC might serve as a good plasma source for magnetized target fusion, as we will see below (Friedberg 2007, p. 344ff).

FRC is closely related to a spheromak, with the important difference that a spheromak has an internal toroidal field while the FRC plasma has not. The plasma has the form of a self-stable torus, similar to a smoke ring. It is normally more elongated than a spheromak plasma, having the overall shape of a hollowed-out sausage rather than the roughly spherical spheromak (Dolan 2000, p. 296).

Private companies now study FRCs for electricity generation, including General Fusion, Tri-Alpha Energy (now TAE Technologies) and Helion Energy. We will come back to them in Chap. 15.

14.3.3 Spheromak

Spheromaks are FRC-like configurations with a finite toroidal magnetic field. These devices are sometimes also called magnetic vortices, magnetic smoke rings or plasmoids. The latter name, coined in 1956 by the American physicist Winston H. Bostick (1916–1991), is more commonly used to denote coherent structures of plasma and magnetic fields and would also include field-reversed configurations and dense plasma focuses. The choice of the name spheromak, proposed in 1978, specifically wants to stress its relationship with the tokamak, because of the similar torus-shaped plasma that eventually forms, but it is in fact more like a stabilized pinch configuration with a trapped toroidal field that is generated by plasma currents alone, complemented with an external field that is purely poloidal. With the spherical tokamak discussed in Chap. 11 it has a small aspect ratio in common and a short definition would be: a tight-aspect-ratio pinch with the toroidal field generated solely by plasma currents.

A spheromak is a toroidal plasma in a chamber without a hole in the middle. So, there is no central conductor and hence there cannot be any coils to generate a toroidal magnetic field. Thus, spheromaks manage to have a toroidal field (generated by the plasma currents) without having toroidal field coils and without an ohmic transformer through the middle.

This first of all allows for a compact design, a small aspect ratio. Unlike the two leading magnetic confinement concepts—the tokamak and stellarator—the dominant component of the magnetic field in a spheromak is generated by internal plasma currents, as opposed to external field coils, although there can be coils for generating a poloidal magnetic field. The device operates through an innovative procedure for driving steady-state currents coupled with an externally applied poloidal magnetic field (Fig. 14.3). As the current increases it eventually reaches a critical value above which the poloidal magnetic field lines break and reconnect forming closed flux surfaces. The current inside the closed surfaces produces ohmic heating, thereby forming the spheromak plasma (Friedberg 2007, p. 373).

Plasma with embedded fields is injected into the chamber from external sources e.g. by a plasma gun, and the device then produces a plasma in MHD equilibrium mainly through the self-induced plasma currents, as opposed to a tokamak device which depends on large externally generated magnetic fields.

The main point is that the force exerted by a current in a magnetic field is perpendicular to the magnetic field. These forces will move the plasma around until a certain balance has been reached, which is the case when the current is everywhere parallel

Fig. 14.3 Toroidal and poloidal magnetic fields in a spheromak (*from* Chen 2011, p. 380)

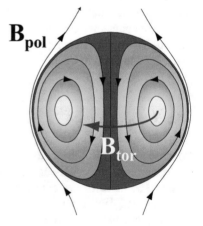

to the magnetic field, so that there is no perpendicular force (Chen 2011, p. 380). The advantages are a simple magnet set (circular coils); internal magnetic fields and plasma current, and a compact less expensive power plant (possibly 1/10 the size of a comparable tokamak).

One of the first spheromak experiments in the US was CTX (Compact Torus Experiment) which ran at Los Alamos until the 1980s. It was at the time the largest and most powerful device, generating spheromaks with currents of 1 MA, temperatures of 100 eV, and peak electron *betas* of over 20%. Another early device was the Spheromak Experiment (SPHEX) at the University of Manchester in the UK. Germany had an early spheromak experiment with the Heidelberg Spheromak Experiment (HSE) at Heidelberg University. The spheromak was converted into a spherical tokamak by inserting a central rod in the experiment. Through this a vacuum toroidal field was superimposed on the spheromak and a spherical torus was created by non-inductive current drive (Bruhns et al. 1987). It operated in this configuration for a very short time in 1986–1987, had an extremely low aspect ratio of 1.1 with a major radius of just 8 cm. In the Heidelberg device it was found that a small current in the rod stabilized the tilt instability of the spheromak.

Spheromaks had some successes during the 1970s and 1980s, but by the late 1980s the tokamak had surpassed the confinement times of the spheromaks by orders of magnitude. This was probably the reason that funding dried up when the entire US fusion programme was dramatically curtailed in fiscal year 1986, resulting in de defunding of most alternative approaches, and spheromak research was mostly abandoned. In the late 1990s research demonstrated that hotter spheromaks have better confinement times, which led to a second wave of spheromak machines.

At Lawrence Livermore National Laboratory the Sustained Spheromak Physics Experiment (SSPX) has been running since 1999 and at least until 2006, but no longer today, to investigate spheromak plasmas, with particular attention to energy confinement and magnetic fluctuations. The device is small with a major radius of 0.31 m and a minor radius of 0.17 m.

A second spheromak programme in the US is the Swarthmore Spheromak Experiment (SSX) at Swarthmore College in Pennsylvania, which started in 1994 and has now been running for 25 years. It studies fundamental plasma physics phenomena such as magnetic reconnection using rings of spheromak plasma. The aim of the experiment is to explore magneto-inertial fusion. Magneto-Inertial Fusion is an intermediary approach between inertial and magnetic confinement fusion in an attempt to lower the cost of fusion devices. It aims to generate small parcels of hot magnetic plasma and then compresses them, similar to the compression of petrol in a car engine. In SSX a volume of plasma is formed into a twisted structure using magnetic fields, accelerated to high speeds, then stagnated and compressed into a suitable fusion target.[1]

Although these programmes aim in the long run to build a nuclear fusion power plant, they reflect the attitude still current in the US to see fusion more as a scientific

---

[1] https://www.scientia.global/professor-michael-brown-literal-sun-jars-shrinking-stars-energy-pro duction/.

research subject than as a technology project for energy production, so that despite its promise as a confinement concept, the spheromak has only been explored at the levels of basic plasma science and concept exploration.

14.3.4   Dynomak

This purely research attitude certainly does not apply however to the dynomak, developed at the University of Washington in Seattle. It is a refinement of the spheromak design, and the alleged breakthrough was the experimental discovery in 2012 of a physical mechanism called Imposed-Dynamo Current Drive (IDCD) (Jarboe et al. 2012) (hence the name "dynomak").

The concept of dynamo action in nuclear fusion is, as is more often the case, borrowed from astrophysics. It is the mechanism by which a celestial body such as Earth or a star generates a magnetic field: currents of fluid metal in the Earth's outer core driven by heat flow from the inner core create circulating electric currents which in turn generate the magnetic field. So, dynamo action involves the self-generation of currents from fluid (or plasma) motion.

In the IDCD mechanism the fluctuations required for dynamo current drive are imposed, and not generated by the plasma. This is claimed to be a step forward towards achieving efficient current drive in magnetically confined plasmas without a central solenoid. The discovery was made on the spherical tokamak HIT-SI (Helicity Injected Torus with Steady Inductance) at the University of Washington. By injecting current directly into the plasma, IDCD lets the system control the fields that keep the plasma confined. The result is that steady-state fusion can be achieved in a relatively small and inexpensive reactor (Sutherland et al. 2014).

The dynomak has not yet been built, but grand plans for it have nevertheless been made. A virus that seems to have affected most fusion practitioners. It is claimed to be smaller, simpler, and cheaper to build than ITER, *and* will produce more power. Since the plasma heats itself, the fusion reaction is self-sustaining as excess heat is drawn off by a molten salt blanket to boil water to run a steam turbine to generate electricity. According to its engineers, a dynomak reactor would cost one tenth of ITER and produce five times as much energy at an efficiency of 40 percent. That means that a one gigawatt dynomak power plant would cost US\$2.7 billion (with total development costs including construction of the power plant amounting to US\$4 billion) against US\$2.8 billion for a coal plant.[2] Such a statement based on almost nothing is apparently enough for the press to proclaim "Fusion Reactor Could Be Cheaper than Coal".[3] How silly can you get!

Big plans were formulated with a full-scale prototype by 2030 and a commercial reactor of 1000 MW electricity production envisaged for 2040. At that time the comparison with coal will almost certainly and conveniently have been forgotten.

---

[2]https://newatlas.com/dynomak-fusion-reactor-university-washington/34174.
[3]https://www.sciencedaily.com/releases/2014/10/141008131156.htm.

The first step in the realization of this programme will be the HIT-SIX experiment to demonstrate IDCD at higher temperatures with sufficient confinement (Sutherland 2014). A private company, CTFusion[4], a spin-off from the University of Washington, was formed in 2015 to bring these plans to fruition, but so far not much has come out of them.

The greatest difference of these three configurations (RFP, FRC and spheromak) with tokamaks and stellarators is that they all have much smaller or zero toroidal fields. A possible reactor based on these concepts would lead to a compact, higher-power-density design reducing capital costs. The smaller toroidal field however results in poorer plasma performance. Transport losses are higher than for tokamaks and each configuration would be macroscopically unstable without the presence of a perfectly conducting wall near the plasma surface. The clear advantage compared to the tokamak is however that no high-field (superconducting) toroidal magnets are needed which in a tokamak are required to provide MHD stability, a problem that is even aggravated in the stellarator. On the other hand, the current drive problems for FRC, RFP and spheromak are more severe than in a tokamak. Each of the alternative concepts improves one aspect of the tokamak concept but suffers with respect to others, and therefore does not obviously offer better prospects.

14.3.5   Rotamak

In Australia so-called Rotamaks were operated from 1987 to 2001, first at Lucas Heights (Sydney) and later at Flinders University in Adelaide. In 2001 the Flinders machine was disassembled and moved to Prairie View A&M University in Texas. The Rotamak is a sort of hybrid device that can form and sustain a field-reversed configuration, in which case it does not have an external toroidal magnetic field. It can also function as a spherical tokamak configuration, when of course it does have such a field. It has a compact torus configuration with the distinctive feature that the toroidal plasma current is driven in a steady-state, non-inductive fashion by means of the application of an external rotating magnetic field (Jones et al. 1998).[5] The plasma current is kept in equilibrium by an externally applied "vertical" field. If the plasma current is large enough, the "vertical" field is reversed on the symmetry axis, resulting in a typical FRC. When a steady toroidal magnetic field is added, the Rotamak is able to operate as a spherical tokamak (Zhou et al. 2015; Petrov et al. 2010).

Research focuses on comparing FRC and spherical tokamak discharges with regard to the magnitude of a driven plasma current, studying active plasma shape control, and investigating instabilities. For the Prairie View device, the vacuum vessel

---

[4]https://ctfusion.net/, the first sentence you see on this website is "the cheapest and most reliable approach to fusion energy" (accessed 23 June 2020).

[5]https://www.pvamu.edu/pvso/research-and-education-activities/fusion-plasma-research-project/, accessed 16.10.2019.

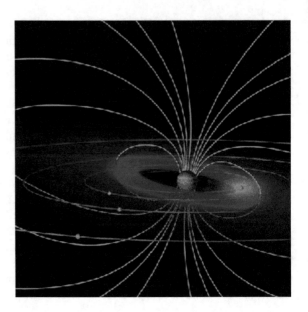

is completely different from an ordinary tokamak setup, made of Pyrex glass and 80 cm long and 40 cm in diameter. The electromagnetic coils can produce a magnetic field up to 0.023 T at the centre of the vessel. Another electromagnetic coil running through the axis of the vessel can produce the magnetic field necessary to make the apparatus into an ST. Some time between 2015 and 2017, all personnel left the project and there has been no activity since.

14.3.6   *Levitated Dipole*

The levitated dipole, which forms a separate category among the toroidal magnetic confinement devices, is a relatively new fusion concept, the first ideas dating from the middle or late 1990s. It was initially motivated by astrophysical observations, in particular the existence of a stable, long-lasting, plasma ring confined in the dipole magnetic field surrounding Jupiter, as shown in Fig. 14.4.

The magnetic dipole[6] field forms a shape with magnetic field lines passing through the planet's centre, reaching the surface near the poles and extending far into space above the equator. Charged particles entering the field will tend to follow the lines of force, moving north or south. As they reach the polar regions, the magnetic field lines begin to cluster together, and this increasing field can cause particles below a certain energy threshold to reflect and start travelling in the opposite direction. Such particles bounce back and forth between the poles until they collide with other

---

[6]A **dipole magnet** is a magnet with two poles (North and South poles) that form a closed field loop. The simplest example of a dipole magnet is a *bar magnet*.

particles. We have already discussed this phenomenon in Chap. 3 in connection with mirror devices (Fig. 3.10). The levitated dipole has thus great affinity with such mirrors. These mirrors used a solenoid to confine the plasma in the centre of a cylinder, and two magnets at either end to force the magnetic lines closer together to create reflecting areas (Fig. 3.9). In some sense the levitated dipole can be thought of as a toroidal mirror. The confinement area is not the linear area between the mirrors, but the toroidal area around the outside of the central magnet, similar to the area around the planet's equator. Particles in this area that move up or down see increasing magnetic density and tend to move back towards the equator area. This gives the system some level of natural stability. Particles with higher energy, the ones that would escape a traditional mirror, instead follow the field lines through the hollow centre of the magnet, recirculating back into the equatorial area (*Wikipedia*; Friedberg 2007, p. 335ff).

The goal of the Levitated Dipole Experiment (LDX), whose construction was completed at MIT in 2005, was to reproduce such stable, long-lasting plasma, on a laboratory scale on Earth. The magnetic configuration of the device basically consists of a single ring current that produces a dipole magnetic field. The coil current and plasma current are purely toroidal, producing a purely poloidal magnetic field. There are no toroidal fields. The plasma forms a hollow toroidal shell surrounding the dipole coil.

A critical feature of LDX is the need to levitate the coil in the vacuum chamber. The reason is that if any type of mechanical supports is used, the plasma must intersect these supports at some location. This would cool the plasma and heat the supports, an unacceptable situation in an experiment, and certainly in a reactor. In practice, long-time levitation of the coil is possible if it is constructed of a superconducting material, which adds to the technological complexity of the concept, but is feasible. A cross section of the experiment is shown in Fig. 14.5.

The advantage of LDX is that it possesses a very simple magnetic geometry, requiring only a single coil for plasma confinement. Also advantageous is the fact that complete ideal MHD stability can be achieved at high *beta*. The absence of any toroidal field coils is an important technological advantage compared to the tokamak.

There are two important disadvantages. First, there is the obvious problem of levitating the superconducting coil, a problem that becomes increasingly more challenging technologically as the experiments grow in size, and the plasma environment becomes more hostile (i.e., increased density and temperature). While levitation itself is feasible, the challenge is protection against failure. Second, for the coil to remain superconducting, it must be shielded from fusion by-products. The magnetic field can shield against charged particles, but not against the large number of high-energy neutrons produced by the D–T reaction. Thus, to reduce the number of high-energy neutrons, a levitated dipole reactor will preferably be based on one of the more difficult D–D or D-^{3}He fusion reactions (Garnier et al. 1999).

LDX was funded by the US Department of Energy and was run in a collaboration of MIT with Columbia University. Although it was successful in determining the underlying physics for plasma confinement in a dipole field and in establishing a levitated dipole as a unique approach to a fusion-based energy source, funding was

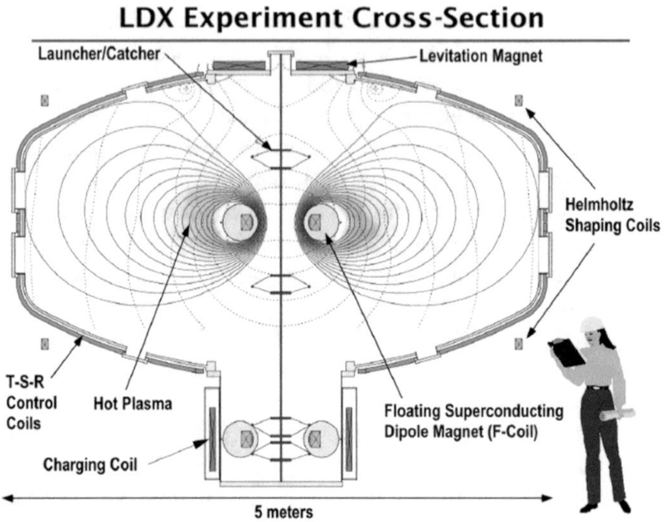

LDX Experiment Cross-Section

Fig. 14.5   Conceptual drawing of cross section of the LDX experiment (*from* LDX website)

ended and the effort stopped in November 2011, to concentrate resources on tokamak designs, so it became a victim of the ITER monster's insatiable appetite for cash.

The D20 dipole experiment at Berkeley and the RT-1 experiment at the University of Tokyo use similar designs.

14.3.7   Dense Plasma Focus (DPF)

The final concept among the non-mainstream magnetic fusion approaches that we have to mention briefly is the Dense Plasma Focus, also called *high-intensity plasma gun device* (HIPGD), or just *plasma gun*. In Table 14.1 it has been classified as a pinch device, as similar to pinch devices it runs large electrical currents through a gas to cause it to ionize into a plasma and then to pinch down on itself to increase the density and temperature of the plasma. It is one of the oldest devices invented to create fusion and was discovered in 1954 by Nikolai Filippov (1921–1998), who worked on early pinch machines in the USSR and noticed that certain arrangements of the electrodes and tube would cause the plasma to form into new shapes. The principle of a plasma focus device is schematically shown in Fig. 14.6.

The device consists of two cylindrical metal electrodes nested inside each other enclosed in a vacuum chamber. The inner cylinder, often solid (brass in the figure), is physically separated from the outer by an insulating disk (the Pyrex insulator) at one end of the device. It is left open at the other end. A low-pressure gas fills the space between the electrodes. When a high voltage is applied between the two electrodes, an intense current flows for a few millionths of a second from the outer to the inner

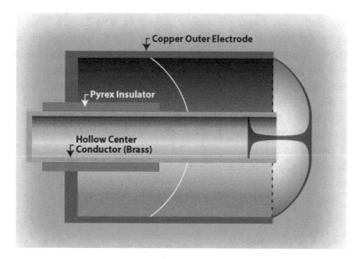

Fig. 14.6 Schematic picture of a dense plasma focus device (*from* www.plasma-universe.com)

electrode through the gas. This causes the gas in the area along the insulator to rapidly ionize and form a plasma, and current begins to flow through it to the outer electrode. The current creates a magnetic field that begins to push the plasma down the tube towards the open end. An ionization curve (shown by the white curve in Fig. 14.6) travels rapidly to the end at the right, which it reaches in microseconds. There the plasma sheet bows out into a shape not unlike an umbrella or the cap of a mushroom. At this point further movement stops, and the continuing current begins to pinch the section near the central electrode. Eventually this causes the former ring-shaped area to compress down into a vertical post extending off the end of the inner electrode (hence the name plasma gun). In this area the density is greatly increased and for a brief instant fusion conditions can be reached. The dense plasma column rapidly pinches, undergoes instabilities and breaks up. Intense electromagnetic radiation and particle bursts occur during the dense plasma and breakup phases. When operated using deuterium, intense bursts of X-rays and charged particles are emitted, as are nuclear fusion by-products, including neutrons (Chen 2011, p. 394–395; *Wikipedia*).

Several groups have claimed that DPF could prove viable for fusion power, even producing temperatures high enough for p-^{11}B fusion, and that the powerful magnetic field can reduce electron-ion collisions and thus reduce bremsstrahlung losses (radiation produced by charged particle collisions). But, both diagnostics and theory are difficult for DPF and understanding is not perfect. Another advantage claimed is the capability of direct conversion of the energy of the fusion products into electricity, with a potential efficiency above 70%. So far only minor experiments and computer simulations have been performed to investigate the DPF prospects for fusion power.[7]

DPF is however quite popular at present and a network of ten identical DPF machines operates in eight countries around the world, coordinated by the Asian

---

[7]https://www.plasma-universe.com/dense-plasma-focus/#DPF_for_nuclear_fusion_power.

African Association for Plasma Training (AAAPT). The International Centre for Dense Magnetised Plasmas (ICDMP) in Warsaw Poland, operates several plasma focus machines. Above we have already mentioned the private company LPPFusion as one of the companies intending to use hydrogen-boron as a fusion fuel. To this end they use the DPF concept (calling it Focus Fusion). In Chap. 15 we will pay attention to this company. According to the LPPFusion website[8] research with DPF devices is carried out by at least 44 groups around the world, with Iran hosting the most groups with eight operating in 2014. Many of the devices are relatively small. The largest facilities are PF-1000 at the ICDMP in Warsaw; PF-3 at the Kurchatov Institute, Moscow; KPF-4 in Sukhumi; Gemini at NSTec in Las Vegas; and FF-1 at LPPFusion.

14.4   Cusp Confinement

The idea of cusp confinement, so named because of the cusp shape of the curve in which opposing magnetic force lines meet, goes back to the early days of fusion. At the 1954 conference at Princeton, and especially as a result of Edward Teller's talk at that conference (see Chap. 3), it was concluded that any device with convex magnetic field lines towards the plasma would likely be unstable, resulting in kink instabilities and the like, and that a configuration with field lines curving away from the plasma should be inherently stable. Could adequate stability be achieved with any of the approaches then under consideration?

A plasma consists of a soup of moving charged particles, which due to this motion create their own magnetic field. In 1955, the American applied mathematician Harold Grad (1923–1986) had theorized that a high-*beta* plasma combined with a cusped magnetic field would improve plasma confinement. The internal magnetism of the plasma can be used to reject or block an applied outside cusped field and plug the cusps. Such a system would be a much better trap. The basic geometry of a **biconic cusp** is shown in Fig. 14.7a and a cross section of this configuration in Fig. 14.7b. The cusp magnetic fields are formed by two circular coils with opposite currents. It is a mirror-like setup with the fundamental difference that the currents have the opposite direction, while in a mirror they are in the same direction (see Fig. 3.9). The magnetic fields are blocked from entering the high-pressure plasma by the plasma's internal magnetic field and the plasma is kept confined in the middle. The plasma presses against the outside cusped magnetic fields. This means that the magnetic pressure is equal to the plasma pressure and *beta* will automatically be 100%. An ideal situation.

At the centre there is a null point in the magnetic field (a zero-field point). There are three classes of particles moving about in the plasma. The first class moves back and forth far away from the null point. Close to the poles of the electromagnets these particles are reflected as in a mirror. These are very stable particles, but their motion

---

[8]https://lppfusion.com/international-research-on-plasma-focus-devices/.

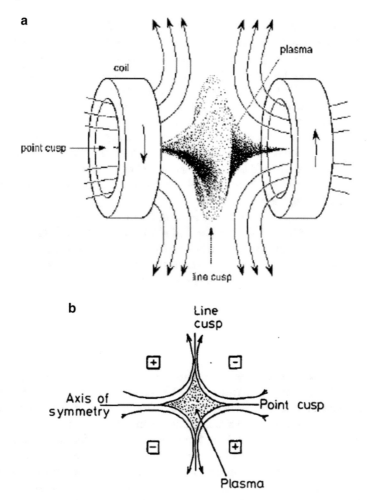

Fig. 14.7 a Cusped magnetic field, produced by two coils carrying currents in opposite directions (*from* Post 1987, p. 1601). **b** Cross section of the geometry shown in Fig. 14.7a (*from* Haines 1977)

changes as they radiate energy over time. This radiation loss arises from acceleration or deceleration by the field. The second type of particle moves close to the null point at the centre. For particles passing through locations with no magnetic field, the motion can be straight. This straight motion causes the particle to make a more erratic path through the fields. The third class of particles is a transition between these two types.

The important advantage of the cusped configuration is its stability due to the favourable curvature of the external magnetic field towards the confined plasma system in the centre.

The chief disadvantage is the large rate of particle loss, as can be imagined, by diffusion along the magnetic field; particles squeeze out between the magnetic force lines where they meet at the cusps. This loss increases with temperature. Nevertheless, the confinement of a plasma in magnetic fields with such a geometry is an improvement over the simpler magnetic mirror discussed in Chap. 3 (Haines 1977).

Papers on cusp confinement were already presented at the 2nd Geneva Conference in 1958 (Berkowitz et al. 1958) and it was fairly popular it seems in the 1970s. However, most cusped experiments failed and disappeared from national programmes by 1980.

Biconic cusps were recently revived by two companies using similar geometries for designs of fusion reactors, the polywell by the company EMC2 (Energy/Matter Conversion Corporation) and Lockheed Martin's Compact Fusion Reactor. We will discuss the polywell below under inertial electrostatic confinement and come back to these companies in the next chapter.

To confine a plasma with a magnetic field, it is not necessary to fill the entire volume with magnetic fields—a surface field would do. A surface magnetic confinement (**surmac**) configuration is characterized by a low magnetic field in the main volume and a high magnetic field on the surface. Such a configuration provides a natural magnetic well and a high-*beta* environment (Chen 1979).

14.4.1 Picket Fence

There exists a large variety of three-dimensional stable cusped configurations, suggested by the prototype depicted in Fig. 14.7. Various modifications were explored at Los Alamos and at New York University. One of them is the "picket-fence" consisting of current-carrying wires laid out in parallel in a planar array of equally spaced cusps, with current alternating from coil to coil. The coils are arranged such that the resulting magnetic field lines bend convexly away from the plasma at all points, thus giving a stable configuration according to Teller's criterion. To create high magnetic fields high currents are passed through the coils. Possible arrangements are shown in Fig. 14.8. Instead of a linear setup, a circular arrangement is also possible leading to a "ringed picket fence". The idea behind these arrangements is

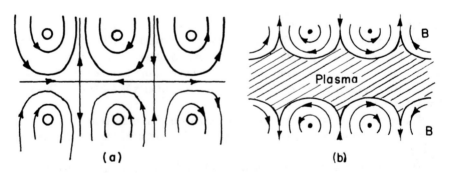

Fig. 14.8 Picket fence arrangements (*from* Berkowitz et al. 1958)

to stop the leakage of particles. As far as I can tell, this setup was the brainchild of James Tuck who already in 1954 had been examining calculations for the picket fence (Bromberg 1982, p. 58). Some time later he apparently said about this: "For the first time I see in this device faint glimmerings of a possibility of making a thermonuclear reactor" (Time Magazine 1960). He was not declared insane but kept busy working on equally unpromising projects.

Although particles losses were reduced, they remained the most important problem for these arrangements. It was calculated that one seventh to one tenth of the particles were lost in the time it took for an ion to bounce from one side of the plasma volume to the opposite side (Bromberg 1982, p. 60). In the meantime, the stability problems with other approaches were seemingly successfully tackled and there seemed little justification in continuing studies with cusp geometries.

14.4.2 Tormac

The Tormac is a concept that combines the favourable stability properties of a cusp geometry with the good particle confinement inherent to a closed field geometry. The advantage of this combination is that a Tormac can contain a high-*beta* plasma, near 85%, as in a mirror, and still have the low loss characteristics of a tokamak. Figure 14.9 shows its plasma and field configuration (Levine 1978; Brown et al. 1978).

The main (bulk) plasma is confined in a toroidal region of closed flux surfaces similar to a tokamak configuration. Completely surrounding this bulk toroidal plasma is a layer (sheath) of plasma confined on open magnetic field lines, containing both poloidal and toroidal field components. The sheath plasma is subject to loss by scattering collisions. A particle flow balance requires that this sheath loss rate be

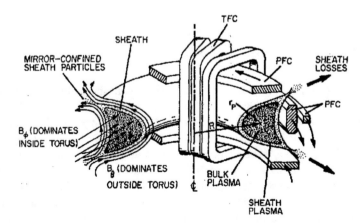

Fig. 14.9 Schematic representation of the Tormac plasma/field configuration (*from* Miller et al. 1979)

offset by diffusive flow of plasma from the bulk plasma into the sheath. The bulk plasma would be sustained by an external refuelling source in the assumed steady-state operating mode. Since the bulk plasma volume can be made large compared to the sheath plasma volume, the overall confinement time of the device can be enhanced over that predicted for simple mirror confinement. It promises high-*beta*, steady-state operation in a device with relaxed technological problems (Miller et al. 1979). Since the late 1970s nothing further has been heard about this arrangement.

14.5 Non-thermal

Since for most aneutronic fuels (such as D-^{3}He, p-^{6}Li, p-^{11}B) thermal plasmas seem to be ruled out, although not everyone is convinced of this, the possibility left is to consider plasma systems which are substantially out of (local) thermodynamic equilibrium, with the electron temperature much higher than the temperature of heavy species of particles (ions and neutrals). This concerns devices normally based on some sort of colliding beams. In all cases of non-thermal plasmas, some power should be spent to maintain the non-Maxwellian distributions of one or more species of particles. This is due to the fact that collisions by the particles will slowly relax their distribution towards a Maxwellian one. This power, which in general is relatively large, has to be added to the power needed to maintain the operating conditions. Thus, for such devices more input power is needed, making it more difficult to reach and exceed breakeven (Rider 1997; Santini 2006).

Quite big claims for non-thermal proton-boron fusion have quite recently been made by Heinrich Hora (b. 1931) and collaborators from the University of New South Wales in Sydney (Australia) (Hora et al. 2017). An Australian start-up company, HB11 Energy, holds the patents for this process and will be considered in the next chapter.

14.6 Inertial Electrostatic Fusion

Amasa Bishop in his 1958 book on *Project Sherwood*, the early American fusion effort says about confinement by electric fields (Bishop 1958, p. 15): "While not unequivocally ruled out, the use of electric fields for confinement does not appear to be feasible. The major deterring factor is that an electric field exerts oppositely directed forces on the electrons and positive ions of the plasma; if made to confine one component, the other would tend to escape. Another factor is the limited density achievable. Numerous proposals have been made for plasma confinement by electric fields, but none appears to have sufficient merit to warrant serious consideration."

He seems to be right, but in spite of this we will consider here its history and continuing fascination for a number of people. Inertial electrostatic fusion concerns fusion devices that use electric fields rather than magnetic fields to confine the plasma, an idea that was also proposed by Oleg Lavrentiev in his 1950 letter to the Central Committee of the Communist Party and led to Sakharov and Tamm proposing the tokamak, as related in Chap. 3, by chucking out the electric fields and replacing them by magnetic fields. Most inertial electrostatic fusion devices directly accelerate their fuel to fusion conditions, thereby avoiding the energy losses seen during the longer heating stages of magnetic fusion devices. In theory, this makes them more suitable for using alternative aneutronic fuels, which as we have seen offer a number of practical advantages.

When accelerated in an electric field, the negatively charged electrons and the positively charged ions in the plasma move in different directions and the field has to be arranged in some fashion so that the two species of particles remain close together. In most designs this is achieved by pulling the electrons or ions across a potential well, beyond which the potential drops and the particles continue to move due to their inertia. Fusion occurs in this lower-potential area when ions moving in different directions collide. The motion provided by the field creates the energy level needed for fusion, not random collisions with the rest of the fuel. Therefore, the bulk of the plasma does not have to be hot and the system as a whole works at much lower temperatures than magnetic fusion devices.

Some detailed theoretical studies (Rider 1997) have pointed out that the inertial electrostatic fusion approach is subject to a number of energy loss mechanisms that are not present if the fuel is evenly heated, i.e. in thermal equilibrium or Maxwellian. These loss mechanisms appear to be greater than the fusion rate in such devices, meaning they can never reach fusion breakeven and cannot be used for power production. These mechanisms are more powerful when the atomic mass Z of the fuel increases, which suggests inertial electrostatic fusion does neither have an advantage with aneutronic fuels. The assumptions made in these studies have been criticised and attempts are still being made by various (private) companies to develop inertial electrostatic fusion devices, but with scant success.

14.6.1 Fusor

The best known and one of the simplest and earliest inertial electrostatic fusion devices is the fusor. Its basic mechanism is explained in Fig. 14.10. It consists of two concentric metal wire spherical grids. The outer grid has a positive voltage compared to the inner grid. A fuel, typically, deuterium gas, is injected into the space between the two grids. When the grids are charged to a high voltage, this gas ionizes. The field between the two grids then accelerates the ions inwards and heats them to fusion conditions. When passing the inner grid, the field drops and the ions continue inwards towards the centre. If they impact with another ion, they may undergo fusion. If they do not, they travel out of the reaction area and again into the charged area, where they are re-accelerated inwards. Over time, a core of ionized gas can form inside the inner cage. Ions pass back and forth through the core until they strike either the grid or another nucleus and possibly fuse. Fusors are popular with amateurs, because they

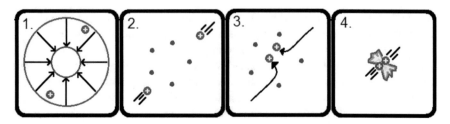

Fig. 14.10 Illustration of the basic mechanism of fusion in fusors. 1. The fusor contains two concentric wire cages. The cathode is inside the anode. 2. Positive ions are attracted to the inner cathode. They fall down the voltage drop. The electric field works on the ions heating them to fusion conditions. 3. The ions miss the inner cage. 4. The ions collide in the centre and may fuse. (*Wikipedia*)

are easy to construct, can regularly produce fusion and are a practical way to study nuclear physics, but need energy to work. They suffer from high conduction losses, and no fusor has come close to energy breakeven.

They were invented in the 1950s[9] by Philo Farnsworth (1906–1971), who also was the inventor of electronic television and apparently coined the name inertial electrostatic fusion.[10] In his work on vacuum tubes for television he observed that electric charge would accumulate in certain regions of the tube. Today, this effect is known as the multipactor effect. The multipactor was also an invention of Farnsworth and named as such by him. He filed a patent application for it in 1935. A particular variant of the multipactor featured two hemispherical electrodes that were placed near the perimeter of a glass sphere, and another electrode–a wire grid–was placed at the centre. When Farnsworth operated his device, he observed a tiny, star-like anomaly suspended within the inner grid at the centre of the tube. When increasing the power level, the point of light became brighter still and, even more impressively, it never touched the walls of the tube itself (Schatzkin 2002, p. 142). Apparently, without realizing it at the time, Farnsworth had created his first fusor. He came back to it in the early 1950s, when he was working for International Telephone & Telegraph, and reasoned that, if the concentration of ions were high enough, they could collide and fuse. In 1962, he filed a patent on a design using a positive inner cage to concentrate plasma, in order to achieve nuclear fusion. During this time, Robert L. Hirsch, whom we have met as head of the American fusion programme in the early 1970s, joined the company and began work on what became the fusor. Hirsch patented the design in 1966 and published it in 1967. The Hirsch machine was a 17.8 cm diameter device with a 150 kV voltage drop across it and used ion beams to help inject material.

Already in 1959 William Elmore and his collaborators, among whom James Tuck of Perhapsatron fame (see Chap. 3), discussed the device and concluded on theoretical

---

[9]A similar device was constructed at around the same time by W.H. Wells at the Bendix Aviation Corporation.

[10]Farnsworth's life and work has been described in Schatzkin 2002. On page 217 Schatzkin claims that Farnsworth called his approach "inertial electrostatic confinement". Two websites (http://fus or.net and http://farnovision.com/) are devoted to Farnsworth's legacy.

grounds that "[a]lthough it is of doubtful utility as a thermonuclear reactor, it may be possible to produce in this way small regions of thermonuclear plasma for study. The device appears to be unstable at economic densities" (Elmore et al. 1959).

Indeed, no fusor has come close to producing a significant amount of fusion power. A fusor can be built at home and produces fusion reactions, but hardly any power and must be connected to a power supply to work. "Plus, the fusor just looks totally cool. An eerie purple-blue glow emanates from the reactor, and a really well-made fusor can produce a mesmerizing phenomenon called a "star in a jar"." They can however be dangerous if no proper care is taken because they require high voltages and can produce harmful radiation (neutrons and X-rays).[11]

14.6.2 Periodically Oscillating Plasma Sphere (POPS)

As already noted above, theoretical studies indicated that non-thermal systems like the fusor cannot scale to net energy generating devices. However, these studies have several approximations, and a more complete study indicated that, if the ion distributions are close enough to thermal, net energy gains are possible, although the fusion power densities are small. The underlying problem is that for nonthermal systems, the Coulomb scattering cross sections are larger than the fusion cross sections. Thus, it can take more energy to maintain the nonthermal distributions than the device produces in fusion power. As a response to this, at Los Alamos National Laboratory a Periodically Oscillating Plasma Sphere (POPS) was built, a machine with a wire cage, in which ions are oscillating around (move at steady state). Such a plasma can be in a local thermodynamic equilibrium. The ion oscillation is predicted to maintain the equilibrium distribution of the ions at all times, which would eliminate any power loss due to Coulomb scattering, resulting in a net energy gain. This reactor concept becomes increasingly efficient as the size of the device shrinks (Park et al. 2005).

Several schemes, to be discussed below, attempt to combine magnetic confinement and electrostatic fields with inertial electrostatic confinement. The goal is to eliminate the inner wire cage of the fusor, and/or the problems that result from it.

14.6.3 Polywell

The polywell is a cross between a fusor and a magnetic mirror/magnetic cusp configuration, so it combines cusp confinement with inertial electrostatic fusion.

Electrons are confined magnetically by a cusp field while ions are confined by an electrostatic potential well produced by electron beam injection. The potential well accelerates ions to high energies for fusion and confines them. It was developed in the

---

[11] https://makezine.com/projects/make-36-boards/nuclear-fusor/ (23.12.2019). This website tells you how to make a fusor.

1980s by the American physicist Robert Bussard (1928–2007), who worked most of his life in nuclear rocket propulsion (Bussard 1991). In 1985 he founded the private company EMC2 (Energy/Matter Conversion Corporation) to develop the polywell. The main problem with the fusor is that the inner cage conducts away too much energy and mass. The solution, suggested by Bussard, was to replace the negative cage with a "virtual cathode" made of a cloud of electrons. As the name suggests, unlike the fusor's inner grid this virtual cathode has no physical barrier, meaning that ions are not lost through ion bombardment losses. This increases the polywell's efficiency by several orders of magnitude and has raised the hope for a net energy gain to be achieved.

A polywell consists of several parts placed inside a vacuum chamber. A set of electromagnetic coils arranged in a polyhedron generates a magnetic field that traps electrons. The most common arrangement is a six-sided cube. The six magnetic poles are pointing in the same direction towards the centre of the cube, where the magnetic field vanishes by symmetry, creating a null point. Electron guns shoot electrons into the centre of the ring structure. Once inside, the electrons are confined by the magnetic fields. Electrons that have enough energy to escape through the magnetic cusps can be re-attracted to the positive rings. They can slow down and return to the inside of the rings along the cusps. This reduces conduction losses and improves the overall performance of the device. The electrons act as a negative voltage drop attracting positive ions. This is a virtual cathode. The negative voltage attracts positive ions. As the ions accelerate towards the negative centre, their kinetic energy rises heating them to fusion conditions, and ions that collide at high enough energies can fuse. Ions are electrostatically confined raising the density and increasing the fusion rate, while the magnetic energy density required to confine electrons is far smaller than required to directly confine ions, as is done in other fusion projects such as tokamaks.

The use of an electron beam provides two critical advantages for the polywell reactor over other magnetic cusp devices. The excess electrons from the beam form an electrostatic potential well. By utilizing an electrostatic potential well, the polywell reactor is highly efficient in accelerating ions to high energies for fusion. In addition, the potential well reduces ion kinetic energy as the ions travel outward towards the cusp exits. This results in electrostatic confinement of ions as well as a reduction in the ion loss rate. In a polywell reactor, the main issues of high-temperature plasma containment and plasma heating are thus reduced to the confinement property of the injected electron beam (Park et al. 2015; *Wikipedia*).

Bussard called the type of confinement provided by a cusped magnetic field (Fig. 14.7a and b) wiffle-ball confinement.[12] He used the analogy to describe the trapping of electrons inside the field. Marbles can be trapped inside a wiffle ball; if they are put inside, they can roll and sometimes escape through the holes in the sphere. The magnetic topology of a high-*beta* polywell acts similarly with electrons. The confinement concept has been explained in Fig. 14.11.

---

[12]A wiffle ball is a hollow, perforated, light-weight, resilient plastic ball used in a variation of baseball, called wiffle ball and mainly played in the US.

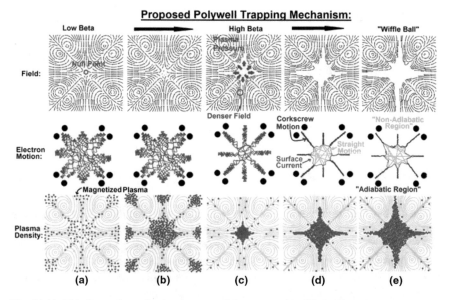

Fig. 14.11 This figure shows the development of the proposed "wiffle ball" confinement concept. The three rows of figures show the magnetic field, the electron motion and the plasma density inside the polywell. **a** The field is the superposition of six rings in a box. At the centre is a null point— a zone of no magnetic field. The plasma is magnetised, meaning that the plasma and magnetic field intermix. **b** As plasma is injected, the density rises. **c** As the plasma density rises, the plasma becomes more diamagnetic, causing it to reject the outside magnetic field. As the plasma presses outwards, the density of the surrounding magnetic field rises. This tightens the corkscrewing motion of the particles outside the centre. A sharp boundary is formed. A current is predicted to form on this boundary. **d** If the pressures find equilibrium at *beta* = 1, this determines the shape of the plasma cloud. **e** At the centre, there is no magnetic field from the rings. This means that its motion inside the field free radius should be relatively straight or ballistic. (*Wikipedia*)

In 2015 EMC2 published a paper (Park et al. 2015) providing evidence that the effect is real, based on X-ray measurements and magnetic flux measurements during its experiment. According to Bussard, the typical cusp leakage rate is such that an electron makes 5 to 8 passes before escaping through a cusp in a standard mirror confinement biconic cusp; 10–60 passes in a polywell under mirror confinement (low *beta*), which he called cusp confinement; and several thousand passes in wiffle-ball confinement (high *beta*).

In February 2013, Lockheed Martin Skunk Works (see next chapter) announced a new compact fusion machine, the high-*beta* fusion reactor, which is related to the biconic cusp and the polywell, and works at *beta* = 1. A third company working on a "polywell fusion reactor" is Progressive Fusion Solutions, based in Vancouver (Canada).

14.6.4   Insulated Fusors

Theoretical studies for a gridded inertial electrostatic confinement fusion system, like the fusor, have shown that a net energy gain is possible if the grid is magnetically shielded from ion impact (Hedditch et al. 2015). This might be a good idea but is unlikely to make the case for inertial electrostatic fusion. No experimental setups are known to date.

14.6.5   Penning Trap

Magnetic and electric fields can be used to increase the time that a charged particle (ion or electron) remains within a discharge. A particle can be trapped 'indefinitely' in a combined homogeneous magnetic field and electrostatic quadrupole field. Such ion traps were developed by the German-American physicist Hans Georg Dehmelt (1922–2017) and the German physicist Wolfgang Paul (1913–1993), who shared half of the 1989 Nobel Prize in physics for this work. Dehmelt named the trap after the Dutch experimental physicist Frans Penning (1894–1953), who studied low pressure gas discharges at the Philips Laboratory in Eindhoven.

Ion traps are used in mass spectroscopy, basic physics research, and controlling quantum states. The two most common types are the above-mentioned Penning trap, which forms a potential via a combination of electric and magnetic fields, and the Paul trap, which forms a potential via a combination of static and oscillating electric fields.

In a possible Penning-trap fusion reactor, first the magnetic and electric fields are turned on. Subsequently, electrons are emitted into the trap and caught. The electrons form a virtual electrode similar to that in a polywell, described above. These electrons are intended to then attract ions and accelerate them to fusion conditions. In the 1990s, researchers at Los Alamos National Laboratory built a Penning trap to do fusion experiments (Penning fusion experiment (PFX) (Mitchell et al. 1997).

14.7   Magneto-Inertial Fusion (MIF)

Before proceeding to discuss specific devices let us first elucidate the general concepts of magnetized target fusion (MTF) and MIF. These two concepts are not completely identical, although they are often used interchangeably. The term MIF encompasses a wider variety of arrangements. That is why it heads the column in Table 14.1 and is MTF classified as a form of MIF, but there is quite some arbitrariness in such classification. They are similar as both combine features of the two main approaches to fusion: magnetic confinement fusion and inertial confinement fusion.

The essential ideas behind MIF have existed for a long time. It involves the creation of a small hot magnetized plasma target and the subsequent rapid implosion as in inertial confinement fusion. During the implosion the magnetic flux is compressed and the intensity of the magnetic field increases. The intense magnetic field suppresses cross-field thermal diffusion in the plasma during the compression, and thus facilitates the heating of the plasma to thermonuclear fusion temperatures. The parameter space for MIF is between conventional magnetic confinement and inertial confinement, in particular, higher densities than a tokamak, but lower than inertial confinement. The mechanism to create fusion power is however the same as in inertial confinement fusion, for which reason I consider this approach closer to inertial confinement than magnetic confinement. MIF uses magnetic fields to confine an initial warm, low-density plasma, then compresses it to fusion conditions using an impulsive driver or "liner." These approaches were mainly developed in an attempt to lower the cost of fusion devices.

There are two main classes of MIF, the class of high-gain MIF and the class of low-to-intermediate gain MIF. An example of high-gain MIF is magnetized liner inertial fusion (MagLIF), to be discussed below. Both attempt to make use of a strong magnetic field in the target to suppress electron thermal transport in the target and thus rely upon the same scientific knowledge base of the underlying plasma physics. However, their strategies for addressing the two challenges of inertial fusion energy, suitable targets and drivers, are different. In the US, MIF is currently being pursued as a science-oriented research programme in high energy density laboratory plasma (HED-LP). Solid and liquid shells have been proposed as liners for compressing various types of magnetized target plasma for low gain MIF, which aims at a fusion gain in the range of 10-30. MIF is a pathway to create and study dense plasmas in ultrahigh magnetic fields (Thio 2008).

14.7.1 Magnetized Target Fusion (MTF)

Magnetized target fusion aims for a plasma density of 10^{25} m^{-3}, intermediate between magnetic confinement fusion (10^{20} m^{-3}) and inertial confinement fusion (10^{31} m^{-3}). At this density, confinement times must be of the order of 1 μs, again intermediate between the two main approaches. Magnetized target fusion uses magnetic fields to slow down plasma losses, and inertial compression is used to heat the plasma.

In the early 1970s a series of experiments were carried out within the framework of the LINUS fusion power project at the United States Naval Research Laboratory. The goal of the project was to produce a controlled fusion reaction by compressing plasma inside metal liners. The reactor design was based on the mechanical compression of magnetic flux (and therefore plasma) inside a molten metal liner. A chamber was filled with molten metal and rotated along one axis. This spinning motion created a cylindrical cavity into which plasma was injected. Once the plasma was contained within the cavity, the liquid metal wall was rapidly compressed raising the temperature and density of the trapped plasma to fusion conditions. Several experimental machines

(SUZY II, Linus-0, HELIUS) were constructed throughout the LINUS project, to gather data and demonstrate various aspects of the system concept. Experiments were largely unsuccessful due in part to delays incurred in the design, fabrication, and assembly phases. The machines were eventually disassembled and placed in storage. One major problem that LINUS encountered was the occurrence of Rayleigh-Taylor instabilities in the liquid liner, if the liquid wasn't uniformly compressed.

These concepts were recently revived by the Canadian private company General Fusion[13] and by Compact Fusion Systems in Santa Fe (New Mexico), which works on the technical development of a Fusion Power Core with commercial availability claimed to start in 2030 (see next chapter).

14.7.2   Magnetized Liner Inertial Fusion (MagLIF)

MagLIF is a very recent method of producing controlled fusion. It is very close to inertial confinement fusion as it uses the inward movement of the fusion fuel to reach densities and temperatures where fusion reactions can take place. Inertial confinement experiments usually use laser drivers to reach fusion conditions, whereas MagLIF uses a combination of lasers for heating and Z-pinch for compression. This latter aspect, in combination with the use of a powerful magnetic field to inhibit thermal conduction and contain the plasma, represents its connection to magnetic confinement fusion. A variety of theoretical considerations suggest such a system will reach the required conditions for fusion with a machine of significantly less complexity than the pure-laser approach. In Chap. 12 we saw that the billion-dollar National Ignition Facility at Lawrence Livermore National Laboratory, which used lasers to compress the fuel, failed to reach its goal of ignition by the 2012 deadline. The first MagLIF experiments have now started and the eventual goal is to create a burning plasma, i.e. ignition.[14] The setup of the experiment is shown in Fig. 14.12.

The fuel (deuterium or deuterium/tritium) is contained in a metal cylinder called a liner (like the hohlraum of Chap. 12), which is magnetized by external field coils. A 100-nanosecond pulse of electricity is run through this cylinder creating an intense Z-pinch magnetic field that crushes the fuel in the liner inwards. Just before the cylinder implodes, a laser is used to preheat the fusion fuel held within the cylinder and contained by the magnetic field. The potential of this method is being explored by Sandia National Labs. For this it uses the Z-machine (short for Z pulsed power facility), the largest high frequency electromagnetic wave generator in the world and designed to test materials in conditions of extreme temperature and pressure (*Wikipedia*).

A computer simulation published in 2012 showed the prospect of a spectacular energy return of 1000 times the expended energy. A 60 MA facility would produce a 100 times yield. The currently available Z-machine facility at Sandia is capable of

---

[13] https://generalfusion.com/.

[14] https://www.sandia.gov/z-machine/research/fusion.html (23.12.2019).

Fig. 14.12 Setup of the liner for the Sandia MagLIF experiment. It uses a heating laser, a stabilising magnetic field and a Z-pinch magnetic field to implode a cylinder of hydrogen fuel. (*Source* Sandia Nat. Lab.)

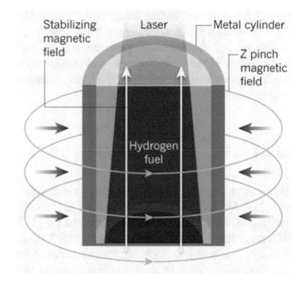

27 MA and may be capable of producing slightly more than breakeven energy (Slutz and Vesey 2012). Breakeven experiments with D-T fuel were planned for 2018, but this deadline was not met.

It has now been established that the liner will not break apart too quickly under the intense energy, which was the biggest concern regarding MagLIF following its initial proposal, that laser preheating is able to correctly heat the fuel and that the magnetic fields generated by a pair of coils above and below the hohlraum can serve to trap the preheated fusion fuel and importantly inhibit thermal conduction without causing the target to buckle prematurely.

Following these experiments, an integrated test started in November 2013. The test yielded about 10^{10} high-energy neutrons. With the aim to achieve scientific breakeven, the facility has gone through a 5-year upgrade to achieve a magnetic field of 30 T, 8 kJ power for the laser, the full power of 27 MA of the Z-machine and D-T fuel handling. In 2019, after encountering significant problems related to mixing of imploding foil with fuel and helical instability of plasma, the tests yielded up to 3.2 $\times 10^{12}$ neutrons.

These results obtained so far represent the first significant experimental evidence to support the claim that intermediate-density magnetized fusion approaches could be significantly lower in cost than magnetic confinement fusion or inertial confinement fusion (Nehl et al. 2019).

With a budget of $5 million MagLIF is a small and cheap experiment compared to the $3.5 billion National Ignition Facility and the €20 billion plus ITER experiment. (Sandia has about $80 million to operate the Z-machine each year, but it also serves other experiments in addition to MagLIF).

Apart from Sandia, MagLIF experiments are currently being carried out at the University of Rochester, using the OMEGA laser system, which is one of the most powerful and highest-energy lasers in the world. The target used will be roughly a factor 10 smaller than used at Sandia. These smaller targets will not achieve magnetic confinement of charged fusion products, and will suffer from greater thermal losses and a greater reduction in magnetic-field compression, but provide the first experimental data on MagLIF scaling and a higher shot rate with better diagnostic access than at Sandia (Davies et al. 2017).

14.7.3 Plasma Liner Experiment (PLX)

The Plasma Liner experiment (PLX) is a magneto-inertial fusion experiment going on at Los Alamos National Laboratory, in collaboration with a half dozen other US institutions. Like the other MIF approaches, the PLX machine combines aspects of both magnetic confinement fusion and inertial confinement. The hybrid approach, although less technologically mature than pure magnetic or inertial confinement concepts, may offer a cheaper and less complex fusion reactor development path, but will likely come too late to divert the world from the route taken, unless it will soon be seen that this route will lead to failure. Like in tokamaks, the fuel plasma is magnetized to help mitigate losses of particles and thermal energy. Like in inertial confinement machines, a heavy imploding shell (the plasma liner) rapidly compresses and heats the fuel to achieve fusion conditions. Instead of high-power lasers driving a solid capsule, PLX relies on supersonic plasma jets fired from plasma guns.

PLX has an additional advantage. Because the fusion fuel and liner are initially injected as a gas, and the plasma guns are located relatively far from the imploding fuel, the machine can be fired rapidly without damage to the machine components or the need for replacement of costly machined targets.

Assembly of PLX at Los Alamos National Laboratory is well underway with the installation of 18 of 36 plasma guns. The plasma guns are mounted on a spherical chamber, and fire supersonic jets of ionized gas inward to compress and heat a central gas target that serves as fusion fuel. (Hsu and Thio 2018).[15]

14.8   Conclusion

This finishes the discussion of alternative non-mainstream fusion approaches. It is difficult to judge the merits of most of these ideas; too few experiments have so far been carried out and no consensus on their (lack of) value has yet been reached. Each has its own idiosyncrasies, advantages and disadvantages, but their development lags

---

[15] https://www.sciencedaily.com/releases/2019/10/191021082828.htm (23.12.2019).

so far behind the tokamak (and even the stellarator) that they are unlikely to occupy the main stage any time soon, even if the tokamak were abandoned. In the next chapter we will pay attention to a few private companies that have recently revived some of the ideas considered in this chapter.

Chapter 15
Privately Funded Research into Fusion

15.1 Introduction

The current situation in nuclear fusion research, contrary to the first decades of the fusion effort, is that most non-mainstream fusion approaches, both as regards devices and alternative fuels, are privately funded. Virtually all public money (globally about $2 billion per year) goes into tokamaks, especially into cash-hungry ITER and other toroidal devices that can show to have some 'ITER-relevance'. Consequently, the opportunities to develop new ideas and start alternative projects in fusion energy research at national and/or university laboratories are dwindling. In view of the fact that it apparently seems to take forever for fusion to become a reality, it is not surprising that governments are not rushing to fund all kinds of wild, or even sensible ideas or projects in this field. So, venture capital can play a useful role here at no extra cost to the taxpayer. For venture capitalists it is just a wager, a little money invested in a project that has little chance of working, but promises a huge payoff if it does. The chance of success should not be zero, though, else it would just be foolish to invest.

The business model of the private companies that are working in the fusion field is to do good research with less overhead than universities or national labs and with the bold intention to produce commercial fusion power within a couple of decades at the most, while drawing on contributions from venture capitalists and at the same time competing for the same government grants, e.g. from the US Department of Energy, taking advantage of money earmarked for small businesses. Quite a few of these companies are spin-offs from universities or national labs, using ideas that have come to fruition at these labs and now need funding to test them in the marketplace. Innovative technologies can only have an impact on society if they emerge from the lab and enter the marketplace. Governments and universities fund basic research but stop well short of the level of development required for commercialization. Industry and investors are reluctant to pick up that technology when the pay-off is longer than a few years and when substantial technical and economic factors are unresolved. This

L. J. Reinders, *The Fairy Tale of Nuclear Fusion*,
https://doi.org/10.1007/978-3-030-64344-7_15

is what MIT calls the "valley of death".[1] Having ideas that are technically sound is not enough—there must be a viable pathway to widespread, profitable deployment in order to succeed. Technologies, like fusion energy, that require substantial investments and take many years to reach fruition face a particularly deep and broad valley. It is claimed that investments in fusion by the US government have been incommensurate with its potential to meet the existential threat posed by climate change, even in its funding of basic research on fusion science and technology. That claim is not correct, if only because the potential of fusion to meet this threat has by no means been proved, in spite of all the funding that the US government has sunk into fusion research in the last 70 years. That funding has gone on for far too long and now it is time for private capital to waste some of its money on this impossible venture. On a path determined by current levels of federal research funding, fusion energy would be unlikely to come on stream much before the end of this century. The new model is to employ private companies stuffed with private capital and academics full of expertise gained at government and university labs—reaching across the valley of death. This new model seeks a much more rapid development path, claiming even that it could develop fusion power in time to address the growing threat that humanity faces from climate change.

Most of these companies have sprung up in the fusion landscape in North America, especially in the United States, where the surroundings of Seattle in Washington State are especially well represented. The reason probably is that funding for fusion research at universities and national labs has been squeezed more in the US than in Europe and that fusion was traditionally not seen as a practical source of energy, but mostly as a research project. In Europe there are far fewer private fusion companies. There are some in the UK (an example is Tokamak Energy which we met in Chap. 11) and there are one or two on the continent. A single Australian and one Chinese company complete the picture.

As is common in fusion land, newspapers have already been raising overblown expectations on what these companies will achieve. 'Limitless' and 'inexhaustible' in respect of the promised energy are favourite epithets in this respect. The Guardian Weekly newspaper of 16 March 2018, for instance, announced 'fusion power to be on grid in 15 years', when describing the collaboration of scientists at MIT with the private company Commonwealth Fusion Systems performing experiments with a tokamak, which is going to make its benevolent appearance just 'in time to combat climate change'. What a relief that we no longer have to worry about that. There can be no doubt though that the best contribution to combating climate change from these endeavours would be to immediately stop the experiments, since their carbon footprint is just gigantic. There are stories of this kind for almost all the alternative approaches tried by private companies. They have all been tried in the past and judged to be lacking promise and now suddenly there are solutions all over the place. These start-ups all want fusion quicker and cheaper (who wouldn't?) and go for small devices, much smaller than the ITER colossus, which by the way is not even a

---

[1] Brochure SPARC–The high Field Path to Fusion https://www.psfc.mit.edu/files/psfc/imce/research/topics/sparc/MITSPARCbrochure.pdf.

power plant. Spherical tokamaks and/or high-temperature superconducting magnets among other things must do the trick. They all have several features in common. The first one is that none has ever produced a single watt of power from nuclear fusion reactions, but they still claim to have 'power on the grid' within 10 to 20 years or even less, and secondly that they all claim to be ahead of one another and of all other fusion energy technologies including the ones that actually have produced fusion reactions. The peculiar thing about all this is that the press is just printing these silly claims without asking any further questions. History is repeating itself where their websites dangle in front of our eyes the prospect of working fusion power plants within an impossibly short term, while at the same time casually mentioning a solution to climate change. This can never hurt as it may entice panicky governments in loosening the purse strings and dole out some taxpayers' money as a bonus. The phrases 'virtually unlimited energy' or 'inexhaustible source of energy' are also useful ones in this respect, but a bit stale as they have been fulfilling their function for close to seventy years by now. And all this while nothing fundamental has changed, and there is no reason whatsoever to think that they will succeed where others have failed for so long. Would the plasma suddenly stop its unruly behaviour because of a relatively small change in the geometry of the torus (more spherical) and start fusing like hell in an orderly fashion and/or click into place at the command of a set of more powerful magnets? Some say that advances in supercomputing and complex modelling are the breakthrough that will do the trick. Modelling plasma behaviour and using supercomputers to grind through the hideously complex models can indeed help to bring within reach long overdue knowledge on plasmas and can speed up the design process, but it cannot do much else. It doesn't change the fundamentals. It is very unlikely in my view that it will help much. An ELM or two will most probably be enough to teach the necessary lesson.

But whatever the motivation, fact is that private capital invested in this kind of projects has soared in the last decade with over $1bn of private venture capital put into fusion projects in the United States alone. This money is also encroaching on mainstream fusion projects, while the tendency of diverting all public funding into the mainstream tokamak approach seems to be changing somewhat; the major example being the UK government investing quite heavily in a spherical tokamak programme, called STEP. In Chap. 11 we have already considered private efforts in spherical tokamaks, especially those made by Tokamak Energy in Britain, whose expertise mainly derives from its scientists having previously worked at JET and other tokamaks at Culham. Although Tokamak Energy with its spherical device can still be considered non-mainstream, albeit just marginally, the private company Commonwealth Fusion Systems, founded in 2017 as a spin-off from MIT, uses a conventional tokamak design with high-temperature superconductors to take the fast route to fusion. Here, in this chapter, we will first pay some more attention to this company and then to the various non-mainstream approaches to nuclear fusion embarked on by private companies with alternative devices. The names of many of these companies have already been dropped in the previous chapter.

15.2 Commonwealth Fusion Systems (CFS)

CFS is a spin-off from MIT and headed by former students and scientists of the MIT Plasma Science and Fusion Center. Like MIT it is based in Cambridge, Massachusetts, and it funds $30 million of research at MIT. MIT's motivation to follow the route of collaboration with private capital has been formulated in its SPARC brochure and has to do with "reaching across the valley of death" as explained above. It is the private company Commonwealth Fusion Systems that collects money from venture capitalists, who apparently value its commercial potential, and uses this money to pay for research carried out at MIT. Over the long term, CFS seeks to commercialize fusion and lead the world's fusion energy industry. So, a grand vision indeed. For the time being, MIT and CFS will collaborate on the research and development of high-temperature superconducting (HTS) magnets and on the SPARC experiment.

In 2018, the year of its launch, CFS announced that it had obtained $50 million in funding from the Italian energy company ENI. In 2019 it raised $115 million from a number of venture capital funds, and in May 2020 a further $84 million.[2] This is not such a big surprise, given that CFS is (and will continue to be) one of the heavyweights in fusion start-ups. It has a large team of more than 60 people; MITs' full weight and by extension the rest of Boston is supporting them. The Alcator C-MOD Tokamak has a long and rich scientific history and when tokamaks are combined with advances in superconductors, it could be a potent mix.

Its timeline is to produce net energy by 2025 and generate power on the electrical grid by 2036, the magical two decades from its inception date. It is unimaginable that they can adhere to that timeline, considering that currently, early in 2020, they are still developing and testing the HTS magnets and no plasma has been produced yet. Even a coal plant would have difficulty meeting that deadline.

In collaboration with MIT the SPARC demonstration plant is currently being built. It is claimed to be the world's first fusion device that produces plasmas which generate more energy than they consume, becoming the first net-energy fusion machine. The SPARC goal is to exceed $Q = 2$, in a significantly smaller device than JET. Whether that means net energy, depends on what its energy consumption will be and what is included on the energy balance sheet. SPARC will have a plasma volume of 15 m^3, the same as a mid-sized fusion experiment—similar to many machines already in operation. It is the new HTS magnet technology that must do the trick. These magnets can be run at much higher field strengths, roughly doubling the magnetic force on the fuel. In this respect CFS follows the same path as Tokamak Energy. Like the Tokamak Energy machines, the CFS machines are high-field, high-density devices. There is no experience with such devices, and it would be very surprising indeed if the plasma would not come up with some surprises.

SPARC is claimed to be about the minimum size experiment that could make a net-energy plasma using this magnet technology. While the SPARC magnets will be superconducting and run in steady state, the plasma will be pulsed—lasting about

[2]https://cfs.energy/.

10 s—to simplify many aspects of the device. This pulse length is long enough for all plasma-related processes to reach steady state.

CFS estimates that its burning-plasma device, SPARC, will cost around $400 million, which is just small change compared to the more than $20 billion construction costs of ITER. But such a comparison is of course not fair, as ITER intends to achieve Q > 10 and is based on old technology, in particular ordinary instead of high-temperature superconductors. If ITER had to be designed now, it would be a completely different device. A possible successor to ITER will consequently also be vastly different.

The development of SPARC heavily relies on data from decades of research on dozens of experiments around the world, including data compilation and analysis during the ITER design stage. This gives CFS a major advantage in the early development stages, making fast progress possible so long as no new ground has to be covered. At a certain point it will enter uncharted territory and things might start to look less rosy. It would be the first nuclear fusion device that does not run into unexpected complications.

Following the SPARC demonstration, which is scheduled to start in 2021, CFS will construct the "world's first fusion power plant", based on the ARC (Sorbom et al. 2015) tokamak concept. ARC (short for affordable, robust, compact) is a theoretical design for a compact fusion reactor developed by the MIT Plasma Science and Fusion Center and has a conventional advanced tokamak layout. It is not a spherical tokamak, although the design is compact. The advanced tokamak aims to increase tokamak performance by using active control of the detailed cross section shape (D-shape) and internal profiles to simultaneously optimize both *beta* and confinement in a manner consistent with steady-state operation. JET and ITER, for that matter, are also advanced tokamaks. The ARC design also uses HTS magnets of course.

ARC is a tokamak with a physical size slightly larger than any of the big tokamaks (JET, TFTR, JT-60), but with a magnetic field that is two to three times larger. It is designed to produce ~200–250 MW in electricity (525 MW of thermal fusion power). ARC is significantly smaller in size and thermal output than most current reactor designs, which typically want to generate ~1 GW in electricity.

The HTS technology is key to CFS's approach. The claim is that "advances in superconducting magnets have put fusion energy potentially within reach", so without these HTS no fusion, and with HTS only potentially so. It is always sensible to hedge one's bets. ITER plans to use low-temperature superconducting (LTS) magnets (that have to be cooled to −269 °C) and such magnets have been tried on tokamaks since 1979, when the Russian T-7 was the first all-superconducting tokamak, followed more recently by the Japanese advanced superconducting tokamak JT-60SA. In these cases superconducting magnets were included to make steady-state operation possible. Now the most important advantage is that ReBCO superconductors can sustain much stronger magnetic fields. The confinement time for a particle in a plasma varies with the square of the size of the machine and the fourth power of the magnetic field, so doubling the field offers a four times larger performance for a same size machine. The smaller size reduces construction costs, although this is offset to some degree by the expense of the ReBCO magnets.

The second important point is that conventional LTS magnets as used in ITER would need thick shielding (of about 1 m) to prevent the high-energy neutrons produced in the fusion reactions from heating the superconductors. HTS magnets would potentially provide a solution for this and ReBCO HTS are especially promising in this respect. ReBCO becomes superconducting at temperatures of around 90–100 °K (−183 to 173 °C), so quite a difference, although there is still a fair amount of shielding needed, reason why ARC cannot utilise an ST design.[3] But it means that liquid neon, hydrogen or even nitrogen can be used for cooling rather than liquid helium, making cryogenics simpler and cheaper.

A third important and desirable point of ARC is that power is produced in a much smaller device than in reactors of the same output, like ITER. But it also means that more power is confined in a smaller space and more heat must be get rid of, else the machine would tear itself apart.

We will have to wait and see if CFS can make good on its promises.

15.3   Privately Funded Non-mainstream Approaches

Several times in the previous chapter we mentioned companies that try to exploit non-mainstream approaches, already tried and discarded in the past, for making fusion work in a *commercially* viable way. One of the surprising things in all these new attempts is that they immediately go for *commercial* energy production, as if net energy production by nuclear fusion as such is a well-established process that has been well and truly proved. Suddenly, within the last twenty years or so, there seem to have sprung up dozens of promising, fast routes to *commercial* energy production from nuclear fusion where before there were none and all the efforts of the last 70 years were essentially in vain. It just does not make sense.

Worldwide there are now a few dozen such companies, taking on the challenge in which public funded endeavours have so miserably failed. Some are very young, others are older, but most were founded in this century. Some disappeared after a few years. The world is awash with cheap money that has to be spent somewhere, so why not redo some of the failures of the past, to remind us that history had a reason for unfolding as it did. They all are very bold in their claims, outrageously so, insofar as they reveal anything about their intentions, and have in common that they want to revolutionise things by promising quick, abundant, clean, inexhaustible energy at a fraction of the cost spent on tokamaks, and that they will probably all fail. If you read the blurb on their websites one after the other, it is quite silly sometimes to see this mantra of clean, safe and abundant energy repeated all the time. Many of these companies have the characteristic that they are pushed by older, or even very old, seasoned nuclear fusion researchers who have spent their professional career at a university or (national) laboratory, have been pensioned off, but in their twilight

---

[3] Tokamak Energy, as we have seen in Chap. 11, is also struggling with this problem. They hope to develop HTS magnets that can be used in a ST design.

years still want to make a mark. The company **EMC2** (Energy/Matter Conversion Corporation) is a case in point. It was founded by Robert Bussard to develop the polywell, but since he died in 2007 the company is in dire straits, has no longer an operating website[4] and is now seeking funding.

Quite a number of these companies have joined forces in the **Fusion Industry Association**[5], whose usefulness (apart for publicity purposes) is not all that clear, but it claims to "advocate for policies that would accelerate the race to fusion energy", as if policies can speed up the laws of nature. Most are also a member of the **American Fusion Project**[6], which is an initiative of the American Security Project,[7] a nonpartisan organization created to educate the American public and the world about the changing nature of national security in the twenty-first century. The American Fusion Project is "an educational organization that informs the public about the benefits of accelerated development of fusion as a solution to our energy, economic, and environmental challenges".

Not all companies have the goal of creating energy to sell on the grid but want to use nuclear fusion for other purposes such as rocket propulsion. An example is **Princeton Satellite Systems**, which licenses patents from Princeton University for space propulsion applications. These patents originate from the Princeton Field Reversed Configuration (PFRC) – a novel plasma heating method invented in 2002 at the Princeton Plasma Physics Laboratory that can lead to a very small fusion reactor. With one end open to space, PFRC becomes the Direct Fusion Drive (DFD), a fusion-powered rocket engine that could enable new robotic and human space missions. DFD is a conceptual fusion-powered spacecraft engine, with the ability to produce thrust from fusion without going through an intermediary electricity-generating step. The designers think that this technology can radically expand the science capability of planetary missions. DFD uses a novel magnetic confinement and heating system, fuelled with a mixture of helium-3 and deuterium.[8]

Let's now turn to discuss these 'fusion pioneers' in more detail. For this we will first break the various companies into groups according to the route they intend to follow.

---

[4]Clicking on its logo on the members' page of the Fusion Industry Association brings you to the website of the non-profit organisation EMC2 Fusion Development Corporation, also founded by Bussard, but a different entity. This website too seems to be defunct.

[5]https://www.fusionindustryassociation.org/.

[6]http://americanfusionproject.org/.

[7]https://www.americansecurityproject.org/.

[8]https://federallabs.org/successes/awards/awards-gallery/2018/princeton-field-reversed-configuration-fusion-reactor-for-space.

15.3.1 Magnetized Target Fusion

General Fusion, Helion Energy, Compact Fusion Systems and MIFTI

This first group consists of companies that work on some form of magneto-inertial or magnetized target fusion (MTF), combining magnetic and inertial confinement. Important names in this respect are the Canadian company General Fusion, founded in 2002 and based in Vancouver, British Columbia; Helion Energy, founded in 2013 and based nearby in Redmond, close to Seattle, Washington State; Compact Fusion Systems, founded in 2017 in Santa Fe, New Mexico; and MIFTI (Magneto-Inertial Fusion Technologies), a Los Angeles-based company founded in 2008 which spun out of the University of California at Irvine.

Especially General Fusion[9] has been around for quite some time now. It was founded by the Canadian physicist and entrepreneur Michel Laberge, has more than 70 employees and managed to attract funding from various venture capitalists among whom Amazon founder Jeff Bezos through his investment vehicle Bezos Expeditions.

The fusion device under development at General Fusion[10] is rather original; it has no vacuum chamber but uses a sphere (approximately 3 m in diameter) filled with a molten lead-lithium mixture. By rotating the metal mixture, a vortex (a form of turbulent flow revolving around a straight or curved axis line) is created at the centre of the sphere. A pulse of magnetically-confined D-T plasma fuel is then injected into the vortex. Around the sphere, an array of pistons drives a pressure wave into the centre of the sphere, compressing the plasma to fusion conditions, releasing energy in the form of fast neutrons. This process is then repeated, while the heat from the reaction is captured in the liquid metal and used to generate electricity via a steam turbine.[11] The thick liquid metal would also protect outer structures and the environment from neutrons, thus solving one of the most serious problems in D-T fusion.[12]

In contrast to most magnetised target fusion systems, which use magnets to compress the plasma, the General Fusion design instead uses a large number of steam-driven pistons to mechanically compress the vortex of liquid metal. Such an MTF system has many advantages compared to tokamaks. There are no superconducting coils, neutral beams, lasers, manufactured targets, fuelling, ash removal, impurity accumulation, current drive, diverter heat load, etc. On the other hand, it introduces new difficulties, not present in tokamaks. Not much is known about confinement at high energy density; other challenges to be solved are liquid metal vaporization, impurities from the liquid (forming an initial spherical liquid surface), Rayleigh–Taylor instability of the fluid surface, kink instability of the liquid shaft,

---

[9]https://generalfusion.com/.

[10]https://generalfusion.com/wp-content/uploads/2018/11/aps-2018-magnetized-target-fusion-overview.pdf.

[11]https://generalfusion.com/technology-magnetized-target-fusion/.

[12]Leonid Zakharov, private communication.

symmetry of implosion, flux diffusion in the liquid, pulsed power requirements, etc. As of 2017 General Fusion was developing subsystems for use in a prototype to be built in three to five years. So, 2022 should be the crucial year.

In 2018 the company published several papers (Laberge 2019) on a new design that intends to use a spherical tokamak as the plasma source, instead of a compact toroid. It is not (yet) clear if this implies a change of direction.

Helion Energy[13] wants to build its Fusion Engine on deuterium-helium fusion, to avoid the troublesome high-energy neutrons produced in D-T fusion. The technology is based on the Inductive Plasmoid Accelerator (IPA) experiments performed from 2005 through 2012 at Helion Energy and at MSNW LLC, a company founded in 1991 that is or was developing revolutionary space *propulsion* technologies. (See e.g. Slough Votroubek and Pihl 2011.)[14] Deuterium and helium are heated until a plasma is formed, after which pulsed magnetic fields accelerate the plasma into the burn chamber at over 1 million miles per hour. A strong magnetic field compresses the plasma to fusion pressure and temperature. The deuterium and helium nuclei fuse, releasing charged particles (helium nuclei) that push back on the compressing magnetic field. The expanding plasma is then directly converted into electricity. The helium is then used for starting up the next cycle. The device they intend to develop according to this schedule, "The Fusion Engine", is supposed to produce 8 times as much energy as put in. It hopes to produce 50 MW of power in modules the size of a shipping container.[15] The aim is to have a commercial plant operational in six years (from 2020).

The next company to be mentioned in this category is the very young Compact Fusion Systems,[16] which has been awarded a grant by the Department of Energy Advanced Research Projects Agency for Energy (DOE ARPA-E) to support the technical development of a *Fusion Power Core* with commercial availability starting in 2030. The development is still very much in the design stage.

Finally, there is MIFTI[17] at Los Angeles which wants to develop a thermonuclear fusion-based stabilized Z-pinch. The firm has recently managed to overcome the instability problems of the Z-pinch by compressing the pinch in stages (staged Z-pinch). The lighter hydrogen plasma is surrounded with a heavier gas (i.e. argon, krypton or xenon), which slows the instability growth rate. The main feature is that when the heavier gas component in the cylinder collapses around the lighter part, a shock front develops that travels faster than instabilities can grow, allowing the plasma to remain stable, long enough for fusion to occur. The shock collides with the outer surface of the low-mass hydrogen (target) plasma, accelerating it inward. The shock transit time in the target is short and the plasma is rapidly heated to fusion conditions. The idea must still be shown to work in practice.

---

[13]https://www.helionenergy.com/.

[14]http://www.iccworkshops.org/icc2007/uploads/203/icc07_ipa_poster.pdf.

[15]https://www.helionenergy.com/; https://arpa-e.energy.gov/sites/default/files/05_KIRTLEY.pdf.

[16]https://www.compactfusionsystems.com/.

[17]http://miftec.com/.

15.3.2 *Inertial Electrostatic Fusion*

Progressive Fusion Solutions, Fusion One Corporation, Lockheed Martin, EMC2, Horne Technologies and Convergent Scientific Inc

Quite a number of companies seek to revive Philo Farnsworth's Inertial Electrostatic Fusion, in combination with cusp confinement and/or other methods. I have identified the ones listed above.

The current difficulties of EMC2 (Energy/Matter Conversion Corporation) founded in 1985 by Robert Bussard have already been mentioned. Work has continued until 2014, but the company seems to be dormant now due to a lack of funds. The device they are working on, called the polywell, has been discussed in the previous chapter. It combines cusp confinement with inertial electrostatic fusion. It uses an electric field to heat ions to fusion conditions. A set of electromagnets generates a magnetic field that traps electrons. This creates a negative voltage, which attracts positive ions. As the ions accelerate towards the negative centre, their kinetic energy rises and ions that collide at high enough energies can fuse. The device is closely related to the fusor, Lockheed Martin's high-*beta* fusion reactor, the magnetic mirror and the biconic cusp.

Progressive Fusion Solutions[18], a very young company from Vancouver (Canada), has the fantastic ambition to develop a car-sized fusion energy generator to replace fossil fuel power plants and end climate change. Why not? You have to think big in this business. It also intends to do this by following the Inertial Electrostatic Fusion path. One of its major achievements seems to be that its founder and CEO completed a desktop fusion reactor in 2017 at age 17: "the sun in a jar". In the previous chapter it has been pointed out that this is no big deal and 17 seems indeed the right age to do this. Anyone can do it and building a fusor is actually quite popular with amateurs. Instructions can be found on the Internet.[19] Apart from some general statements about "implementing fusion energy into society and ending climate change" the company has not yet anything to show for.

Convergent Scientific Inc. or CSI was founded in 2010 and is based in Huntingdon Beach (California).

It has developed a variation of the Convergent Ion Focus (CIF), i.e. polywell, approach to fusion, which is capable of producing extremely energetic non-neutral plasmas at much lower relative input energy requirements. They tested their first polywell design on steady-state operations in 2012. The intention is to retrofit coal-fired power plants with their technology. Its website convsci.com does no longer exist and no further information is readily available. The company seem to have gone under.

---

[18]https://www.progressivefusionsolutions.com/.

[19]https://makezine.com/projects/make-36-boards/nuclear-fusor/.

Started in 2008, Horne Technologies,[20] based in Denver, is a company that develops technology to optimize containment of plasma for the purpose of improving the energy balance of fusion devices. In 2017 it has developed the Horne Hybrid Reactor (HHR), which they claim is the world's first continuous-operation enabled, superconducting, high-*beta* fusion research device. In spite of this momentous claim it has attracted surprisingly little attention, much less in any case than its closest relative Lockheed Martin's Compact Fusion Reactor. Its prototype demonstration device uses a combination of fusion technologies with high-temperature ReBCO superconductors in a high-*beta* style magnetic configuration. Back in 2011, Horne was one of the first companies to apply superconducting wire to the problem of fusion plasma.

Heating is achieved through the use of inertial electrostatic confinement, improved by a magnetically-shielded grid, and a high-*beta* fusion core (in the form of a cusp confinement configuration). As we have seen in the previous chapter, in fusors, the potential well is made with a wire cage; because most of the ions and electrons fall into the cage, fusors suffer from high losses. By using a grid that is magnetically shielded it is theorized that these losses can be reduced. Funding has recently been obtained for a second-generation device, for optimization experiments which are currently in progress, while for stage III a full-scale net energy producing device is being planned.

Fusion One Corporation was a US organization in business from 2015 to 2017. The lead physicist of EMC2 was one of its founders. The company developed a magneto-electrostatic reactor named "F1" that was partly based on the polywell, and some novel additions to stem particle losses through the magnetic cusps. The results yielded an energy distribution that was non-thermal, but more Maxwellian than monoenergetic. The input power required to maintain the distribution was calculated to be excessive and ion-ion thermalization (i.e. reaching thermal equilibrium through mutual interaction) was a dominant loss channel. With these additions, a pathway to commercial electricity generation seemed no longer feasible.

That leaves Lockheed Martin, one of the largest companies in the aerospace, defence, security, and technologies industry, formed in 1995 by the merger of Lockheed Corporation and Martin Marietta. As of December 2017, it employed approximately 100,000 people worldwide. It is the world's largest defence contractor with revenues of more than US$ 50 billion (2018). It has ample experience with research into nuclear fusion as the former owner and operator of Sandia National Laboratories from 1993 to 2017. As of May 2017, Sandia National Laboratories is being managed by National Technology and Engineering Solutions of Sandia, a wholly owned subsidiary of Honeywell International.

Lockheed Martin Skunk Works intends to build the Lockheed Martin Compact Fusion Reactor (CFR), a high-*beta* compact reactor to be achieved by combining cusp confinement and magnetic mirrors to confine the plasma. Skunk Works is an official pseudonym for Lockheed Martin's Advanced Development Programs. It is responsible for a number of aircraft designs, used in the air forces of several countries. The designation "skunk works" or "skunkworks" is widely used in business,

---

[20]https://www.hornetechnologies.com/.

engineering, and technical fields to describe a group within an organization that is given a high degree of autonomy and is unhampered by bureaucracy, with the task of working on advanced or secret projects; an environment that is intended to help a small group of individuals design a new idea by escaping routine organizational procedures. The term originated with Lockheed's World War II Skunk Works project. (*Wikipedia.*)

On the website of its compact fusion endeavour[21] Lockheed does not reveal much, only that it intends to build a "reactor small enough to fit on a truck [to] provide enough power for a small city of up to 100,000 people". It is in general very secretive about its efforts and apart from a poster at an APS conference in 2016 nothing substantial has been published (McGuire et al. 2016).[22] Why is not clear, and its secretive attitude has been criticized on various Internet fora. There is hardly any danger that a competitor would run away with the idea and an open discussion in the literature might yield useful criticism. Its approach to fusion is a high-*beta* concept, implying that the ratio of plasma pressure to magnetic pressure is equal or close to 1 (compared to 0.05 for most tokamak designs). By using a high fraction of the magnetic field pressure, or all of its potential, they claim to be able to make their devices 10 times smaller than previous concepts, to fit on the back of a truck.

The project started in 2010 and in 2014 Lockheed Martin announced a plan to "build and test a compact fusion reactor in less than a year with a prototype to follow within five years". This period has been over for some time now. Since 2014 they have built four different test reactor designs, as well as a number of subvariants. The latest one is a device called T5 that will be followed by various others.

Their approach is based on the idea of cusp confinement considered in the previous chapter whereby the internal magnetism (diamagnetism) of the plasma presses against an applied outside magnetic field and creates a diamagnetic cusp trap, the "perfect plasma trap", with a "*beta* = 1" plasma. The drawback is that at the cusps, where the field is sharply bent, plasma can leak out. Other companies, notably EMC2, have failed to bring this concept to fruition, but Lockheed is of course quite a different company, a very serious company indeed, and with its team and funding it is perfectly positioned to exploit the concept to the full.

Its T4B device, still a small device of just 2 m long and a diameter of 1 m, was announced in 2016 at the APS conference mentioned, including a poster with some details. The basic reactor configuration is shown in Fig. 15.1. As can be seen from the figure, the basic design system is completely different from the magnetic toroid designs currently in use.

It basically consists of a canister with a set of superconducting magnetic coils inside and other electromagnets on the outside (Fig. 15.2a). The reactor is no longer as compact as it set out to be with its core being 18 m long and 7 m in diameter.

Plasma confinement is achieved in magnetic wells with self-produced sharp magnetic field boundaries. Neutral beam heating is used to heat the plasma to an ignited state. The dominant losses are ion losses through the axial cusps (along the

---

[21] https://www.lockheedmartin.com/en-us/products/compact-fusion.html.

[22] https://web.archive.org/web/20171225092237/http://fusion4freedom.us/pdfs/McGuireAPS.pdf.

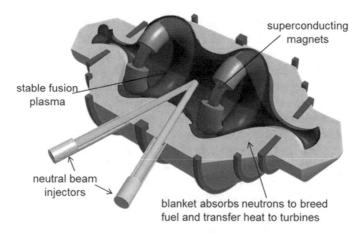

Fig. 15.1 Diagram showing Lockheed's basic reactor configuration used in its CFR programme

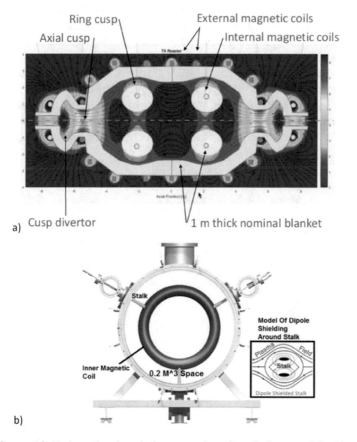

Fig. 15.2 a and **b** Horizontal and vertical cross sections through the core of Lockheed's T4B experimental reactor (*from* McGuire 2016)

axis of the plasma; the dotted line in Fig. 15.2a), the ring cusps into the stalks [the stalk is a piece of material that sticks into the plasma to hold up the magnetic coil (Fig. 15.2b)]. This is a problem for many researchers in this type of fusion device. Robert Bussard had problems with mass being lost through the stalks holding the polywell and through the sheath, i.e. the skin of the trap. To stem the loss through the stalks Lockheed is trying to shield them using a dipole field, a very innovative method. The models made to simulate these losses have many assumptions, which need confirmation in further experiments.

At the start of a shot, the vacuum is pumped down and the magnets are fired by using supercapacitors. This creates the fields for the plasma coming from two sources. A source of electrons at the end of the tank (a lanthanum hexaboride (LaB_6, an inorganic chemical) cathode, whose principal use is precisely for such hot cathodes). It is heated, electrons come off and are directed into the chamber. The other source of plasma is the neutral beam, which is the source of hydrogen ions. The plasma stays trapped after the sources are shut off; a promising sign that cusp confinement is happening. The plasma geometry with the magnetic coils is shown in Fig. 15.3.

The first claim Lockheed made for its T4B experiment is that its plasma inflated to high *beta* (without specifying the actual value), meaning that it is pressing against the magnetic field, as expected from a cusp confinement trap. Secondly, everything is very sensitive to plasma density; a change by a factor of 10 from 10^{19} to 10^{20} particles per cubic metre (hundred times less than in JET) changes things dramatically.

Halfway through 2019 Lockheed announced to be in the process of constructing its newest experimental reactor, known as the T5 and that, despite slower than expected progress, it remains confident that "the project can produce practical results, which

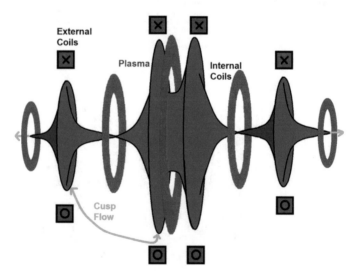

Fig. 15.3   A sketch of the plasma geometry and magnetic coils inside the T4B model. This design has since been superseded with a model using only two main cusps (*Wikipedia*)

would completely transform how power gets generated for both military and civilian purposes". The T5's main job will be to further test whether Skunk Work's basic reactor design can handle the heat and pressure from the highly energized plasma inside, which is central to how the system works.

15.3.3 Hydrogen-Boron Fusion

LPPFusion, HB11 Energy, TAE Technologies

The third group of companies tries to find the holy grail of fusion in alternative fuels, especially proton-boron plasmas. These companies include LPPFusion,[23] TAE Technologies and the Australian company HB11 Energy.

Fusion of boron with hydrogen (p-^{11}B fusion) is well known to be extremely difficult, five orders of magnitudes more difficult than the easiest D-T fusion under conditions of local thermal equilibrium at about 100 keV (one billion degrees) temperature. Proton-boron fusion is a factor of 100,000 less efficient, reason why this fusion option is usually excluded. This can be changed by igniting the plasma under conditions of local non-thermal equilibrium, which is the method that all these companies use, but then each in its own way.

TAE Technologies[24] (formerly Tri Alpha Energy), founded in 1998 and based in Foothill Ranch (California) claims to be the world's largest private fusion reactor company. It is indeed one of the larger players in fusion in the US, but as a company it is dwarfed by Lockheed Martin or General Atomics, for which nuclear fusion (reactors) is just a side issue. TAE grew out of the physics department at the University of California, Irvine. With its $600 million (others say $800 million) in funding from private investors, it has received a big chunk of total private venture capital put up in the US. One of its investors was Microsoft co-founder Paul Allen, who died in 2018. His former colleague Bill Gates apparently invests in one of TAE's competitors, Commonwealth Fusion Systems, but it seems that Gates, with good reason, has more confidence in power from nuclear fission, judging by his company TerraPower.[25] TAE is currently testing a full reactor and raising funds to build a next-generation reactor. The reactor will use a combination of high-energy radio-frequency pulses, particle-accelerator beams, and magnets to control a proton-boron plasma. They intend to demonstrate net energy generation by 2024, which is very soon indeed. Earlier they made the even more outlandish claim to start commercialization of fusion by 2023.[26]

---

[23]LPP stands for Lawrenceville Plasma Physics. Lawrenceville Plasma Physics Inc. is the parent company of LPPFusion.

[24]https://tae.com/company/.

[25]https://terrapower.com/; TerraPower is developing a class of nuclear fast reactors called the travelling wave reactor (TWR).

[26]https://www.nextbigfuture.com/2019/01/nuclear-fusion-commercialization-race.html#more-153607.

This claim was soon corrected into starting fusion commercialization 'efforts' by 2023, whatever that may mean. It illustrates in any case how one should weigh every word in the fusion business, but it also implies that nothing will probably come of it.

Its latest prototype machine, C-2 W, which was renamed Norman in honour of the company's co-founder Norman Rostoker (1925–2014), is a field-reversed configuration (FRC). The averaged *beta* value of FRCs is near unity. Its setup is shown in Fig. 15.4, which also gives an idea of the considerable size of the machine (from the figure of a man at the front), given that it is only a preliminary experimental device. In Norman both ends of the cylindrical colliding beam fusion reactor (CBFR) heat hydrogen gas to form two smoke rings of plasma. These rings are then shot at each other at supersonic speeds, merging in the middle of the machine, forming a spinning plasma ball held together by self-created magnetic fields (very similar to the merging compression method used by Tokamak Energy described in Chap. 11). Neutral beams, consisting of neutral particles with no electric charge, are then used to stabilize the plasma and help make it hotter and last longer. First plasma was obtained in 2017 and a temperature of about 20 million degrees in 2018, far below the required value for proton-boron fusion (Gota et al. 2019). In March 2019 TAE reported a fusion yield of 40 microjoules, the only company apart from LPPFusion (see below) to do so, but the yield cannot have originated from proton-boron fusion.

LPPFusion started operations in the 1980s but incorporated only in 2003. It is based in Middlesex, New Jersey. Their mission is the usual one: "to provide environmentally safe, clean, cheap and unlimited energy for everyone through the development of Focus Fusion technology, based on the Dense Plasma Focus device and

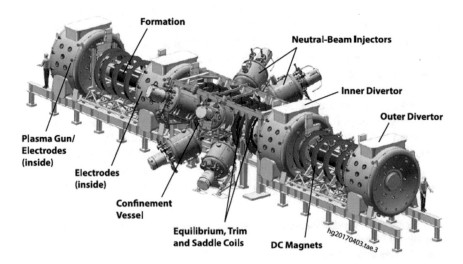

Fig. 15.4 Illustration of the C-2 W experimental device, Norman (*from* Gota et al. 2019)

hydrogen-boron fuel."[27] President and Chief Scientist of the company is the controversial popular science writer and independent plasma researcher Eric Lerner (b. 1947).[28] Between 1994 and 2001 he received funding for his research from NASA's Jet Propulsion Laboratory, to explore whether his approach based on Dense Plasma Focus might be useful for advanced fuel fusion and space propulsion.

The company has revived the Dense Plasma Focus (DPF) concept (see previous chapter), which it calls Focus Fusion. Their device consists of two cylindrical metal electrodes nested inside each other. The outer electrode is generally no more than 6–7 inches in diameter and a foot long. The electrodes are enclosed in a vacuum chamber with a low-pressure gas filling the space between them. A pulse of electricity is discharged across the electrodes. For a few millionths of a second, an intense current flows from the outer to the inner electrode through the gas. This current starts to heat the gas and creates an intense magnetic field. Guided by its own magnetic field, the current forms itself into little whirlwinds of hot plasma. This plasma travels to the end of the inner electrode where the magnetic fields produced by the currents pinch and twist the plasma into a tiny, dense ball only a few thousandths of an inch across called a plasmoid. See also Fig. 14.6 and the description in that chapter. All of this happens without any external magnets. The magnetic fields very quickly collapse, which induces an electric field that causes a beam of electrons to flow in one direction and a beam of ions in the other. The electron beam heats the plasmoid to extremely high temperatures, the equivalent of billions of degrees (particle energies of 100 keV or more, which are needed to eventually fuse hydrogen and boron). Ions collide and fuse, which adds still more energy to the plasmoid. So, in the end, the ion beam contains more energy than was put in by the original electric current. (The energy of the electron beam is dissipated inside the plasmoid to heat it.) This happens even though the plasmoid only lasts a few billionths of a second or so; because of the very high density in the plasmoid, which is close to the density of a solid, collisions are very likely and to occur extremely rapidly.[29]

A detailed quantitative theory of DPF device operation has been developed and tested against experiments in collaboration with the University of Illinois and Texas A&M University. The claim is that this theory, including important refinements that include magnetic effects, gives the ability to show in advance that hydrogen-boron fusion is feasible.[30]

In 2016 an ion temperature of 2.4 billion degrees was achieved in the FF-1 experimental device, which also gave a fusion yield of 1/8 joule (Lerner et al. 2017). The fusion yield is still tiny, but the temperature is adequate for proton-boron fusion (although no proton-boron fuel has yet been used). As we have seen many times before in this book, temperature alone is not sufficient for a burning plasma, density and confinement time are also important. Their latest device is called FF-2B, where

---

[27]https://lppfusion.com/mission/, last accessed 27 June 2020.

[28]Among other things he wrote the book *The Big Bang Never Happened*, which rejects mainstream big bang cosmology. See his *Wikipedia* page for more details.

[29]https://lppfusion.com/technology/dpf-device/.

[30]https://lppfusion.com/, an extensive commendable website with lots of information.

2B stands for beryllium electrode and boron fuel. Its electrodes are made of beryllium, while before tungsten was used. This will reduce impurities from vaporised electrode material as beryllium is a low Z material; $Z = 4$ compared to $Z = 74$ for tungsten.

Experiments with hydrogen-boron fuel were said to start in 2019, which however did not happen. Troublesome rapid oscillations in the current caused delays. Different parts of the current arrive at different times at the pinch, where the plasmoid forms. This means that the region that contains all the current never gets very small, preventing high compression, high density and thus high fusion yield. Instead a song contest was held, which must have been fun too. They are now hurrying up, I suppose, to show net energy production in 2020 and present a first prototype 5 MW generator in 2023.

The company HB11 Energy,[31] a spin-off from the University of New South Wales in Sydney (Australia) makes very big claims indeed for laser driven non-thermal proton-boron fusion. It is a type of inertial confinement fusion but with proton-boron fuel, which is the reason for its classification in this group. The claims on their website beat all others and are simply amazing, including that HB11 Laser Boron Fusion Reactors, so far only existing on paper, will produce electricity at a quarter of the price of electricity from existing coal fired power stations and are much cheaper to build; the technology already exists and the first prototype could be built in just 5–10 years; only modest investment is required for further research to clarify some of the scientific methods and engineering requirements; the reactors can be built anywhere and need no water as they produce electricity directly, so there is no need for steam turbines; they can be built to scale, meaning that smaller communities can build smaller reactors, while larger reactors can be built for large scale urban and industrial usage; they can be used on ships and submarines, for industry and manufacturing, isolated communities or interconnected to major power grids. In short, an almost free lunch for all, where Laser Boron Fusion Reactors offer an inexpensive, clean, long term solution, with no carbon emissions, no radioactive waste and a very small environmental footprint.

In Chap. 12, when discussing D-T fusion via inertial confinement, we have described how the National Ignition Facility uses 196 laser beams to irradiate a small hohlraum at the centre of a 10 m diameter sphere. At the centre of the hohlraum sits a small sphere with D-T fuel, whose implosion causes the fuel to fuse. The laser amplifiers cover most of the size of three football fields, collect the 196 beams into the sphere and focus them at its centre. For laser proton-boron fusion, HB11 Energy also proposes a reactor of spherical shape (diameter 1 m), but with the basic simplification that the ignition of the reaction is produced by a single laser beam. Moreover, they seek to implode a small cylinder filled with fuel, in their case a proton-boron mixture.

Proton-boron (p-^{11}B) fusion has of course the advantage that no neutrons are produced, but only charged alpha particles, and is therefore much more environmentally friendly. Under conditions of local thermal equilibrium, temperatures must reach

---

[31] https://www.hb11.energy/. Last accessed 27 June 2020.

about 600 keV for proton-boron fusion to be feasible and for laser ignition a compression to 100,000 times solid-state density must be achieved. However, according to HB11 Energy, when the plasma is not in local thermal equilibrium, ignition can be achieved by using extremely short (picosecond (10^{-12} s)) laser pulses with more than 10 petawatt of power (one petawatt being 10^{15} W, i.e. one billion megawatt). The laser beam accelerates plasma blocks at nearly solid density without heating within a picosecond to velocities above 10^9 cm/s. Such ignition by lasers had been tried before with nanosecond (10^{-9} s) laser pulses with thermal compression, but this turned out to be extremely difficult. With picosecond laser pulses the ignition threshold for proton-boron fusion increased by five orders of magnitude arriving at the level of D-T fusion due to the non-thermal transfer of the laser energy directly into the ultrahigh acceleration of the plasma blocks (Hora et al. 2017). The approach is based on the possibility of triggering an avalanche reaction in laser-driven p-^{11}B fusion by collisional energy transfer from alpha particles to protons. A very high number of proton-boron fusion reactions observed (Eliezer et al. 2016) in experiments at the Prague Asterix Laser System (PALS) has been interpreted as indicating such avalanche multiplication, occurring through the generation of three secondary alpha particles from a single primarily produced alpha particle. This interpretation has however been questioned (Belloni et al. 2018; Shmatov 2016).

The only problem seems to be that petawatt lasers cannot (yet) be bought off the shelf in a hardware store.

15.3.4 Inertial Confinement Fusion (ICF)

First Light, Proton Scientific, Fusion Power Corporation

Private enterprise has been active in ICF in the past. In the 1970s the company KMS Fusion, a subsidiary of KMS Industries, was the only private company to pursue controlled thermonuclear fusion research using laser technology. KMS successfully demonstrated ICF, achieving compression of a D-T pellet by laser energy in 1973, and in 1974 carried out the world's first successful laser-induced fusion (although net energy production was still off by a factor of more than 10 million).[32] KMS Industries went under, in part due to heavy opposition from the Atomic Energy Commission and large federal weapons laboratories. Many people in both government and scientific sectors were bitterly opposed to the operation of such a fundamental and important energy programme in the private sector. This undoubtedly also contributed to the untimely death of its founder Kip Siegel (1923–1975).

The three companies mentioned above apply more or less conventional ICF techniques to achieve fusion, but of course they think to have something extra to solve the problems that prevented the National Ignition Facility to achieve ignition, as related in Chap. 12.

---

[32] *Wikipedia*, https://en.wikipedia.org/wiki/KMS_Fusion.

First Light Fusion[33] spun out of the University of Oxford in 2011 and is one of the very few private companies known to research energy generation via ICF. It is working towards demonstrating "first fusion" before the end of the year 2019, in their Oxford-based laboratory, but have obviously failed to do so. By 2024 they hope to build a device in which the energy created outstrips the energy required to start the reaction.[34]

First Light Fusion calls its approach "projectile fusion." Their machine launches a copper disk at high velocity to collide with a pellet of fuel they call the "target," which has a hydrogen bubble inside. At the point of collision, immense pressure is put on the bubble, forming a pressure wave which travels through the target, forcing the hydrogen cavity to collapse. For that brief fraction of a second, the plasma that is created is very hot and denser than lead. The company already demonstrated plasma in 2012, and hoped to demonstrate fusion for the first time in 2019.

For First Light Fusion, the target is key. The rest of their set up predominantly uses existing technology. And here lies the business model opportunity: consumables. The plan would be to manufacture and sell the targets to power plant operators—the ultimate Nespresso capsule, as it were.

According to the critical website *New Energy Times*, First Light Fusion is actually using the same deception technique as JET and the other tokamaks have done and are doing with their clever ways of defining various sorts of breakeven. You may remember that JET claimed to have gained fusion power equal to 67% of the power put into the plasma, while in actual fact is was only 2% if you take account of all the power needed to run JET. First Light Fusion does a similar thing. They are also hiding the actual power required to operate the device. It claimed that, by 2024, its experimental fusion reactor would "achieve gain—generating more energy than is used to spark a reaction." In inertial fusion science terminology, the phrase "sparked a reaction" indicates only the final amount of power that arrives at the fuel target. It does not include the vast majority of the power required to operate the fusion device. The company's reactor is actually designed to consume more electric power than it produces by fusion, resulting in a net energy loss.[35] But, nonetheless, if they achieve what they are claiming, it would be a great result. Wouldn't they then have beaten the National Ignition Facility? We will have to wait and see if First Light will indeed be the dawn for ICF.

Proton Scientific[36] was established in 2012 in Champaign (Illinois) and states as its goal to further develop and commercialize the technology for fusion energy production using an electron beam as driver. Websites tend to change rather quickly though and now, mid 2020, its far grander mission is "to develop a commercial process that produces clean and affordable fusion power which will meet global

[33] https://firstlightfusion.com/.

[34] https://www.forbes.com/sites/gemmamilne/2019/08/29/how-this-unconventional-oxford-sta rtup-plans-to-win-the-fusion-energy-race/#156f10cc29f8 (29 August 2019).

[35] https://news.newenergytimes.net/2019/02/12/first-light-fusions-fake/.

[36] http://protonscientific.com/.

energy demands in the twenty-first century and beyond". So, still in this century, we will all benefit from the fruits of their labour.

A pulsed power-generated electron beam as a driver for nuclear fusion has been considered for more than four decades (see e.g. Lubkin 1977). Proton Scientific wants to achieve its goals via a pulsed electron beam generator, which goes under the name IVR-3 and was constructed in Kiev, Ukraine, and transferred to Champaign in 2014 (Adamenko et al. 2015). At the end of that year, the pulsed power generator, now called "Thunderbird", became operational. The task of this generator is to produce a relativistic electron beam which is then used to strike a small metallic target. Proton Scientific's claim is that its Thunderbird pulsed power generator can produce a concentrated burst of power strong enough to achieve fusion and ignition. An extremely high power, up to 20 GW, comparable to the average annual electricity consumption of an entire industrial country, is delivered into the microscopic volume of the solid target, equal in size to the tip of a pin. When such huge power focuses into a micron size area, its density reaches a level of 10 TW/mm^2. That such power density level is needed for ignition in electron beam fusion was already indicated in the early 1970s (Yonas et al. 1974). For a long time, it was believed that only a large pulsed power machine can achieve this goal. The mid-scale Thunderbird device is claimed to be able to bring the power density up to these extreme levels.

Additionally, because the fuel pellet size does not need to scale up to achieve ignition, the overall device will be smaller, meaning the eventual cost to build a commercial power plant based on this approach should be much less.

In the original direct-beam driver fusion scheme high uniformity of the electron beam had to be maintained in order to inhibit the growth of hydrodynamic instabilities during the target implosion. The best efforts to address these challenges were implemented in the Particle Beam Fusion Accelerator-II (PBFA-II) project at Sandia National Laboratories, which used a light-ion beam as driver. Proton Scientific sees the effect of the "self-pinching" of the electron beam as an advantage, making it possible to achieve the extreme power densities necessary for fusion in the small volume of a target. This region of extreme energy deposition under certain conditions can spark the thermonuclear burn to sustain it through the whole volume of the target. Such an approach is similar to the fast-ignition concept in laser-driven ICF research.

Two major technological challenges are: (1) meeting the high energy and power requirements for the target compression, and (2) minimizing the growth rates of the instabilities during target compression.

A conceptual design of the power plant is supposed to be ready in 2020–2021, while the end goal is to develop and build a full-scale electron-beam fusion power plant. It is not clear how far things have proceeded.

Fusion Power Corporation[37] from Sacramento, California, was founded in 2008. In 2011 they promised electricity-generating nuclear fusion power stations within ten years with a process known as RF Accelerator Driven Heavy Ion Fusion, but nothing

---

[37]http://www.fusionpowercorporation.com/.

has been seen yet. Its website has disappeared and the company is now probably defunct.

15.3.5   Miscellaneous

ZAP Energy[38] was incorporated in 2017 as a spin-off, like many others, from the University of Washington and Lawrence Livermore National Laboratory. Zap Energy's website does not reveal much, only that magnetic coils are too expensive, inertial confinement is too inefficient and conventional Z-pinches are too fleeting, so the most promising path is sheared-flow-stabilised Z-pinch, meaning that its technology stabilizes plasma using sheared flows—that is, with layers of plasma flowing at different velocities at different radii—rather than magnetic fields. The traditional Z-pinch, as we have seen earlier in this book, is known to be plagued by instabilities that prevent conditions for net fusion energy output to be achieved. Sheared axial flows have been shown to stabilize disruptive Z-pinch instabilities at modest plasma conditions. Driving electric current through the flow creates the magnetic field, which confines and compresses the plasma. The higher the current, the greater the pressure and density in the plasma. By stabilizing the plasma with a sheared flow, the high-temperature, high-density reactive medium can be confined long enough for fusion reactions to occur.

The US Department of Energy has provided funding for this venture since 1998, that is including the work by the FuZE team who pioneered the technology at the University of Washington and Lawrence Livermore National Laboratory (Shumlak and the FuZE team 2017). In 2015, ZAP Energy was awarded a grant from ARPA-E's ALPHA programme and it exceeded all of its aggressive programme milestones. What these are is however not explained.

It sees its reactor as the least expensive, most compact, most scalable solution with the shortest path to commercially viable fusion. Based on progress to this point, it is said to be on track to reach $Q = 1$ energy breakeven plasma conditions. By the end of 2019 the goal for Zap Energy's next step device is to achieve 600 kA of plasma current, with plasma density and temperature predicted to approach conditions of scientific breakeven, when fuelled with a 50–50 D-T mix (Wang 2019).

Another American company is **AGNI**[39], based in Washington State, a start-up pursuing an unconventional version of scaled-down fusion in which a beam of high-energy deuterium atoms is fired at a target of lithium and tritium (Fig. 15.5). In their own words they are expected to achieve: 16 million times the efficiency of coal, 10 times the efficiency of fission, no waste and zero emissions. Isn't it fabulous? The following is from its website:

Deuterium and helium-3 are injected into the chamber where they are ionized and accelerated into an ion beam. Magnetic focusing lenses serve to excite the ion beam

---

[38]https://www.zapenergyinc.com/, last accessed 27 June 2020.
[39]https://www.agnifusion.org/.

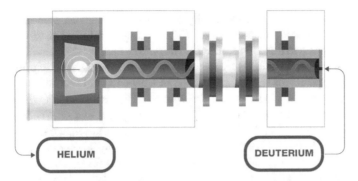

Fig. 15.5 The Agni fusion reactor. The AGNI fusion reactor uses both electric fields and magnetic fields, giving the nuclei a very short flight time before they hit the solid target, so the nuclei don't need to be controlled for very long before the fusion occurs

and control it to prevent it from straying off course on the way to the target. The lenses also serve to periodically increase the density to induce ionic heating through deuterium-helium-3 fusion. The target is made of lithium borotritide ($LiBT_4$), which has the highest density of tritium by volume. Here, the excited deuterium hits the target with a large amount of kinetic energy and fuses with tritium, creating a vast surplus of energy, in the form of helium-4 and neutrons. The neutrons carry most of the energy, 14.1 MeV, that will be used to power a steam turbine, generating electricity.

The AGNI reactor uses several types of fusion fuel in order to take advantage of different energies and the fusion-ion heating of aneutronic fusion, in which neutrons carry only about 1% of the total released energy, as opposed to 80% in traditional fusion reactions.

In the previous chapter we have discussed the dynomak, a fusion device developed by the University of Washington in Seattle. The spin-off private company **CTFusion**,[40] formed in 2015, tries to further develop the concept of imposed-dynamo current drive (IDCD) and the dynomak, a power-generating spheromak reactor. Halfway through 2019, CTFusion was awarded several million dollars from ARPA-E to develop the dynomak concept. It has enough similarities with the tokamak that much of the theory, codes and expertise of the tokamak can be applied. A key innovation of the dynomak is the use of helicity injection, whereby a wobble is imposed on a self-sustaining loop of plasma. By imposing a wobble, the company can eliminate a spheromak plasma instability, effectively heat the material and reduce the number of magnets needed in the machine. The progress they make can be easily followed as they tend to publish the results of their research in the scientific literature, which most of the companies discussed here don't do.

---

[40]https://ctfusion.net/.

Type One Energy[41] is a University of Wisconsin-Madison spin-off, applying innovations in additive manufacturing (3D printing), analytical theory, and high-field HTS magnets to drive the stellarator fusion concept towards commercialization as a compact and cost-effective power plant. It builds on the results of the Helical Symmetric Experiment (HSX), mentioned in Chap. 13, and has laid out a 15-year path to a pilot power plant, named the Spitzer-1 stellarator, which will have 800-1200 MWe power generating capacity, is about twice the size of Wendelstein 7-X, and should be ready around 2035. Type One Energy does not intend to use tritium as a primary fuel, but to utilize the catalysed D-D fusion cycle (see previous chapter) by recycling the tritium produced in one of the two D-D fusion reactions.

In spite of being at the very beginning of its 15-year development path and without having yet constructed any power-generating stellarator, Type One Energy already sees massive revenues "generated through the mass production and sale of stellarator power units produced from Type One gigafactories fabricating a high fraction of pre-assembled components to expedite construction for a 2-year on-site install target." They see themselves having a 15% market share of energy-generating capacity in 2050 by capturing 56% of new energy consumption. An incredible feat of wishful thinking.

HyperJet Fusion Corporation[42] is also a member of the Fusion Industry Association and probably wants to achieve fusion in one way or another. How is as yet unclear. Their website only informs us that for them "Hypervelocity Plasma Guns [are the solution] for Clean Energy, Industry & Space". The only thing that I have been able to find out is that the "company is developing an advanced hypervelocity plasma armature railgun for use as a driver in a new type of impact fusion." Impact fusion is when a solid hypervelocity projectile impacts D-T fuel contained inside a fuel target to create a fusion reaction, so very similar to ICF. Apparently in 2017 the company merged with **HyperV Technologies Corporation**, a private fusion energy research and development company, founded in 2004 and located in Chantilly, Virginia. The company specializes in the development of ultra-high performance plasma guns for use in fusion energy, plasma physics research, and industrial applications. The name HyperV comes from the word "HyperVelocity" and refers to the extremely high velocities achieved by plasma when formed and fired from these plasma guns. The fusion technology concerns magneto-inertial fusion driven by a plasma jet.[43]

Helicity Space[44] is located in Berkeley, California, and was formed in 2018. The company focuses on designing a fusion-driven rocket engine. Their approach is to inject helical filaments of plasma[45] into a magnetic nozzle. When the plasma reaches the centre, the magnetic fields reconnect—releasing energy into the plasma

---

[41] https://www.typeoneenergy.com/, well-presented, informative and clear website.

[42] http://hyperjetfusion.com/.

[43] See https://www.titansofnuclear.com/dougwitherspoon for an interview with the founder of HyperV Technologies Corp.

[44] https://www.helicityspace.com/.

[45] The company calls them plectonemes; a plectoneme is a loop of helices twisted together such that they cannot be separated without breaking them. The word is especially used for nucleic acids.

and kicking off fusion reactions. The resulting plasma is heated by the magnetic reconnection and by fusion events, and escapes out of the back of the rocket. No results have yet been reported.

Renaissance Fusion[46] is one of the very few fusion start-ups in continental Europe. It was founded in 2019 and is based in Granada, Spain. Renaissance Fusion aims to develop subsystems for stellarators and tokamaks. It belongs to the small minority of companies that see prospects in stellarators. Their aims are to demonstrate a new manufacturing technique of high-temperature superconductors with potential benefits for medical imaging and stellarator simplification; to build the first high-field (10 T) HTS stellarator with thick, flowing liquid metal walls; and to put fusion electricity in the grid within 13 years from today. Grand aims indeed and no results have yet been reported.

Finally, I mention a company simply called **fuse**.[47] It is a member of the Fusion Industry Association with a website, which however cannot be accessed without a password, and nothing further is known about this company.

Apart from the companies considered above there are also some involved in the 'nuclear fusion industry' that are not necessarily focused on building a nuclear fusion power plant. These include:

General Atomics, which we met before in Chap. 5 and has been involved with nuclear fusion from a very early stage; **Phoenix** (founded in 2005 and formerly known as Phoenix Nuclear Labs) manufactures compact intermediate and high-yield neutron generators, based on both D-D and D-T fusion; **Woodruff Scientific** is a company that since 2005 manufactures magnetic field coils, pulsed power systems, and scientific instrumentation and supplies them to the nuclear fusion industry; **Sorlox** (part of the German company HBM) makes plasma by taking deuterium gas and ionizing it using a high-strength magnetic field; **Shine Medical Technologies**, that spun out of Phoenix, is dedicated to being the world leader in the safe, clean, affordable production of medical tracers and cancer treatment elements; **NSD Fusion** (based in Luxemburg, one of the few European companies) also makes neutron generators.

15.4 Conclusion

This chapter comprises an overview, insofar as information was available, of the efforts in nuclear fusion by private companies. Almost all of these companies are based in the United States, with only very few in Europe or elsewhere. The most important of the European companies is undoubtedly the British company Tokamak Energy, which has been dealt with in Chap. 11. The reason for the proliferation of such companies in the US is, among other things, that venture capital is much more readily available there. Most of the companies considered here have in common that they promise a quick fix to the difficulties that have beset nuclear fusion research now

---

[46]https://stellarator.energy/.

[47]http://www.f.energy/.

for close to seventy years. The surprising thing is that they apparently do this with all the confidence in the world and that the venture capitalists are easily convinced of the merits of all the proposals and overblown expectations. We will see in Chap. 18 that Daniel Jassby brands most of these start-ups pursuing alternative fusion concepts as Voodoo Science.

All this private activity also had its effect on the public sector. In general, nuclear fusion is characterised by the fear of losing out. Imagine, however unlikely this may be, that a country or company cracks the fusion nut in the next decade or so. Then we, of course, also want to benefit. This is undoubtedly one of the motives for the British government to start the STEP programme, and for the US and India, and perhaps other nations too, to continue to participate reluctantly, but on the cheap in the ITER programme.

The private efforts of the last years have also convinced the US Department of Energy that the private sector has important contributions to make in the quest for fusion energy. Why they think so is not clear as the private sector has so far not contributed anything of any significance. In 2019 the INFUSE program[48] (Innovation Network for Fusion Energy) was announced, an effort to pair national laboratories with private companies. To apply for a grant a private company must partner with a national laboratory. The funding provided is rather low compared to the amounts put in by venture investment funds. A typical starting contribution is about $100,000 in funding, with the private company getting about 40% of that money. However, the INFUSE program will allow companies to get major experiments done at the national laboratories. It will allow them to do big experiments on large scale equipment. In October 2019 the Department of Energy announced funding for 12 projects with private industry to enable collaboration with DOE national laboratories in fusion energy development, the first awards to be granted within the INFUSE program. The program does not provide funding directly to the private companies, but instead provides support to the partnering DOE laboratories to enable them to collaborate with their industrial partners, which include Commonwealth Fusion Systems, which got four such awards, TAE Technologies (three awards), Helicity Space (two awards), HyperJet Fusion Corporation and Proton Scientific Technologies.

The DOE has also announced a new programme called Breakthroughs Enabling Thermonuclear-Fusion Energy (BETHE) as part of ARPA-E. It says, as if nothing has happened in the last seventy years, that BETHE aims to support the development of timely, commercially viable fusion energy. Based on numerous studies examining the cost challenges facing advanced nuclear energy, ARPA-E believes that a commercial fusion power plant should aim at an overnight capital cost (OCC)[49] of US$2 billion and $5/W. It adds that "if a grid-ready fusion demonstration can be realized within approximately twenty years (*sic!*) while satisfying these cost metrics, then, as a firm low-carbon energy source, fusion can contribute to meeting global, low-carbon energy demand and cost-effective deep decarbonization in the latter half of the century". This book has failed if it has not made clear that they are fooling

---

[48]https://infuse.ornl.gov/.

[49]See the Glossary for an explanation of this term.

themselves. Energy production by nuclear fusion is and remains a pipe dream. Private enterprise will not be able to change that. It is of course not excluded, and actually quite likely, that this research will lead to useful results in other areas, such as neutron generators and high-temperature superconductors, and that in the end the investment into the private company will reap high returns, but these returns will not be earned from power generation.

Chapter 16
Engineering and Materials Issues

16.1 Introduction

Energy generation by nuclear fusion faces numerous technological challenges. These are in the first place connected with the availability of suitable materials. Materials are required that do not easily become highly radioactive after exposure to the high-energy neutrons generated in the reactor and maintain good mechanical and thermal properties under extreme conditions (considering that temperatures within a fusion reactor range from millions of degrees down to a few hundred degrees below zero (if high-temperature superconductors are used) or even just a few degrees above absolute zero (if ordinary superconductors are used, like in ITER)).

The neutrons produced in D-T fusion reactions have an energy of 14.1 MeV, compared to <2 MeV on average for a neutron emitted in a nuclear fission reaction. The fusion neutron spectrum is consequently much harder than the spectrum of fission neutrons, and, since neutrons carry no charge, they cannot easily be stopped and are able to penetrate several metres into materials before coming to a standstill. As they move within the materials, they produce gamma rays, secondary particles and radioactive nuclei, and create microscopic changes in the material structure which may cause degradation of physical and mechanical properties. Especially the first wall of the reactor vessel will be affected by this. Once the damage has accumulated, the reactor must be shut down, the vacuum vessel vented, the highly activated first wall replaced by remote control, a new one installed, impurities pumped out, etc., before the reactor can be restarted. This kind of interruptions in running a fusion reactor cannot be avoided and could well be more costly than the proceeds from electricity production from the reactor.

Hence, for fusion materials, such radiation-induced damage (neutron activation) is a huge problem and its study necessarily has to go far beyond the damage level in fission materials. Fission materials have always been tested in experimental fission reactors, whereas no facility exists that offers the suitable flux and neutron spectrum required for fusion materials' research. For this we need a fusion relevant neutron source, which is an indispensable step towards the possibly successful development of

© The Author(s), under exclusive license to Springer Nature Switzerland AG 2021
L. J. Reinders, *The Fairy Tale of Nuclear Fusion*,
https://doi.org/10.1007/978-3-030-64344-7_16

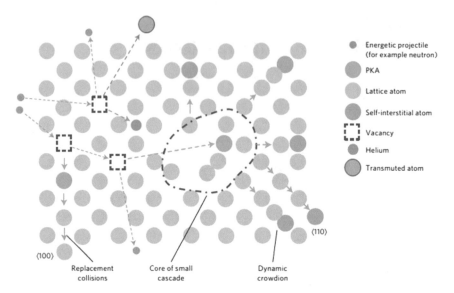

Fig. 16.1 Schematic illustration of radiation damage. In the case of elastic scattering, projectiles with energy E (<14.1 MeV for fusion neutrons) scatter at atoms of the impacted solid, creating primary knock-on atoms (PKAs) in different directions and with different energies. The PKA atom loses its energy by damage production as well as by ionization. The displaced atom, called a self-interstitial atom, can 'annihilate' with another vacancy or can share a regular lattice site with another atom (resulting in a 'crowdion'). Significant amounts of protons and helium particles are created by transmutation reactions, leading to irradiation embrittlement (*from* Knaster et al. 2016)

fusion. None of the currently available neutron sources is adequate for testing neutron activation of fusion materials. Since materials are very sensitive to the specifics of the irradiation conditions, material tests require the neutron source to be comparable to a fusion reactor environment. This a very serious problem and the failure to master the challenges of structural and functional materials would be a further reason for stable burning plasmas for electricity generation to remain but a dream for humankind.

Neutrons of a given energy spectrum and flux induce structural changes in materials. The flux is the number of neutrons that go through each square metre per second. Apart from the flux there is also the fluence, which is how many neutrons have gone through a square metre during the whole life of the material, i.e. the flux integrated over time. In fusion power plants the neutron flux will be of the order of 10^{18} particles $m^{-2}s^{-1}$ and their energy is, as said, a staggering 14.1 MeV, resulting in a neutron wall loading (energy flux)[1] of about 2 MW/m^2. Each neutron impact will give rise to a cascade of collisions which will displace many atoms from their original positions in the material. This has been illustrated in Fig. 16.1 (Knaster et al. 2016). This structural change or damage is expressed in displacements per atom (dpa). Its definition is the number of times that an atom is displaced for a given

---

[1]10^{18} m^{-2}s^{-1} × 14.1 MeV = 14.1 × 10^{18} × 10^6 × 1.6 × 10^{-19} J m^{-2} s^{-1} = 2.25 × 10^6 J m^{-2} s^{-1} = 2.25 MW/m^2.

fluence, whereby one dpa means that, on average, every atom within the material has been displaced once. In both elastic and inelastic collisions, a significant part of the neutron energy is transferred to the recoiling atom (primary knock-on atom, PKA), which is left in an excited state and can in turn displace secondary knock-on atoms.

In addition, the neutrons will generate transmutation reactions in the material (conversions into another chemical element). This damage can be as important as displacement damage. In such transmutations significant amounts of hydrogen and helium will be released and lead to a presently undetermined degradation of structural materials after a few years of operation. These reactions and some of the consequences have also been illustrated in Fig. 16.1. The large helium production (about 40 times more than in fission reactors) is an example of the problems that are encountered in fusion reactors because of the high-energy neutrons striking the first wall, the part of the reactor vessel facing the plasma and bearing the brunt of the onslaught of these neutrons. Even low concentrations of He have a significant impact on mechanical properties. He-induced embrittlement, for instance, already a concern for fission materials, becomes even more critical for fusion materials (Knaster et al. 2017).

The new tritium campaign in JET, which is expected to start in 2020, will also in particular address these issues. JET will generate neutron yields large enough to cause easily measurable activation in materials and degradation in their physical properties. The aim is to reduce the risks and uncertainties associated with ITER operation and maintenance by measuring these quantities in JET and comparing them with simulations and numerical predictions. JET will obtain the first complete and consistent "nuclear case" for a tokamak using the D-T fuel cycle. This includes the accurate measurement of the neutron source and radiation field in the device and the surrounding areas as well as the effects on materials exposed to 14 MeV neutrons, the tritium inventory in plasma-facing materials, the amount and type of waste produced, and the occupational radiation exposure (Batistoni 2016).

Whereas in ITER structural damage in the steel of the device will not exceed 2 dpa at the end of its operational life, damage in a fusion power plant, which will have to deal with hard neutrons and tritium on a continuous basis, is expected to amount to 15 dpa per year of operation. After many dpa the material will swell or shrink and become so brittle as to be useless. Therefore, a commercial fusion reactor will require materials capable of withstanding 150 dpa.

Especially the choice of the material for the reactor vessel's first wall, which directly faces the plasma, and for the divertor is one of the toughest engineering problems faced by fusion research. Before turning to a discussion of these problems, we will first pay attention to the International Fusion Materials Irradiation Facility (IFMIF), the fusion development facility proposed by Europe and Japan, to produce fusion-equivalent neutrons for testing such new materials.

The second toughest problem is the breeding of tritium. As noted before, future power reactors will have to breed their own tritium and ITER will be a testing ground for this. This problem will also be discussed in this chapter. Further problems that deserve attention are the avoidance of plasma-killing disruptions and the mitigation of Edge-Localised Modes (ELMs).

Finally, we will consider in some detail high-temperature superconducting magnets. The magnet system is one of the most important and most expensive parts of any tokamak. In the last decades, after the design of ITER was completed, important progress has been made in such HTS magnets, which could drastically change the fusion playing field. ITER will not much benefit from this development, but future devices undoubtedly will.

16.2 The International Fusion Materials Irradiation Facility (IFMIF)

In Chap. 10 when discussing the squabble about where to locate ITER, we have noted that, for softening the blow of not having secured ITER, Japan would host a €1 billion support facility—IFMIF. It has been in the planning for a very long time now and, in spite of the fact that it is urgently needed, it is still not clear when actual construction of the facility will begin.

Much useful information on materials can be obtained from the experience of fission reactors and from computer simulations. For ITER such information is sufficient, but for the design and construction of any fusion reactor subsequent to ITER a neutron source with a suitable flux and spectrum is an indispensable facility, and results must be available before construction of such a follow-up plant can start. IFMIF will carry out testing and qualification of advanced materials under conditions similar to those of a future fusion power plant.

The IFMIF project was started in 1994 as an international scientific research programme, carried out by Japan, the European Union, the United States, and Russia. Since 2007, the project has been pursued by Japan and the European Union under the Broader Approach Agreement, an agreement for complementary fusion research and development between Euratom and Japan.[2] The agreement was concluded for an initial period of ten years. Recently in March 2020, Euratom and the Japanese government signed a joint declaration to extend the agreement. It does not comprise any new activities, but just continues the projects agreed upon earlier.

Within the Broader Approach three gigantic projects have been set in motion: the design and construction of IFMIF, the foundation of a fusion energy research centre for advanced plasma experimentation and simulation (IFERC), and the completion of an upgrade of the JT-60 tokamak in Japan to an advanced superconducting tokamak (JT-60SA, see Chap. 8). IFERC, like IFMIF based in Japan at Rokkasho, in the very north of Japan's main island Honshu, has been busy designing a DEMO plant since 2007, holding conferences and producing large numbers of papers and reports, which all, I am sure, is very enjoyable, but also very premature as ITER has not even been constructed yet, and will not produce any plasma for close to a decade. Apparently,

---

[2] Agreement between the European Atomic Energy Community and the Government of Japan for the Joint Implementation of the Broader Approach Activities in the Field of Fusion Energy Research, *Official Journal L* 246, 21/09/2007.

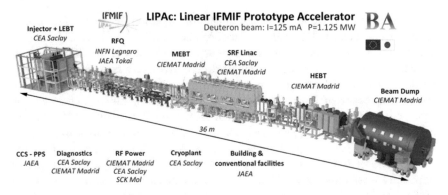

Fig. 16.2 The Accelerator Facility LIPAc for the IFMIF (*from* https://www.iter.org) (The abbreviations in the figure stand for: MEBT/HEBT/LEBT is Medium/High/Low Energy Beam Transport; RFQ is Radio Frequency Quadrupole; SRF is Superconducting Radio Frequency; CCS is Central Control System; PPS is Personnel Protection System.)

it is felt that ITER and its outcome, whatever it will be, has not much relevance for DEMO, but intermediate technological developments will surely make any DEMO design obsolete before any construction can be contemplated. IFERC activity can be followed on its website.[3]

The IFMIF materials testing facility is currently in its EVEDA phase, where EVEDA stands for engineering validation and engineering design activities. The IFMIF/EVEDA project has the task to produce an integrated engineering design of IFMIF and the data necessary for future decisions about the construction, operation, exploitation and decommissioning of IFMIF, and also to validate the continuous and stable operation of each IFMIF subsystem. It consists of two parallel mandates: the engineering design activity (EDA) and the engineering validation activity (EVA). The EDA has by now been completed and the engineering design report of the IFMIF plant has been released (Knaster et al. 2015).

To minimise the risks in constructing IFMIF, prototypes of the Accelerator Facility, the Target Facility and the Test Facility, being the three systems that face the main technological challenges in establishing a fusion relevant neutron source, have been designed and constructed in the validation phase. The Accelerator Facility is a prototype of the IFMIF accelerator device, see Fig. 16.2, completed in 2017 at Rokkasho. IFMIF's basic tool will be an accelerator-driven neutron source, producing a high intensity fast neutron flux with a spectrum similar to that expected at a fusion reactor's first wall. This neutron source is based on Li(d,n) nuclear reactions, where Li(d,n) is shorthand for the reaction, demonstrated theoretically by Rober Serber (1909–1997) already in 1947, of a beam of deuterium nuclei (here indicated by d, the common symbol for a deuterium nucleus which also goes under the name deuteron) impinging on a lithium target and stripped of their proton, while the neutron continues on its way. It is equivalent to the reaction Li + d → n + ?, where the question mark

---

[3] https://www.iferc.org/.

stands for anything possible, including in any case a proton (the proton left over of the deuterium nucleus). It is very similar to spallation whereby a nucleus is hit by a particle with sufficient energy and momentum to knock out several small fragments or smash it into many fragments. Such neutron sources have been developed since the 1970s. The IFMIF tool will consist of two 40 MeV steady-state linear accelerators, creating beams of deuterium nuclei that strike a flowing lithium target under an angle of 20°, and in this way provide an intense neutron flux of about 10^{18} neutrons $m^{-2}s^{-1}$ with a broad peak near 14 MeV, exactly as required to look like nuclear fusion neutrons. The energy of the beam (40 MeV, which means that the deuteron velocity is about 20% of the speed of light, so about 60,000 km/s) and the current of the parallel accelerators (2×125 mA) have been tuned to maximize the neutron flux (10^{18} $m^{-2}s^{-1}$) while creating irradiation conditions comparable to those in the first wall of a fusion reactor. Damage rates higher than 20 dpa per year of full operation are expected to be reached in a volume of 0.5 l of its High Flux Test Module (see below). Note how small this volume actually is, just half a litre.

The LIPAc (Linear IFMIF Prototype Accelerator) in Fig. 16.2 is a 1:1 prototype of the accelerator for the future IFMIF. As can be seen in the figure various labs in Europe are heavily involved in its design and construction.

The second prototype facility, the Target Facility, is a lithium test loop (the EVEDA Lithium Test Loop or EliTe), integrating all elements of the IFMIF lithium target facility. It was completed in 2010 at Oarai in Japan, where the Japan Atomic Energy Agency operates a research centre with a number of nuclear research reactors. It is a small town about 120 km northeast of Tokyo and close to Naka where the JT-60 tokamak is located. The facility was damaged during the Great East Japan Earthquake of 11 March 2011, the most powerful earthquake ever recorded in Japan and also the cause of the Fukushima disaster, but it remained in operation until October 2014 and successfully demonstrated the long-term stability of IFMIF's lithium target operation. In the actual IFMIF lithium loop, the high-energy deuterium nuclei from the accelerator will hit a fast-flowing film of liquid lithium, generating high-energy neutrons which will in turn be directed at small samples of material to be tested. In the test loop, the capacity of the system will be tested to generate a steady 25-millimetre thick "liquid screen" of molten (250 °C) lithium, flowing over a "backplate" at a speed of 15 m per second. The lithium screen fulfils two main functions: to react with the deuterons to generate a stable neutron flux in the forward direction and to dissipate the beam power in a continuous manner. The test loop will also validate several diagnostics systems. Its work is complemented by corrosion experiments performed at a lithium loop at the Brasimone ENEA Research Centre in Italy.

The third facility is the Test Facility consisting of the High Flux Test Module (HFTM) in which the radiation tests will actually be carried out. Given the limited available irradiated volume (just 0.5 litre), the testing samples required are small (typically about 25 mm long). This has been a severe point of criticism, as it is argued to be completely inadequate for testing the large components of ITER and DEMO. A full-scale prototype of the High Flux Test Module, with two instead of four irradiation compartments, was studied in the HELOKA loop facility at the Karlsruhe Institute of Technology in Germany. The Test Facility will provide high, medium and low flux regions ranging from > 20 dpa/(full power year) to <1 dpa/(full power year) with the available irradiating volumes increasing from 0.5 l to 8 l, which will house different metallic and non-metallic materials potentially subject to different irradiation levels in a power plant. It is complemented by the Creep Fatigue Test Module (CFTM) manufactured and tested at full scale at the Paul Scherrer Institute, part of the ETH in Zurich and located in Villigen (Switzerland). The purpose of the CFTM in IFMIF is to perform creep fatigue experiments under intense neutron irradiation. Creep and fatigue are actually two different things, but very similar. The first can be defined as the slow and progressive deformation of a material with time under a constant stress and at high temperature (like the gradual elongation of an elastic band when a weight is hung on it), while fatigue is the weakening of a material caused by cyclic loading that results in progressive and localized structural damage (crack growth and fracture). To limit creep ITER uses several types of so-called austenitic stainless steel for various components. The name refers to the crystalline structure of the metal which is called austenite. It is corrosion resistant, non-magnetic, relatively inexpensive and can operate at temperatures up to about 600 °C, but for fusion reactors something better is needed. Reduced activation ferritic/martensitic (RAFM) steel is the best candidate and is currently the benchmark structural material for the in-vessel components of a fusion reactor (Dolan in Dolan 2013, p. 423). European researchers have been developing such type of steel for several decades now, called Eurofer. And although this would seem to be a perfect and obvious subject for cooperation, China has been developing its own version of this material, called CLAM (Chinese Low Activation Martensitic steel) (Huang et al. 2014), and so have India (Raj et al. 2010) and Japan.

Apart from the three main facilities mentioned above the Post-Irradiation Examination (PIE) facility has to be mentioned whose aim will be to minimize the handling operations of irradiated samples.

Figure 16.3 presents a cut-open view of IFMIF.

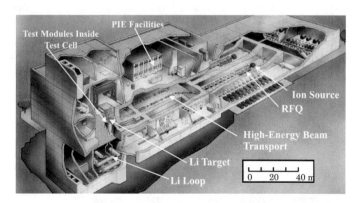

Fig. 16.3   Three-dimensional view of IFMIF (*from* IFMIF Comprehensive Design Report)

Once all this has been validated, the construction of IFMIF can start. When this will be has not yet been decided. According to the IFMIF Comprehensive Design Report[4], after the EVEDA phase a Construction, Operation and Decommissioning Activities (CODA) phase will start. The EVEDA phase was planned to be complete in 2009, but this has now, early 2020, apparently not yet been achieved. Construction of IFMIF is supposed to start immediately after approval of CODA, and will last 7 years for the first half-power operation phase using one accelerator only. The first operation phase will last about three years and the second accelerator will then be built during that same period. A first prototype LIPAc has apparently been completed in Rokkasho, and a second one is soon to be commissioned. After the commissioning of the full-power operation using both accelerators, the steady mode operation will continue for 20 years, with a possible extension for an additional 10 years. It is clear that the construction of IFMIF is way beyond schedule, more than a decade. The CODA has not yet been approved and construction has not yet started, which means that it may take another 20–30 years before IFMIF will make its expected contributions to the nuclear fusion community, including data for the engineering design for DEMO and a multitude of other relevant information.

But of course, in spite of the fact that IFMIF has not yet come off the ground, a successor has already been thought of. It is impossible to escape the relentless pressure of thinking ahead. Don't we always do that? Dreaming about the dessert when the main course has not yet been served? DONES, which stands for DEMO Oriented Neutron Source, is a future version of IFMIF and will take over when the latter's operation has come to an end. It will be designed to mimic the conditions of neutron irradiation in DEMO and allow scientists to test materials in an environment mimicking DEMO conditions and characterize candidate fusion materials. Keen to have some flows of money directed to their countries, Croatia, Poland and Spain have early expressed interest in hosting the facility. Poland has apparently dropped out of the race and, although the location has not yet been decided, Spain and Croatia have now agreed to propose Granada, Spain, as host. Should this not be possible for technical reasons, the project would be hosted in Moslavačka Gora, Croatia. Like IFMIF, the new facility will also be built within the framework of the Broader Approach agreement between the EU and Japan.[5] It is not clear to me though why a completely new facility has to be built for this; one would think that IFMIF can easily be upgraded (the neutrons released in DEMO are very similar to those of ITER, certainly energy-wise) to do the job.

---

[4]IFMIF Comprehensive Design Report (International Energy Agency 2004).

[5]https://fusionforenergy.europa.eu/mediacorner/newsview.aspx?content=1197.

16.3   The Fusion Nuclear Science Facility and Materials Plasma Exposure Experiment

To bridge the large gap between ITER and DEMO, the Fusion Nuclear Science Facility (FNSF) has been proposed in the US, partly as an alternative and partly as a complement to IFMIF. It has grown out of an earlier proposal for a Fusion Development Facility (FDF), a tokamak specially proposed by General Atomics for technology tests. FNSF too will be a tokamak, unlike IFMIF. It will ultimately operate with a D-T plasma and is designed to achieve major advances in plasma duration and fusion nuclear environment through a phased programme reaching power plant operating regimes over a period of ~30 years (Rowcliffe et al. 2018). The device would precede the DEMO, which is supposed to provide routine electricity production and operations. The motivation for a smaller first step, like FNSF, has been based on concerns over the behaviour of materials and components in an actual fusion environment, which cannot be tested in advance of FNSF. FNSF serves as the platform to establish a new database of material and component behaviour. It is strongly anticipated that the actual fusion environment will give rise to behaviour not seen before. A conceptual design for the facility has been presented which, when approved, is scheduled to come on stream in the early 2030s and will be in operation until the mid 2050s (Kessel et al. 2018), but later publications already mention a later timespan from 2040 to 2070. It is unlikely though that the US government, already not showing much generosity in funding nuclear fusion research, will be prepared to sanction expenditure of undoubtedly ITER-like proportions for FNSF. There is consequently not much point in discussing the details of the device in this book.

As a precursor to a possible FNSF, Oak Ridge National Laboratory is designing the Material Plasma Exposure eXperiment (MPEX), a steady-state linear plasma device with superconducting magnets, to address the plasma material interactions of fusion reactors. This development actually receives funding from the federal government, which has been increased from $15 million in 2019 to $21 million in 2020. In the device a steady-state plasma will be generated and confined with superconducting magnets and the experiment will cover the entire range of expected plasma conditions in the divertor of a future fusion reactor (Duckworth et al. 2017).

This route to DEMO proposed by the US differs from the one taken by Europe and Japan, for instance, which are jointly building JT-60SA and IFMIF as complements to ITER. They consider the results from this partnership to be sufficient to be able to reliably build a DEMO plant. In this respect it is also relevant to note that after the shutdown and dismantlement of TFTR, no big tokamak is currently in operation in the US. The DIII-D tokamak, operated since the late 1980s by General Atomics in San Diego, is at present the largest magnetic fusion experiment in the US, together with the NSTX-U spherical tokamak at Princeton, which however has been in the doldrums since 2016. In the next chapter the routes taken by the various ITER members towards DEMO and an actual power plant will be charted in greater detail.

16.4   The First Wall

As said, one of the first and greatest challenges that must be met is to find a suitable material for the first wall, the inner wall of the vacuum vessel that faces the plasma and has to bear the brunt of the neutron onslaught. In ITER the first wall consists of detachable panels. Depending on their position inside the vacuum vessel the panels are subject to different heat fluxes. Two different kinds of panels have been the object of a multi-year qualification programme that has included the fabrication of semi- and full-scale mock-ups and testing: a normal heat flux panel designed for heat fluxes of up to 2 MW/m^2 and an enhanced heat flux panel that can withstand heat fluxes of up to 4.7 MW/m^2. Within ITER's operational lifetime, these panels will be replaced at least once.

Figure 16.1 illustrates various processes that can take place when neutrons hit the first wall. In early tokamaks stainless steel was used, but that is clearly not a high-temperature resistant material. Current tokamaks use carbon fibre composites (CFCs), which are light, strong and high-temperature resistant, but carbon cannot be used in fusion reactors that burn deuterium-tritium as it absorbs tritium (just forming a hydrocarbon as carbon does with other forms of hydrogen), depleting this scarce fuel and weakening the CFC. The same applies to carbon fibre-reinforced graphite (C/C). Tungsten was proposed as a material, but as a high Z (atomic number) material it has so many electrons (74 to be precise) that it cannot be completely ionized and the remaining electrons will radiate energy away, cooling the plasma.

In ITER the first wall will consist of 440 detachable panels made of high-strength copper and stainless steel with a beryllium coating on top. They will cover the 610 m^2 surface of the vacuum vessel. Beryllium has low Z (just 4 electrons), but a low melting point and must therefore be aggressively cooled. ITER will be the first fusion device to operate with such cooling (first tested on JET, which also used beryllium). But beryllium cannot be used in a DEMO device or a real fusion reactor. It melts too easily, and its retention of tritium and its toxicity would be problematic. In short, the conditions that a first-wall material must meet are that it does not absorb tritium, has a low atomic number, takes high temperatures and is resistant to erosion, sputtering (the ejection of microscopic particles from the surface of a solid material when bombarded with highly energetic particles) and neutron damage. Various other proposals have been made, including alloys of tungsten with e.g. iron, nickel, titanium, vanadium and zirconium, or tungsten doped with various oxides (e.g. lanthanum-oxide and yttrium-oxide), but no fully satisfactory material has yet been found. A materials testing facility for testing such materials under fusion equivalent conditions is therefore especially needed for the comparatively large step between ITER and DEMO. The neutron damage in DEMO is expected to be about 80 dpa (compared to < 3 in ITER) and 150 in a real reactor. The difference is due to the fact that ITER is just an experiment, while in a reactor the first wall should last for some 15 years of almost continuous operation before it is replaced (Chen 2011, p. 313ff; https://www.iter.org/mach/Blanket).

A liquid lithium first wall is one of the proposals that has now been abandoned. Promising materials that are still being investigated are silicon carbide (SiC) composites. The advantages of this material include low activation, high temperature capability, relatively low neutron absorption, and radiation resistance. Since the use in combination with tungsten for first-wall, blanket structure and divertor may be desirable, a key challenge is to develop a robust joining technique for SiC and tungsten. One drawback is that at present there is no known method for manufacturing SiC in large quantities. It can be produced by so-called chemical vapour infiltration, a method to form fibre-reinforced composites. However, several technical challenges must be addressed before SiC or SiC composites can be considered for use in fusion energy systems. Apart from the joining technique mentioned above, these include understanding radiation effects (high-dose radiation and transmutation effects) and chemical compatibility with coolant and/or breeding materials (Koyanagi et al. 2018).

The conclusion of this section is that the solution to the problem is still wide open.

16.5   The Divertor

More important still is the material to be used for the divertor, since sixty per cent of the plasma exhaust is designed to go into the divertor, taking the major part of the heat load away from the first wall. In JET a fully tungsten first wall has been tested, but as mentioned above this metal will not be used for the first wall in ITER. The ITER divertor will however be fully tungsten,[6] mainly because it has the highest melting point of all metals, but also to minimize tritium trapping. Although it retains less tritium than beryllium or carbon, this factor is less important, and carbon could in fact also be used, as the divertor parts are easier to replace than parts of the first wall and the tritium can be removed periodically. The heat load on the divertor surfaces is huge, about 20 MW/m^2 (ten times higher than that of a spacecraft re-entering Earth's atmosphere), and cracking and melting under such high loads may damage the tungsten surface and shorten its lifetime, so cooling is essential. In ITER water cooling, which is limited to about 170°C, is possible and will use huge volumes of water for this purpose, but in DEMO or a fusion power plant helium cooling would have to be used. Tungsten alone is a poor structural material, so alloys will be used as a surface layer on RAFM steel.

The Chinese-French tokamak West is equipped with a full ITER-grade tungsten divertor and will start to study these issues in 2021 (see Chap. 6).

---

[6]The first choice for the divertor material was CFC (carbon fibre composite), but budget restrictions forced the ITER Organization to reconsider this and change to a full tungsten variant (Pitts et al. 2013). This change has also consequences for the eventual power output and may well result in ITER being unable to achieve its projected 500 MW of fusion power.

16.6   Tritium Breeding

ITER will obtain the tritium it needs from the market, but future devices are supposed to breed their own tritium as the global supply of tritium will be insufficient for their demands. A tritium-breeding blanket ensuring self-sufficiency in tritium is a compulsory element for DEMO, the step after ITER. DEMO will need 300 g of tritium per day to produce 800 MW of electric power.[7] ITER will test tritium-breeding concepts by testing pilot models of breeding blankets and a special section, the Tritium Breeding Blanket Systems Section, has been set up for this in the ITER Organisation.

A blanket is a thick, massive and complex structure that serves three major purposes: capture the neutrons generated in the fusion reactions, shield the magnets from these neutrons and use them to produce tritium. For ease of replacement the blanket is composed of modules and in a reactor there could be hundreds of blanket modules. ITER already will have 440 blanket modules, each one measuring 1×1.5 metres, weighing up to 4.6 tonnes and covering, like the first wall, the entire inner walls of the vessel. The entire blanket must be inside the vacuum vessel and is the first layer after the first wall (see Figs. 10.7 and 10.8). The neutrons first strike the first wall and then go into the blanket, where their energy is captured. The heat is taken to heat exchangers outside via hot gas or liquid coolants (ITER will be the first fusion device to operate with an actively cooled blanket), and in future reactors will be turned into electricity. The tritium will be bred in the blanket from lithium, as related in Chap. 10, and about 300 kg of ^6Li will be needed per reactor per year.

Tritium-breeding blankets will contain lithium, lead and beryllium, in addition to a structural material. The latter will be the same as used for the first wall, while the lithium can be in the form of solid pebbles or a lithium ceramic, a liquid mixture of lead and lithium, or a lithium and beryllium containing molten salt. In ITER six proposals for tritium-breeding solutions will be operated and tested, with different proposals coming from the EU, Japan, Korea, China and India. They are still in the design stage. The main problem is cooling and taking out all the heat for generating electricity. Blanket designs differ in the way they are cooled, with the main coolants being water, liquid metals and helium.

Figure 16.4 gives a graphic representation of the basic reaction of the tritium-breeding process. One of the problems with the proposed designs is that they can barely breed enough tritium to keep a D-T reactor going. The one neutron created in a fusion reaction is not enough as it can create only one tritium nucleus by hitting a lithium nucleus. Those that hit something else or nothing at all do not give rise to tritium. The beryllium in the blanket however acts as a neutron multiplier. When a high-energetic neutron strikes a beryllium nucleus, the latter breaks up into two helium nuclei and two neutrons, which in turn can react with lithium to produce tritium.

The *helium-cooled-ceramic breeder* (HCCB), the breeder blanket proposed by China, uses solid material (ceramic pebbles) with the beryllium multiplier and the

---

[7]https://www.iter.org/mach/TritiumBreeding.

Fig. 16.4 Tritium
production in the breeder
blanket: a neutron strikes a
lithium nucleus which
disintegrates into a helium
nucleus and a tritium
nucleus. Neutrons are blue
and protons are red in this
figure

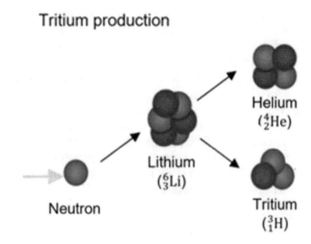

Tritium production

Helium
($^4_2$He)

Lithium
($^6_3$Li)

Neutron

Tritium
($^3_1$H)

lithium breeder in separate compartments. This technology may also be tested in a planned Chinese experimental tokamak, the China Fusion Engineering Test Reactor (CFETR). The *helium-cooled lithium lead* (HCLL) blanket, one of the proposals from the EU, uses a molten alloy of lithium and lead, called a eutectic, meaning 'easily melted' in Greek. A eutectic has the property that it melts at a lower temperature than its constituents. The preferred eutectic is Pb-17Li, which contains 17% lithium enriched to 90% Li^6 and melts at 234 °C, in between the melting points of lead (328 °C) and lithium (181 °C). Lead is a neutron multiplier like beryllium and in this blanket proposal the multiplying and breeding are done in the same liquid.

The Indian proposal, called a *lithium-lead ceramic breeder* (LLCB), uses both a lithium-lead liquid eutectic and a lithium-containing ceramic. The liquid metal cools the tritium breeding zone, and helium cools the plasma-facing wall of the blanket module. Originally developed in collaboration with Russia, this proposal is claimed to give higher tritium production, without an actual number having been given.

The number of tritium nuclei created in the blanket for each incoming neutron is the tritium-breeding ratio (TBR). It has so far not been possible to design a blanket with a TBR larger than 1.15, so no bigger margin than 15%. Since only a small percentage of the tritium injected into the plasma actually fuses (this percentage is called the *fractional burnup*) and produces a neutron that can generate a tritium nucleus in the blanket, the implication is that tritium self-sufficiency can only be achieved after several decades, which is obviously not good enough for a real reactor. The fractional burnup is only a few per cent, in ITER it is expected to be only 0.3%, so the step between ITER and a real reactor is still huge. In earlier tokamaks without a divertor this percentage was much higher, perhaps 30%, because of recycling. Ions of the plasma that would hit the first wall and recombine into neutral gas were re-ionised, re-heated and returned to the plasma. In tokamaks fitted with a divertor such ions are prevented from hitting the wall, led to the divertor and pumped out, so not recycled. ITER will use up most of the tritium available in the world, so there is some urgency to develop breeding blankets with higher TBR (Chen 2011, p. 320 ff). Advances with breeder-blanket design are being made, and as said six will be tested in ITER, but there is no guarantee that any of the proposals will be satisfactory. The failure to develop a suitable tritium-breeding blanket would actually be a showstopper for D-T fusion, as there would simply not be enough tritium to do the job.

An added complication is that tritium is mildly radioactive, as its two neutrons in the nucleus do not sit well with a single proton. It decays with a half-life of 12.3 years by emitting an electron (β-decay) and converting into helium-3. So, it is continually lost, 5.5% per year. The emitted electron has very low energy and cannot penetrate the skin, and even in air it can go only 6 mm, but it can be a radiation hazard when inhaled, ingested via food or water, or absorbed through the skin, so special precautions must be taken.

16.7   Disruptions

The phenomenon of disruptions was already introduced in Chap. 2 of this book, and elaborated on in Chap. 8 when discussing such events in JET and TFTR. A disruption occurs when an instability grows in the tokamak plasma with a rapid loss of the stored thermal and magnetic energy. It is a violent event that causes the current in the tokamak to abruptly terminate, resulting in the loss of temperature and confinement. The amount of heat in a large experiment like ITER will be about 400 MJ, equivalent to the explosive yield of 80–90 kg of TNT, that means the power of a fairly large bomb. In addition, another 400 MJ of energy is held by the poloidal magnetic field (created by the tokamak current). The toroidal magnetic field actually holds much more energy, but as long as the toroidal field coils remain undamaged that energy will not be released in a disruption. The big tokamaks JET and TFTR suffered from such violent events. Fortunately, not every disruption is violent and JET actually experienced a few thousand disruptions in the last decade of its operations. They are however much more powerful in ITER and therefore extremely dangerous, capable of causing considerable damage. The large thermal and magnetic energy contained in a full-performance discharge means that it is essential for ITER to have an exceedingly low rate of disruptions by the time it reaches D-T operation. Unmitigated disruptions at plasma currents above 8.4 MA may be so severe that they can only be allowed to happen once or twice in the machine's lifetime, and a disruption at ITER's planned current of 15 MA must consequently be avoided at all cost. Since the cause of disruptions is still poorly understood, prevention or mitigation is essential. If this turns out to be impossible, the entire structure of the device has to beefed up to be able to absorb all the energy that is released in a disruption. This of course would considerably add to the costs.

The damage caused by a disruption is threefold. First, the plasma's heat is deposited into the walls and causes them to vaporize in spots (this is called thermal quench). Even if most of the heat can be channelled into the divertor, there is no time for it to be conducted away and the tungsten and carbon in the divertor will also vaporize. Secondly, the plasma current decreases very rapidly (current quench) and causes a counter current to be driven in the conducting parts of the confining vessel. This current will exert a tremendous force on the vessel, and, if it is not sturdy enough, move or deform it. Thirdly, there are the runaway electrons; electrons accelerated to high speed in the toroidal electric field that never find an ion to collide with and literally 'run away'. They can amount to 50–70% of the original tokamak current and are dangerous for the plasma facing components (Chen 2003, p. 291).

It is an extremely serious problem and as recently as May 2018 the Science and Technology Advisory Committee (STAC) of the ITER Council characterized disruptions as "a serious threat to ITER's mission." No wonder that the disruption mitigation system is considered one of ITER's key systems to ensure its reliable and successful operation. In that context and the advanced state of ITER's construction it is a little surprising that only in 2018 a special Disruption Mitigation Task Force was established for designing such a disruption mitigation system. The strategy of tackling the problem had apparently first to be agreed upon before the Task Force could get off the ground. The design of such a system has proved challenging because of the complex physics involved in stopping runaway electrons. During plasma disruptions in ITER massive generation of such electrons is expected. The disruption mitigation system has to protect the plasma-facing components against the heat and the forces that arise during the disruption, and at the same time it must tame the runaway electrons, which could cause melting of the first wall and leaks in the water cooling circuits.

The concept chosen for the ITER disruption mitigation system is based on so-called shattered pellet injection (SPI), a technique developed at Oak Ridge National Laboratory (Baylor et al. 2019) and pioneered at General Atomics' DIII-D tokamak in San Diego. The SPI pre-empts an abrupt termination of the plasma, i.e. a disruption, by shooting frozen deuterium-neon pellets into the plasma which brings the temperature down significantly and increases the density (which suppresses the runaway electrons). It is a safe way to dissipate plasma energy and to minimise the damage to in-vessel components from the disruption. The pellets are shattered into small pieces just before entering into a disrupting plasma. The largest pellets are larger than a wine cork, with a diameter of 28 mm. Despite this "enormous" size for a cryogenic pellet, several of these have to be fired at the same time to reach the required quantities to stop the worst-case runaway electron beam in ITER. Several tokamak experiments are planned and in preparation to gain experience with this system, e.g. at JET and KSTAR.[8] Recently (October 2019) JET obtained an important result in countering (the adverse consequences of) disruptions; a result that is also relevant to ITER. Using the SPI-technology on JET offers a better and more complete picture of what to expect when deployed on ITER. The deuterium pellet shoots out at a staggering 250 metres per second, nearly the same speed as for a jet airliner travelling at full altitude.[9]

---

[8]https://www.iter.org/newsline/-/3183.

[9]http://www.ccfe.ac.uk/news_detail.aspx?id=488.

It should be emphasized though that such techniques do not prevent the disruption, but only mitigate the consequences. The result is still a machine that has to be emptied and started up anew (TFTR for instance needed more than a month for recovery after big disruptions), not something you want to happen too often when running a power station. It is also clear that the collapse of the plasma occurs on such a fast time scale (the energy sprays out in a matter of 10 ms) that the pellet delivery time must be very precise and that a few milliseconds difference in delivery times can be detrimental to the mitigation achieved. Figure 5.2 illustrates what happens in a disruption, while also showing the timescale on which this happens.

Prevention of disruptions, which must be the eventual aim for a power generating device, is at a much less advanced stage, although some progress is being made (Strait et al. 2019). As often in nuclear fusion science, the basic knowledge is lacking and the search for a solution proceeds backwards. Disruptions occur when the system is pushed beyond or too close to its limits and control over the plasma is lost, in particular the density limit and the pressure or *beta* limit. The mechanism behind these limits is not fully understood. Impurities can also give rise to instabilities and disruption as they affect pressure and current density. An average over all tokamaks shows that in 13% of the pulses a disruption has been suffered (Chen 2011, p. 292), while JET achieved an average rate of 3.4% unintentional disruptions in the 2008–2010 period, but the disruption rate increased significantly following the change from a carbon wall to an ITER-like metal wall (Strait et al. 2019, p. 9). In ITER nearly disruption-free operation will be required, in addition to highly reliable mitigation of any disruptions that do occur, whereby disruption mitigation should be a rarely-used last resort.

In addition to the above, the alpha particles produced in the fusion reactions can also be a cause of disruptions, a problem that cannot be studied before ignition has been achieved. As the alpha particles cool down, they transfer their energy to the plasma keeping it hot. Before this happens, they stream in the form of beams along the magnetic field lines and can excite so-called Alfvén wave instabilities, electromagnetic waves that become so strong that they disrupt the plasma (Chen 2010, p. 342).

16.8   Suppression of Large ELMs

Edge Localized Modes (ELMs) have shown up before in this book. See e.g. Chap. 2, where they were described as a way for the plasma to let off steam when the pressure is building up, and Chap. 6. ELMs are very nasty instabilities and could easily spoil confinement in ITER. A tokamak requires strong and carefully engineered magnetic fields, but even then part of the plasma escapes. As we have seen, in 1982 the "high-confinement" mode (H-mode) was discovered by chance in the German ASDEX tokamak. As the plasma was heated, plasma losses abruptly dropped and turbulences at the plasma edges practically disappeared. Confinement improved and energy was retained more effectively. But, for this more stable confinement a price had to be paid as pressure built up at the plasma's edge and a new type of quasiperiodic instability appeared. The H-mode plasma abruptly experiences "storms" amid the calm, akin to solar flares on the Sun. These are called ELMs, instabilities at the edge of the plasma, as their name suggests, and ejecting a jet of hot material. With each ELM, the surface of the vessel faces a sudden increase of temperature by thousands of degrees and a large ELM can represent 5–10% of the total energy stored in a fusion plasma. Especially the divertor and the plasma facing components will suffer from the onslaught of ELMs. At ITER thousands of ELMs are expected and it was found that ELMs are more damaging than originally thought. Type-I (giant) ELMs can deposit enough energy to melt some of the beryllium wall, so their occurrence must be minimized by stimulating smaller Type-III ELMs (Dolan 2013, p. 422).

The spherical tokamak MAST at Culham has been using an ELM mitigation technique called resonant magnetic perturbation, which consists of applying small magnetic fields around the tokamak to punch holes in the plasma edge and release the pressure in a measured way. This technique has been successful in curbing ELMs on several tokamaks. A resonant magnetic perturbation is an externally-induced small perturbation of the equilibrium magnetic field. It shakes the plasma and throws particles off course as they move around the magnetic field lines in the plasma, changing their route and destination. Changing the shape of a small area of the plasma in this way lowers the pressure threshold at which ELMs are triggered. This should produce a stream of smaller, less powerful ELMs that will not damage the tokamak. [10]

To utilize this mitigation technique in ITER, its design was changed to include additional electromagnetic coils that will attempt to control the ELMs by providing such magnetic perturbations. These coils will be very close to the inner wall, thus they must resist large thermal expansions due to the high temperatures, while ensuring to be reliable for 20 years or so. The final ITER design now includes such ELM coils, but of course practice must show whether they indeed cure the problem (Campbell et al. 2019).

Another way of trying to calm down ELM instabilities was dubbed the "snowball in hell" technique. It consists of throwing a small pellet of cold fuel directly into the plasma, which produces a minor instability that, somehow, prevents the much larger ELMs. It is similar to the frozen-pellet injection discussed above for mitigating disruptions.

Several other ways to tame ELMs are being investigated. For instance, at PPPL a device is tested on the DIII-D tokamak operated by General Atomics in San Diego to diminish the size of ELMs. It injects granular lithium particles into tokamak plasmas to increase the frequency of the ELMs. The method aims to make ELMs smaller and reduce the amount of heat that strikes the divertor.[11]

It is clear that the final and definitive solution has not yet been found. ELM control is required for the achievement of fusion energy in ITER. Two of the schemes

---

[10]https://www.iter.org/doc/www/content/com/Lists/Stories/Attachments/1229/ELM%20lobes%20PR.pdf.

[11]https://www.iter.org/newsline/-/2010.

mentioned above to minimize the impact of ELMs are currently preferred, pellet injection and in-vessel ELM control coils.

16.9 Superconducting Magnets

ITER will use ordinary superconducting magnets (low-temperature superconducting (LTS) magnets), which have to be cooled by liquid helium to a few degrees above absolute zero, which obviously is no mean feat, knowing that temperatures just a few metres away will reach millions of degrees. At the time when ITER was designed there was no choice, as high-temperature superconductors (HTS) had yet to be discovered. So, the recent advances made in HTS magnets are a considerable step forward, not just for fusion, but also for countless other applications.

A superconducting magnet is an electromagnet made from coils of supercon-ducting wires. The important advantages of superconducting magnets are that they can produce greater magnetic fields (20–50 T) than ordinary non-superconducting magnets (usually limited to about 16 T) and are cheaper to operate as no energy is dissipated as heat in the windings. To cool such magnets liquid helium is mostly used, even for magnets with critical temperatures far above helium's boiling point of $-$ 269 °C. This is because the lower the temperature, the better superconducting wind-ings work—the higher the currents and magnetic fields they can withstand without returning to their non-superconducting state.

The breakthrough in HTS (a real one; not the spurious ones often encountered in nuclear fusion research) was the discovery of high-temperature superconductors by IBM researchers J. Georg Bednorz and K. Alex Muller in 1986, who went on to win the 1987 Nobel Prize in physics for their work (with the shortest interval ever between a discovery and a Nobel award). The definition of high-temperature superconductor is unfortunately not unique; preferably the term should be reserved for superconductors with critical temperature above the boiling point of liquid nitrogen, but that would even exclude the original discovery.

Bednorz and Müller discovered their superconductivity in a new class of ceramics: the copper oxides or cuprates. The resistance of the particular oxide they studied dropped to zero at around -238 °C. While this development was of course extremely exciting for physics, the physical form of the new compound, small black crystals, was not obviously suited for practical purposes, and it took a decade or two to turn the discovery into engineering materials and commercial products.

Cuprate superconductors are made of layers of copper oxides (CuO_2) alternating with layers of charge reservoirs (CR), which are oxides of other metals. These copper oxides include the family of rare-earth barium copper oxides (ReBCO), which we already encountered in Chaps. 11 and 15. The rare-earth metals form a set of seven-teen chemical elements in the periodic table (see Fig. 1.3), specifically the fifteen lanthanides (with $Z = 57$–71 and collectively denoted by Ln), as well as the transition metals scandium Sc ($Z = 21$) and yttrium Y ($Z = 39$). Any rare-earth element can be used in a ReBCO and made into tape to be wrapped around a conductor; popular choices include yttrium (YBCO), lanthanum (LBCO), samarium, neodymium and gadolinium. The cuprate of barium and yttrium (YBCO) was the first superconductor found above liquid nitrogen boiling point (-195.8 °C). The significance of this is the much lower cost of the refrigerant needed to cool the material below the critical temperature. YBCO can actually be made with tools and equipment available in a high-school science laboratories (Grant 1987).

Other HTS are made of bismuth strontium calcium copper oxide, or BSCCO, also a type of curate superconductor, discovered in 1988 and the first HTS which did not contain a rare-earth element. BSCCO, with YBCO the most studied cuprate superconductor, was the first HTS material to be used for making practical super-conducting wires. There is an analogous thallium family of thallium barium calcium cuprates, referred to as TBCCO, and a mercury family (HBCCO).

Research in HTS is vibrant and new superconducting materials are discovered all the time. Figure 16.5. shows a timeline of discoveries. Other promising candidates for HTS are iron-based superconductors discovered in 2006, which contain layers of iron and a pnictogen (such as arsenic or phosphorus or any other element in group 15 of the periodic table, also known as the nitrogen family) or a chalcogen (the elements in group 16 of the periodic table, also known as the oxygen family).

Since May 2019 the record for HTS stands at a mere -23 °C, achieved with a material, called lanthanum hydride H_3La (not shown in Fig. 16.5), under very high pressure (more than one million times Earth's atmospheric pressure, something not easily achievable in a nuclear fusion reactor) by the same people who held the

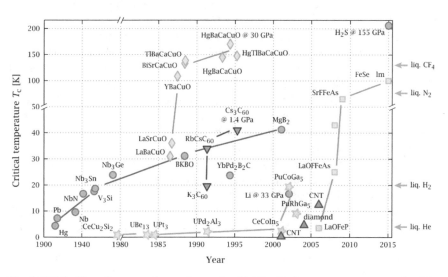

Fig. 16.5 Timeline of superconductor discoveries. The liquid nitrogen temperature on the right usually divides high-temperature from low temperature superconductors. Cuprates are displayed as blue diamonds; iron-based superconductors as yellow squares. Low-temperature superconductors are displayed for reference as green circles (P. J. Ray, CC BY-SA 4.0, https://commons.wikimedia. org/w/index.php?curid=46193149)

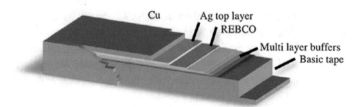

Fig. 16.6 Layout of a ReBCO tape. The width of the tape depends on the fabrication possibilities (typically 12 mm) (*from* Fietz et al. 2013)

previous record of −70 °C with hydrogen sulphide (Drozdov et al. 2019).[12] The race is now on for room-temperature superconductors whose discovery would revolutionise all kinds of technical applications, e.g. electrical efficiency, vastly improving power grids, high-speed data transfer, and electrical motors.

ReBCO HTS is now available in thin steel ribbons, with an ultra-thin layer of superconducting material deposited on it. The principal layout of second-generation ReBCO tape is shown in Fig. 16.6. This is available in long length and with high critical currents, the critical current being the current above which the material is normal and below which the material is superconducting. The total thickness of the tape is typically 0.1 mm, with the ReBCO layer being only 1.5 μm = 0.0015 mm.

Because of their superior superconducting properties ReBCO is the only appropriate candidate for higher temperatures and magnetic fields (Fietz et al. 2013).

HTS magnets have been built with fields up to 42 T, proving that the superconductor can operate at the fields needed for fusion. Only in the last three to five years the performance of the superconductor, and the ability to manufacture it in sufficient quantities, has been adequate for fusion applications. Still, significant R&D will be needed to enable the fabrication of large-volume, high-field magnets that will be useful for fusion. Although "rare earths" is the name given by chemists to the particular group of elements used in ReBCO superconductors, they are actually not so rare that scarcity of their supply would pose a problem for a fusion energy economy. Thousand fusion power plants a year could be produced with YBCO superconductors before it would even have a 1% impact on the current rare-earths market.

One of the further important advantages of HTS magnets is that their operating temperature need not be controlled precisely (temperature control can be relaxed to a few degrees rather than a few tenths of a degree in conventional LTS), greatly simplifying the cryogenic system. One of the goals of the search for HTS magnets to be used in nuclear fusion devices is to build magnets that can be cooled by liquid nitrogen alone (to about −190 °C).

Currently (2018) the prices for ReBCO tape are in the range of $15–30,000 per kg, which is still a factor of 5–10 higher than the price for Nb_3Sn wires as used in ITER. The delivery of long-length tapes, longer than 200–300 m, is still challenging. Improvements in increasing current-carrying capability and fabrication process may

---

[12]https://www.nature.com/articles/d41586-019-01583-y.

continue to bring down the price. Commercial tape manufacturers also predict significant cost reduction when a market for fusion devices develops. In view of the ITER project similar prospects and predictions were also made for Nb_3Sn in the 1990s, but the market price remained unchanged while worldwide production scaled up by over an order of magnitude. In the ITER project, the cost of the magnet system is estimated to be about 28% of the overall cost of the plant components and the cost of the LTS material (NbTi and Nb_3Sn) is almost half of the magnet cost, i.e. about 14% of the total. If the ITER TF coils would be made from HTS materials, the overall cost of ITER would double and the magnet system would cost about 70% of the total. From this it is clear that the use of HTS magnets would further increase the costs of a fusion power plant, although for ReBCO tapes, a drastic cost reduction may be possible if the production process could be simplified. For BSCCO wires, even if the process and material cost were reduced, the use of silver in its production has a significant impact on raw materials costs and sets a quite high lower limit. In specific cases, such as current leads, the use of expensive HTS can be economically justified. HTS leads cost up to ten times more than conventional copper leads but pay back these costs with savings in cooling power. Similarly, in the central solenoid of the EUROfusion DEMO and CFETR, some parts can be made from HTS, reducing the size of the solenoid and thus paying back the higher capital cost.

In conclusion it is fair to say that HTS technology is very promising for fusion applications and, for that matter, for other applications as well. New high-temperature superconductors are being discovered all the time. The time scale for the development, qualification and demonstration of HTS-magnet technology for fusion is of the order of 10 years if adequate funding is available. So far, no showstoppers have been identified, but the major objection to their use remains the high cost, further enhancing the already staggering capital costs of fusion devices (Bruzzone 2018).

16.10 Conclusion

In this chapter we have identified some obstacles, and potential showstoppers, on the way to a practical energy generating fusion reactor. In the first place the development of materials that can withstand the flux of high-energy neutrons originating from fusion reactions. Such materials are especially required for the first wall and divertor of the device. Far too little progress has been made in this respect and neutron sources with a fusion-relevant neutron spectrum must urgently be built for a reliable design of a DEMO plant or pilot fusion plant to be possible.

Secondly, there are the difficulties in breeding sufficient tritium to make a fusion reactor self-sufficient in tritium. Various breeding-blanket proposals will be tested in ITER. It is however very unlikely in my view that tritium self-sufficiency will ever be achieved. The tritium-breeding ratio is bound to remain too low and the scarcity of some materials (beryllium, lithium; see Chap. 19) may also play a disruptive role in this respect.

In addition there are the long-standing physics issues of disruptions and instabilities, in particular large ELMs. Instabilities and disruptions are part and parcel of fusion plasmas, and have to be lived with. It does not help of course that the physics of these phenomena is not very well understood either. The question is whether mitigation techniques can be developed that can sufficiently moderate and mitigate such events without shutting down the power generation of the plant.

We ended with the positive story of the development of high-temperature superconductors, which mainly took place outside fusion, but has nevertheless a positive impact on fusion, as it will in future greatly reduce the cost of the magnet system of fusion reactors.

Chapter 17
Post-ITER: DEMO and Fusion Power Plants

17.1 Introduction

After ITER the path to a commercial fusion reactor is still a very long one, even if ITER turns out to be the greatest success in nuclear fusion the world has ever seen. Although not all approaches are the same, the route is fairly clear and consists of two further steps before one can start thinking about building a full-size nuclear fusion power plant, to be operated by industry.

The first step consists of the construction of one or several large machines like IFMIF (or FNFS) and DONES for solving the engineering and materials problems discussed in the previous chapter. The toughest engineering problems are the material of the vacuum vessel's first wall and tritium breeding, added to the physics problems of disruptions and instabilities (ELMs). The solutions presented in the last chapter for disruptions and ELMs (the controlled injection of fuel (gas puffing), internal correction coils) are too crude to be suitable for DEMO or power plants. Nevertheless, some believe that such a first step is not necessary and that the ITER experiments will give enough information to design DEMO, but in that case these ITER experiments have to be waited for, which in the current impatient climate is apparently not what one wants to do either.

The second step will then be the construction of DEMO, a proto-type reactor built to run like a real reactor but not producing full power. DEMO designs are already well underway, as we have seen, even before the above-mentioned first step has been made and even before ITER has been completed and carried out any experiments. The most amazing thing in this whole business is that all DEMO designs aim to burn deuterium and tritium, while the only experience with tritium is a few minutes of experiments with JET and TFTR.

In spite of all the hullabaloo and optimistic stories around it, ITER has turned out to be a cumbersome project and hideously expensive, so expensive that one would think it unlikely for any one country to be able to go it alone to a power plant or wanting to run the associated risks of a dismal failure. Nevertheless, several of the current partners in the ITER consortium, which are also the only serious

L. J. Reinders, *The Fairy Tale of Nuclear Fusion*,
https://doi.org/10.1007/978-3-030-64344-7_17

players in the nuclear fusion power game, seem intent to do so. The consortium that joined forces to build ITER is unlikely to continue into the post-ITER period. The spirit of collaboration that characterised the early activities on ITER has apparently not whetted the appetite for more of the same for subsequent stages of the fusion effort, despite the fact that future devices will probably be even more costly than ITER and might greatly benefit from collaboration. What is the point of six or seven almost identical DEMO plants? With multiple countries already now starting on their own (premature) design for a DEMO or power plant, one actually wonders whether there is any point in continuing with ITER after its construction is completed. The construction will teach engineers many lessons that are useful for building a DEMO, but the ITER design itself dates back thirty years and for DEMO or a real power plant much more advanced technology will be used. A DEMO, if ever built, should be a very different machine, although the first glimpses of the DEMO designs available show a very ITER-like device. The versions that are being designed now cannot benefit much from experiments to be carried out on ITER after 2035 or so, when according to the current schedule its most important operations, those with deuterium and tritium, will start. ITER will probably continue as long as the EU continues to pay almost 50% of the cost, as it guarantees a cheap ride for the other ITER Members. That may change when construction finishes and the EU contribution will drop to 34% and if the cost of running ITER turns out to be higher than currently expected. In view of the different routes the various countries seem to take, it will be very hard in future to forge an ITER-like collaboration for DEMO or for a real power plant, which, if the EU Roadmap is any guide in this, must be built by or in close cooperation with industry, i.e. private companies, while in view of the complexity and cost of such a project collaboration between nations would be the only way to possibly achieve success. Industry will of course always be prepared to 'contribute' if money can be earned. For operating a nuclear fusion device, it will be very hard though to find enthusiasts among currently operating utilities companies that, being used to the simple process of burning coal, oil or gas, are prepared to run such a complicated device.

17.2   Plans of the Various ITER Members

Already late in 2013 the ITER Members presented their own projects for DEMO during the 2013 Monaco ITER International Fusion Energy Days (MIIFED 2013). All have the objective of building their machine by 2050. The various members are exploring different routes towards DEMO and some different regroupings have taken place, whereby some countries (Russia, South Korea, India) seem to go it mostly alone or not at all (US). Japan, Korea, India, Europe and Russia stated their intention to begin building DEMO in the early 2030s in order to operate it in the 2040s. This time schedule seems currently a little too ambitious, but in fusion land time limits were never sacred.

The conceptual designs all sketched out a machine that is larger than ITER with a major radius—which determines the size of the machine—ranging from 6 to 10 m. In comparison, ITER's major radius is 6.2 m and JET's half of that. As far as power is concerned the designs varied—from an electricity output of around 500 MW for the European DEMO to 1500 MW for the early Japanese DEMO (which has now been scaled down).[1] The ambitions for DEMO also differ. For some Members, DEMO will be a pre-industrial demonstration reactor; for others, it will be a quasi-prototype that requires no further experimental step before the construction of an industrial-scale fusion reactor. In this range of possibilities, the Russian DEMO project stands out from the others, as it is a hybrid combining fission and fusion within the same device.[2]

As mentioned, unlike ITER most work on DEMO has and is being done without much international collaboration although Europe and Japan are cooperating on DEMO design work as part of the Broader Approach agreement. Within that framework they are building the powerful JT-60SA tokamak in Naka, Japan, as a complement to ITER. In addition, the joint programme includes IFMIF/EVEDA and IFERC (see previous chapter). Part of the latter is the DEMO Design and R&D Coordination Centre, which coordinates design and R&D activities for DEMO and ITER. Construction of DEMO is however not part of the Broader Approach agreement, which does not go beyond the ITER construction phase. ITER serves as the school where physicists and engineers will learn how to build DEMO. Joint construction of DEMO by Japan and the EU is not foreseen at present.

In 2018, the Japanese authorities reviewed the Japanese strategy for the development of a DEMO fusion reactor. Since 2005 Japan based its fusion energy development programme on the "Future Fusion Research and Development Strategy." But in the years thereafter the fusion landscape changed, especially due the change of public opinion towards nuclear power in general, including fusion, after the Fukushima disaster, as well as by developments in Broader Approach activities and increased interest in renewable energy sources. The new guideline states that the decision to transition to a DEMO reactor phase will be taken in the 2030s when ITER will demonstrate D-T burning plasmas, with the added condition that the economic feasibility of a commercial reactor must be foreseeable when proceeding to the DEMO phase. As objectives of the DEMO reactor it mentions the realization of a steady and stable electricity output beyond several hundred megawatts (the earlier target for this was in the gigawatt range) and fuel self-sufficiency by tritium breeding. In addition,

---

[1] A 1500 MW electric power plant would be equivalent to a third-generation fission reactor of the EPR type (European Pressurized Reactor) as under construction in Flamanville, France, or Olkiluoto, Finland. It is interesting to note that the costs for the Olkiluoto reactor was initially (in 2005) estimated at €3.7 billion, which has now gone up to more than €8 billion (in 2012). Regular electricity generation is due to start in March 2021. The construction of the EPR unit at Flamanville (which already operated two older units) started in 2007 and is scheduled to start operation in 2022. Its cost ballooned from €3.3 billion (2007) to €12.4 billion (2019). Both plants took 16 years to construct at enormous cost, reflecting the sad fact that everything nuclear seems to be doomed.

[2] https://www.iter.org/mag/3/22.

safety, an acceptable level of construction costs, and a flexible blanket and divertor are important issues.[3]

Although the decision on DEMO has now been postponed to the 2030s, Japan is nevertheless currently working on the design of a steady-state JA DEMO with a major radius of 8.5 m, minor radius 2.42 m and fusion power of 1.5–2 GW (three or four times more than ITER, corresponding to an electricity output of roughly one third of this) based on a water-cooled solid tritium-breeding blanket and niobium-tin superconducting magnets like ITER (no HTS). A Joint Special Design Team for Fusion DEMO is in charge of this, a collaboration of 23 institutions, universities and industrial enterprises, with a conceptual design stage until 2027, an engineering design stage until 2032 and after that the manufacturing design stage until 2036.[4] Compared to earlier designs there is a change from more compact DEMOs producing high fusion power (about 3 GW) to lower fusion power and less compact (larger radius) devices. From these earlier DEMO studies it was learned that handling such a high power with existing or foreseeable technology, especially in divertor heat removal, would be difficult and led to the compact options being abandoned, a point that might also be relevant for other compact tokamaks, such as spherical tokamaks (Tobita et al. 2019).

According to the European Roadmap DEMO will mark the very first step of fusion power into the European energy market by supplying between 300 and 500 MW of net electricity to the grid, which is approximately 3–4 times less than an average nuclear fission reactor. Europe's DEMO would be a "demonstration power plant" to be followed by the first-of-a-kind fusion power plant. It will largely build on the ITER experience and will have (1) to breed its own tritium (closed fuel cycle); (2) to demonstrate materials capable of handling the flow of neutrons produced in the fusion reactions; (3) to demonstrate safety and environmental sustainability, and (4) sufficient technology to allow a first commercial power plant to be constructed. Very little of this has been accomplished and the experience with ITER, although useful, will not help that much. The demands differ very little from the Japanese demands for DEMO.

Figure 17.1 shows an artist's impression of a future EU DEMO plant, which looks very much like ITER, except for the blue clad building at the centre of the illustration, where the thermal power generated by the tokamak will be converted into electric power that will leave the site via the attached power lines. An interesting question is whether, calculated over the lifetime of the plant, the power leaving the site via these lines will exceed the power coming in through the power lines on the left. Only in that case, it will be a net energy producing device.

At first, according to the 2012 Roadmap, the conceptual design of DEMO was expected to be completed by 2017, with construction beginning around 2024 and the first phase of operation commencing from 2033. Following the delays in ITER, this has since also been delayed with early conceptual design(s) of a European DEMO

---

[3]https://www.mext.go.jp/component/b_menu/shingi/toushin/__icsFiles/afieldfile/2019/02/18/140 0137_02.pdf.

[4]https://www.fusion.qst.go.jp/rokkasyo/en/project/reactor-sp.html.

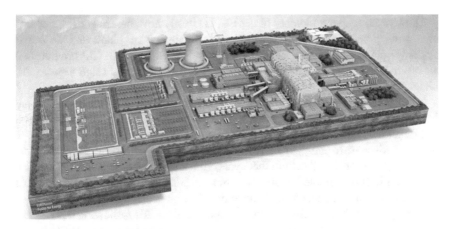

Fig. 17.1 The European DEMO, very similar to ITER. On the left the power comes in and in the middle at the top the generated power leaves the site (*from* EUROfusion and Fusion for Energy)

expected around 2027 and construction now planned for after 2040. After high power burning plasmas have been demonstrated in ITER, DEMO will be operational for around 20 years, i.e. until after 2065 if there are not too many more delays (Fig. 17.2) (Federici et al. 2019).

To achieve fusion electricity in the second half of the century, a European DEMO construction has to start in the early 2040s, shortly after ITER is supposed to achieve its $Q = 10$ milestone, if it ever does so. Therefore, engineering design for a DEMO will become a major activity after 2030.

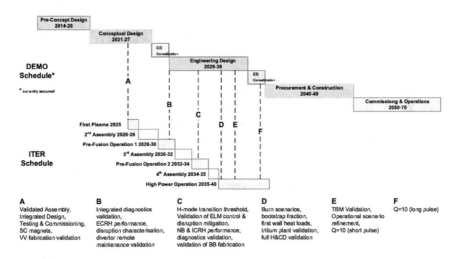

Fig. 17.2 Overview of phasing and key technical inputs from ITER and DEMO schedule (*from* Frederici et al. 2019)

The UK, no longer a member of the EU, is already now, before ITER has shown any results, going ahead with its own plans for fusion power plants and has (rather suddenly) placed all its eggs in the spherical tokamak basket (STEP programme), as reported in Chap. 11. All other designs are still conventional (high aspect ratio) tokamaks.

Within the framework of SIFFER (Sino-French Fusion Energy Center) China and France[5] have joined forces and made a planning for the China Fusion Engineering Test Reactor (CFETR), which aims to bridge the gap between ITER and DEMO. Its main objectives are to complement ITER; produce 200 MW of fusion power (less than ITER); demonstrate a full tritium fuel cycle, with a tritium-breeding ratio above 1.0; explore options for remote handling techniques; address the physical and technical solutions for achieving steady state operation. It took nearly four years to complete the conceptual design of CFETR, which was achieved in 2017, showing that China is in a hurry. A preliminary estimate of the total project cost of a fully superconducting tokamak, which CFETR will be, was about $4 billion, an astonishingly low figure when compared to ITER. A comparison was also made with the cost of a tokamak project with copper magnets, which was about $2 billion. The electricity cost per annum in the D-T burning operation stage of the copper magnet tokamak option is expected to be more than twice that of the superconducting tokamak option ($600 million compared to $250 million per year (based on 2009 electricity prices in China)) (Chen et al. 2015). China has now progressed to the detailed engineering stage of CFETR and construction is planned to start in the early 2020s. It will be upgraded to a full DEMO in the 2030s (see Chap. 7 and Fig. 7.3 for an artist's impression of the future CFETR) to obtain fusion power production of 1000 MW, a fusion gain (Q) higher than 12 and tritium self-sufficiency. The programme implies that China's DEMO might be (close to) working before ITER has started its D-T operations. In any case it shows that at least for China ITER is apparently not essential for designing a DEMO reactor. The construction of a Prototype Fusion Power Plant (PFPP) (about 1 GW of electric energy) will then be completed in 2050–2060 and be the final step in the Chinese roadmap towards a commercial fusion power plant (Wan et al. 2017; https://www.neimagazine.com/features/featurechina-fusion-roadmap-7436879/.)

India has announced plans to start building a device called SST-2 (Steady State Tokamak-2) (Srinivasan et al. 2016). It will be a new D-T machine with the aim to test and develop components for a demonstration plant around 2027. The SST-2 reactor is a medium-sized device with low fusion gain ($Q \sim 3$–5). The fusion power output can be from 100 to 300 MW. It is apparently thought that ITER ($Q = 10$ and a similar power output) cannot answer all the questions and SST-2 is required to fill the technological gaps between ITER and a demonstration plant. One of its objectives is to show the capability of a tritium fuel cycle, which is beyond the scope of ITER. The SST-2 will also allow the testing of technological components developed in India,

---

[5]The French like to have fingers in all pies. They of course also participate in the Broader Approach between the EU and Japan, and have recently concluded a separate five-year cooperation plan that includes ITER with Japan in advance of the G20 summit in Osaka in July 2019; www.iter.org/of-interest/885.

like the breeding blanket. Construction of a demonstration plant will then start as soon as 2037. According to the latest (2018/2019) report of the Indian Department of Atomic Energy the SST-2 is still under development. The situation with DEMO is difficult to assess, but it is clear that India wants to pursue an independent route towards such a device and has currently no intention to start any collaboration beyond ITER.

South Korea initiated a conceptual design study for a K-DEMO (Kim et al. 2015) in 2012, with as target the construction of the device by 2037 with potential for electricity generation starting in 2050. In its first phase (2037-2050), K-DEMO will demonstrate a self-sustained tritium cycle, and develop and test components. Then, in the second phase after 2050, a major upgrade is planned, replacing in-vessel components in order to show net electricity generation of the order of 500 MW (roughly 1500 MW of thermal fusion energy).

The main parameters of K-DEMO have already been defined. It is expected to be a 6.65 m major radius tokamak, a bit larger than ITER. South Korea plans to invest about KRW 1 trillion (US$ 941 million) in the project.[6] This probably only covers the design stage. Construction can then start and is supposed to be completed in 2037.

Russia plans the development of a fusion-fission hybrid facility called DEMO fusion neutron source (FNS), a conventional tokamak (fairly small with major radius $R = 3.2$ m) with a superconducting magnet system that would harvest the fusion-produced neutrons to produce (through interaction of the neutrons with heavy elements like thorium or depleted uranium) fissile material and to break down radioactive waste (Shpanskiy et al. 2019). Especially the latter application is interesting as it could be used to promote a revival of ordinary nuclear fission plants by solving the nuclear waste problem. It can also be used for the production of tritium.

Development of the device started in 2013 (Kuteev et al. 2017) at the Kurchatov Institute and construction is expected to start in 2033 (it was earlier reported as being planned as early as 2023,[7] which can hardly have been realistic as it is currently still in the modelling stage and it would in any case be sensible to wait for some ITER results). Nevertheless, it is still part of what is called Russia's fast-track strategy to a fusion power plant by 2050.

The Russian strategy (Kuteev et al. 2015) is shown in Fig. 17.3. If the DEMO-FNS project is successful, an experimental industrial scale hybrid facility (Pilot Hybrid Plant (PHP)) with a fusion power of 40 MW and total thermal power of 500 MW (including the fission part) will be built to demonstrate the possibility of producing electric power from nuclear fuel and offering radioactive waste disposal services on a commercial scale (Ananyev et al. 2015).

---

[6]https://newenergytimes.com/v2/sr/iter/public/WORLD-NUCLEAR-20171222.pdf.

[7]Fusion Energy: For Peace and Sustainable Development, IAEA (2018); ITER Newsline / 14 May 2018.

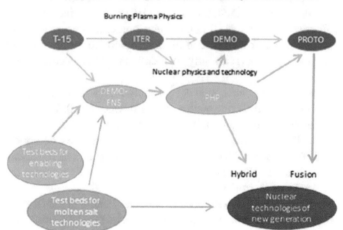

Fig. 17.3 Russian strategy combining pure fusion, hybrid systems and molten salt technologies. PHP stands for pilot hybrid plant (*from* Kuteev et al. 2015)

The US has not yet officially engaged in a DEMO project, but plans have been developed for a so-called Fusion Nuclear Science Facility (FNSF)[8] as an intermediate step between ITER and DEMO, in addition to the Materials Plasma Exposure eXperiment, and to be used for the development and testing of fusion materials and components for a DEMO-type reactor (see Chap. 16). Plans call for operation to start after 2030, and construction of a DEMO after 2050.[9] It is unlikely though that the US government will provide funding for these projects. The Trump administration requested an almost 30% cut in the 2020 budget for fusion research compared to 2019 from $564 million to $403 million. Congress however eventually awarded $671 million which actually means a rise of 19% compared to 2019 (but all of the increase going to ITER). The total budget for the Department of Energy's Office of Science rose $415 million, or 6%, to $7 billion. Particularly notable is a near doubling in construction funding for ITER to $242 million, reflecting renewed congressional confidence in its management (the Trump administration had proposed to lower it by about 20%).[10]

Very recently, in March 2020, a new report[11] was issued by the US fusion community, hailed as a consensus report put together by a diverse group of researchers from academia, government labs, and industry. The top energy-related priorities in the

---

[8] https://sites.nationalacademies.org/cs/groups/bpasite/documents/webpage/bpa_185482.pdf, accessed 25 February 2020.

[9] https://www.iter.org/newsline/-/3009; https://www.iaea.org/newscenter/news/charting-the-intern ational-roadmap-to-a-demonstration-fusion-power-plant.

[10] https://www.aip.org/fyi/2020/final-fy20-appropriations-doe-office-science.

[11] A Community Plan for Fusion Energy and Discovery Plasma Sciences, Report of the 2019–2020 American Physical Society Division of Plasma Physics Community Planning Process.

report include: development of a shared neutron source facility that can be used for development of critical materials and power plant designs; continued cultivation of burning plasma physics knowledge through ongoing participation in the ITER programme and expanded public–private collaboration in the United States; and immediate pre-conceptual design of a new US tokamak facility, which would begin operation by the end of the decade and support work on power extraction from exhaust heat and plasma sustainment. In these recommendations it does not differ much from an earlier 2018 report (National Academy of Sciences 2018), already referred to in Chap. 11, which had recommended that "the United States should remain an ITER partner as the most cost-effective way to gain experience with a burning plasma at the scale of a power plant and (…) should start a national programme of accompanying research and technology leading to the construction of a compact pilot plant that produces electricity from fusion at the lowest possible capital cost". This pilot plant should be operational in the 2040s. It was explicitly stated in that report that the attractiveness of participation in ITER is that "as a partner, the United States receives full benefit from the technology developed for ITER while providing only a fraction of the financial resources". In other words, for all ITER members, apart from the EU, participation in ITER is a cheap ride.

The new 2020 report seems to work out the second recommendation of the 2018 report, but is rather optimistic in assuming that, when starting a pre-conceptual design of a, presumably fairly large, tokamak now in the early 2020s, a tokamak can be running by the end of the decade. It supports the earlier recommendation for the US to remain a partner in ITER and to begin a science and technology programme leading to the construction of a fusion pilot plant that would operate as early as the 2040s. It states that "remaining a full partner in ITER remains the best option for US participation in a burning-plasma experiment. However, ITER and other planned/existing facilities will not be able to fully address the high heat flux and neutron fluence conditions that will be present in a fusion power plant." Therefore "the conceptual design of a new US tokamak facility capable of handling power exhaust at conditions typical of a fusion power plant while simultaneously demonstrating the necessary plasma performance should begin immediately, with the goal of beginning research operations on the new facility before the end of the decade." A tough task indeed, as funding for such a project is unlikely to be provided any time soon.

17.3 Prospects

In various studies in the final decade of the last century the potential for fusion power was already investigated and this many times over, but always by people for whom a lot was at stake; insiders in the field whose jobs and research opportunities depended on a favourable outcome for the prospects for fusion. Already in those years and based on very little hard data it was concluded that "fusion power has *very promising* potential to provide inherent safety and favourable environmental features, to address global climate change and gain public acceptance. In particular, fusion energy has the potential of becoming a clean, zero-CO_2 emission and inexhaustible energy source". Knowing that that on itself was not enough it was added that "the cost of fusion electricity is likely to be comparable with that from other environmentally responsible sources of electricity generation".[12] Especially objectionable in such a conclusion are the words 'very promising' as these were not based on any facts. JET and TFTR had just had some very limited success with the first D-T operations (less than expected as both were supposed to achieve breakeven) and the rest was just wishful thinking. The second conclusion is in any case completely false, as the cost of nuclear fusion power plants will in all probability be exorbitant and the costs of 'environmentally responsible sources' like solar power and wind have substantially decreased. The problem with all non-carbon energy sources at present is that electricity from fossil fuel (oil, gas, coal) is dirt cheap and that there is no shortage in fossil fuels for a long time to come. Predictions of peak oil, peak gas and peak coal to be reached in the first decade of this century have all turned out to be false by a big margin. Unless a (long overdue) sizable carbon tax is imposed on the use of fossil fuels, electricity from nuclear fusion will never have a chance of becoming competitive any time soon. And if such tax is then finally imposed, it will be shown in no time that the capture of carbon dioxide released in fossil-burning plants is a much cheaper option, leaving nuclear fusion again standing in the cold.

17.4   Power Plant Studies

Designing real nuclear fusion power stations has already been a pastime for decades. The early argument given in the 1960s was that the question of power plant design is as important as the question of plasma confinement, since what is the good of plasma confinement if we don't have an idea that a reactor system is feasible? Although that may be a valid question, it was equally clear that the question of the feasibility of a nuclear fusion power plant could not be settled in those early days. It is a bit like the Wright brothers fantasizing about jet aircraft before they had managed to get their first contraption off the ground. It can be argued that even today that question can still not be answered.

As recalled in Chap. 4, September 1969 saw the first international conference on fusion reactors held at Culham in the UK. This date is traditionally taken by the fusion community as marking the beginning of sustained and serious interest in fusion reactor design studies and of the encroachment of nuclear engineers upon the fusion territory. In such designs no hardware is produced, but design work is carried out with computer codes. Since this conference, more than 50 conceptual power plant studies have been conducted in the US, EU, Japan, Russia and China (El-Guebaly 2009, 2010). Several groups sprang up in the US resulting in an increasing number of nuclear engineers becoming involved in fusion. These people very much formed a separate group. They had so far not been concerned with nuclear fusion and were

---

[12] A Conceptual Study of Commercial Fusion Power Plants, EFDA (April 2005), EFDA-RP-RE-5.0.

distinct from both the fusion physicists and the fusion hardware engineers, who were concerned with engineering problems in fusion research, not with power generation from a fusion plant. But they soon made their presence felt. Their studies had an effect on the strategy of the entire fusion programme (recall the Hirsch episode related in Chap. 4 in which fusion became a mission-oriented endeavour (deliver a power plant) instead of just a research programme) and they started to make demands upon the fusion engineers, where before it had been assumed that they would work with whatever plasma the physicists came up with (Bromberg 1982, pp. 178–183.)

Desirable characteristics of a practical fusion reactor were already listed in a NASA technical memo in 1976 (Roth 1976) and include:

a. Steady-state operation (power interruptions associated with cyclic operation are awkward for utility companies);
b. High *beta* (implies smaller reactor size, lower capital investment e.g. in the magnet system);
c. Self-Sustaining fusion reaction (without this characteristic commercial power generation will be impossible);
d. Possibility of advanced fuel cycles (meaning other fuel than D-T, since such fuel cycles release more of their energy in the form of charged particles);
e. Direct conversion into electric power (related to d., as energy of charged particles can be converted into electric power by direct conversion schemes);
f. No neutrons or activation of structure (related to d. and e.; the advanced fuel cycles either do not generate neutrons or at least minimize the neutron generation and/or activation of the reactor structure);
g. Environmentally safe (i.e. minimizing possible radiation hazards);
h. High capital and resource productivity (capital expenditure should not be too high and the return on the investment should be adequate).

Desirable does of course not mean that all these characteristics are necessary and it will not be possible to design a plant that has all these characteristics, but it is striking that the tokamaks (spherical or otherwise) that have been or are currently being constructed, do not naturally meet a single one of these criteria, and nonetheless it is virtually the only approach that is being pursued.

A fusion power plant will be built up in a series of concentric layers—like the layers in an onion—as shown schematically in Fig. 17.4. The *burning plasma* forms the core, which is surrounded by the *first wall,* followed by the *tritium-breeding blanket*, a *neutron shield*, the *vacuum vessel*, the *magnetic coils* (in the case of magnetic confinement), and, finally, a second shield to reduce the radiation down to the very low level acceptable for personnel working nearby (McCracken and Stott 2013, p. 167).

Figure 17.5 displays the timeline of the large-scale magnetic power plant designs that have been made in the US from 1970 until the present day. As can be seen from the figure, in addition to the dominant tokamak designs, six other magnetic fusion concepts were developed since the 1970s: stellarators, reversed-field pinches, field-reversed configurations, spherical tokamaks, tandem mirrors and spheromaks. These alternative design studies were especially numerous in the 1980s. All this has led to

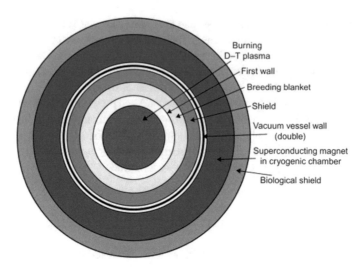

Fig. 17.4 Cross-section of a conceptual design for a fusion power plant. The power is extracted from the blanket and is used to drive a turbine and generator (*from* McCracken and Stott 2013, p. 168)

an impressive body of knowledge, information and understanding about power plant design even before the basic question whether a burning plasma can be sustained for any length of time has been answered.

In the 1970s a group at the Fusion Technology Institute of the University of Wisconsin, led by Robert W. Conn (b. 1942), currently President and CEO of the Kavli Foundation, was especially active in this field and turned out various tokamak reactor studies, called Premak (1971), UWMAK-I (1973), UWMAK-II (1975) and UWMAK-III (1976). The latter design was unusual as it was one of the first to feature a non-circular plasma cross section (Bromberg 1982, p. 237). Later these early studies were followed by designs based on the mirror and stellarator concepts. Noteworthy is also the STARFIRE design which was the first to promote steady-state current drive. In total the University of Wisconsin was actively involved to a greater or lesser extent in 44 magnetic fusion energy designs[13] and 26 inertial confinement reactor designs[14] (see below).

17.4.1   ARIES Program

From around 1990 the ARIES program (Advanced Reactor Innovation and Evaluation Study) in the US was leading in designing fusion reactors. It was a multi-institutional national research project, with among others involvement from

---

[13]https://fti.neep.wisc.edu/ncoe/timeline/mfe.

[14]https://fti.neep.wisc.edu/ncoe/timeline/ife.

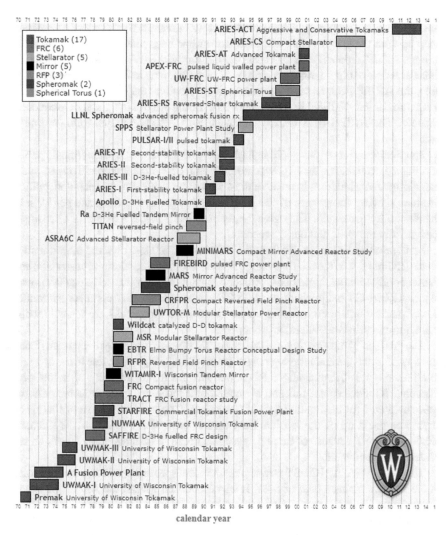

Fig. 17.5 Timeline of US magnetic fusion power plant designs (*from* https://fti.neep.wisc.edu/ncoe/timeline/mfe/US)

Wisconsin, at the University of California at San Diego to conduct advanced fusion systems research and to explore the potential for fusion development. It continued working until around 2013. The purpose of the program was to develop fusion reactors with enhanced economic, safety, and environmental features. The program has produced a whole series of conceptual studies of fusion systems, including the tokamak studies ARIES-I, ARIES-II, ARIES-III, ARIES-IV, ARIES-RS (reversed shear; the benefits of the reversed-shear configuration being high *beta* and large bootstrap current fractions), ARIES-AT (advanced tokamak), ARIES-ST (spherical tokamak), ARIES-ACT (advanced and conservative tokamak), and ARIES-CS (compact stellarator plant). When one design was finished, they could start again to keep up with the progress that had in the meantime been made in physics and technology. ARIES-I dates from 1990 and the last study, ARIES-AT, was completed in 1999. It is interesting to note that, as new physics and new technology became available, the ARIES designs became progressively smaller (major radius R decreasing from 8 m for the ARIES-I to a little over 5 m for the ARIES-AT), and consequently cheaper. The ARIES-ACT (2015) and ARIES-CS (2008) studies followed much later.

Practical considerations such as high safety, environmental friendliness, public acceptance, reliability as a power source and economic competitiveness were paramount considerations in these studies, which are in general very detailed, putting a value on every component of the plant and even predicting the price per kWh of the electricity generated. They give the feeling that it is just a matter of choosing a design from the shelf and build a reactor.

Stellarator studies in the 1980s and 1990s had led to large size and mass for such plants (mainly due to the relatively large aspect ratio and other design constraints) and consequently to much higher cost projections than for the advanced tokamak power plant. ARIES-CS (Najmabadi et al. 2008) demonstrated that compact stellarator power plants that are comparable in size to advanced tokamaks are feasible while maintaining desirable stellarator properties.

The ARIES-ST design (Najmabadi et al. 2003) was inspired by the work of Peng and Strickler and the encouraging results from START. High *beta,* due to the increase in allowable plasma elongation, of course was the major property of a spherical tokamak. ST devices are tall and elongated and the low aspect ratio restricts the space available for the inboard legs of toroidal field coils. In fact, most of this space should be taken up by the toroidal field central post. This central post is a central challenge for ST because of limited space and the high field and large forces on the central post conductor.

The 1000 MWe ARIES-ST power plant has an aspect ratio of 1.6, a major radius of 3.2 m, and attains a *beta* of 50%. It was found that overall the ARIES-ST study showed that the ST concept leads to attractive fusion power plants. The cost of electricity from ARIES-ST is comparable to that of the advanced tokamak ARIES-AT. In the physics area, most of the ARIES-ST research has borrowed from tokamak data bases extrapolated to low aspect ratios.

A pertinent remark, also already alluded to in Chap. 16 and applying to all designs presented here (or elsewhere for that matter), is that the physics of burning plasmas and alpha-particle dynamics remain unresolved issues for fusion research. Obtaining advanced tokamak modes in the presence of dominant alpha-particle heating is a critical issue that can only be addressed in a long-pulse, burning plasma experiment. In other words, not enough is currently known to design a nuclear fusion power plant with any confidence.

17.4.2 Power Plant Studies Outside the US

Figure 17.6 displays the timeline of power plant studies conducted in Europe, Russia, Japan and China, including the more recent DEMO designs. It is remarkable that apart from two early Russian designs, no design studies were carried out outside the US until the 1990s and, when discounting the DEMO designs, there are in total only a dozen, with the majority coming from the fossil-fuel starved countries in the Far East, reflecting a different attitude towards nuclear fusion, I suppose. While US studies have ended about fifteen years ago with the ARIES-CS, the rest of the world is still very busy designing DEMOs and other reactor types.

Four of the power plant studies in Fig. 17.6 originate from the EU. While the US was, and perhaps still is, highly motivated to obtain a fusion power plant that is economically competitive with other electric power sources, Europe and Japan also lay great store on environmental aspects and safety and would be prepared to take a somewhat higher electricity price for granted. The Safety and Environmental Assessment of Fusion Power (SEAFP) was commissioned by and undertaken for the

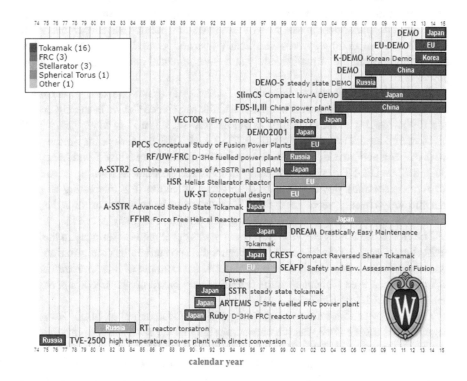

Fig. 17.6 Timeline of magnetic fusion power plant designs outside US including DEMOs (*from* https://fti.neep.wisc.edu/ncoe/timeline/mfe/intl)

European Commission in the framework of the Fusion Programme 1990–94.[15] At the time, when the design of ITER was in full swing, it was quite sensible to investigate the safety and environmental issues of fusion power. The assessment started with the development of two conceptual power plant designs, with 3,000 MW of fusion power each, using, where possible, the database for the design of ITER or reasonable extrapolations thereof. Since SEAFP is a safety-directed study, safety considerations dominated the concept and the conclusion is that fusion reactors have a great potential for safety. There is no possibility of uncontrolled power runaway. Even in the event of a total loss of active cooling, the low residual heating excludes melting of the reactor structures.

As regards waste management the study concludes that over their lifetimes, fusion reactors would generate, by component replacement and decommissioning, activated i.e. radioactive material similar in volume to fission reactors, but qualitatively different in that the long-term radiotoxicity is very much lower. After about a hundred years (still a considerable time), radiotoxicity indices (relating to ingestion and inhalation) for the total activated materials from both plant models considered fall to levels comparable with the ashes from coal-fired plants. The study indicates that fusion waste would not constitute a burden for future generations. The inclusion of the latter conclusion indicates, in my view, that some politics was involved as such a conclusion can in no way be scientifically justified. Overall, it is concluded that the studies confirm the attractive safety and environmental characteristics of fusion power.

The UK-ST design study (Voss et al. 2000, 2002) took the spherical tokamak concept as its basis and was initiated at the Culham Science Centre in the UK in the late 1990s, roughly at the same time as the ARIES-ST. It is a 1224 MW plant with $R = 3.4$ m and aspect ratio 1.4. Both the UK-ST and ARIES-ST were inspired by the work of Peng and Strickler, and of course by the results from START and MAST at Culham. The UK-ST design study sees the spherical tokamak as a strong candidate for future economic power generation. In this connection it is interesting to note that currently (twenty years later) the UK government has chosen for the spherical tokamak as the basis for its new STEP research programme (see Chap. 11).

The Helias Stellarator Reactor (HSR) study was developed in Germany in the late 1990s based on the Wendelstein7-X experiment. The stellarator configuration can also be produced using continuous helical coils. An example of this approach is the Japanese Force Free Helical Reactor (FFHR) study based on the LHD experiment (El-Guebaly 2010, p. 1074).

The EU Power Plant Conceptual Study (PPCS) (Marbach et al. 2002) from the beginning of this century was the successor study to the safety and environmental assessment mentioned above. It focussed on five models that spanned a wide range of physics and technology options. All these models produce about 1.5 GW of electricity, but like the ARIES models become progressively smaller and use less power. Safety and environmental issues were carefully considered. The study highlighted the need to establish the basic features of DEMO, a device to bridge the gap between

---

[15] Safety and Environmental Assessment Report, European Commission, EURFUBRU XII-217/95.

ITER and the first-of-a-kind fusion power plant. It also gave an estimate, for what it is worth, for the cost of electricity generated by fusion and concluded (in 2004) that fusion compares favourably with other renewable sources, such as solar and wind.

Japan, as can be seen from the figure, has been quite active since the early 1990s. Including the recent DEMO designs, Fig. 17.6 includes 11 Japanese design studies. The start is made around 1990 with the Ruby and ARTEMIS studies, two deuterium-helium fuelled field-reversed configurations. The 1990s witnessed the emergence of deuterium-tritium and deuterium-helium fuelled FRC designs with steady-state operation. One of the challenging physics issues for field-reversed configurations is to sustain the plasma current during the steady-state operation. Some other issues (plasma stability, energy confinement, and an efficient method for current drive) have hindered the progress of the FRC concept. In the US such configurations had been studied earlier with the SAFFIRE, TRACT, LANL-FRC and FIREBIRD studies, and were taken up again in the late 1990s. Russia has also conducted an FRC study with its RF/UW-FRC. The conclusion of the latter study is that "the total mass of the reactor is much smaller than of D-T or fission power plants, which together with simplicity and easy maintenance make a D-3He FRC power plant the most handsome and powerful candidate for a fusion reactor." (Khvesyuk et al. 2001). The rest of the fusion community apparently does not share that conclusion.

A series of pioneering steady-state tokamak reactor (SSTR) studies was developed in Japan in the 1990s to achieve high power density through high magnetic field strength. The successor studies DREAM and CREST promoted the approach of easier and faster power core maintenance to achieve high overall availability. The most recent VEry Compact TOkamak Reactor (VECTOR) is very compact with a major radius of 3.75 m, aspect ratio 2, and > 16 T superconducting toroidal-field magnets operating at 20 °K (−253 °C). VECTOR's design features were incorporated in the design of a compact Demo (SlimCS) with low aspect ratio and slim central solenoid.

The Chinese fusion design study (FDS) included in Fig. 17.6 is one of a series of studies conducted in the last 15 years and covering a broad range of tokamak concepts, including a hybrid tokamak to transmute fission products and breed fissile fuels (FDS-I), an electricity generator (FDS-II), hydrogen producer (FDS-III), and spherical tokamak (FDS-ST) (El-Guebaly 2010, p. 1073).

17.5 Inertial Fusion Power Plants

There also has been some (fairly extensive) activity in designing IFE (Inertial Fusion Energy) power plants. As we have seen in Chap. 12, the concept of inertial confinement fusion (ICF) is based on the uniform heating of the surface of a spherical pellet by laser light or high-speed particles. For such pellets ICF uses microspheres filled with a mixture of deuterium and tritium. They act as miniature hydrogen bombs, which are made to explode and deliver their energy to a wall and cooling medium.

Figure 17.7 shows a schematic arrangement of a laser-fusion reactor. After the

Fig. 17.7 Simplified
schematic representation of a
laser-fusion power reactor
(*from* Murray and Holbert
2020, p. 535)

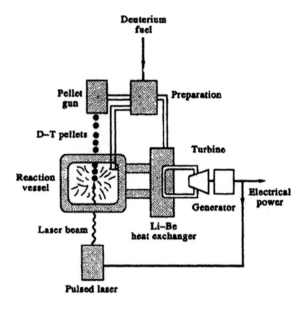

D-T fuel pellet has been prepared, the pellet gun shoots the pellet into the reaction vessel where a pulsed laser beam delivers the energy to the pellet causing it to implode and generate fusion reactions. In the Li-Be heat exchanger the heat is extracted and converted into electric power. At the same time the materials are recovered from the target debris and recycled into new fuel pellets.

The operation of an IFE reactor will consist of a rapid repetition of the following four steps: (a) intake of the fusion fuel pellet into the reactor chamber; (b) compression of the microcapsule in order to initiate the fusion reactions; (c) explosion of the plasma created during the compression stroke, leading to the release of fusion energy; (d) exhaust of the heat to be converted into electricity and reaction residue to be treated afterwards to extract all the reusable elements, mainly tritium, but also other metals. For instance, in the LIFE design (see below) each target contains approximately 3 g of lead, which with more than 1 million targets used per day would amount to a daily throughput of about 4 tonnes. This material would also be collected and recycled into future targets.

To accomplish these four steps the plant consists of four major, separate, but interconnected elements. The first three are (a) the driver, either a laser or particle accelerator, which converts electric power into short pulses of light or particles and delivers them to the fuel pellet to cause implosion, ignition and thermonuclear burn; (b) the target or pellet factory, which manufactures the fuel capsules or pellets, fills them with D-T fuel and sends them to the reactor to be injected into the reaction chamber (in this respect we have met the British company First Light Fusion in Chap. 15 which envisages to sell such pellets as a kind of Nespresso capsules to power stations); (c) the reaction chamber, in which the injected fuel pellet (target) is tracked, i.e. its position, flight direction and velocity are measured precisely. Driver beams are directed at the target to implode it and to produce thermonuclear energy with a repetition rate of a few times per second. The entire target disintegrates and a portion of the D-T fuel is burned. Each capsule explodes with an energy equivalent to several kilograms of high explosive and the wall of the reaction chamber will have to withstand the blast waves of such explosions. Target debris is deposited in the chamber gas and/or in the blanket fluid. The thermonuclear emissions are captured in a surrounding structure called a blanket, and their energy is converted into heat. Tritium is also produced from the lithium in the blanket.

The fourth element is the remainder of the plant, whose components (blanket, heat transfer system, neutron shielding and electricity generators) are broadly similar to the equivalent parts of a magnetic confinement power plant. Here the electricity is generated and materials are recovered.

These components will have to withstand the repetitive thermal and mechanical stresses. More than in a continuously operating system, the necessarily pulsed operation of an IFE power plant can lead to metal fatigue and additional stress. At present inertial confinement experiments shoot less than one target per hour (in some cases only one per day), where a power plant will require about ten shots every second. An important advantage claimed for IFE power plants is the possibility of physically separating the target factory and driver from the radiation environment of the reaction chamber, resulting in ease of maintenance. However, some components, like target handling systems, have to be mounted on the target chamber, where they will be subject to blast and radiation damage (McCracken and Stott 2013, pp. 178–179).

Tritium and other target materials are extracted from the re-circulating blanket fluid material and from the reaction chamber exhaust gases. These extracted materials are then recycled to the target factory for use in new target fabrication. The thermal energy in the blanket fluid is converted into electricity, part of which is used to power the driver. This recirculation power fraction should be less than 25% for economic reasons. If it is too large, the cost of electricity rises rapidly because much of the plant equipment is then used simply to generate electricity for the driver itself. Just as a tokamak power plant, an IFE power plant must be safe and have a minimum impact on the environment (Nakai and Mima 2004).

IFE power plants that accomplish all this are still only figments of the imagination as ignition (plasma burn) has not yet been reached in ICF either (as related in Chap. 12) and ICF faces continued technical challenges in reaching the conditions needed for ignition. Even if these were all solved, there are a significant number of other problems that seem just as difficult to overcome. Given the low efficiency of the laser amplification process (about 1–1.5%), and the losses in electricity generation (steam-driven turbine systems are typically about 35% efficient), fusion gains would have to be on the order of 350 just to energetically break even. These sorts of gains appear to be impossible to achieve, and work on ICF turned primarily to weapons research, although some hope has recently been raised as will be related below. A safe conclusion is though that there is at present no clear path to energy production via inertial confinement fusion, whatever driver one may use.

17.5.1   IFE Power Plant Design in the US

In spite of these bleak prospects, a number of conceptual inertial fusion reactor designs has been developed by national laboratories, universities, and companies, as can be seen from Figs. 17.8 and 17.9, which show the timelines of inertial confinement fusion power plant design in the US and abroad. A large majority of devices uses lasers as driver.

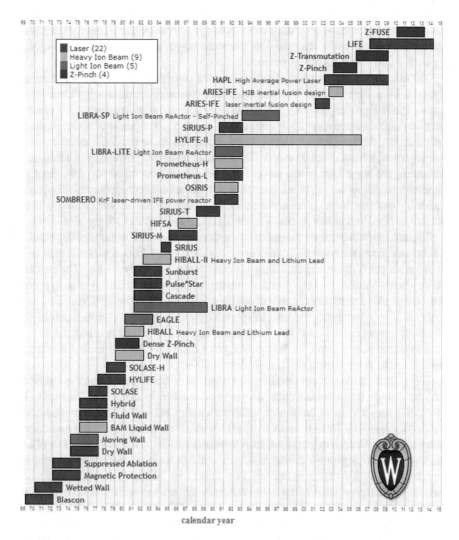

Fig. 17.8 Timeline of US inertial confinement fusion power plant designs (*from* https://fti.neep. wisc.edu/ncoe/timeline/ife/US)

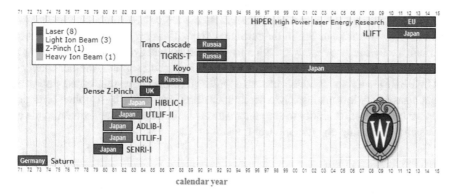

Fig. 17.9 Timeline of ICF power plant designs outside US (*from* https://fti.neep.wisc.edu/ncoe/timeline/ife/intl)

American activity in this field started already in 1969, even earlier than for magnetic confinement fusion. The designs are intended to achieve power outputs comparable to those of fission reactors. Examples are HIBALL-II (University of Wisconsin), HYLIFE-II and CASCADE (Lawrence Livermore), Prometheus (McDonnell Douglas), and OSIRIS and SOMBRERO (W.J. Shafer Associates).

It does not make much sense to discuss all these studies here in detail. The stage of development is far behind magnetic confinement fusion and many comments have already been made in Chap. 12. Here I will therefore restrict myself to some remarks.

It can be seen from Fig. 17.8 that most early US activity was in laser-driven plants, while from the early 1990s onwards there has been a shift towards heavy and light ion beam and Z-pinch devices.

One of the primary concerns in an IFE reactor design is how to successfully remove heat from the reaction chamber without interfering with the targets and driver beams. Another serious concern, similar to magnetic confinement fusion devices, is that the huge number of neutrons released in the fusion reactions react with the materials of the plant, causing them to become radioactive, as well as mechanically weakening metals. Fusion plants built of conventional metals like steel would have a fairly short lifetime and the core containment vessels will have to be replaced frequently. Proposed designs include dry walls, magnetically protected walls, gas-filled chambers, wetted walls, thick liquid walls, or thick-flowing ceramic-powder walls. The reactor designs Wetted Wall, Magnetic Protection, Moving Wall, BAM Liquid Wall, Fluid Wall, Dry Wall, HYLIFE, HIBALL, LIBRA and CASCADE are some of the various proposals made in this respect (Perlado and Sanz in Velarde, Ronen and Martinez-Val 1992, p. 616). Thick liquid walls essentially eliminate the "first wall" problem, and lead to a faster development path, as no new neutron test facilities are required. One more recent application of this concept is in the HYLIFE-II design, which uses a thick liquid wall of a molten mix of fluoride salts of lithium and beryllium (FLiBe), to both protect the chamber from neutrons and carry away heat. In this design a thick molten FLiBe-salt layer is injected between the reaction chamber walls and the explosions. The FLiBe is then passed into a heat exchanger where it heats water for use in the turbines. The tritium produced by the fission of lithium nuclei can also be extracted. This will close the fuel cycle and is a necessity for perpetual operation because tritium, as we know, is in short supply. Another concept, SOMBRERO, uses a reaction chamber built of carbon-fibre-reinforced polymer, which has a very low neutron cross section. Cooling is provided by a molten ceramic, chosen because of its ability to stop the neutrons from traveling any further, while at the same time being an efficient heat transfer agent.

17.5.2   Z-Pinches

Of particular interest are the various recent Z-pinch concepts (Z-pinch, Z-Transformation, Z-Fuse). As can be seen from Figs. 17.8 and 17.9, there had already been some renewed interest in Z-pinches in the late 1970s to early 1980s, both in the US and the UK (Dense Z-Pinch in the figures).

In the Z-pinch a current is passed through a cylinder of a magnetically confined plasma parallel to the axial or z-direction, producing an inward force. This force can be employed to collapse a hollow plasma cylinder or to confine the plasma through the pinch effect. Unlike most other magnetic confinement schemes for fusion studies, there is no obvious limit to the strength of the magnetic field. Magnetic fields in the 100–1000 T range are routinely produced in present-day Z-pinch experiments.

This clear and remarkable shift of interest back to perhaps the oldest concept in fusion physics, which had been abandoned because of its basically unstable character, has arisen from the spectacular increase in X-ray power and efficiency from the collapse of cylindrical wire arrays mounted between electrodes, especially at the Z Pulsed Power Facility (Z-machine) at Sandia National Laboratories. Z-pinches are driven by the Z-machine which typically delivers 20 MA of current through more than 300, about 7 μm diameter, tungsten wires arranged in a 2 cm radius, 1 cm tall cylindrical ring (a hohlraum). The wires vaporize, forming a very uniform plasma sheath that implodes under the force of its own radial magnetic field onto a low-density foam or ring-shaped foil. This compression heats the interior of the foam to temperatures as high as 230 eV (2.7 million degrees). The thermal X-rays emitted during the course of the implosion contain up to 1.8 MJ of total energy and radiate for about 10 ns. This fast Z-pinch[16] technology therefore can create high-level radiation environments on time scales similar to those created in indirect-drive laser

---

[16]The word 'fast' in this context refers to a class of Z-pinches where a sub-microsecond electrical pulse (rise time of ~100 ns on the Z-machine) implodes an annular, cylindrically symmetric plasma at supersonic velocities. When Z fires, it releases the stored energy so fast, that the generated electric power is more than all the power plants in the world combined for that brief moment in time (~100 ns). Currently, the Z-machine can fire once a day. A fusion energy machine would have to fire around six times per minute, capture the energy from the shots, and then transmit it to a power-producing system.

hohlraums or ion-beam ICF drivers, and can be used to indirectly heat an ICF fuel capsule (Derzon et al. 2000).

It has resulted in quite dramatic results in the form of high yields of soft X-rays (1–2 MJ) with peak powers greater than 200 terawatt in a pulse of 4.5 ns duration and, as Sandia National Laboratories have demonstrated, with temperatures and time scales nearly appropriate for driving ICF capsules. The advantages of the Z-pinch are that larger volumes of plasma can be produced of very high energy density, and with greater efficiency than e.g. in laser-produced plasmas.

The next step would then be the Z-IFE (Z-inertial fusion energy) test facility, the first true Z-pinch driven prototype fusion power plant. Major problems include producing energy in a single Z-pinch shot, and quickly reloading the reactor after each shot. In 2005 a roadmap was set out which envisages the construction of a Z-Pinch IFE Demo in the 2030s. For this the entire Z-machine has been refurbished into the ZR-machine (R for refurbished) (Olson et al. 2005).

The newer Z-machine can shoot around 26 MA (instead of 18 MA previously) in 95 ns. The radiated power has been raised to 350 terawatt and the X-ray energy output to 2.7 MJ. The ultra-high temperatures reached in 2006 (2.66–3.7 billion degrees) are much higher than those required for D-T fusion, for which about 150 million degrees is needed. At least theoretically, they would allow the fusion of deuterium with heavier elements such as lithium or boron, which do not produce any neutrons and thus no radioactivity or nuclear waste (aneutronic fusion). Sandia's roadmap includes another Z-machine version called ZN (Z-Neutron) to test higher yields in fusion power and as the first step towards a transmutation plant that can be used to transmute transuranic waste produced from nuclear fission plants. It is planned to give between 20 and 30 MJ of fusion power with a shot per hour by using a Russian Linear Transformer Driver (LTD). The LTD is a ring-shaped parallel connection of switches and capacitors designed to deliver rapid high-power pulses, designed at the Institute of High Current Electronics (IHCE) in Tomsk. The next step would then be the Z-IFE power plant. It is not clear how far the programme has advanced. The Sandia website does not give any information about ZR, ZN or Z-IFE.

17.5.3 The Lawrence Livermore LIFE Project

At Livermore National Laboratories, which also house the National Ignition Facility, thoughts turned to designing an ICF power plant when NIF started to operate successfully and seemed to fulfil its promises and achieve ignition by the end of 2012. It failed, but between 2008 and 2013 the LIFE (Laser Inertial Fusion Energy) project aimed to develop the technologies needed for converting the laser-driven fusion concept of NIF into a practical commercial power plant. A typical example of embarking head over heels on a project for which there was no good scientific basis whatsoever. No wonder that it was criticized throughout its development for being based on physics that had not yet been demonstrated. That does not necessarily mean that it was not or could not be an interesting project, it just was and still is a bridge too far for this kind of research.

It was a very ambitious project as was also reflected in report titles like "LIFE: The Case for Early Commercialization of Fusion Energy" (Anklam et al. 2011), although such overblown titles are by no means uncommon in fusion science. The report contains a design and build schedule with 2035 as the start of commercial fusion operations and a demonstration power plant (LIFE.1) generating 400 MW to be operational by the mid 2020s. Some fusion researchers have criticized LLNL for having seriously understated the challenges to be overcome in order to build an ICF power plant, and overpromised and oversold the project.

LIFE used the same basic concepts as NIF, but aimed to lower costs by using mass-produced fuel elements, simplified maintenance, and diode lasers (see Chap. 12) with higher electrical efficiency. It is based on indirect-drive targets injected into a xenon-gas-filled chamber. It first explored a fusion-fission hybrid system, like the Russian DEMO discussed above, and then switched to a pure fusion system.

In the hybrid concept the fast neutrons from the fusion reactions are used to induce fission in *fertile* nuclear materials, i.e. in material that is not itself fissionable by thermal neutrons, but can be converted into a fissile material by neutron absorption and subsequent conversions of nuclei. The concept was designed to generate power from both fertile and fissile nuclear fuel and to burn nuclear waste. In H-bombs the yield of the fusion section of the bomb is increased by wrapping it in a layer of depleted uranium,[17] which undergoes rapid fission when hit by the neutrons from the fusion bomb inside. The energy of the 14 MeV neutrons produced in fusion reactions is sufficient to cause fission in uranium-238, as well as in many other transuranic elements. The same basic concept can also be used with a fusion reactor like LIFE, using its neutrons to cause fission in a blanket of fission fuel. Unlike a fission reactor, which stops burning fuel once the uranium-235 drops below a certain threshold value, these fission–fusion hybrid reactors can continue to produce power from the fission fuel as long as the fusion reactor continues to provide neutrons. Since the neutrons are highly energetic, they can potentially cause multiple fission events, resulting in the reactor as a whole producing more energy, a concept known as *energy multiplication*, which has been around at least since the 1970s (see e.g. Lidsky 1975). Even left-over nuclear fuel taken from conventional nuclear reactors will burn in this fashion. This is potentially attractive because this burns off many of the long-lived radioisotopes in the process, producing waste that is only mildly radioactive and without most long-lived components. In most fusion energy designs, fusion neutrons react with a blanket of lithium to breed new tritium for fuel. A major issue with the fission–fusion design is that the neutrons causing fission are no longer available for tritium breeding, which as we have seen is necessary to close the fuel cycle. The designer has to choose which is more important; breeding tritium or providing power through self-induced fission events.

The economics of fission–fusion designs has always been questionable. The same basic effect can namely be obtained by replacing the central fusion reactor with a

---

[17]I.e. with a lower content of the fissile isotope uranium-235 than natural uranium.

specially designed fission reactor, and using the surplus neutrons from the fission to breed fuel in the blanket. These fast breeder reactors have proven uneconomical in practice, and the greater expense of the fusion systems in the fission–fusion hybrid has always suggested that they would be uneconomical too unless built in very large units.

After the further design of the hybrid option stopped at Livermore in 2009, it was decided to redirect the project towards a pure fusion design with a net electrical output of about 1 gigawatt. One of LIFE's goal was to improve the inefficiency of NIF's lasers by using diode lasers and to greatly reduce the size of NIF while making it much easier to build and maintain. Whereas an NIF beamline for one of its 192 lasers is over 100 m long, LIFE was based on a design about 10.5 m long that contained everything from the power supplies to frequency conversion optics. Each module was completely independent, allowing the units to be individually removed and replaced while the system as a whole continued operation.

A considerable amount of LIFE's effort was put into the development of simplified target designs and automated manufacture that would lower their cost, compared to NIF's expensive pellets. Working with General Atomics, a concept was developed involving the use of on-site fuel factories that would mass-produce pellets at a rate of about a million a day. It was expected that this would reduce their price to about 25 cents per target, although other references suggest the price was closer to 50 cents, and LLNL's own estimates range from 20 to 30 cents.

A further advantage of LIFE is that the amount of tritium required to start up the system is greatly reduced over magnetic fusion concepts. In the latter a relatively large amount of fuel is prepared and put into the reactor, just for start-up, while only a fraction of this tritium is burned. LIFE, by virtue of the tiny amount of fuel in any one pellet, can begin operations with much less tritium, on the order of 10%.

By 2012, the baseline design of the pure fusion concept, known as the Market Entry Plant (MEP), aka LIFE.1, had stabilized. A total of 384 lasers would provide 2.2 MJ of ultraviolet light, producing a Q of 21. Per second 15 targets would be fired into the target chamber, a two-wall structure with liquid lithium or a lithium alloy filling the space between the walls. The lithium captures neutrons from the reactions to breed tritium, and also acts as the primary coolant. The plant had a peak generation capability, or nameplate capacity, of about 400 MWe, with possible expansion to as much as 1000 MWe. MEP was not intended to be a production design, and would be able to export only small amounts of electricity. It was projected to be operational 10–15 years following ignition on NIF at a total build cost of between \$4 billion and \$6 billion. It would serve as the basis for the first production model, LIFE.2, which would produce 2.2 GW of fusion energy and convert that to 1 GW of electric power at 48% efficiency. Over a year, assuming a rather ideal 90% capacity, this plant would then produce 8 billion kWh, for which 425 million fuel pellets would have to be burned. At a cost price of 50 cents per pellet as suggested above, it would cost over \$200 million per year to fuel the plant, apart from the other costs. The average rate for wholesale electricity in the US as of 2018 was around 3 cents/kWh, so the 8 billion kWh might be sold for about \$240 million, and fuel costs would eat most of this, making it already clear that such a plant would never be competitive (barring a sizeable carbon tax to be levied on fossil fuels).

There are of course also other costs, especially capital costs for which the following rough calculation can be made. Capital expenditure (CAPEX) for the plant was estimated to be $6.4 billion. Financing the plant over a 20-year period would add another $5 billion, assuming a 6.5% unsecured rate. Considering only CAPEX and fuel, the total cost of the plant in this 20-year period would therefore be $15.4 billion.[18] Dividing the total cost by the energy produced over the same period (20 times 8 billion kWh) results in 9.6 cents/kWh as a rough estimate of the cost of electricity for a 20-year lifetime operation. A 40-year operational lifetime would reduce this by half. There are still other (operating) costs that have not been included and the calculation presented above only applies to the rather ideal situation that no mishaps occur, nor major maintenance has to be carried out. So, it is safe to conclude that LIFE.2 would be unable to compete with modern renewables or other sources.

LLNL estimated that further development after widespread commercial deployment might lead to further technological improvements and cost reductions, and proposed a LIFE.3 design of about $6.3 billion CAPEX and 1.6 GW nameplate capacity with a projected electricity cost of 5.5 cents/kWh, which is competitive with offshore wind as of 2018, but unlikely to be so in 2040 when construction of LIFE.3 would start.

The saga of NIF has been told in Chap. 12. Construction was completed in 2009 and design capacity reached in 2012. During this period, NIF started running the National Ignition Campaign, with the goal of reaching ignition by 30 September 2012. Ultimately, the campaign failed as unexpected performance problems arose that had not been predicted in the simulations. In the years since, NIF has run a small number of experiments with the aim of edging closer to ignition, but progress has been slow and it is expected that a number of years of additional work are required before ignition can be achieved, if ever. When it became clear that NIF would fail to solve the problem of ignition, there was no reason to continue the LIFE project and at the end of 2013 it was quietly cancelled.[19]

17.5.4   Efforts Outside the US

Let us now turn to some of the developments outside the US, depicted in Fig. 17.9.

The introduction of fast ignition (see Chap. 12) and similar approaches have dramatically improved the gain of laser-driven inertial confinement as mentioned above. At the High Power laser Energy Research facility (HiPER) gains of 100 are predicted. HiPER is not a conceptual design, but as related in Chap. 12, a proposed experimental fusion device which is currently undergoing preliminary design. Given a gain of about 100 and a laser efficiency of about 1%, HiPER will produce about the

---

[18]$6.4 billion + $5 billion financing + 20 x $200 million for fuel.

[19]https://en.wikipedia.org/wiki/Laser_Inertial_Fusion_Energy.

same amount of fusion energy as the electrical energy needed to create it. As of 2019, the effort appears to be inactive, but has so far not been cancelled. Brexit will neither be of help in this respect. A 2013 report by the US National Academy of Sciences (National Academy of Sciences 2013, p. 214) concluded that at this time fast ignition appears to be a less promising approach for IFE than other ignition concepts, because of issues surrounding low laser-target energy coupling, a complicated target design, and the existence of more promising concepts (such as shock ignition). Since none of the approaches proposed are in any way or sense promising for (commercial) energy generation, ICF as an energy generating option can very much be considered a lost cause.

None of this nor other conclusions of the above-mentioned report like "the lack of understanding surrounding laser-plasma interactions remains a substantial but as yet unquantified consideration in ICF and IFE target design" do deter fusion scientists from doggedly pursuing this course.

The Japanese have launched a design effort called LIFT, Laser Inertial Fusion Test, or iLIFT, based on a single laser system. It will first aim at repetitively producing fusion reactions in a burst mode (100 consecutive shots) while deepening the understanding of the interaction of the fusion products with a solid surface, then produce the first megawatts of electric power and lastly clarify IFE reliability (through continuous operation for half a year) and economics (Jacquemot 2017). At Osaka University the Fast Ignition Realization Experiments (FIREX) are being carried out. Before LIFT can start, FIREX has to successfully demonstrate ignition and burn based on fast ignition, which was planned to be achieved around 2019 but has so far not been done. In 2013 it was reported (National Academy of Sciences 2013, p. 202) that the plans for LIFT include operation from 2021 to 2032, start of the engineering of the demonstration plant in 2026, and operation of a single-chamber system from 2029 to be expanded to a four-chamber commercial plant operating at 1.2 MJ at 16 Hz in 2040.

As can be seen from the figure, the KOYO project in Japan has been ongoing since the early 1990s. This project also relies on fast ignition, which is considered attractive because a high gain can be achieved with smaller lasers. No breakthroughs have so far been obtained with this system.

The other projects shown in Fig. 17.9 go back even further into the last century and have not yielded any results that are noteworthy.

17.6 Heavy-Ion Fusion

Before finishing this chapter, we have to say a few words about heavy ion fusion (HIF).

Investigations in HIF emerged in 1976, when this concept was presented at the first ERDA Workshop on Heavy Ion Fusion in Oakland, California. The perspective was to use the success story of particle accelerators to solve the energy problem without the risks inherent to nuclear fission. A typical HIF scenario consisted of

482 17 Post-ITER: DEMO and Fusion Power Plants

two major steps: 1) an accelerator to accelerate high intensity beams of heavy ions to the energy required for a single target ignition, accompanied by a compression scheme to compress the beam energy to the typical time scale of 10 ns; and 2) an inertial fusion target to absorb the pulsed energy and thus enable fast heating and shock-induced compression of the target (pellet) such that fusion conditions of high temperature and density are met.

On the other hand, it was soon realized that due to the extremely short pulse structure, the challenges of HIF on accelerator design would significantly exceed those of nuclear fission applications. In all cases, whether fission or fusion, driver accelerators for a typical 1000 MW power station would have to deliver of the order of 10 MW average beam power. Inertial fusion, however, would in addition have to compress the total pulse energy of initial linear accelerator pulses by a factor of a million, and to repeat this several times per second (Hofmann 2018).

After its first emergence, HIF soon turned into a research field that continued to fascinate researchers, but as yet without large funding in the US, and neither in Europe or Japan. In the US, LBNL (Lawrence Berkeley National Laboratory) was especially active and remained the leading lab on the linear accelerator (linac) concept for more than three further decades, with a number of accompanying activities in other laboratories and universities. In Germany, HIF studies for energy production were initiated in 1979 within a government-funded research programme focussing on the linear accelerator storage ring concept, in which a continuous or pulsed particle beam injected into the ring by a linear accelerator can be kept circulating for many hours. Noteworthy in this respect are the HIBALL Study (Heavy Ion Beams and Lead Lithium, 1981–1985) in a collaboration between German research groups and the University of Wisconsin; and the later HIDIF Study (Heavy Ion Driven Ignition Facility, 1995–1998) elaborated by a European Study Group under the leadership of CERN and GSI (Gesellschaft für Schwerionenforschung) in Darmstadt, Germany. The early work on HIF triggered studies in all kinds of other fields. In a variety of other countries different studies have been carried out and are still continuing: in Japan, with the HIBLIC (Heavy Ion Beam and Lithium Curtain) study in 1985, a design of a heavy ion fusion driver, and a variety of HIF-related target physics studies; at ITEP (Institute for Theoretical and Experimental Physics, Moscow, Russia) since the late 1980's until today; and more recently at IMP (Institute of Modern Physics) in Lanzhou, China, with experiments at their nuclear physics heavy ion facility.

Similar to laser driven targets, in case of heavy-ion directly driven targets the outer shell of the fuel capsule is heated by the incident heavy ions and ablated. The radial pressure from the ablation compresses the fuel in the same way, heats it up to ignition temperature and ignites by a converging sequence of shock waves. For heavy ions, the number of beam lines is limited by geometrical constraints due to their large bending radius, thus the required spherical symmetry is most difficult to achieve in a directly driven target scheme, and more difficult than with lasers. Indirect drive is advantageous. In that case the spherical fusion capsule is enclosed in a cylindrical hohlraum of high-Z material. The energy of the heavy ion beam is absorbed by two converters on opposite sides and transformed into soft X-ray radiation. The spherical fusion capsule is radiatively imploded at high symmetry. It was shown that by some further technical improvement the radiation asymmetry could still be considerably reduced and kept down theoretically to the required level of about 1%.

Heavy ion accelerators can hardly be competitive with lasers developed for single shot ignition purposes, but their potential has always been seen in terms of possible future application in commercial energy production, where efficiency, repetition rate and reliability are crucial issues. So far there are only model calculations and code simulations to give any comfort to the idea that HIF is a viable concept. There are two complementary accelerator scenarios as potential inertial fusion energy drivers: the RF linac & storage rings and the induction linear accelerator concept. In an RF linear accelerator particles are accelerated by radio frequency (RF) ion sources. The concept of RF linac & storage rings benefits from the large operating experience with RF linear accelerators, synchrotrons and storage rings. This scheme has been the focus of the two system studies coordinated by GSI Darmstadt, the HIBALL as well as the later HIDIF study. The induction linear accelerator concept was invented in the 1950s by Nicholas Christofilos, whom we met before in Chap. 3 when discussing the Astron device, for high-current (kiloamperes) electron beams. In 1976, at LBNL a novel technology was proposed for inertial fusion drivers with heavy ions. The idea was to have a device, where a single heavy ion bunch (or several parallel bunches) was at the same time accelerated and longitudinally compressed. Key experiments were carried out until the programme was terminated in the US in 2011/12.

In 1993, a number of European laboratories set up a European Study Group to research the idea that an accelerator had the potential of being competitive with the laser based NIF. The efforts of the group merged into the European HIDIF Study to demonstrate the accelerator needs for target ignition with significant energy gain. HIDIF was conceived as the heavy ion driven "ignition facility" based on the expectation that the goal of an experimental "single-shot" facility (with a few shots per day) would relax the driver requirements. In the course of the study, it was, however, realized that no particular advantage could be drawn from the single-shot assumption, and it almost goes without saying that ignition was not achieved. HIDIF ended in 1998 and no further feasibility studies have been carried out in heavy ion fusion (Hofmann 2018).

17.7 Conclusion

It may be obvious but I feel that it is still worth noting that power plant design studies (magnetic confinement and inertial confinement) have now been carried out for more than half a century where most design reports ended with something like: "These studies have shown that attractive combinations of environmental, safety, and economic characteristics for IFE can be found with a robust variety of driver and fusion chamber concepts, enhancing the probability of ultimate IFE success." This citation has been randomly chosen from a paper published in 1994 about inertial fusion power plants (Logan 1994), but equally optimistic and in hindsight completely unwarranted conclusions can also be found in reports of magnetic fusion power plant designs. It goes almost without saying that a considerable gap remains between performance required in these designs and that obtained in the laboratory to date. It will still take probably at least half a century before the first fusion power plant will see the light (if it ever happens). By then a century will have passed since the first design. This fact alone is probably sufficient for the reader to draw their own conclusion, and it certainly does not bode well for the future of energy from nuclear fusion.

Part V
Concluding Chapters

Chapter 18
Criticism of the Fusion Enterprise

18.1 Introduction

Since the dream of abundant and cheap energy associated with nuclear fusion is slow in coming, many have wondered whether it is worth all this effort and money. On the Internet you can read an unlimited amount of hostile comments about the waste of money that research in nuclear fusion might imply. Even books have been written to belittle the fusion effort and cast doubt on the sincerity and integrity of the scientists involved. An example is Charles Seife's book "Sun in a Bottle: The Strange History of Fusion and the Science of Wishful Thinking" (Seife 2008), who accuses scientists of deception and cheating.

Such criticism is certainly not without ground. Although it probably goes too far to accuse fusion scientists of outright dishonesty in the pursuit of their dreams, some deception and cheating is certainly going on. When reading some of the careless statements by people calling themselves scientists, of which we have seen a few examples scattered throughout this book (see e.g. the discussion about the term breakeven in Chap. 8) and of which we will see more examples below, one indeed wonders whether they know how science ought to work. Some see it as an exercise in publicity, it seems, with the simple goal of raking in as much money as possible to be able to continue on this endless road. If a little dishonesty is needed to achieve this, so be it; once the splendid dawn of cheap and abundant energy is upon us, all will be forgotten. This dawn however refuses to break and it can indeed be said without hesitation that the result of the huge effort spent in fusion research in the last 60–70 years, is so far deeply disappointing, to the extent that no one can be blamed for no longer believing in the promise ever to be fulfilled.

The argument often used by proponents of nuclear fusion is that we do not have a choice. That fusion is necessary not only to combat climate change and provide a carbon-free energy source, but also to be able to meet the global demand for energy.

© The Author(s), under exclusive license to Springer Nature Switzerland AG 2021
L. J. Reinders, *The Fairy Tale of Nuclear Fusion*,
https://doi.org/10.1007/978-3-030-64344-7_18

The number of people on Earth is soon to reach a staggering 10 billion[1] who all want to use electricity in ever greater amounts. Burning fossil fuels is no option. Not only is there not enough in the long run, the climate change consequences become ever more pressing. Nuclear fission energy has been banned in many countries because of its waste problems and the risk of man-made (Chernobyl) and natural (Fukushima) disasters. In the long term the only resources capable of producing energy on a massive scale will be solar and wind in one form or another and fusion. Solar and wind energy need breakthroughs in production and storage before they can be deployed on the scale needed, which leaves nuclear fusion energy as the only alternative to be pursued.[2] The pros, if it works, are obvious: abundant, high-energy density fuel (deuterium plus lithium (to produce tritium)), producing per gram of fuel about 10 million times more energy than gasoline or coal, no greenhouse gases, safe, minimal "afterheat", no nuclear meltdown possible, small residual radioactivity with short-lived and immobile products, minimal proliferation risks, minimal land and water use, no energy storage issue and no seasonal, diurnal or regional variation. Some of these advantages are undoubtedly real, but not quite as real as the proponents want us to believe, as we will see in this and the next chapters. The cons are only high capital costs and, most importantly, the fact that it does not work and probably will never work. This latter fact is of course lethal. From the description of the fusion effort so far in this book, it is clear that, although recent developments give some hope, an awful lot has still to be done to make it work. Success is in no way guaranteed, very unlikely even, and will surely not be achieved within the foreseeable future. It is safer to wait for the breakthroughs necessary for solar and wind than for fusion. No sooner than around 2050 there might be any certainty about whether or not large-scale energy generation from fusion is *in principle* possible or should be abandoned and left to the stars. And even if it works and power plants can be built, the costs will probably be prohibitive, not to speak of the engineering problems and complexity of a fusion reactor, which will make power companies very reluctant to run them. They may be forced to take them on in order to survive, as other power plants will no longer be tolerated. It may even have to come to a situation in which there is no longer any place for private energy generating companies and (inter)governmental enterprises will have to take over power generation.

In any case, the technical and scientific challenges faced by fusion are impossible to overcome in the short time the world has available for making the switch to carbon-free power generation. Advances in renewable energy (and its storage) will probably manage to solve problems with future energy needs faster and at lower cost than fusion will ever be able to do, if at all. Fusion will take far too long to reduce greenhouse gas emissions to avoid unacceptable global warming. Even if green energy growth

---

[1] Although recent research has suggested that global population is to peak at 9.7 billion around 2064, before falling down to 8.8 billion by the end of the century (Vollset et al. 2020; https://doi.org/10.1016/S0140-6736(20)30677-2).

[2] For a convincing plea in favour of decarbonization through a mix of power resources, including nuclear power (just fission power though, no contribution from fusion is envisaged), solar, wind and other renewable resources, as well as conventional power plants with carbon capture and storage, see Sivaram 2018.

were too slow to save the planet, nuclear fission could be revived by employing safer devices and the use of natural gas could be ramped up. Energy from nuclear fusion in any amount that could make a difference cannot be expected in this century. If all goes well, and that is a big if indeed, a first experimental power plant may be available at staggering cost at the very end of the century. We will discuss this in some more detail later in this chapter.

First, I will consider the cost factor of fusion research, followed by the rosy pictures painted by scientists and the media about fusion's prospects. The rest of the chapter will be devoted to some of the (serious) criticism made of the fusion enterprise as a whole, of its unachievable goals, or of specific projects, such as ITER. The criticism mainly comes from (former) scientists. There are very few active fusion scientists who vent their critical observations, but a growing number of retired fusion scientists are now starting to speak out against the fusion folly. It seems that as long as fusion pays for their mortgage and the schooling of their kids, they behave as ardent supporters of fusion and refrain from speaking their mind. It may also be dangerous for their career as an early critic, Lawrence Lidsky, found out at his peril as we will see below. Only in retirement they can forgo the pretence and really speak their mind.

18.2 Cost

A much-heard point of criticism has to do with cost. In this respect Seife speaks of the 'unfathomable wealth' spent in vain on fusion. That criticism does not wash. In actual fact, the amounts spent on fusion, currently annually just a couple of billion of dollars for all countries taken together, are not that large when compared with other relevant pursuits of mankind. It is of course a large amount of money and, if well spent (on trying to understand the science behind fusion for instance), could do a lot of good,[3] but compared to the amounts sunk into energy research and exploration of fossil fuels, e.g. by the big oil companies (for Royal Dutch Shell alone an amount in the order of 20–30 billion dollars annually (at least until recently)) they are actually fairly small, not to speak of the costs due to oil spills and other disasters with ordinary fossil-based energy, or the sums spent in US oil subsidies. US tax breaks for oil and gas companies still amount to $4 billion a year (Sivaram 2018, p. 268). The BP oil spill in the Gulf of Mexico in 2010 is estimated to have cost about $40 billion and dwarfs the entire fusion effort, while the price of the Chernobyl and Fukushima disasters is indeed unfathomable. Nobody has any idea how high that price eventually will be or how to calculate it, but it will certainly be in the order of trillions of dollars.

Moreover, other renewable energy sources are not particularly cheap either. In Germany subsidies for solar and wind energy have amounted to on average €25

---

[3]If such amounts were spent in research on carbon capture technology (old fashioned technology, not very sexy), directly tackling the problem of carbon emissions, better results can be expected (see Sivaram 2018).

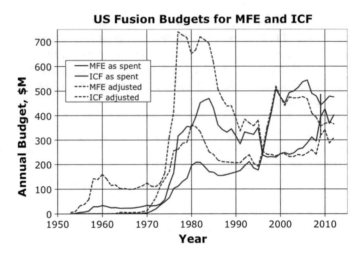

Fig. 18.1 US fusion budgets from 1950 to 2012. MFE stands for Magnetic Fusion Effort, ICF for Inertial Confinement Fusion. Adjusted means adjusted for inflation (*Data source* Department of Energy via Fusion Power Associates)

billion per year in recent years. Wind and solar contribute nowadays about 40% of the German electricity mix but had a negligible effect on CO_2 emissions from electricity generation. This seemingly surprising result is due to Germany's foolish decision to phase out nuclear fission power stations and replace them by coal-fired plants or even worse by lignite burning plants. That does not help, of course. How foolish can you be?

As can be seen from Fig. 18.1, the total amount spent on fusion energy in the US, since the fusion programme began in 1953 until 2012 is $24.1 billion dollars (adjusted for inflation to $30.4 billion). That is an average of $400 million a year, adjusted to $515 million per year. It includes the expenditure on inertial confinement fusion (e.g. the National Ignition Facility), as well as on tokamaks and alternatives.[4] It is arguable that the expenditure on inertial confinement fusion should (at least partly) be deducted, as it involves weapons research and not so much energy generating technology. The fact that ICF mainly concerns weapons research is also the reason that expenditure on it has actually gone up through the years, as can be seen in the figure.

The total 2016 US fusion budget was $951 million; by 2019 it had gone down to a mere $564 million and for 2020, thanks to action by Congress, it went up again to $671 million. As a comparison, NASA's budget in 2016 was eight billion. The funding for fusion is now roughly split in half between ICF and magnetic (tokamak) fusion, the latter being mainly the US contribution to ITER.

As can also be seen from the figure, US spending has been rather erratic, reflecting the many policy changes by the US government over the years. It is argued that these

---

[4]Source *Fusion Power Associates*, from the website https://aries.ucsd.edu/FPA/OFESbudget.shtml, accessed 13 March 2020.

policy changes, often caused by changing priorities as a new administration took office, have regularly stymied attempts by the fusion community to capitalize on fusion's scientific successes in the past. This argument is rather shaky in my view, but it is obvious that, as it stands at present, the US fusion budget is not on a track that will ever lead to a US demonstration plant (Dean 2013, pp. 233–234).

18.3 Painting Rosy Pictures

It is common practice in the fusion business to paint rosy, if not dishonest, prospects. For instance, in the abstract to a 2005 "Status report on fusion research" (IFRC 2005) published in a *scientific* journal by the International Fusion Research Council (IFRC), an august body created by the International Atomic Energy Agency (IAEA) in 1971 to advise the agency on its activities in the field of nuclear fusion, it is stated: "Fusion is, today, one of the most promising of all alternative energy sources because of the vast reserves of fuel, potentially lasting several thousands of years and the possibility of a relatively 'clean' form of energy, as required for use in concentrated urban industrial settings, with minimal long term environmental implications. The last decade and a half have seen unprecedented advances in controlled fusion experiments with the discovery of new regimes of operations in experiments, production of 16 MW of fusion power and operations close to and above the so-called 'breakeven' conditions." This is on the verge of outright lying, and in any case grossly misleading and unscientific, as fusion in its current state is in no way promising (is promising a scientific term? Moreover, should I trust a scientist who calls his own work promising? What would he actually mean?) and the advances of the last decade or so may have been unprecedented (in the field of nuclear fusion not much is needed to be unprecedented, as there simply is no precedent), but certainly not promising or even encouraging. The fact that there are vast reserves of fuel is completely irrelevant if this fuel cannot be made to 'burn'! If anything, the whole fusion enterprise looks hopeless. Breakeven conditions are still very far away. The use of the word 'breakeven' in the quote above is aptly (but unscientifically) qualified by 'so-called', as 'breakeven' in nuclear fusion, as we have seen in Chap. 8, is a completely different concept from what a normal sensible person would mean by it, and is used by fusion proponents in a very misleading way, probably with the intent to mislead, a magnificent case of doublespeak if there ever was one. It is in any case not true that there have been operations above breakeven. In the report itself this is qualified by saying that "present day fusion experiments have already exceeded *conditions equivalent* to a $Q = 1$ operating power plant", without of course explaining what '*conditions equivalent*' actually means. From the discussion in Chap. 8 we know that it means by extrapolating D-D results to D-T results, so-called extrapolated breakeven, which has nothing to do with reality. Moreover, a $Q = 1$ device cannot be called a power plant, as it is just a power consuming device. The statement made above may arguably be included in a publicity or advertisement leaflet but has no place in a scientific journal.

They make matters worse by stating in the body of the report that it "appears realistic that fusion power plants delivering electricity will be available for commercial use towards the middle of this century". The statement dates from 2005 when construction of ITER had not even started (the squabble about the ITER site had just ended) and the authors must have known that there was no scientific basis for making such a statement; they cannot themselves have believed it!

The rosy pictures painted by the leaders of the various fusion enterprises have in general unquestioningly been accepted by the media, as is often the case with science. Scientists still tend to be trusted for no other reason than that they are scientists. The media often even tries to surpass them in screaming headlines of great progress made towards unlimited energy or similar nonsense. There have been pitifully few who did not fall for these fairy tales. Especially noteworthy in this respect is Steven Krivit who runs the *New Energy Times* website and was at first mainly concerned with debunking the cold fusion fiasco, one of these unfortunate cases of pathological science, where specialists from other fields lacking basic knowledge about nuclear fusion processes claim to have found a shortcut to fulfilling the dream of cheap and inexhaustible energy. After having done this rather thoroughly, from 2006 Krivit also started to cast a critical look at the statements of leaders of hot fusion projects like JET and ITER. What he discovered was really shocking: a marvellous and almost endless collection of false statements, some so blatantly untrue that they, assuming that the people who made them are competent in the field, must have been made with the clear intent to lie and/or mislead the people they were addressing. I believe Krivit started with "The ITER Power Amplification Myth"[5] in which he pointed out that the ITER management and communications office have led journalists and the public to believe that, when completed, the reactor will produce 10 times more power than goes into it. We know by now that that is not true, but that is only so because Steven Krivit did not tire of pointing this out. The true picture (from his website) is shown in Fig. 18.2. It shows that the amount of *electric* power that goes into the reactor will be 300 MW and that 536 MW may be produced in *thermal* power. When converted into electricity at 40% efficiency (which is high), it will result in 214 MW electric power output, leaving a shortage of 86 MW.

This stands in glaring contrast to what the ITER website stated in 2017: "The goal of the ITER fusion program is to produce a net gain of energy and set the stage for the demonstration fusion power plant to come. ITER has been designed to produce 500 MW of output power for 50 MW of input power—or 10 times the amount of energy put in. The current record for released fusion power is 16 MW (held by the European JET facility located in Culham, UK)."

Thanks to Krivit's prodding the text on the website has now been changed into: "ITER has been designed for high fusion power gain. For 50 MW of power injected into the Tokamak via the systems that heat the plasma it will produce 500 MW of fusion power for periods of 400 to 600 s. This tenfold return is expressed by $Q \geq 10$ (ratio of thermal output power to heating input power). The current record for fusion power gain in a tokamak is $Q = 0.67$ held by the European JET facility located

---

[5] https://news.newenergytimes.net/2017/10/06/the-iter-power-amplification-myth/.

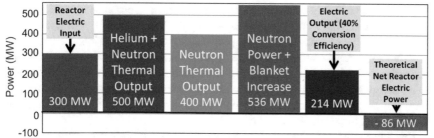

Fig. 18.2 Showing the electric power that goes into the ITER reactor, and the output in thermal power, resulting in net power usage, not power gain (from https://news.newenergytimes.net/)

in Culham, UK, which produced 16 MW of thermal fusion power for 24 MW of injected heating power in the 1990s."[6]

In Chap. 8 we have already seen what JET's 16 MW of produced power actually means for the efficiency of the machine (just 2%, as also dug out first by Krivit, but kept under the rug and only released when asked for, while it is the only number that really counts!). And in Chap. 10 we already have related what ITER's 500 MW actually means and that the website of the Japanese JT-60SA has never thought it necessary to hide the truth.

The new text on the website is still not a model of clarity. No member of the public will understand it without further explanation. It nonetheless is an improvement, although still untrue. ITER has not been designed for *high* fusion power gain, but for pitifully *low* power gain. After all, it is, as the JT-60SA website states, a zero (net) power reactor. It remains common policy among fusion managers not to mention how much electricity ITER requires to operate, and it is very difficult to figure this out. For instance, the 300 MW mentioned in Fig. 18.2 does not include ITER's non-interruptible power usage; power that will be needed regardless whether the reactor is operating or not. As we will see below, Daniel Jassby has a lot to say on this subject.

Krivit has waged a crusade against these false statements, not only by the ITER organisation, but by EUROfusion and other organisations and officials as well. They were all involved in the "Great ITER Power Deception".

A wrong statement on a website is one thing but providing confusing or misleading information to the US Congress is quite another. But nonetheless, as recent as March 2018 Congress was shamelessly misled by ITER director Bigot and the acting associate director of the Department of Energy's Office of Fusion Energy Sciences Van

---

[6]https://www.iter.org/factsfigures.

Dam, as Krivit reports.[7] Such statements tend to propagate through the community. Others start to repeat these inaccuracies on their own website. The wrong information spreads as a highly contagious virus and is almost impossible to eradicate, as a collection on Krivit's website shows.[8] Even reputable journals like *New Scientist* do not seem capable of getting it right or providing correct information.

Krivit's comments and criticism are of course belittled with statements like "this discussion is irrelevant in the case of ITER since its purpose is not to produce as much energy as possible but to demonstrate the technological feasibility of fusion."[9] If so, why do the fusion leaders mention these numbers and don't they tell Congress to stop asking these irrelevant questions?

Krivit has much more to tell about the probably deliberately deceptive way ITER is presented to the public. Further details can be found on the website of the *New Energy Times* (https://news.newenergytimes.net/). I will restrict myself here to repeat his conclusion:

"Given the preponderance of misrepresentations of the ITER power values on prominent Web sites, in news outlets such as the *New York Times, Bloomberg,* and the *BBC*, in science publications such as *Nature*, in major worldwide Web references such as *Wikipedia*, EUROfusion, and World Nuclear Association, and in a publication of the European Parliament, logical conclusions are that: (1) The fusion representatives who created the misrepresentations had to have known of the effects of their public relations efforts; (2) A significant number of fusion scientists who were not directly responsible for the creation of the misrepresentations must have read about their project in the news media and known of the falsehoods—yet for at least five years before October 2017, they corrected none of the falsehoods.

Even the director-general of ITER, Bernard Bigot, had to have seen the falsehood in [an article in *Nature*[10]], which says that ITER "is predicted to produce about 500 megawatts of *electricity*." He added a comment to the article after it [was] published.

Was the public broadly misled? The list above shows that, yes, it was. Was the Q-switch intentional on the part of the scientists? In most cases, this is difficult to prove. Was the Q-switch intentional on the part of the people—like Neil Calder and Laban Coblentz[11]—whose job it was to create, manage and track the worldwide public

---

[7] https://news.newenergytimes.net/2018/03/28/experts-testify-before-congress-on-future-of-u-s-fusion-energy-research/.

[8] https://newenergytimes.com/v2/sr/iter/Effects-Misleading-Statements-Fusion.shtml.

[9] Claessens (2020), p. 165. In his book Claessens also mentions Krivit very briefly in just a few lines on p. 103–104, stating that he pointed out some false statements on websites, without giving a further opinion. He apparently did not dare to publish in his book what he wrote to Krivit and can be read on the *New Energy Times* website about this matter (https://news.newenergytimes.net/2017/01/19/former-iter-spokesman-confirms-accuracy-of-new-energy-times-story/).

[10] https://www.nature.com/news/us-advised-to-stick-with-troubled-fusion-reactor-iter-1.19994.

[11] Former and current head of communications at ITER.

communications of their respective organizations? Clearly, they were in control of the public messaging about ITER and fully aware of the results."[12]

When entering Krivit's website please be forewarned that it is at your own peril, the peril of losing your belief in scientists, the only people on the planet you still thought could (sometimes) be trusted.

18.4 Lawrence Lidsky

The first criticism that aroused some attention goes back to a 1983 paper (Lidsky 1983)[13] by the fusion insider Lawrence Lidsky (1935–2002), at the time a professor of nuclear engineering at MIT and associate director of the MIT Plasma Science and Fusion Center. He also was the founding editor of the *Journal of Fusion Energy*. Lidsky wrote a paper, called *"The Trouble with Fusion"* in the MIT Technology Review. At the time he said that he wrote it because "I couldn't get an internal discussion going. Some didn't care and some didn't want to know."[14] I am afraid that is still the case. He actually was of the opinion that the fusion programme had come prematurely under the sway of machine-builders and that as a consequence science was suffering. An explicit goal was established, generating commercially competitive electricity from D-T fusion early in the twenty-first century. Once such a goal has been set, it is not easy to change. Producing net power from fusion is a valid scientific goal, but generating electricity commercially is an engineering problem, Lidsky said.

In the article he discussed all the problems that are still harassing nuclear fusion reactor designs based on the tokamak concept and trying to burn deuterium and tritium. He pointed out that the engineering problems involved in a fusion reactor are so immense that, even if a reactor were eventually built, no one will want to buy it, no power company would ever want to work with a reactor with such complex engineering. "The costly fusion reactor is in danger of joining the ranks of other technical "triumphs" such as the zeppelin, the supersonic transport, and the fission breeder reactor that turned out to be unwanted and unused." "A chain of undesirable effects ensures that any reactor employing d[euterium]-t[ritium] fusion will be a large, complex, expensive and unreliable source of power." He pointed out that deuterium-tritium as a fuel had undesirable effects, including a tritium supply problem and

---

[12]https://news.newenergytimes.net/2017/12/11/evidence-of-the-iter-power-deception/;   *Q*-switch refers to the confusing use of other definitions of breakeven, in this case making it look as if the actual *Q* achieved was engineering breakeven, instead of scientific breakeven.

[13]A short time after the article appeared, Lidsky resigned his position, and the US Congress reduced funding for the fusion program by five percent the next year. An adapted version entitled '*Our Energy Ace in the Hole is a Joker: Fusion Won't Fly*' was published in the Washington Post of 13 November 1983. Lidsky had already written a critical paper in 1982 (Lidsky 1982) which includes a plea for fission-based systems.

[14]https://archive.thinkprogress.org/ny-times-funding-for-fusion-better-spent-on-renewable-sou rces-of-energy-that-are-likely-to-be-c6ea8398eaba/.

the release of high-energy neutrons whose bombardment would weaken the reactor structure and make it radioactive, problems that still need solving now in the early twenty-first century when we should have been enjoying commercially attractive electric power from fusion reactors. He also pointed out that nuclear fusion reactors would inevitably be very complex as they had to handle enormous heat flows and huge temperature gradients over a distance of a few metres (from the 150 million degrees of the plasma to a few degrees above absolute zero for the superconducting magnets). When he formulated his criticism, tokamaks were still fairly small, with a couple of large ones (TFTR and JET) in the process of being built, and they have become ever larger still, with costs going through the roof.

At the time he was not thanked for his message, on the contrary the reaction was swift and substantial, leading to Lidsky being quietly 'purged' in almost Soviet style, made a pariah in the fusion community and stripped of his title as an associate director of the MIT Plasma Science and Fusion Center (Herman 1990, p. 191; Clery 2013, p. 304).[15] Especially Harold Fürth, the then director of PPPL, went to great length seeking to destroy every argument Lidsky was making and trying to nip in the bud any damage his article might have caused. TFTR was being constructed, but its successor the Toroidal (or Tokamak) Fusion Core Experiment (TFCX) was being planned. It would be the first device to achieve ignition, but was never built, as Congress slashed the fusion budget. The decrease was not that much, just five per cent, but devastating for these plans for which an increase was needed. It may well be that Lidsky's dissent affected the vote in Congress (Marshall 1984).

Reading Lidsky's paper now, more than thirty-five years after it was published, it is clear that his arguments are still as fresh and apt as they were at the time, even more so as solutions are more urgently needed and in spite of the alleged 'great progress' that has since been made, as has abundantly been shown in the preceding chapters of this book. A fusion reactor will indeed be large, complex and expensive. The neutrons are a problem and tritium breeding is not that simple. Since no reactor has yet produced any power to speak of, it is not yet possible to say whether it will be unreliable, but the signs are certainly not favourable.

After his rather early death in 2002 at age 66, one of Lidsky's colleagues observed: "He was one of the earliest engineers to point out some of the very, very difficult engineering challenges facing the program and how these challenges would affect the ultimate desirability of fusion energy. As one might imagine, his messages were not always warmly received initially, but they have nevertheless stood the test of time."[16]

---

[15]Lidsky was asked to resign (a questionable request for an academic institution to make) by Ronald C. Davidson (1941–2016), the first director of the MIT PSFC, who later became director of PPPL.
[16]MIT News, March 5, 2002.

18.5   Robert L. Hirsch

Hirsch, whom we have encountered earlier in this book, held important management positions in the US fusion research programme, e.g. director of this programme during the 1970s in which he aggressively pursued the tokamak option. As related in Chap. 4, he even predicted that commercial fusion power could be on the grid by the year 2000. However, it took not very long for him to drastically change his mind. Already in 1985, he infuriated the American fusion community by denouncing the tokamak as an impractical reactor design (Hirsch 1986. See also Herman 1990, pp. 210–211 about this.) The fatal flaw of the tokamak is, he said, that "it is inherently a complex maze of rings and a toroidal chamber inside of other rings. In my view, this complex geometry will not be acceptable to the utility world, where power plants must be maintained and serviced rapidly at low cost. In that world, simple geometries are essential." With this he was essentially making the same point of unacceptable complexity that Lawrence Lidsky had made earlier.

Twenty-five years later in a book on the history of the US fusion effort (Hirsch in Dean 2013, pp. 217–218), Hirsch elaborates on his earlier remarks by stating that the current effort to achieve commercial fusion energy will almost certainly fail for three main reasons: (1) fusion power plants as currently envisaged will be extremely large and complicated, and consequently very expensive while low cost is a major requirement of electric power generation (see the ITER saga about this in Chap. 10); (2) the radioactivity problem will require complicated remote handling operations, expensive maintenance and disposal of large volumes of radioactive waste (this aspect will be discussed in Chap. 20); and (3) the currently envisaged fusion reactors are not inherently safe (also to be discussed in Chap. 20). They require large high-field superconducting magnets, which contain large amounts of stored energy. There is a small, but nevertheless possible risk for this energy to be explosively released, if one or several of the magnets were suddenly to become non-superconducting.

And Hirsch has stuck to this view as a few years back he wrote in *Science and Technology*: "tokamak fusion power will almost certainly be a commercial failure, which is a tragedy in light of the time, funds, and effort so far expended." (Hirsch 2015). He even wants to learn lessons first from the tokamak experience before embarking on another method, quite a different attitude from the one he himself displayed when directing the AEC fusion programme. Such cavalier approach seems pervasive in US fusion research–we have seen it earlier in statements by Trivelpiece– and has damaged the fusion enterprise more than anything else, I believe. Everybody knows that science cannot achieve everything you would like it to, and certainly not on command or by just throwing a lot of money at it. Certain things are just not or not yet possible, because of physics or engineering problems or simply for lack of knowledge and understanding of the physics behind them. It makes sense to try to find out first what the basic problems of a certain approach are before embarking on large-scale and hideously expensive ventures. Of course, if Hirsch's prediction of fusion power on the grid by the year 2000 mentioned above had come true, he would have been hailed as a visionary and the saviour of mankind, so for him perhaps it was worthwhile the risk although he must have known that the odds were heavily against him.

Let us now turn to criticism of the ITER project itself.

18.6   Criticism of ITER

Let us start this section with Hirsch who of course had also something to say on ITER. In a book he wrote with two others in 2010, entitled *"The Impending World Energy Mess"* (Hirsch et al. 2010), in which as the title suggests impending doom is prophesized (mainly as regards an oil shortage), it is said "the outlook for success with the physics aspects of ITER is good, but the likelihood of commercial success is near zero. Thus, when ITER operates 'the operation might well be a success, but the patient will in effect be dead'. As a result, the world will have wasted decades and tens of billions of dollars on a dead-end concept. Sadly, the ITER waste could have been avoided." For the latter remark Hirsch is advised to look in the mirror and see the man who is (partly) to blame for this. As regards the prophecy in the title of his book, it applies that like most such prophecies also this one has turned out to be completely false. The world is no messier today than it was in 2010.

Hirsch however gets too much credit as a critic. There are much more capable people, very capable people indeed, on a par with Lidsky, who have joined the critics: the French physics Nobel Prize winners Pierre-Gilles de Gennes[17] (1932– 2007) and George Charpak (1924–2010) and the Japanese physics Nobel Prize winner Masatoshi Koshiba (b. 1926), who have in particular criticized ITER as a useless and overpriced reactor. The first two have conveniently died by now, but their criticism is still very much to the point. De Gennes spent part of his career with the French Atomic Energy Commission (CEA) and can be considered an expert in all things nuclear. He feels that too much money is spent on actions that are not worth it and mentions nuclear fusion as an example. "European governments, as well as Brussels, have rushed into the ITER experimental reactor without having carried out any serious reflection on the possible impact of this gigantic project. (…) Before building a 5-tonne chemical reactor, one must have fully understood the operation of a 500-L reactor and have evaluated all the risks that it entails. However, this is absolutely not how we have proceeded with the ITER experimental reactor. We are unable to fully explain the instability of plasmas or the thermal leaks of current systems. So, we are embarking on something which, from the point of view of a chemical engineer, is heresy. And then, one last objection. Being fairly well informed about superconducting metals, I know that they are extraordinarily fragile. So, to believe that the superconducting coils that are used to confine the plasma and are subject to a fast neutron flux comparable to an H bomb will be able to survive such flux during the entire lifespan of such a reactor (ten to twenty years), sounds crazy. The ITER

---

[17]https://www.lesechos.fr/2006/01/recherche-le-cri-dalarme-dun-prix-nobel-1068633.

project was supported by Brussels for reasons of political image, and I find that it is a fault."

Georges Charpak was a member of the CERN staff since 1959 where he invented and developed the multiwire proportional chamber, a device for detecting charged particles and photons that can give positional information on their trajectory by tracking the trails of gaseous ionization. Charpak's main criticism concerns the enormous costs of ITER, in combination with the fact that it in all likelihood will not solve the problems fusion is faced with ("it will only study the stability of its own plasma"). He had earlier vented his concern that ITER's cost would eat into the science budgets of EU member countries. In a last interview (Charpak et al. 2010) with the journal *Libération* in 2010 just before his death and just before construction of ITER started in earnest, he and his colleagues Treiner and Balibar said that what they had feared was actually happening: the estimated cost of the construction of ITER has gone up from 5 to 15 billion euros (*it is even more now (LJR)*), and there is talk of passing on the consequences to the European budget for scientific research. "This is exactly the catastrophe we feared. It is high time to give it up. (…) For France, the expense will represent more than the total amount (excluding salaries) made available to all physics and biology laboratories for twenty years! Much more important research, including for the energy future of our planet, is thus threatened." They continued by stating the three major problems for fusion that nobody has managed to solve for more than 50 years: to keep the plasma inside the reactor vessel, as it is unstable; to produce tritium in industrial quantities and to invent materials to enclose this plasma under ultra high vacuum in a vessel of a few thousand cubic metres. The most formidable problem they say is the third one: violently irradiated by very energetic neutrons (14 MeV) emitted by fusion reactions in the plasma, the material of the vessel will lose its mechanical strength. No matter how often it is said that materials can be imagined that will resist irradiation, they remain sceptical. The development of a prototype power plant, let alone a commercial power station, is still very far away. It is in no way justified to take money away from other research projects on the pretext that an almost infinite source of energy is on the horizon. Plasma physics must be funded on par with the other major fields of fundamental research. They then say: "If we continue, all areas of research will suffer. This situation is reminiscent of the construction of the International Space Station, the ISS. Another pharaonic project, the ISS cost $100 billion and our colleague astrophysicists still remember the budget cuts that its construction entailed. What was the ISS used for? For virtually nothing. To observe the Earth or the Universe, it is better to send robots into orbit that are more stable and cheaper.[18] In fact, the astronauts are bored up there. So, they spend their time studying their own health! ITER is likely to be comparable: if built, this large machine will only be used to study the stability of ITER's plasma. Isn't it a little expensive to spend 15 billion euro on that? (…) So, rather than masking an initial bad decision by an even worse escalation, it would be better to finally admit that the gigantism of the project is disproportionate to the expectations, that its management

---

[18] A view shared by Robert Park in his book *Voodoo Science* (see below).

appears deficient, that our budgets do not allow us to pursue it, and transfer that money to useful research."

In his book[19] on ITER, Michel Claessens, ITER's former communications director, has only a feeble and rather unconvincing reply to this, mainly of course as there is no way around their arguments. Particularly unconvincing is his argument that, since ITER is now funded directly from the EU budget, French and European researchers would be immune from budgetary restrictions caused by ITER. A very weak argument indeed as money can be spent only once and the money spent on ITER cannot be spent on other (energy) research or climate-change combating measures,[20] of which there are currently pitifully few. Moreover, he immediately fatally weakens his argument by stating that there is no direct impact on national research priorities and funding in the EU, but that his argument does not necessarily hold for the other ITER Members. Neither does it hold for EU members. The Dutch Institute for Fundamental Energy Research, DIFFER, for instance, carries out a lot of ITER-related work and is, in any case partly, funded by the Dutch government and the Dutch Research Council (NWO), money that, if there were no ITER, would be spent on other scientific projects. The same is most probably true for all other EU countries involved in any way with ITER. There may be a division between ITER funding and funding for other research at EU level, but such division does not exist at EU-member state level, so spending for ITER undoubtedly affects other scientific research in EU-member states. And indeed, as regards the other ITER Members, it does not hold for the US, which explains in part why the US, although contributing comparatively little, is especially sensitive about ITER's costs. ITER is eating the larger part of the budget of the Office of Fusion Energy Sciences. Because of ITER's thirst for cash, many other concepts have been strangled or shut down. Researchers often need to "show relevance" to ITER, otherwise they run the risk of not obtaining funding or getting closed down. The United States has so far contributed about $1 billion to ITER, and had been planning to contribute an additional $500 million through 2025. Although it promised to behave well when it re-joined the collaboration in 2003, it has several times failed to fulfil its obligations. The latest is that it cut its 2017 contribution from a scheduled $105 million to $50 million and had planned to cut its 2018 contribution from $120 million to $63 million, which would have resulted in further delays to the project. But at the last minute a $122 million in-kind contribution for ITER was approved (De Clercq 2018).

Koshiba drew attention to possible hazards in connection with the presence in the reactor of four kilograms of tritium, which is radioactive, and although his worries in

---

[19]Claessens 2020, p. 98ff, whose discussion of criticism of the ITER project is in general rather disappointing. In spite of this Claessens, who as former science communications officer at ITER must have been used to singing the praises of the ITER project or at the very least presenting it to the public as a palatable venture, has written an admirably balanced book, which clearly brings to the fore the difficulties involved in managing the ITER project.

[20]The idea, often heard nowadays, that fusion is actually needed to combat climate change is just laughable. The fusion effort has a gigantic carbon footprint, and climate change with all its adverse consequences will be all over us long before any electricity from fusion will flow through the grid. If anything, the fusion effort and in particular the building of ITER will make climate change worse.

this respect seem to be merely theoretical, there is no need to call him a discredited scientist, as has foolishly been done by someone whose bread is, no doubt generously, buttered by fusion money (Claessens 2020, p. 102).[21] Koshiba is a very serious scientist and should be treated as such. His criticism does not only pertain to tritium but encompasses much more. Just like Charpak and his colleagues, he issued very apt warnings about the highly energetic neutrons that will be released in huge quantities in the fusion reactions and scientists do not know how to handle. How to absorb them by the walls surrounding the reactor, without the material becoming radioactive and forcing a biannual replacement for which the reactor will have to be shut down, an expensive and uneconomical solution? No wonder Koshiba was very happy when it was decided to build ITER in France and not in Japan.

It is interesting to see that a growing number of retired fusion scientists are now starting to speak out against the ITER folly. Only in retirement they can forgo the pretence and really speak their mind. A remarkable example is Daniel Jassby, who was a principal research physicist at the Princeton Plasma Physics Laboratory until 1999 and worked for 25 years in fusion energy research. During that time, he was an ardent promotor of fusion, like many others. In Chap. 11 he has been credited with having stood at the cradle of the spherical tokamak. He clearly is not someone who can or should be dismissed out of hand, not a 'mere' journalist. In a few publications in the online version of the Bulletin of the Atomic Scientists (Jassby 2017, 2018) and on the website of the American Physical Society[22] he made short shrift of ITER and the entire fusion enterprise. He is worth quoting in some detail. Most of the problems and arguments he brings to the fore have been discussed and identified as such in the previous chapters of this book, but it is worthwhile reading them in the succinct and pertinent manner put by Jassby. In his criticism of ITER he especially focuses on power yield and usage, which we have already referred to above and in Chap. 10. Water usage, radiation and radioactive waste will be dealt with in Chap. 20.

A vast amount of energy will be expended in operating the ITER project—and indeed every large fusion facility. Its non-interruptible electric power drain varies between 75 and 110 MWe, i.e. power that has to be drawn from the main grid continuously, day and night, also when the plant is not operating (Gascon et al. 2012). When ITER actually operates, its plasma requires at least 300 MWe, the output of a small power station,[23] for tens of seconds for heating and generating the necessary plasma currents. During the 400-s operating phase, about 200 MWe will be

---

[21]It is with some apparent relish that Claessens quotes alleged foolish statements or remarks by ITER critics, without paying much attention to the criticism itself. Apart from Koshiba he also derides the French physicist Jean-Pierre Petit, who may be a maverick, but that does not necessarily invalidate his criticism of ITER. He also says without giving evidence that the ITER project is supported by the vast majority of scientists. I doubt this to be true. Fusion scientists will of course support it, but others who see their budgets slashed because of ITER will not be so enthusiastic. Moreover, it doesn't matter how many support it; it is abundantly clear that it is a flawed project and may well be a case where even scientists won't listen to the arguments.

[22]https://www.aps.org/units/fps/newsletters/201904/voodoo.cfm.

[23]The International Atomic Energy Agency (IAEA) defines 'small' as under 300 MWe, and up to about 700 MWe as 'medium'. The bigger power stations yield more than 1600 MWe.

needed to maintain the fusion burn and control the plasma's stability. Even during the coming years of plant construction, the on-site power consumption will average at least 30 MWe. The expected 500 megawatts of output refer to fusion power (thermal power)—and has nothing to do with electric power. How much electric power this could be depends on the efficiency of the installation in converting the thermal power into electric power, and ITER is not going to do that. The 50 MW of heating power injected into the plasma helps sustain its temperature and current, and is only a small fraction of the overall electric input power to the reactor, which varies between 300 and 400 MWe. This agrees with what the JT-60SA website states about ITER's power production, as we have seen in Chap. 10: virtually nothing. One wonders what the great difficulty is for the ITER Organisation to state such matters clearly on its website, instead of (falsely) boasting of being "the first of all fusion experiments in history to produce net energy gain".

This input electric power of 300 MWe and more is indisputable, but a fundamental question is whether ITER will produce 500 MW of anything, a query that revolves around the vital tritium fuel—its supply, the willingness to use it, and the campaign needed to optimize its performance. Fusion practitioners, so Jassby states, are in fact intensely afraid of using tritium for two reasons: first, it is somewhat radioactive, which implies that there are safety concerns connected with its potential release to the environment; and second, there is unavoidable production of radioactive materials as fusion neutrons bombard the reactor vessel, requiring enhanced shielding that greatly impedes access for maintenance and introduces radioactive waste disposal issues. Assuming that the ITER project is able to acquire an adequate supply of tritium and is brave enough to use it, nobody knows whether it will actually achieve 500 MW of fusion power. ITER's current schedule (Fig. 10.10) envisages deuterium and tritium use in 2035. But there is no guarantee of hitting the 500 MW target. During the unavoidable teething stage through the early 2040s, it is likely that ITER's fusion power will be only a fraction of 500 MW, and that more injected tritium will be lost by non-recovery than actually burned. The permeation of tritium at high temperature in many materials is not understood to this day (Causey et al. 2012). The deeper migration of some small fraction of the trapped tritium into the walls and then into liquid and gaseous coolant channels cannot be prevented. Most of such tritium will eventually decay, but there will be inevitable releases into the environment via circulating cooling water. In designs of future tokamak reactors, it is commonly assumed that all the burned tritium will be replenished by fusion neutrons reacting with the lithium in the blanket surrounding the reacting plasma, i.e. that they are self-sufficient in tritium. But even that fantasy totally ignores the tritium that is permanently lost in its globetrotting through reactor subsystems. ITER will demonstrate that the total of unrecovered tritium may rival the amount burned and can be replenished only by the costly purchase of tritium produced in fission reactors. The conclusion must be that tritium self-sufficiency is a fantasy, not only for ITER, for which self-sufficiency in tritium is not a requirement, but also for DEMO or real fusion power plants. In Chap. 10 we stated that there will be no commercial development of fusion energy if self-sufficiency in tritium cannot be achieved, so the conclusion from this is clear, with the proviso of course that it only applies to D-T burning reactors.

Whether ITER performs well or poorly, according to Jassby, its most favourable legacy is that, like the International Space Station, it will have set an impressive example of decades-long international cooperation among nations both friendly and semi-hostile. But why cooperation for the sake of cooperation is a good thing, is beyond me. And if you at any price want to stimulate international cooperation, then choose something that is bound to be useful, the eradication of some ghastly diseases for instance. Moreover, the ITER cooperation would not have come off the ground if the EU had not foolishly agreed to pay about half the cost, letting the other participants enjoy a cheap ride. The international collaboration and the resulting immense management problems have greatly amplified ITER's cost and timescale. It can be said that all nuclear energy facilities—whether fission or fusion—are extraordinarily complex and exorbitantly expensive. Other large nuclear enterprises, such as recent nuclear power plants in the United States (Summer and Vogtle, which were greatly affected by the bankruptcy of Westinghouse) and Western Europe (Olkiluoto and Flamanville), and the US MOX nuclear fuel project in Savannah River, whose construction was cancelled in 2018 by the Trump administration because it would cost tens of billions of dollars more than originally estimated, have experienced a tripling of costs and construction timescales that ballooned from years to decades. ITER will however beat the lot with an estimated $60 billion, as shown in Chap. 10. The only sensible conclusion from this can be that all such projects should be abandoned as soon as possible. They cannot be the solution.

ITER may allow physicists to study long-lived, high-temperature burning D-T plasmas. As such ITER will be a havoc-wreaking neutron source fuelled by tritium produced in fission reactors and powered by hundreds of megawatts of electricity from the regional electricity grid, demanding unprecedented cooling water resources. In ITER structural damage will not exceed 2 dpa at the end of its rather short operational life, but that is altogether different in any subsequent fusion reactor that will attempt to generate enough electricity to exceed all the energy sunk into it. That reality alone should be enough to abandon fusion as an energy generating option. As Jassby states, rather than heralding the dawn of a new energy era, it's instead likely that ITER will perform a role analogous to that of the fission fast breeder reactor, whose blatant drawbacks mortally wounded another professed source of "limitless energy" and enabled the continued dominance of light-water reactors in the nuclear arena.

Let me end this section with a further quote from Hirsch, the man who is partly to blame for the rush into tokamaks in the early 1970s, but has now lost all faith in it: "One can only guess at why ITER continues to be built. Did the researchers ignore the engineering warnings associated with "sufficient"? Perhaps they chose to circle the wagons and hide the realities of their chosen concept. Where were the government officials who were supposedly responsible for overseeing fusion research? The media must not have been paying attention, either. When the truth regarding current tokamak fusion research is recognized, embarrassment and repercussions may well be widespread" (Hirsch 2017).

Doubts are also being voiced by people still working in the fusion enterprise, albeit still rather timidly and mainly by people who have another interest. In this respect Leonid Zakharov, a leading plasma physicist at PPPL and currently a proponent of some type of spherical tokamak, is a fairly vocal critic who stated at the Physics Colloquium at Princeton University in December 2000[24] that "tokamak fusion devices (...) are now in an eventual state of defeat and possible shutdown in the US. Despite much better understanding of the tokamak plasma now, many fundamental problems on the way to the tokamak reactor remain unresolved even at the conceptual level. These problems include stability and steady state plasma regime control, power extraction from both the plasma and the neutron zone, activation and structural integrity of the machine under 14 MeV fusion neutron bombardment, maintenance of future reactors, etc." This he wrote in 2000 and, as nothing much has changed since then in respect of the problems he mentions, he has consistently propagated the same message. In a talk in 2018[25] in answer to the self-posed question "Can expectations (i.e. the expectations for fusion to happen) be converted to reality?" he states that "from the mid 1980s to 1990s insufficient attention to science was shown for addressing fusion reactor problems. Then TFTR and JET failed. The leaders disappeared, the program became complicated and unmanageable, and progress was lost. Nobody was capable of understanding that the failure of TFTR and JET indicated that the adopted approach was exhausted. ITER is the implementation of the same failed approach. There are many indications that the program is in the stage of degradation being insensitive to science and experimental data."

Finally, in a very recent paper (Zakharov 2019) he writes: In 2020 the world best tokamak JET will perform the second D-T experiment in the same high recycling regime, which already failed in 1997. With no way of getting $Q = 1$ or even 0.6 as in 1997, it will provide the experimental evidence that the currently adopted approach to fusion is incapable of making progress and is hopeless. It was exhausted 20 years ago. The JET D-T experiment will be proof of the failure of the entire fusion crowd, including management, who were ignorant of science and instead relied on interpretations, scaling, cooked explanations and fake understandings. Since the mid 1990s the science leaders of the program have disappeared and science itself has become unwelcome in the program.

All this is pretty devastating. His statement confirms the assertion stated often in this book that the whole enterprise went down hill when the fusion project was 'upgraded' from a scientific research project to an engineering project that tried to take a shortcut towards a power generating facility, a shortcut that has now landed it in a cul-de-sac. One of the main culprits in this respect is Robert L. Hirsch, currently one of the most vociferous critics of the project.

---

[24]https://w3.pppl.gov/~zakharov/PU001207.pdf.

[25]Talk at the Seminar of the Institute of Physics and Technology Research, People's Friendship University of Russia, Moscow (24 October 2018). For some reason this talk cannot be found on the Internet, e.g. not on Zakharov's website (https://w3.pppl.gov/~zakharov/), and was kindly sent to me by Professor Vladimir Voitsenya from KIPT in Kharkov. The website has however several other critical talks.

18.7 Criticism of Other Fusion Endeavours

Magnetic fusion research today focuses almost exclusively on the tokamak concept. The stellarator concept has not been abandoned but is far behind in development, and other concepts (mirrors, pinches) get little to no attention.

At present privately funded companies display most activity in such alternative fusion concepts. Jassby has branded many of these private efforts as being Voodoo Science. This expression for scientific research that falls short of adhering to the scientific method[26] has been termed by the American physicist Robert L. Park (1931–2020) in his book *Voodoo Science: The Road from Foolishness to Fraud* (Park 2000). It lists cold fusion and the International Space Station as examples of such science, meaning that Charpak is not alone in deriding this venture. So, how can such an exalted exercise as trying to "bottle the Sun" be Voodoo Science?

Daniel Jassby has the answer. He especially directs his scathing sarcasm at the many fusion start-ups that have sprung up in the "valley of death" after ITER construction started.

As we have seen in Chap. 15, many of these start-ups promise to develop practical fusion electric power generation within 5 to 15 years, and claim to do so by surpassing ITER's planned performance in a fraction of the time at 1% of the cost. According to Jassby, these projects are nothing more than modern-day versions of Ronald Richter's Argentinian arc discharges of 1951 (already mentioned in Chap. 3), the first scam in fusion history. In that year Argentinian president Perón announced to the international press that "the Argentine scientist Richter"—a German who couldn't speak a word of Spanish—had achieved the controlled release of nuclear-fusion energy and, although it soon became evident that the claims were spurious, it jolted other countries in starting fusion projects (e.g. Spitzer's stellarator in the US Project Matterhorn). Just as Richter's contraption was unable to generate a single fusion reaction, none of the current projects has given evidence of more than token fusion-neutron production, if any at all. Park demolished "cold fusion" but never mentioned any of the failed "warm plasma" fusion schemes of his era in either his book or one of his columns. Most of these plasma-based fusion attempts can nonetheless be classified as voodoo technology, voodoo fusion, defined for present purposes as those plasma systems that have never produced any fusion neutrons, but whose promoters claim will put net electric power on the grid in just a couple of years from today. The absence of fusion neutrons apparently works like a Voodoo incantation, putting a spell on journalists, investors and politicians making them believe that commercial fusion electric power generators are just around the corner. The messianic incantations of the voodoo priest-promoters, so Jassby says, invoke the aura of "the energy source that powers the sun and stars" as well as the myth that terrestrial fusion energy is

---

[26]The scientific method implies careful observation, applying rigorous scepticism about what is observed, to avoid that cognitive assumptions can distort how one interprets the observation. It involves formulating hypotheses, via induction, based on such observations; experimental and measurement-based testing of deductions drawn from the hypotheses; and refinement (or elimination) of the hypotheses based on the experimental findings.

"clean and green" in order to cast a spell over credulous investors and politicians. The point is that more than 90% of fusion concepts have never produced measurable levels of fusion neutrons, which means those systems may have little practical value.

Jassby excludes some companies, like Tokamak Energy and Commonwealth Fusion Systems from the voodoo class despite what he calls their preposterous and insupportable claims of near-term electric power production, solely because their schemes are based on tokamaks. Tokamaks have demonstrated for 50 years that they can produce fusion reactions, which is of course quite another thing from producing net energy that can be turned into electricity to feed the grid. LPPFusion whose Dense Plasma Focus device does produce meaningful levels of D-D fusion neutrons is ranked similarly.

Let us hear a bit more from Jassby. "Journalists and promoters rarely mention neutrons, because most journalists have never heard of them, while the promoters assume that neutrons can eventually be made to issue from their contraptions by the appropriate voodoo recitation. The permanent fusion R&D organizations, mainly government-supported labs, are the silent spectators of the parade of naked emperors, only occasionally challenging their insupportable assertions and predictions. One feature that voodoo fusion schemes do share with their neutron-producing rivals is that while they will never put electricity onto the grid, all of them take plenty of energy *from* the grid. The *voracious consumption* of electricity is an inescapable feature of all terrestrial fusion schemes."

18.8   Conclusion

All this criticism is of course being ignored, presented as false claims or miscommunication, or belittled as nothing new. The juggernaut just thunders on, crushing everything and everybody on its path. In that respect it is very convenient that much of the criticism comes from old(er) people after their retirement from active research. Fusion is a long-term project and these critics conveniently die and can then be safely ignored. I bet that hardly anybody now working in the fusion community will for instance have heard of Lidsky's criticism, even though it is still very much to the point and was hitting the nail on its head close to forty years ago. Even more than that: he has been completely vindicated. In spite of that, he is not considered worth a mention in most of the more recent books on fusion I read, an exception being Daniel Clery's 2013 book "*A Piece of the Sun*".[27] Most of these books continue to sing the praises of this deeply flawed venture, scientifically and technically. When finally all come to their senses and realise the folly of it all, it will hit science hard. It will cause permanent damage to the entire science enterprise and bring home to

---

[27]To show how quickly things change, I note that in his book Clery says: "But even many of the most ardent fusion enthusiasts concede that commercial fusion is unlikely before 2050." This was printed in 2013, just seven years ago. Nowadays it will be hard to find people who believe that just feasibility of fusion, let alone commercial fusion, will have been shown by 2050.

society that scientists are as unreliable as the rest of us, and just put their mouth where the money is. It has been made abundantly clear that the fusion enterprise as currently conducted is a hopeless one. Although the Dutch saying has it that "you never know who a cow catches a hare", it seems unwise to put much confidence in such eventuality.

Let me finish this chapter with a citation from a 2018 interview[28] with Chris Llewellyn Smith, former director-general of CERN and former director of the JET facility at Culham (UK), and currently Director of Energy Research at Oxford University. He has been quoted at other places in this book and has always been a staunch proponent of nuclear fusion, but now seems to be slowly changing his mind and is becoming more critical: "Ten years ago, I would have replied that I'm reasonably confident that we will be able to make a fusion power plant, although we need ITER to be sure, and the real question is can we make one that's reliable and competitive? The question of reliability will remain unanswered until we try, although operation of ITER will provide clues. I used to think there was a reasonably good chance that fusion could compete with other low carbon sources of power, but, while I would not say that it is impossible, the situation has changed. The cost of wind and solar power has decreased faster than anyone could have dreamt. Meanwhile ITER has gone way over budget, partly because of the way that the project was set up and because it's the first of a kind, but probably also because fusion reactors will be intrinsically more expensive than we thought a decade ago. I think we need to finish ITER and establish once and for all whether fusion really is a viable option. We will then have to reassess the likely cost of fusion power in the light of the experience gained with ITER and in comparison with the cost of alternatives before deciding whether to go ahead and build a real fusion power station." We are now two years later, and it seems to be a foregone conclusion what the result of the reassessment referred to here will be.

---

[28]https://scgp.stonybrook.edu/archives/24923#_ftn6.

Chapter 19
Economic and Sustainability Aspects of Nuclear Fusion

19.1 General Considerations

All the expense and effort for fusion experiments will only be justified when in the end a viable reactor emerges that produces electricity reliably and at a competitive rate. The current situation in the energy market is that electricity from fossil fuel sources is very cheap. In this connection it is not helpful that the world is awash with oil and gas, and that in the current wasteland of the free market this will be so for a very considerable time to come. It is clear though that these resources are finite and will not last forever.

Renewable sources, wind and solar, are on the march, but have so far only been shown to be viable with the support of large subsidies, although there are signs that things are rapidly changing in that respect. Climate change has brought home the urgency of switching from fossil fuels to carbon-free energy sources. A carbon tax is seen as one of the most efficient ways to achieve this. Some sort of price on carbon has now been adopted by more than 40 governments worldwide, either through direct taxes on fossil fuels or through cap-and-trade programmes. The European Union Emissions Trading System (EU ETS) is such a cap-and-trade system. It is the biggest greenhouse gas emissions trading scheme in the world and sets a maximum on the total amount of greenhouse gases installations may emit. "Allowances" for emissions are then auctioned off or allocated for free, and can subsequently be traded. The system has been criticized for several failings, including over-allocation, windfall profits, price volatility, and in general for failing to meet its goals. It cannot be considered as a proper means to deal with the problem. A more positive example is Britain where coal use plummeted after the introduction of a carbon tax in 2013. However, in practice, most countries have found it politically difficult to set prices that are high enough to spur truly deep reductions. Carbon pricing programmes are mostly fairly modest and partly for that reason, carbon pricing has, so far, played only a supporting role in efforts to mitigate global warming.[1]

---

[1] https://www.nytimes.com/interactive/2019/04/02/climate/pricing-carbon-emissions.html, accessed 30 March 2020; Plumer and Popovich 2019.

An analysis of the International Monetary Fund has recently concluded that such a tax should be imposed immediately around the world and should rise to $75 per tonne of CO_2 by 2030 (Parry 2019). If imposed, it would add about $0.17 to the price of a litre of gasoline.[2] The average carbon emission allowance price in the EU ETS increased from €5.8 per tonne in 2017 to €15.5 per tonne in 2018, still well below the price stated by the IMF report.[3] In the end such a tax is the only way to have users of carbon fuels pay for the climate damage they cause by releasing carbon dioxide into the atmosphere. It will also be a sure and just way to make renewable energy sources competitive and factor in all the costs in a proper way. Now we are essentially paying twice. First with our health for the carbon pollution caused by burning fossil fuels, which are now allowed to spew out their waste for free, and second through taxes for the subsidies needed to make wind and solar competitive.

Even if such a tax were levied, it would make little difference for fusion as it will not play a role of any significance in energy generation, at any rate not in this century. If anything, the various gigantic machines that are being built or planned to be built will consume huge amounts of electric power generated for a great part by fossil-fuel burning power stations. The carbon footprint of these machines is gigantic. It would for instance be interesting to know how much electric power in kWh the JET tokamak has drawn from the public grid in the 20–30 years of its existence, but these figures are unfortunately not made public. We know that when JET runs, it consumes 700–800 MW of electric power (the equivalent of 1–2% of the UK's total electricity usage). Although a huge amount, it is only used for a very short time (the duration of the plasma pulse) and does not say much about total electricity usage. In this respect EUROfusion says: "The power required to keep a reactor working is an interesting question." And then the usual story[4] follows about keeping the plasma hot, but the question, although apparently interesting, is not answered, while it would have been easy to give an idea about energy usage by revealing JET's electricity bill. How can we ever know whether fusion will be worthwhile when the question of how much power is needed to start the reactor and keep it running is not even answered? The much more difficult question when fusion power will be on the grid has always been answered (wrongly) with much (unwarranted) confidence and undue haste.

In this chapter we will assume that ITER will be a success (meaning that indeed a burning plasma will be achieved by 2040, heated mainly or to a great extent by the alpha particles released in the fusion reactions), that after ITER one or several DEMOs will be built and made to work successfully, and that in the end indeed net power can be generated in quantities that are sufficient to feed into the grid. It is unlikely that this scenario, roughly corresponding to the schedule presented in Fig. 17.2, can be completed without any further hiccups and delays, if only because

---

[2] One litre of gasoline weighs 0.75 kilo and produces 2.3 kg of CO_2. The proposed tax on 2.3 kg would be 2.3 times $0.075 which equals $0.1725 per litre of gasoline.

[3] https://www.eea.europa.eu/themes/climate/trends-and-projections-in-europe/trends-and-projec tions-in-europe-2019/the-eu-emissions-trading-system, accessed 30 March 2020.

[4] https://www.euro-fusion.org/faq/top-twenty-faq/how-much-power-is-needed-to-start-the-reactor-and-to-keep-it-working/, last accessed 2 July 2020.

of unforeseen and unrelated events as the outbreak of the Covid-19 pandemic, but we will nonetheless assume it to be the case, else there is simply no case for fusion.

Sufficient generation of net power is of course a necessary condition but does not yet mean that it is also economically attractive. One of the first economic impediments to making such a scenario economically attractive is the unprecedentedly high level of investment needed for the proof of principle (the cost of ITER and DEMO) and the long construction time of fusion plants. Within the mainstream scenario of a few DEMO reactors towards 2060 (see Fig. 17.2) and the subsequent construction of a few relatively large reactors, there is no realistic path for fusion to make an appreciable contribution to the energy mix in this century (Lopes Cardozo 2019).

In other words, fusion will not contribute to the energy transition in the time frame of the 2015 Paris climate agreement, i.e. to achieve a near complete decarbonization of energy generation by 2050, and even if this is delayed for a couple of decades the situation for fusion will not improve. In Chap. 9 we have quoted the 2012 EFDA Roadmap which makes the ludicrous statement that "fusion can start market penetration around 2050 with up to 30% of electricity production by 2100"! There we have illustrated why that is a ludicrous statement and that in this century there is no way that fusion will make a sizable contribution to the energy supply, even if the most optimistic scenario of the fusion proponents comes true.

But leaving these details aside we can still attempt to calculate or reason if a fusion power plant (after having been built) can indeed be competitive with for instance wind and solar energy. The latter two energy-generating options are also still in development and several problems have to be solved before they can make a real impact. But, they are actually already generating huge quantities of energy, which is the reason why the investment in photovoltaic[5] solar and wind power has soared to $288.9 billion in 2018, approximately 3% of the world energy market, with the amount spent on new capacity far exceeding the financial backing for new fossil fuel power. It was the fifth year in a row that investment exceeded the $230 billion mark. (*Renewables 2019*, Global Status Report REN21 2019.) In the light of such figures, fusion with a global investment of a mere $2–3 billion per year is just a pitiful exercise. It will never be able to match such numbers as long as it does not produce any net energy. But it also implies that fusion will not be the long-awaited saviour that rescues the world from an impending carbon-dioxide death. When it ever becomes available, it will enter a largely decarbonised market that will have organised itself in one way or another. An intruder or newcomer in such a market must offer something that is not already available, or offer it more cheaply, or be competitive in any other way. Being carbon free, safe, clean and unlimited will not be enough. This is nowadays demanded from any new energy source. Wind and solar are also carbon free and safe, and even more unlimited (no tritium supply problem for instance) and cleaner than fusion (no radioactivity whatsoever). A fusion reactor will be a very complex beast, certainly in comparison with the relatively simple engineering involved in solar panels and windmills. Here Lawrence Lidsky's 'trouble with fusion' discussed in the previous chapter that no power company would ever

---

[5]See the Glossary for an explanation of the term photovoltaics.

want to work with a reactor with such complex engineering becomes very relevant and most probably true. Power companies do not normally hire nuclear or plasma physicists to run their plants, while they are perfectly happy to run a windmill or solar park.

The most basic problem of solar and wind power is that they do not always generate energy at the right time. Reliable ways to store the generated energy must be found, but this is by no means a simple matter. It is at present still necessary to complement basic grid needs with fossil fuel power. The deployment of energy saving technologies such as so-called 'smart grid'[6] solutions might help, certainly in the short run, to enable renewables to become baseload electricity suppliers. These problems are however peanuts compared to the problems fusion is faced with, where a proof of principle is still far away.

Two solutions for the storage problem are currently being explored and both will probably provide part of the solution. The first is the development of more efficient batteries, which is well on the way, but may in the end founder on a shortage of lithium (see below). It will also run up against limits dictated by physics, as battery energy density is limited by the energy that can be stored in an atom, which is of the order of a few eV.

The second solution is using the generated energy to produce hydrogen through the electrolysis of water. Electrolysis is a very old process indeed, going back centuries; it is old hat and moreover chemistry, i.e. just atoms, not something modern physicists much care for. It is a technique that uses a (direct) electric current to drive an otherwise non-spontaneous chemical reaction, in this case the decomposition of water into hydrogen and oxygen. The surplus energy generated by wind and solar can be used to decompose water in its components oxygen and hydrogen. The hydrogen is then stored and can later be used, burnt in a power station for instance, for the reverse process whereby energy is released. Electrolysis plants are springing up at various places in Europe (e.g. in Groningen (Netherlands), the Rhefyne project at the Shell Rheinland refinery in Wesseling, between Bonn and Cologne in Germany, and at the Port of Rotterdam which is planning a 250 MW 'green' hydrogen plant, the largest in Europe). The development of such techniques is much more promising than fusion ever can hope to be. The physical and technological principles have been proved and industrial scale installations are already available. The electrical energy needed to generate one kg of hydrogen is about 50 kWh, using an electrolyser efficiency of 65%, so at $0.06 per kWh would cost $3. Since 1 kg of hydrogen is almost exactly equivalent to 1 gallon of gasoline in energy content, this price would already now be competitive. But, the use of electricity, a very efficient energy carrier, to generate hydrogen, another energy carrier, and then convert it back into electricity is not the best option. Electricity is so valuable as electricity that we may not want to use it for anything other than that. Therefore, various other direct solar processes, converting

---

[6]A smart grid is an electricity network that can cost efficiently integrate the behaviour and actions of all users connected to it—generators, consumers and those that do both—in order to ensure an economically efficient, sustainable power system with low losses and high levels of quality and security of supply and safety (EU definition).

energy directly into hydrogen, are under research and development. For more about this see Sivaram 2018.

19.2 Is Fusion Energy Sustainable?

According to the stories told on the websites of fusion promoting endeavours, fusion power seems like the perfect energy source. They say that it is clean, inexpensive, and can draw from an inexhaustible resource of fuel. If commercial fusion plants were to become a reality, an unlimited, nearly free, clean source of energy would be available. But is this actually true? In the preceding chapters we have on occasion already made some critical comments on this story. It is true that deuterium is almost free, but that is certainly not the case for tritium, and both are needed in the envisaged nuclear fusion power stations. In Chap. 16 the difficulty in procuring enough tritium, or in breeding tritium in the reactor itself, has already been discussed and looks very bleak. The tritium required for ITER (12.3 kg) will be supplied from the production by the CANDU reactor in Ontario, but while this reactor may be able to supply 8 kg for a DEMO fusion reactor in the mid-2050s, it will not be able to provide 10 kg at any realistic starting time. The tritium required to start DEMO will depend on advances in plasma fuelling efficiency, burnup fraction, and tritium-processing technology. In theory it is possible to start up a fusion reactor with little or no tritium, but at an estimated cost of \$2 billion per kilogram of tritium saved, it is not economically sound. If ITER and further fusion development are successful, two or three countries may build their own reactors, giving another major source of uncertainty in tritium requirements. If Canada, Korea and Romania make their tritium inventories available for fusion, there is a reasonable chance that 10 kg of tritium would be available for fusion R&D in 2055, but stocks would likely have to be shared if more than one fusion reactor is built (Kovari et al. 2018).

This reflects the point already made several times in this book that tritium breeding is of paramount importance for fusion ever to become a success. For the tritium-breeding blankets to be used in future reactors, lithium and some other metals are needed, and their availability is not without problems either.

19.2.1 Lithium, Beryllium and Lead

Let us first list some facts on lithium, which as regards fusion is the most important of these elements. The chemical element lithium (Li), with atomic number 3, is the lightest of all metals. In its pure elemental form, it is a soft, silvery-white metal. It is a so-called alkali metal, i.e. belonging to the same group as e.g. potassium (K) and sodium (Na), and widely distributed on Earth. Due to its high reactivity it does not naturally occur in elemental form and is always found bound in stable minerals or salts. It appears in the form of two natural isotopes: ^{6}Li (natural abundance 7.42%)

and ^{7}Li (natural abundance 92.58%). Lithium has many uses, the most prominent being in batteries for cell phones, laptops, and electric and hybrid vehicles (46% in 2019). Lithium is added to glasses and ceramics for strength and resistance to temperature change (26% in 2019), it is used in heat-resistant greases and lubricants (11%), and it is alloyed with aluminium and copper to save weight in airframe structural components. Lithium is used in certain psychiatric medications and in dental ceramics. The use with the greatest potential benefit is in rechargeable batteries, which take advantage of lithium's light weight, compared to nickel–cadmium and nickel–metal hydride cells, and high electrochemical potential (Lithium Factsheet, USGS).

In the Earth's crust (upper 16 km), lithium can be found in the range from 20 to 70 ppm by weight in various forms. In the first place minerals which have concentrations of Li_2O (lithium oxide) from about 4.5 to 7%. The most important mines are in Australia. The second form is brine, which is a high-concentration solution of salt in water. The evaporation process of salt lakes leads to an increased lithium content and, when the lake is completely dried up, the remaining salt can contain 4–6% of its weight in lithium. The lithium is extracted as lithium carbonate (Li_2CO_3). The present most important production site is in Chile. The world's biggest salt flat is located in Bolivia. It holds perhaps 17% of the planet's total lithium reserves and will soon be put into production. In the near future, equivalent brine sources with lower lithium concentration could also be economically exploited. The third and most important potential source of lithium is seawater which has a mean lithium content of 0.17 ppm. The world's oceans contain an estimated 180 billion tonnes of lithium. Up to now, only experimental processes have been tested in order to extract lithium from seawater, but the production price is expected to remain too high to be competitive on the present market (Fasel and Tran 2005).

As we have seen, fusion reactors will use lithium in the tritium-breeding blankets surrounding the plasma. In Chapter 16 we have stated that about 300 kg of ^{6}Li will be needed per reactor per year, which with the envisaged thousands of reactors operating for 20–30 years adds up to quite an amount. It is however not so easy to estimate the required amount of lithium and the required enrichment level, i.e. its ^{6}Li content. To be able to make such an estimate, a relatively detailed knowledge of the blanket and—in case of liquid breeders—the design of the infrastructure system, including all pipework, cooling, cleaning, and tritium extraction systems would be required. The EU fusion roadmap focuses on two main candidate blanket concepts, the water-cooled lithium lead and the helium-cooled pebble bed breeder concept. The first one is a liquid breeder and the second one solid. DEMO liquid breeders are a good starting point for making an estimate. For the water-cooled lithium lead blanket of a 2 GW (fusion power) device, 844 m^3 (or 8200 tonnes) of a eutectic[7] lead-lithium alloy (Pb–Li) is assumed to be needed, enriched to 90% of ^{6}Li. With a mass ratio of lithium in Pb–Li of 6.4×10^{-3}, this implies that 52 tonnes of pure ^{6}Li are needed (26 tonnes per GW). These numbers are the total inventories. The consumption of ^{6}Li depends on the tritium production rate (2 g ^{6}Li needed to produce 1 g tritium) and

---

[7] See Chap. 16.

is small compared to the large total lithium inventory (112 kg of ^{6}Li consumption per full power year and GW). In general, concepts based on liquid breeders can be assumed to have a larger inventory than the ones with a solid breeder. Nevertheless, the amount of enriched lithium needed for any blanket concept is of the same order of magnitude, which means that some ten tonnes of 'fusion grade' lithium will be required. For solid breeders, 'fusion grade' refers to an enrichment level of 30–60%, for liquid breeders up to 90%. This need for such large quantities of enriched ^{6}Li asks for an isotope separation process that allows a minimum output of several tonnes of 'fusion grade' lithium (i.e. lithium with an isotopic composition that can directly be used in the blankets) per full power year. There is currently no facility available world-wide that could satisfy this demand, and it is neither straightforward to build such a plant. If DEMO becomes operational in the 2050s, blanket manufacturing must start in the mid-2040s. Several tonnes of enriched lithium will already then be needed, and a fully operational isotope separation plant must be ready in the late-2030s. Pilot plant update and tuning will take approximately half a decade, implying that activity on the separation plant must be started by the early 2030s. Assuming that facility design and construction take some years, the design of the process equipment and the result of the process development must be available by the mid-2020s at the latest (Giegerich et al. 2019).

For a considerable part the ^{6}Li market has up to now been supplied by stockpiles of lithium produced in the US by the COLEX process, a chemical method based on the use of mercury. It requires a significant amount of mercury and there are many opportunities for leaks into the environment, but this process has so far been the only method that enables industrial scale production of enriched lithium at minimal cost. From 1955 to 1963 the US operated a lithium isotope separation plant in Oak Ridge, Tennessee. The ELEX process (isotope separation by electron exchange) was the first method to be deployed; then came the OREX process (in which an organic solution of lithium is exchanged with a solution of lithium in mercury). The COLEX process turned out to be the most efficient and became the primary source of enriched lithium-6. The COLEX plants in Oak Ridge had a very rough start in 1955 with major problems in this new, complicated, and potentially hazardous technology. Due to environmental concerns because of strong environmental contamination with mercury, the US stopped lithium enrichment operations in 1963. The use of this process is now banned. It seems that China is presently the only country in the world that officially employs the COLEX process to enrich lithium. Consequently China has a near total hold over the market of enriched lithium, followed by Russia, which uses a process that is different from the COLEX process. Although US industry relies heavily on Chinese and Russian enriched lithium, the ecological concerns of the process may impede the future domestic use at industrial scale.

Commercial available ^{6}Li today is only sold in small amounts and at very high prices. For the amounts needed to make fusion successful an enrichment plant is needed with a capacity of several tonnes/day. This would probably also decease the price.

The ITER website paints a very optimistic picture of the availability of lithium. It asserts that lithium from proven, easily extractable land-based resources would provide a stock sufficient to operate fusion power plants for more than 1000 years. What's more, it says, lithium can be extracted from ocean water, where reserves are practically unlimited (enough to fulfil the world's energy needs for about 6 million years).[8] Not a word here about the costs or about other applications demanding a big share of the lithium pie. At another place on the website it is even said that "lithium availability will not be an issue for let's say the next thousand years, as there are approximately 50 million tonnes[9] of proven lithium reserves in the world, which means about 3 million tonnes of ^{6}Li (7.42%), which is the form preferably to be used in tritium-breeding blankets. It takes 140 kilos of ^{6}Li to obtain the 70 kilos of tritium necessary to produce one gigawatt of thermal power for one year. Assuming an availability of 80% and a conversion efficiency from thermal to electric power of 30%, the production of one gigawatt of electric power (the estimated size of an average fusion reactor) will require approximately 500 kilos of ^{6}Li per year, which would bring the total requirement for 10,000 reactors to 5,000 tonnes annually. Obtaining 5,000 tonnes of the precious isotope will require processing (…) approximately 70,000 tonnes of "regular" lithium … still a very small fraction of available resources. Fusion specialists generally consider that, in a world where all energy would be produced by fusion, the quantity of lithium ore present in landmass would be sufficient to provide the required tritium for several thousand years. As for lithium present in oceans, it could last us for millions of years."[10]

The fact that there is no isotope enrichment plant available that can provide such amounts of ^{6}Li is not even mentioned, but also in other respects it is doubtful that the situation is as simple as depicted here. In the picture painted above, resources and reserves seem to be taken to mean the same, which is not the case, and several things have just been swept under the rug, notably the fact that it will not be possible to achieve self-sufficiency in tritium with only ^{6}Li. As has been explained in Chapter 16, a neutron multiplier is needed as ^{6}Li just produces tritium and helium when struck by a neutron. Using ^{7}Li would yield an extra neutron which can again react with another lithium nucleus to produce tritium, but that reaction is endothermic, i.e. requires energy (see Chap. 1). Without such extra neutrons self-sufficiency in tritium cannot be achieved. Therefore, beryllium is proposed as a neutron multiplier, causing further supply problems, as we will see below.

In any case, the 50 million tonnes mentioned is questionable and 70,000 tonnes is not a small fraction of actually annually mined lithium, but close to the total amount mined in 2019 (77,000 tonnes). Estimates of lithium reserves (not counting the lithium in ocean water) vastly differ from around 4 million tonnes to roughly 40 million tonnes (Mohr et al. 2010).

However, when making such estimates a distinction must be made between lithium reserves and lithium resources, whereby the reserve estimates refer to the extractable portion of the resources. Estimates for lithium resources go indeed up to more than 50 million tonnes, even as high as 80 million tonnes, but that would be a wrong number

---

[8]https://www.iter.org/sci/FusionFuels, accessed 23 March 2020.

[9]In this book 'tonne' always means 'metric tonne', i.e. 1000 kg.

[10]https://www.iter.org/newsline/-/3071 accessed 23 March 2020.

to quote for the reserves. In any case it is surprising how large the differences between the various studies of lithium reserves or resources are. One reason for this is that most lithium classification schemes are developed for solid ore deposits, whereas brine is a fluid for which these schemes are less suitable due to varying concentrations and pumping effects. Some of the differences also arise from the use of different numbers for the percentage of lithium contained in the ore or salt. This percentage is in any case rather small, at most a few percent. The in my view thorough and critical analysis by Mohr et al. (2010) states reserves of 23 million tonnes, listing them by country. Their numbers are substantially higher than the amount of 14 million tonnes of economically recoverable lithium reserves estimated in 2018 by the US Geological Survey (USGS) (from over 39 million tonnes of lithium resources worldwide) or the numbers given by statistica.com[11] which reports a total of 15 million tonnes in lithium mineral reserves for the top nine countries in 2019. Chile had the largest lithium reserves worldwide with an estimated 8.6 million tonnes, more than half of total global reserves. Australia came in second with reserves estimated at 2.8 million tonnes (vastly more than the report by Mohr et al. or the USGS figures). Bolivia that does not produce any lithium yet has huge resources quoted at 21 million tonnes. Australia was the top country in terms of lithium mine production in 2019, producing 42,000 tonnes of lithium in that year.

It seems that in terms of lithium ore or salt deposits there is indeed no shortage on earth. Western Australia alone hosts five of the world'sbiggest lithium mines, whose combined resources exceed 475 million tonnes of ore, containing 1 to 3% of lithium carbonate equivalent (LCE),[12] good for a couple of million tonnes of lithium.[13] Knowing deposits of ore or salt and their lithium content, it is indeed easy to calculate reserves, but that does not yet mean that they can be mined in an economically and environmentally sound manner. The higher reserve estimates mentioned above also include reserves that are of no actual or potential commercial value.

No wonder that in the literature doubts have been raised about the adequacy of easily minable lithium deposits. An early paper on the matter states that the "classical" D-T-Li cycle could supply most of mankind's future long-term power needs, but only on condition that the required lithium fuel can be extracted from seawater at reasonable expense. The estimated landbound lithium reserves are too small to that end, they will last for about 500 years at most (*still a long time and quite hilarious that such timescales are considered by an industry that does not even exist (LJR)*), depending on forecasts of future energy consumption and on assumptions about exploitable resources. Recovery of lithium from seawater would extend the possible

---

[11]https://www.statista.com/statistics/268790/countries-with-the-largest-lithium-reserves-worldwide/ accessed 24 March 2020.

[12]Data relating to lithium grades in mineral assays and ore resources and reserves for hard rock and brine projects are reported using a number of differing measurement units, e.g. ppm Li and percentages of Li, Li_2O, or lithium carbonate. To normalise this data, data is often also reported in terms of "lithium carbonate equivalent" or "LCE" so that information can be easily compared on a like-for-like basis.

[13]https://www.mining-technology.com/features/top-ten-biggest-lithium-mines/.

range by a factor of 300 or so, provided that extraction technologies which are at present available in the laboratory, could be extended to a very large and industrial scale (Eckhartt 1995).

The latter statement was made in 1995 when electric cars with lithium-ion batteries were still a fairly distant prospect. However, the times have now changed and the recent predictions of the future demand for lithium-ion batteries are alarming. For rechargeable batteries manufacturers use more than 160,000 tonnes of lithium every year, a number that is expected to grow nearly tenfold over the next decade (Service 2020). It is conceivable that the automobile industry may acquire, and according to some estimates even use up, the terrestrial lithium reserves in the next few decades. Various battery configurations use between about 10 and 22 kg of lithium per car. Just imagine what that implies as regards lithium demand if a large proportion of the close to 100 million cars, 2008 figures, manufactured annually will be electric vehicles. Millions of tonnes annually just for car batteries. Nothing may be left for anything else. A mitigating factor as regards fusion is that the latest conceptual designs for the tritium-breeding blankets in the DEMO fusion reactor have much lower initial lithium requirements than those expected from the European Power Plant Conceptual Study (PPCS) (Bradshaw et al. 2011).

In a 2007 report entitled "*The Trouble with Lithium*" (Tahill 2007) it is stated: "Analysis of lithium's geological resource base shows that there is insufficient *economically recoverable* lithium available in the Earth's crust to sustain electric vehicle manufacture in the volumes required, based solely on lithium-ion batteries. Depletion rates would exceed current oil depletion rates and switch dependency from one diminishing resource to another." In a 2008 follow-up to this report (Meridian International Research 2008) a more comprehensive analysis is carried out which confirms this conclusion: "realistically achievable lithium carbonate production will be sufficient for only a small fraction of future PHEV (Plug-in Hybrid Electric Vehicles) and EV global market requirements; demand from the portable electronics sector will absorb much of the planned production increases in the next decade; other battery technologies that use unconstrained resources should be developed for the mass automotive market; major economically recoverable lithium brine reserves are lower than previously estimated at only 4 million tonnes of lithium; mass production of lithium carbonate is not environmentally sound, it will cause irreparable ecological damage to ecosystems that should be protected and lithium-ion propulsion is incompatible with the notion of the 'Green Car'". The number of 4 million tonnes, the lowest of the numbers in the literature, is arrived at by only including economically recoverable reserves whose exploitation does not do irreparable damage to the environment.

Global demand for lithium is projected to go up from 307,000 tonnes in 2019 to 820,000 tonnes of LCE in 2025, while S&P Global estimates that new mines and brine lakes, coupled with expanded output from several existing projects should put global lithium production above 1.5 million tonnes of LCE by 2025.[14] To convert

---

[14] https://www.spglobal.com/en/research-insights/articles/lithium-supply-is-set-to-triple-by-2025-will-it-be-enough.

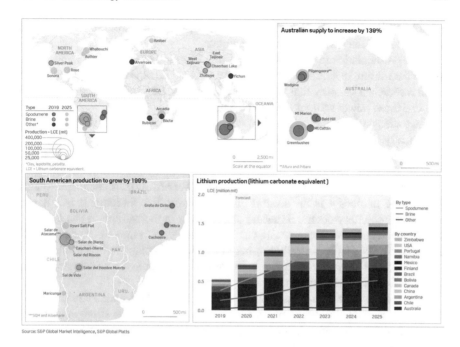

Fig. 19.1   Global trends in lithium availability and production to 2015

the LCE numbers to pure lithium they have to be multiplied by 0.188.[15] There is a strong argument that further out, as momentum builds up especially with the electric automotive industry, demand could outweigh supply. "If forecasts for EV penetration are to be believed (…) then lithium demand is set to increase tenfold over the next decade." In February 2019, 76% more LCE was deployed globally in batteries of newly-sold passenger EVs than the same month the previous year. The current situation with lithium in the world has been summarised in the above figure (Fig. 19.1).

What does this mean for fusion? Can its lithium requirements be met? Let us ask the following (purely hypothetical and rather premature) question: How much lithium would be required annually if fusion were to provide 30% of baseload[16] electricity supply at the end of this century or early in the next century, as the EFDA Roadmap wants? It has been calculated (Bradshaw et al. 2011) that in that case approximately 24,000 TWh would be required from 2760 fusion power stations, each providing 1 GWe. Using the number of 500 kilos of lithium-6 per year as quoted above from the ITER website, they would consume 1380 tonnes of lithium-6, for which about 17,200 tonnes of natural, non-enriched lithium would be required annually, equivalent to about 90,000 tonnes of LCE. The competition from other

---

[15]https://www.savannahresources.com/lithium/lithium-overview/.

[16]Baseload power refers to the minimum amount of electric power needed to be supplied to the electric grid at any given time.

uses like rechargeable batteries may be fierce, but as follows from Fig. 19.1 it may be possible that the fusion demand can be met.

More problematic are the much larger initial lithium loadings (see above). The sum of the lithium inventories for all power plants would represent almost one tenth of the reserves, a quantity that would just not be available in the comparably short time that these power stations have to be built.

If we also include the potential of seawater, there is enough lithium, at least theoretically, for the operation of 2760 power plants for 23 million years! Some progress in lithium extraction from sea water has recently been reported (Service 2020).

As regards beryllium the following applies. This (relatively) rare chemical element with atomic number 4 is the second lightest metal (after lithium). It is a health and safety issue for workers. Exposure to beryllium in the workplace can lead to a cell-mediated immune response and can over time develop into a chronic allergic-type lung disease, called berylliosis; beryllium and its compounds are category 1 carcinogens and are carcinogenic to both animals and humans. The primary risk is the inhalation of beryllium dust.

The metal is mainly used in defence and aerospace applications because of its stiffness, dimensional stability over a wide temperature range, its strength (stronger than steel) and light weight (lighter than aluminium). Globally identified resources of beryllium amount to a little over 80,000 tonnes. The extraction of beryllium from its compounds is a difficult process. Currently the United States, China and Kazakhstan are the only three countries involved in industrial-scale extraction. The United States is the world's largest beryllium producer by far, with production of 170 tonnes in 2019. China came a distant second with 70 tonnes. Global beryllium production in 2019 totalled an estimated 260 tonnes.[17]

In ITER, beryllium will be used as armour for the plasma-facing first wall panels fitted inside the tokamak. For this about 12 tonnes of beryllium will be required to cover a surface of approximately 610 m^2, which will be managed through a detailed beryllium safety programme. As discussed in Chap. 10, ITER will not have a tritium-breeding blanket. It will only test mock-ups of breeding blankets. Hence for ITER there will not be a supply problem, but the beryllium burn-up in the 2760 power plants envisaged above would be 524 tonnes annually and the initial loading 331,000 tonnes (about 120 tonnes per reactor), vastly exceeding the present estimate of resources! (Bradshaw et al. 2011.) This makes it in any case necessary, even for operating just a few reactors, let alone the above postulated 2760 reactors, for beryllium to be recovered from waste beryllium that has become radioactive in the reaction with neutrons. Methods are being developed for this, but the above figures also show that the use of beryllium on a large scale (i.e. for more than one or two reactors) as a neutron multiplier in tritium-breeding blankets is just out of the question. There simply is not enough beryllium around.

A further problem is that most beryllium contains uranium (up to 150 wppm (weight parts per million)). The ITER Organization has demanded that the uranium

---

[17] https://www.statista.com/statistics/264925/world-beryllium-production/.

content in the beryllium to be used in ITER should not exceed 30 wppm, but the designers of the China ITER test blanket module have let it know that the uranium content in the beryllium to be used in the test blanket as neutron multiplier is 100 wppm. Interaction of neutrons with uranium in such a blanket will result in the production of fissile isotopes (enrichment) and fission. It has been estimated that, after 5 years of full power operation of a future European design power plant using beryllium with a uranium content of 30 wppm, about 4.4 kg of plutonium and other fissile isotopes will be produced in the first wall beryllium (Kolbasov et al. 2016). The presence of such amount of plutonium and other fissile isotopes obviously poses a huge problem. For ITER this problem will be manageable or non-existent as its blanket is only a mock-up and not a real tritium-breeding blanket.

The situation for lead is somewhat better: for the 2760 reactors mentioned above the annual burnup would be 8560 tonnes and the initial loading 11.3 million tonnes. Global lead reserves are about 85 million tonnes. Total global consumption of refined lead in 2018 amounted to 11.7 million tonnes, so if fusion comes around that production has to be doubled to provide the initial loading, but the burn up is such that there would be sufficient lead for quite a number of years. The situation is not optimal, but better than for beryllium or lithium.

These are all rather speculative and premature calculations, and it must be said that in the past, for instance in the 1972 report *The Limits to Growth*, doom stories have been spread around about the world running out of resources. These stories never came true; resources just refuse to run out, it seems. Resources do get scarce, but then the price goes up, encouraging people to adapt by conserving it or finding cheaper substitutes. The calculations presented here in any case do show that fusion, if ever realised to the extent assumed here, might make heavy demands on the lithium supply and might come into conflict with the even heavier demand expected for the production of batteries for electric vehicles. Moreover, lithium mining has a considerable negative environmental impact (Meridian International Research 2008) which will further erode fusion's environmentally clean image. The demand for beryllium from fusion cannot be met, not even for one or two reactors, and its large-scale use may give rise to dangerous radioactive materials. In summary, it can be stated that the shortage of these three metals (lithium, beryllium, lead) might raise problems for the fusion enterprise and that other solutions must be looked for.

19.3   Analyses of fusion's Potential Contribution to the Energy Mix

The literature contains several analyses of fusion's entry into the world of energy production, mostly based on some model. As with all models, the outcome greatly depends on the assumptions made and the input data, for instance assumptions on global lithium reserves, on the efficiency of the power plants, and on the timescale for the realisation of fusion. In 2012 a model calculation was made (Cabal et al.

2012) assuming global lithium reserves of 12 million tonnes, which is realistic, even a little low, as we have seen above; plant efficiencies of 42% for basic plants and 60% for advanced plants, about which nothing much can at present be said, whereby basic plants will start operating from 2050 and advanced plants from 2070. This timescale appears to be very unrealistic, since the 'proof of principle' of fusion will not be forthcoming before 2035 when ITER is scheduled to start operations with D-T plasmas. Before a basic power plant can be constructed, one or several DEMO plants must first run their course, which will, according to the schedule of Fig. 17.2, not be before 2070. Hence the schedule used by Cabal et al. will have to be shifted forwards by at least 20 years, probably more as it is unlikely that no further delays will be incurred. This is important as in 2070 full decarbonization of energy generation will probably already have been achieved and the urgency of including fusion will have greatly diminished.

They also assume a scenario for a carbon tax. For OECD regions the tax is assumed to rise from $20 per tonne of CO_2 in 2020 to $50 in 2100, while in non-OECD regions these figures are assumed to be $10 and $25 respectively. Now in 2020 such levels of taxation have not yet been reached. They also stay far below the tax of $75 in 2030 demanded by the IMF report quoted above.

The model results in an almost linear build-up of zero power from fusion in 2080 to 4–5 TW electric power in 2100. This would imply that in the twenty years from 2080 to 2100 about 5000 power plants (more than 10 times the nuclear fission reactors currently operating in the world) with a 1 GW electric capacity will have to be built, so 250 per year. If that is compared with the time of more than twenty years needed for the construction of just one ITER for which the entire world had to be mobilized or with the fact that the present global nuclear fission industry (a mature industry) has the capacity of building about ten reactors per year, then it is abundantly clear that this outcome is far from realistic. Building 250 plants per year with a construction time of 10 years and assuming a price tag of $10 billion per plant (probably too low), would also require investors to be prepared to make an upfront investment of tens of trillions of dollars in today's money before they have seen the first plant work. They would have to provide $2.5 trillion in 2070 for the first reactors to start operating in 2080, and repeat such generosity every year between 2070 and 2080. A further point is that a 10-year construction time implies that there can hardly be a learning effect. The construction of the next batch of plants will have to start before information about the performance of the preceding generation becomes available. To allow an effective learning curve, an innovation cycle much shorter than 10 years, and a unit size and unit cost much lower than presently foreseen, will be crucial (Lopes Cardozo 2019). It will further require that from day one in 2070 an industrial capacity will stand ready to take on the construction of 250 power plants, 25 times the size of the capacity currently available for constructing nuclear fission power stations. Nobody has at present any intention to even think of starting to build up such a capacity. It will be clear to the reader, I suppose, that all this is unachievable, a truly ludicrous outcome! So, how the authors can conclude that "fusion represents a key technological option for the future energy system" is beyond me. Nor is it true that climate change is the key driver for fusion penetration. It will come far too late for that. If it had been available now, i.e. in 2020, it would have stood a chance, but with the ITER/DEMO schedule of Fig. 17.2, no chance at all. In this respect it strikes me that many analyses (see e.g. Han and Ward 2009 and the EFDA Roadmap) are driven by the desire to present a favourable outcome, in other words by wishful thinking, very much in line with the fusion effort itself. All sense of reality seems to be lost and the facts do not seem to matter. See in this respect the quote of Bertrand Russell in front of this book.

Another analysis from 2016 (Lopes Cardozo et al. 2016) developed a model based on economical considerations and compared the road map of fusion power with the historic development and deployment of other energy technologies, fission, photovoltaic solar and wind. A rather generic pattern was found, in which an exponential growth with a doubling time of 3 years is followed by a linear growth that essentially lasts one plant lifetime. Assuming that the same pattern could be followed by fusion, it is concluded that fusion—while having a late start, lagging behind photovoltaic solar by about 50 years—could reach 100 GW of effective installed power by 2070 and impact the world energy system significantly by the turn of the century. To follow the same rate of development as fission, wind and solar, fusion will need to have 3 DEMO reactors operational in the early 2050s, followed by 10 generation one plants in the early 2060s and 100 generation 2 plants in the early 2070s. It is clear that fusion does not follow this pattern. Perhaps 1 GW of power will be installed by 2070, certainly not 100, and at most one DEMO will be operational in the early 2050s. In 2016 it was already clear that ITER was experiencing considerable delays and cost overruns. The authors seem to have just ignored this and continued with a schedule that cannot be met. Just two years later one of the authors of that study conceded, as already quoted above, that there is no realistic path for fusion to make an appreciable contribution to the energy mix in this century (Lopes Cardozo 2019).

We will dwell on that latter paper a while longer as it makes several important points, in spite of the fact that in that paper too the unrealistic starting point is 10 (generation 1) fusion power plants by 2070. That date is too early. Adopting a more realistic timeline and following the reasoning in that paper, I believe it to be unlikely that, even if everything proceeds smoothly with ITER and a successor DEMO plant, 10 generation 1 power plants can be ready before 2100. These first full-fledged power plants will be constructed after DEMO has provided the proof of principle around 2070. Building these 10 plants constitutes a significant technological risk. This first generation will probably not perform very well; have low availability, will probably be prone to unplanned outages, have low plant efficiency and need relatively high capital investment. These 10 plants together will then be good for an average electric power comparable to that of wind power in 2000. "In other words, generation 1 fusion requires an upfront investment of hundreds of billions of euro, which is coupled to a large technological risk, in order to bring a product to the market that is not competitive in performance or price, at a scale that is meaningless in terms of energy generation." Who is going to pay for such plants and why should they do so? This corresponds to what is commonly called the 'valley of death' in product development or innovation. It is used as a metaphor to describe the 'funding gap' between basic research and commercialization of a new product. The development of new and promising products requires large funds. The funding of a company without a working product (and thus without revenue) is not something venture capitalists or other investors are fond of. The idea of the valley of death is that companies acquire or generate innovation and then fail to find a way to nurture it to the stage that it delivers results. This valley can only be crossed with solid financial backing, often with government support of one form or another. In the case of fusion its crossing will require high-risk investment at the level of hundreds or even thousands of billions of euro. Other energy technologies provided proof of technical viability at investment levels orders of magnitude lower than that. This is one of the big economic issues facing the deployment of fusion power, certainly when it will enter an already decarbonised market without having anything special to offer.

A second issue is the so-called technology lock-in. The construction time of fusion reactors is long, and due to their complexity will remain long, while they will have to be built fast and in large numbers, if fusion wants to make an impact on world energy demand. This implies that a next generation of power plants will have to be ordered and financed, before the preceding generation has started operations and before improvements learned from its operations can be implemented. In other words, there will be no major design evolution. A sign of this can already be seen in ITER whose design provides for low-temperature superconducting magnets, while high-temperature superconducting magnets are all the craze these days and if it were designed now, it would deploy such magnets. So, if the first generation applies low-temperature superconductors, so will the second etc., locking in this technology and there is little possibility to test new technologies.

The conclusion of all this must be that there are severe economical constraints, in addition to the shortage problems of materials discussed above, to the rapid deployment of fusion power, due to in particular large unit size, high investment costs and long construction times. This may change if the promise of spherical tokamaks as dangled in front of us is fulfilled. They will be smaller and cheaper with consequently shorter construction times and may offer the possibility of crossing the valley of death at a lower financial risk.

19.4   Price of Electricity from Fusion

In spite of these gloomy prospects for fusion and the fact that it has not even shown to be able to generate electricity, and such demonstration is still at least 15–20 years away, calculations of the price of electricity generated by fusion have already been made. It will be clear from the preceding section that such calculations are largely meaningless, but the predilection for senseless calculations seems hard to suppress. And although Llewellyn Smith and Cowley (Llewellyn Smith and Cowley 2010) for instance write that "these cost figures should not be taken too seriously in detail", a very precise number for the expected price of electricity from fusion is nonetheless given. In the European power plant study it varied, depending on the model, from 9 eurocents per kWh to 5 eurocents per kWh, which, surprise, surprise, happens to fall squarely in the competitive range. The main point, they continue, is that the order of magnitude is not unreasonable. "The conclusion of the power plant study is that economically acceptable fusion power stations, with major safety and environmental advantages, seem to be accessible through ITER with material testing at IFMIF, and intensive development of fusion technologies." It is clear though that if the numbers had been somewhere between 15 and 27 eurocents, which would still have been of the same order of magnitude, this conclusion would not have been drawn and fusion would have been decried as being uncompetitive.

Such calculations are called *ex-ante* economic analysis. One might also call it speculation on a shaky basis, which admittedly is better than mere speculation. Not surprisingly, the outcome of such calculations is always a competitive price for such electricity in relation to other renewable forms of energy. If that were not the case, the results would probably not be published, but sent back to the authors with the request to do better.

When discussing the cost of electricity, capital costs and running costs have to be distinguished. The capital costs are the one-time costs for putting up the power plant, costs of construction, purchase of land and suchlike. They differ vastly between the various electricity generating options and are commonly expressed as overnight costs (i.e. not including any interest to be paid during construction, as if the plant were realised 'overnight') per kW of generating power of the plant to be built. As of 2019 estimated costs for some of the most common generating options are[18]:

- gas/oil power plant—$1000/kW;
- onshore wind—$1600/kW;
- offshore wind—$6500/kW;
- solar PV— $1060–1130/kW (utility), $1800–2000/kW;
- conventional hydropower—$2680/kW;
- geothermal—$2800/kW;
- coal—$3500 to 3800/kW;
- advanced nuclear—$6000/kW.

From this it can be seen that offshore wind and nuclear are the most expensive options, with gas/oil fired stations the cheapest, followed closely by solar and onshore wind. This reflects the huge capital costs of nuclear fission power stations (apparently taken here to be $6 billion for a 1 GWe plant) and the difficulty of installing windmills at sea. It is no surprise to see solar (with its cheap solar panels) and gas/oil as the cheapest options. We don't know of course what the construction cost of a future nuclear fusion power plant will be, but if the experience with ITER is anything to go by and a 1GWe nuclear fusion power plant could be built at the same price (it will probably be vastly higher), fusion power would be by far the most costly with $22,000/kW, more than three times the figure quoted above for advanced nuclear and in a completely different league from most solar and wind options. Running a nuclear fusion power plant must be incredibly cheap for it to be able to make good such high starting costs.

---

[18] *Wikipedia*; https://en.wikipedia.org/wiki/Cost_of_electricity_by_source.

Running costs include the cost of fuel, maintenance costs, repair costs, wages, handling waste, etc. Fuel costs tend to be highest for oil fired generation, with coal being second and gas still cheaper. Nuclear fuel is much cheaper per kWh, and fuel is of course free for solar and wind.

For estimating the costs of electricity, a quantity called the **levelized cost of energy (LCOE)**, or **levelized cost of electricity**, is used. It "represents the average revenue per unit of electricity generated that would be required to recover the costs of building and operating a generating plant during an assumed financial life and duty cycle."[19] It is a measure of the average net present cost of electricity generation for a generating plant over its lifetime. For technologies with no fuel costs and relatively small variable operation and management costs, such as solar and wind technologies, LCOE changes nearly in proportion to the estimated capital cost of the technology. For technologies with significant fuel cost, both fuel cost and capital cost estimates significantly affect LCOE. It is clear that most of these factors are largely unknown for nuclear fusion power plants.

Nevertheless calculations of such nature for fusion go back as far as 1995 (Galambos et al. 1995), and have been continued ever since in all kind of versions, even for heavy-ion fusion (see e.g. Bustreo et al. 2019; Helsley and Burke 2014). The calculations are made on the assumptions that the fuel for fusion is inexhaustible and available at an insignificant price, nuclear safety is inherent, as well as environmental impact negligible, resulting in a possible conclusion that this provides "great scope for reducing investment cost on the basis of technological research and development with a high probability to become the cheapest and cleanest energy source from the end of this century for an unlimited time onwards" (e.g. Entler et al. 2018) or some similar exclamation of great promise. That is quite something: "a high probability to become the cheapest and cleanest energy source (…) for an unlimited time", and that of something that does not even exist! Such calculations, based on very bold and partly dubious assumptions and extrapolations, do not warrant such conclusions. A simple example is the assumed amount of capital investment, i.e. construction costs, needed for a plant to be built early in the next century. In the paper just quoted a figure of just $8.5 billion is used (resulting in capital costs of $8500/kW, to be compared with the number for other energy generating options presented in the table above). Knowing that ITER's construction cost will far exceed $20 billion, that amount is at least a factor 2–3 too low. Fortunately, not everybody in the fusion world jumps to this type of outrageous conclusions.

Inherent nuclear safety is never mentioned for a coal plant or other type of fossil fuel burning plant, as it is obviously irrelevant for such a plant. It is standard for nuclear fusion proponents to add this feature to the advantages of nuclear fusion as they like to compare nuclear *fusion* power stations with nuclear *fission* power stations, but there is not much reason for doing so. If it were not inherently safe, nobody would even look at it. This feature is only relevant when a comparison is made between nuclear fusion and nuclear fission reactors. Since there are no nuclear fusion reactors a realistic comparison is not possible. Although most currently operating nuclear

---

[19] https://www.eia.gov/outlooks/aeo/pdf/electricity_generation.pdf (February 2020).

power plants are not inherently safe, inherently safe or passively safe fission power stations are not impossible. But, certainly in the West nuclear fission power is like a boxer just recovering from a knockout he can hardly be blamed for, lying in his corner of the ring and unable to get up in time to continue the fight. There is no need to give it another kick to the head with the sole aim of acting out a fantasy. Another reason for making the comparison with nuclear fission is that fission, as we know, has insurmountable problems with nuclear waste. The nuclear waste in case of fusion, although certainly not insignificant (see the next chapter), is a much smaller problem than for fission. So, the comparison shows fusion in a favourable light, which would not be the case if the nuclear waste comparison were made with solar or wind power, as they simply have no nuclear waste problem.

The hopeless position for fusion is also illustrated by a 2005 calculation of the LCOE for fusion (Ward et al. 2005). In the publication all kinds of assumption are made on the 'learning factor' of fusion and the LCOE is then compared with wind and photovoltaic solar. Photovoltaic solar, still underdeveloped at the time, always comes out much more expensive than fusion, while fusion is shown to be competitive with wind, already then considered a mature technology. The cost development of solar, even in the few years that have since passed, has however been completely different from the assumption made in that calculation. In 2014 LCOE for solar was already at the level of around €0.20 per kWh, varying between €0.06 for the southern European countries and €0.26 for the Scandinavian countries, far below the values quoted in the paper (Huld et al. 2014). Solar has already advanced to a mature technology and will become even more so in the near future. Another report from 2015 states in this respect: "Solar photovoltaics is already today a low-cost renewable energy technology. Cost of power from large scale photovoltaic installations in Germany fell from over €0.40/kWh in 2005 to €0.09/kWh in 2014. Even lower prices have been reported in sunnier regions of the world, since a major share of cost components is traded on global markets." (Fraunhofer ISE 2015). And all this while fusion can only start its learning curve early in the next century, if at all, when solar and wind will be household products.

A still more recent report (IRENA 2019) states: "In most parts of the world today, renewables are the lowest-cost source of new power generation. As costs for solar and wind technologies continue falling, this will become the case in even more countries" and "Onshore wind and solar PV are set by 2020 to consistently offer a less expensive source of new electricity than the least-cost fossil fuel alternative, without financial assistance". In short, the situation for fusion just seems hopeless.

19.5 Conclusion

In this chapter we have put all our eggs in the ITER basket, so to speak, to see what the hatched-out chicks would look like, but as already hinted a few times in this book, there is a possibility that ITER will be bypassed, even left uncompleted, and that some other countries, notably China, India, South Korea and possibly Japan will go their own way and speed up the race towards fusion by constructing DEMO plants before ITER is ready and running.

Another possibility is that the mainly private companies that promote compact or spherical tokamaks (see Chaps. 11 and 15) manage to solve the problems whose solution has eluded mankind for the last seventy years in the next decade or so and have working, grid-feeding power plants ready by 2030 as their websites tend to cry out.

In 2010 Daniel M. Kammen, a leading researcher in Energy at the Energy and Resources Group of the University of California at Berkeley and director of its Renewable and Appropriate Energy Laboratory, was asked to make a prediction where his field would be in 2020. This is part of what he had to say: "By 2020, humankind needs to be solidly on to the path of a low-carbon society—one dominated by efficient and clean energy technologies. It is essential to put a price on carbon emissions, through either well-managed cap-and-trade schemes or carbon taxes. Creative financing will also be needed so that homes and businesses can buy into energy efficiency and renewable energy services without having to pay up front. (…) Government funding of research is crucial. Several renewable technologies are ready for explosive growth. Energy-efficiency targets could help to reduce demand by encouraging innovations such as net-positive-energy buildings and electric vehicles. Research into solar energy—in particular how to store and distribute it efficiently—can address needs in rich and poor communities alike." (Kammen 2010). It is debatable that humankind is indeed on the path of a low-carbon society, but it is clear from this that he does not see any place for nuclear fusion in the future energy mix.

In spite of all the difficulties we have presented here, many so-called specialists still believe that nuclear fusion will become the major energy source by the end of the twenty-first century, or perhaps somewhat later. They claim that there are no other alternative energy sources of equal size at the disposal of mankind even in the long run, just ignoring solar, wind and nuclear fission. Nuclear fusion, they say, apparently without blushing, has in principle been technologically mastered. The key laws of the process are understood and, although the problem has proved to be much more complicated than initially believed, it is hardly doubted that it will be solved with time. If controlled thermonuclear fusion is available, humanity will have a virtually inexhaustible energy source for thousands of years. It must be very gratifying to repeat such a sentence over and over, I suppose, since it is done so often. Optimists are hoping that after ITER a demonstration fusion power plant will be constructed by the mid-twenty-first century, but even in that case nuclear fusion energy would not be available for the market before the end of the twenty-first century.

In addition to the fairly recent environmentally friendly developments of solar and wind energy, there also seems to take place a hesitating return to nuclear fission solutions, perhaps not so much by and for the western world, but in any case for China and the developing world. This is illustrated by the International Conference on Climate Change and the Role of Nuclear Power, organised by the International Atomic Energy Agency (IAEA) in cooperation with the Organisation for Economic Co-operation and Development (OECD) and the Nuclear Energy Agency (NEA), and held in Vienna in October 2019. No role was assigned to nuclear fusion at this conference; it was not even discussed. Among other possibilities, SMRs (Small Modular Reactors) are being put forward as options for nuclear fission devices that could play a role in the energy transition. Various companies (and countries), e.g. Rolls Royce, Toshiba, (UK, US) are involved in their development. Small modular reactors (SMRs) are defined as nuclear reactors of generally 300 MWe equivalent or less, i.e. approximately one third of current nuclear plants. They have compact and scalable designs, can provide power away from large grid systems, offer a host of safety, construction and economic benefits, as well as manageable capital costs. Such small reactors are already operating in China, India, Russia and Pakistan, with the ones in Russia being very small, the smallest one just 11 MWe. In various other countries SMRs are under construction or in a well-advanced stage of development.[20] Also in this respect nuclear fusion, which aims at large plants (at least 1GWe), is losing out.

We have to wait and see what the future will bring, but it is clear that nuclear fusion will come too late, if at all, and that its case is all but hopeless.

---

[20]https://www.world-nuclear.org/information-library/nuclear-fuel-cycle/nuclear-power-reactors/small-nuclear-power-reactors.aspx.

Chapter 20
Safety and Environment

20.1 Introduction

Let us first make clear that there is no such thing as a perfectly clean energy option, free of any impact on the environment. Nobody will argue that, when actually generating energy, solar, wind and hydropower are very clean indeed, but before they can be put into operation solar panels, windmills and dams have to be manufactured or constructed, which requires investment, materials etc. Most simply it can be stated that they cost money and that everything that costs money will produce waste and/or cause pollution. There is no escape from this. A solar panel for instance is produced in a factory, uses materials of which some may be mined in an environmentally damaging manner. It has to be transported in a (polluting) ship from the production facility, most probably in China, to let's say Western Europe, where it will have a limited lifetime perhaps on a large solar park taking up land and destroying habitats. It will then be discarded, perhaps recycled, but partly dumped or removed in a possibly environmentally harmful manner. Even hydropower requires the construction of dams, which disturb ecosystems, etc. To say it with a quote from Adam Smith (1723–1790), known as the 'Father of Economics' or the 'Father of Capitalism', from his book *The Wealth of Nations*: 'The real price of every thing ... is the toil and trouble of acquiring it', and viewed in that way it can be imagined from what we have learned in this book that the price for fusion is very high indeed.

Nuclear fusion has always been presented as the cleanest energy generating option available. In this chapter we will consider to which extent that is true and concentrate on environmental and safety issues regarding nuclear fusion power stations in their operational phase, meaning that we will not consider any environmental impact from their construction. Before they are in that phase, due to the huge construction costs involved, such power stations have already left a very considerable carbon footprint. As noted in the previous chapter, it would be helpful to know the data of electric power consumption for e.g. JET to be able to estimate such footprints. For ITER, at a price of more than €20 billion, it will be colossal. Whatever happens, nuclear fusion power generation will not be able to erase this footprint in this century by generating its 'clean' energy. For the time being all the activity and effort to bring fusion to fruition only aggravates the problem of climate change, in spite of its alleged 'great promise' to alleviate it.

That nuclear fusion power stations will not be simple facilities as regards possible environmental impact and safety is already borne out by the fact that under French legislation ITER is classified as a nuclear facility. This implies that a special licensing procedure has to be followed and special licenses be obtained. As the ITER website proudly states: "For the first time in the history of fusion research, a fusion device had successfully gone through the process of nuclear licensing. Throughout construction, commissioning and operation, ITER's safety processes will comply with French regulations, as verified regularly through audit and inspection by the French nuclear authorities."[1] ITER is the first nuclear installation to comply with the 2006 French law on Nuclear Transparency and Security and the first fusion device in history to have its safety characteristics undergo the scrutiny of a Nuclear Regulator to obtain nuclear licensing. The fact that all this is necessary shows that ITER is not just a piece of cake. Or has all this hullabaloo been raised for show? It will be clear that, apart from nuclear fission power stations, no other means of generating energy needs supervision by or licensing from a nuclear regulator. ITER needs such licensing as it is the first fusion facility which will have enough radioactive inventory, i.e. radioactive materials on its premises, to be potentially dangerous to the public and the environment. Although the radiotoxicity of radioactive substances (tritium and activated materials) is lower by several orders of magnitude compared to the waste from fission power plants, volumes will be much bigger. For comparison I mention here that in its 60-year-long history up until 2019 the nuclear fission programme in the UK produced 133,000 m^3 of radioactive waste. This waste is classified as high-level waste (HLW), intermediate-level waste (ILW), low-level waste (LLW) or very-low-level waste (VLLW). Three-quarters of this waste is ILW and only 2150 m^3 is high-level waste that needs to be stored for millions of years.[2] High-level waste will not be produced by nuclear fusion plants, but the other categories will. And JET for instance, just a fusion experiment, already produced 3000 m^3 of such waste.

Radioactivity is just one of the potential safety and environmental issues of nuclear fusion plants. These issues include the following (Dolan 2013, Chap. 12):

1. Tritium releases;
2. Disposal of activated structure i.e. radioactive material;
3. Accidental releases of tritium or activated materials;
4. Chemical discharges;
5. Thermal discharge to water or air (i.e. cooling);
6. Stored energy release via liquid metal fire, magnet coils, radioactive afterheat, atmospheric pressure on chamber (e.g. disruptions);
7. Effects of stray magnetic fields;
8. Plant decommissioning;

---

[1] https://www.iter.org/mach/safety.
[2] The Geological Society, *Geological Disposal of Radioactive Waste* (2019).

9. Earthquakes, floods, storms;
10. Aesthetic impact.

We will discuss most of them in this chapter and assess their potential harmful effects.

20.2 Tritium

For some data on tritium see Chap. 1. Tritium is radioactive which makes it hazardous to humans, and its use in nuclear fusion power stations therefore raises both safety and environmental issues. Tritium emits very low-energy *beta* radiation (i.e. electrons), which when inhaled or otherwise taken inside the body is diluted throughout the body tissues and organs, and is eliminated fairly quickly from the human body. These characteristics make tritium one of the least hazardous radioactive materials and it is a potential health risk only when taken inside the body, e.g. via tritiated water (HTO).

Since hydrogen isotopes are very mobile and can diffuse through materials, the containment of tritium is however a major safety concern. In Chap. 18 we have already noted that for this reason scientists are in fact afraid of using tritium. Deuterium-tritium is the preferred fusion fuel, but so far has hardly been used. This is one of the peculiar facts of the entire fusion enterprise. All the hullabaloo and beautiful stories about deuterium-tritium fuel are based on very scant experience indeed. In both TFTR and JET operations with D-T plasma have been severely limited, because of the dangers posed by tritium. The total amount of tritium processed in TFTR was just 99 g and in JET 36 g. The quantity injected into JET in 1991 was only 5 mg and in 1997 35 g. Only a small part of the injected tritium actually burns in the fusion reactions, at most a few per cent. For ITER it is less than 1% and the planned Chinese reactor CFETR calls a burn-up fraction of more than 3% very challenging (Xie et al. 2018). The rest of the tritium escapes from the reaction region, must be recovered (in real time) from the surfaces and interiors of the reactor and its sub-systems, and re-injected ten to twenty times before it is completely burned.

A large fraction of the finally remaining tritium is extracted from the vacuum vessel using a special system and sent for recycling. This is or can be made into a well-controlled operation and will only be a possible hazard to workers and public safety in case of accidents. For ITER, a multiple-layer barrier system has been designed to protect against the spread or release of tritium into the environment. If all works well, this tritium containment system should be adequate to prevent tritium from escaping into the environment. Part of the tritium however remains trapped in the materials of the vacuum vessel, e.g. in the plasma facing components. In the TFTR and JET machines about 13%, respectively 10% of the tritium still remained after various cleaning methods had been applied. This tritium diffuses further into cooling systems and other parts of the plant. All this has caused that only now, in 2020, for the first time since 1997 JET is preparing for a resumption of the tritium operations. For

ITER about 3 kg of tritium will be needed for start-up, and the total amount needed for the basic physics phase has been estimated to be in the range of 15–30 kg, so the problem for ITER will be vastly bigger than for JET (Porfiri et al. 2013). For the extended physics phase the required quantity of tritium is probably still considerably higher. Such amounts of tritium are however never present at the site.

In short, tritium is very unpleasant to work with. To limit the radiological risk to workers and the public, it is a design requirement that the amount of tritium in each part of the plant does not exceed a predetermined maximum amount, i.e. including the amounts that are buried and continue to build up in the plasma facing components. Although the plasma burn on itself is not affected by the tritium in such components, such tritium would be a concern if the tritium inventory allowed in the facility is exceeded, which would force a shutdown of the plant. According to the ITER website the maximum amount of tritium in the facility will be set by the French safety authorities, and will not exceed 4 kg.[3] The quantity of tritium present in the vacuum vessel will be less than or equal to 1 kg.

In spite of the experience with TFTR and JET there are still large uncertainties in quantifying tritium retention in components of the device. The use of mixed materials (carbon, tungsten, beryllium) as foreseen in the plasma facing components of ITER and other reactors under design introduces significant uncertainties in the tritium accumulation, which can only be taken away when the plant is in operation.

The above concerns tritium during normal operations. There is not much to go by, but for ITER it is claimed that during such operations its radiological impact on the most exposed populations will be one thousand times less than natural background radiation.[4]

Accidents are however also possible and non-negligible amounts of tritium could be released into the environment, e.g. from the leaking or breaking of pipes of the tritium system, and people can get exposed. Tritium would probably be released into the air in gaseous form in such accidents, where it would diffuse and disperse rather rapidly, keeping the hazard low. If released in the form of tritiated water, it is much more dangerous to humans. (See Porfiri et al. 2013, p. 326 ff. for an extensive discussion of possible accidents.) Tritium in tritiated water (some of the coolant will get contaminated by tritium) always causes difficulties in nuclear installations, including equipment corrosion. Especially in Japan the health effects of tritium are a major public concern due to the presence of tritium contaminated water (HTO) at the Fukushima Dai-ichi Nuclear Power Plant, although there was not any direct danger to the public. In addition, together with tritiated water, radiolytic decomposition products, from the dissociation of molecules by ionizing radiation, can also be found, e.g. hydrogen peroxide formed during decay, chloride ions produced by degradation of organic seals, and oils used for tightness and pumping, as well as acid pH produced by the excitation of nitrogen in air by the *beta* particles in the tritium radiation (Porfiri et al. 2013, p. 242).

---

[3]https://www.iter.org/faq#What_will_be_the_total_amount_of_tritium_stored_on_site_What_are_the_procedures_foreseen_to_confine_and_control_the_stock_.

[4]https://www.iter.org/mach/safety.

For ITER it applies, as stated on its website, that even in the event of a cataclysmic breach in the tokamak, the levels of radioactivity outside the ITER enclosure would remain very low. For postulated "worst-case scenarios," such as fire in the tritium plant, the evacuation of neighbouring populations would not be necessary, it is said.[5] This does however not agree with other assessments (Perrault 2019) which claim that the theoretically possible maximum release of around 1 kg of tritium for ITER would result in doses to the public of the order of one sievert, making its evacuation necessary within a large perimeter around the installation. According to model calculations one sievert carries with it a 5.5% chance of eventually developing cancer. And thus, to prevent such releases, ITER and any other fusion facility needs, just like a fission reactor, to implement a safety concept. It has been concluded that the safety demonstration for the ITER facility, the first of its kind, still needs additional studies and that all facilities created after ITER will raise even more significant nuclear safety and radiation protection issues. If not properly addressed, they could be an obstacle (in terms of delay or additional cost) or a stop (no licensing) on the way to fusion energy.

The last point to be mentioned in respect of tritium is tritiated waste. As soon as tritium is used as fuel for the fusion reaction, in-vessel components will be contaminated by tritium adsorption (adhesion to surfaces) and permeation. Components of the fuelling system and tritium plant will likewise be tritiated due to tritium permeation. High-level tritiated waste would for instance be produced from the regular replacement of plasma facing components. Disruptions and suchlike will result in the creation of dust that will have the same radiological characteristics as the first few microns of the plasma facing components and will be tritiated. Furthermore, water circulating in the heat transfer systems will corrode the pipes, leading to activated corrosion products in the fluids and pipe circuits (Rosanvallon et al. 2016). All this waste must be detritiated, i.e. the tritium must be removed, before it can be disposed of.

The situation in ITER is still fairly simple as far as tritium is concerned. Future nuclear fusion power plants however will use tritium on a continuous basis and will have to breed their own tritium, probably in a tritium-breeding blanket surrounding the vacuum vessel. They must do so in sufficient quantities to become self-sufficient, which however will be very hard as some of the tritium will escape and cannot be recovered as experience with JET and TFTR has shown. If only one per cent of the unburned tritium is not recovered and re-injected, even the largest surplus in the lithium-blanket regeneration process cannot make up for the lost tritium (Jassby 2017). As we have seen in Chap. 10, for DEMO about 300 g of tritium will be needed per day to produce 800 MW of electric power, while a 1 GW fusion reactor would require about 56 kg/year of tritium (Dolan 2013, p. 632). The quantities of tritium in such a plant will be considerably bigger than in ITER. This, together with the tritium breeding, compounds the problems involved in the safe handling of tritium.

This discussion shows that nuclear fusion with deuterium-tritium as fuel is not such a clean option after all and that the use of tritium raises considerable safety and

---

[5]https://www.iter.org/mach/safety, last accessed 17 July 2020.

environmental issues. That is probably why the term "environmentally responsible" is now becoming popular in this respect. The tritium problem is certainly highly non-trivial, but much work is being done to keep it under control. Renewables like solar and wind energy have however no such problems, nor do they need licensing as a nuclear installation.

20.3 Radioactive Waste

A further important point is radioactive waste (points 2 and 3 in the list above). Activation of the structure will already start with the 2.5 MeV neutrons resulting from the D-D operations in the period preceding the D-T operations, but especially the stream of highly energetic (14 MeV) neutrons from fusion reactions in the D-T plasmas will produce huge volumes of radioactive waste as they bombard the walls of the reactor vessel and its associated components (see Chap. 16 where this has been discussed as an issue for the materials to be used for the plasma facing components). Damage to exposed materials from neutron radiation, which causes swelling, embrittlement and fatigue, is a long-recognized drawback of fusion energy. The total operating time with high neutron production in ITER, which equals the number of D-T shots times the pulse length of 400 s, will be too small to cause any significant damage to structural integrity, but neutron interactions will still create dangerous radioactivity in all exposed reactor components, eventually producing a staggering 30,000 tonnes of radioactive waste.

For fusion power plants, which are supposed to work in steady-state, this will be a very severe problem. The volume of radioactive waste is much larger than for a fission power plant, but fortunately much less dangerous. The waste is classified as very low, low, or intermediate level waste (VLLW, LLW or ILW), but is nonetheless radioactive waste and must be treated as such. At ITER all waste materials (such as components removed by remote handling during operation) will be treated, packaged, and stored on site for the duration of the ITER experiment. Surrounding the ITER tokamak, a monstrous concrete cylinder 3.2 m thick, 30 m in diameter and 30 m tall, called the bioshield, will prevent X-rays, gamma rays and stray neutrons from reaching the outside world. The reactor vessel and non-structural components both inside the vessel and beyond up to the bioshield will become radioactive by activation from the neutron streams. Among other things, it implies that downtimes for maintenance and repair will be prolonged because all maintenance must be performed by remote handling equipment. For the much smaller JET project the radioactive waste volume is estimated at 3000 cubic metres, and the decommissioning cost will exceed $300 million, according to the *Financial Times*.[6] Those numbers will be dwarfed by ITER's 30,000 tonnes of radioactive wastes, which will in turn be dwarfed by the waste

---

[6]Brexit's nuclear fallout: 3000 cubic metres of Oxfordshire waste, *Financial Times* 28 November 2016. Since the future of JET after Brexit has been secured until the end of 2020, this waste problem has been conveniently shifted to the future.

produced by power stations that operate continuously for a considerable period of time. Most of this induced radioactivity will decay in decades, but that does not make the radiation less real. ITER's *Final Design Report* reckons that after 100 years some 6000 tonnes will still be dangerously radioactive and require disposal in a repository (Jassby 2017).

A recent analysis (Zucchetti et al. 2018) comparing total radiotoxicity of pressurized water reactors (the large majority of currently operating fission reactors), Generation IV fission reactors (not yet in operation, but currently being researched) and fusion reactors arrived at the conclusion that fusion reactors have higher radiotoxicity due to short-lived radionuclides, mainly activation products of structural materials and breeders. After some decades of decay, the situation changes completely. For instance, after 100 years of decay, total radiotoxicity of fusion reactors becomes 100 times lower than fission pressurized water reactors and after some 500 years the total radiotoxicity of fusion reactors has fallen to levels close to the natural radiotoxicity contained in fly ashes from coal-burning power plants. From this it is clear that radiotoxicity from nuclear fusion plants is nontrivial, even serious. There is however a large degree of uncertainty in such estimates as in both the blanket and divertor the long-term activity and hence waste classification is largely determined by the specific amount of ^{14}C produced in reactions with the main ^{14}N isotope of nitrogen, which is added to the material composition of the steel used in order to improve its mechanical and thermodynamic properties. This highlights that even small levels of impurities could have significant impact on the severity of the produced radioactive waste (Gilbert et al. 2018).

The 100 years necessary for storage could possibly be reduced for future devices through the development of 'low activation' materials, like RAFM (Reduced activation ferritic/martensitic) steels, which is an important part of fusion research and development today. Some call these 100 years fortunate, since they compare it with the waste from fission plants, which has to be stored for millions of years, but 100 years is still a very long time. Radioactive waste and the accompanying radiation belong to the most important problems fusion will be faced with. Just imagine how much waste the hypothetical 2760 power plants of Chap. 19 would produce! Nobody in France seems to worry that much about this, although according to the ITER agreement France will be responsible for the decommissioning of the device. It is specified that after the final shutdown, and following a five year 'deactivation' period, responsibility for the installation and the waste will be transferred to France with the decommissioning fund generated by the ITER Organization during the period of operation.

20.4 Scarcity of Materials

Many materials are needed for the construction of nuclear fusion power plants. In Chap. 19 we have already discussed that in the future difficulties might arise in connection with in particular the availability of lithium and beryllium. Other elements that may be in short supply include helium, copper, chromium, molybdenum, nickel, niobium, tungsten and some rare-earth elements. Niobium should be avoided in structural materials exposed to neutrons, because of induced radioactivity, but is used in superconducting magnet coils. Future fusion power plants will probably be less prolific users of both niobium and helium than for instance ITER as they will not use niobium-based and helium-cooled low-temperature superconductors but high-temperature superconductors. The ITER cryogenic system requires 24 tonnes of helium; the total amount of helium on site will, on average, be about 27 tonnes. The size of the helium inventory is primarily related to the cooling of the superconducting magnets and in particular to the length and diameter of the required circuitry. Future power plants will probably also need helium for the latter.

Helium is a rare element and a non-renewable resource that plays a vital role in the development of a number of sophisticated technologies, including fusion reactors and various superconducting applications. The cumulative demand up to 2050 for superconducting transmission lines is estimated at about 12 billion cubic metres. This demand could limit the helium resources available for fusion power plants (a similar situation to the one for lithium). The global production of helium in 2019 was 160 million cubic metres. Helium production is traditionally dominated by the US, a bit less than half of the total global production in 2019, 68 million cubic metres, was extracted from natural gas in the US. One of the major issues of helium production is that a large amount of helium is lost, due to the venting of helium rich gases in the natural gas industry (Mohr and Ward 2014). Fusion also produces helium as one the products of the nuclear fusion reactions and of the tritium breeding in the blanket. It is paramount that this very valuable 'ash' is collected for future use (Bradshaw and Hamacher 2013).

Niobium is most commonly used to create alloys. Even as little as 0.01% of niobium markedly improves the strength of steel. It is also used in small amounts in superconducting wires. Global production in 2019 amounted to about 75,000 tonnes, with 65,000 coming from Brazil. World reserves are also located for more than 90% in Brazil and demand is not expected to outstrip supply.

The world's reserves of tungsten are 3.2 million tonnes, mostly located in China (1.8 million tonnes), Canada, Russia, Vietnam and Bolivia. China dominates the market with a share of about 80%. Approximately half of the tungsten is consumed for the production of hard materials—namely tungsten carbide—with the remaining major use being in alloys and steels. Demand is rising, but supply and demand are still roughly in equilibrium. In ITER it will be used as the plasma facing material of the divertor. It is not clear what the effect on world supply will be when tungsten will be applied on a large scale in fusion power stations (Dolan 2013, p. 638).

In summary, shortages in materials for fusion are most probable for lithium, beryllium, and helium.

Shortages of some materials might also pose problems for renewables like solar and wind for that matter. It has been estimated that, for equivalent installed capacity, solar and wind facilities require at least an order of magnitude more concrete, glass, iron, copper and aluminium than fossil fuel or nuclear energy power plants (Vidal et al. 2013).

20.5 Plasma Disruptions and Quenching

This covers point 6 in the above list. In Chap. 10 we have already noted that disruptions and quenching are safety issues for ITER. Accidental disruptions are expensive and dangerous. The heat in a disrupted plasma can be ten times higher than the melting point of the first wall and the divertor. It could also lead to leaks in the water cooling circuits.

A disruption occurs when an instability grows in the tokamak plasma to the point where there is a rapid loss of the stored thermal and magnetic energy. This rapid loss can also accelerate electrons to very high energy (runaway electrons). To safeguard the device against such eventuality a special disruption mitigation system has been envisaged to protect the plasma-facing components against the heat and forces that arise during a disruption, and tame at the same time the runaway electrons. A special task force has been established for this.[7] See Chap. 16 for more details about disruptions and the system chosen to protect ITER. If this system does not work, it will be the end of ITER. Damage will probably be extensive and, in view of the intricacy of the device and the required fine-tuning of the components, repair will be difficult.

A safety analysis has suggested an occurrence of such an event by the failure of a penetration line, e.g. a diagnostic line of which there are many in ITER. Such failure leads to pressure build-up in the vacuum vessel due to gas (e.g. air) inflow. Plasma burning is terminated, and a disruption triggered. The vacuum vessel pressure increases until the pressures inside and outside the vessel are almost equal. After pressure equilibration, the air inside the vessel is heated by hot component surfaces, causing expansion of the vessel atmosphere and the air to flow out of the vessel (Honda et al. 2000). A possible hazard is that in such an event tungsten nanoparticles could be released into the environment and induce occupational exposure via inhalation.

In the literature acronyms are used for possible accidents that can lead to disruptions: LOCA, LOFA and LOVA, respectively standing for loss of coolant accidents, loss of (coolant) flow accidents and loss of vacuum accidents.

The second problem is quenching, the situation when a superconducting magnet suddenly becomes a normal electromagnet and releases its energy. When cooled to around minus 269 °C, ITER's magnets become powerful superconductors. The electrical current surging through a superconductor encounters no electrical resistance. This allows superconducting magnets to carry the high current and produce the strong magnetic fields that are essential for ITER's experiments. Superconductivity can be maintained as long as certain threshold conditions are respected (cryogenic temperatures, current density, magnetic field). Outside of these boundary conditions a magnet will return to its normal resistive state and the high current will produce high heat and voltage. This transition from a superconducting to a resistive state is referred to as a quench. During a quench, temperature, voltage and mechanical stresses increase—not only on the coil itself, but also in the magnet feeders and the

---

[7]https://www.iter.org/newsline/-/3183.

magnet structures. A quench that begins in one part of a superconducting coil can propagate, causing other areas to lose their superconductivity. As this phenomenon builds up, it is essential to discharge the huge energy accumulated in the magnet to the exterior of the Tokamak Building.

ITER's coils contain the same energy as 10 tonnes of TNT. Quenching causes components to overheat and melt; it may even start dangerous fires. Such events may occur as the result of mechanical movements that generate heat in one part of the magnet. Variations in magnetic flux or radiation coming from the plasma can also cause quenches, as well as issues in the magnet cryogenic coolant system. ITER is developing an early quench detection system to protect its magnets.

Magnet quenches are not expected often during the lifetime of ITER, but it is necessary to plan for them. They are actually not accidents, failures or defects, but part of the life of a superconducting magnet and the latter must be designed to withstand them. An early quench detection system must make it possible to react rapidly in order to protect the integrity of the coils, avoid unnecessary machine downtime, and discharge large amounts of stored energy to avoid damage to the first confinement barrier—the vacuum vessel.[8] There is some experience with quench detection at the Large Hadron Collider at CERN, which can be used at ITER, but to have an appropriately working system at ITER is nonetheless a tremendous challenge.

20.6   Water Usage

ITER, its website says,[9] is provided with a cooling water system designed to reject to the environment all the heat generated from the components (nuclear and non-nuclear) using water as coolant. The only exception is the vacuum vessel, whose heat is released, via a separate heat transfer system, to the air coolers. The total heat to be released to the environment during the D-T pulse is about 1100 MW by water plus about 13 MW mainly from the vacuum vessel (10 MW) and other systems (3 MW) by air coolers. Among others, one of the main critical issues of the ITER reactor is the minimization of the release of all the gaseous, liquid and solid effluents to the environment.

Concerning water supply, approximately 3 million cubic metres of water, roughly the annual usage of 300,000 people, will be necessary per year during the operational phase of ITER. This water will be supplied by the nearby Canal de Provence and transported through underground tunnels to the fusion installation. The volume of water needed for ITER represents 1% of the total water transported by the Canal de Provence. The combined consumption of the ITER installation and the adjacent CEA facilities remains below 5% of the total volume of water transported by this canal.

---

[8]https://www.iter.org/newsline/278/1652.

[9]https://www.iter.org/mach/CoolingWater.

It was one of Daniel Jassby's points of criticism (Jassby 2017) that ITER would need torrential water flows to remove heat from various parts and components of the facility. Including fusion generation, the total heat load could be as high as 1000 MW, but even with zero fusion power the reactor facility consumes up to 500 MWe that eventually becomes heat that has to be removed. ITER will demonstrate that fusion reactors will be much greater consumers of water than any other type of power generator, because of the huge parasitic power drains that turn into additional heat that needs to be dissipated on site. (By "parasitic" we mean consuming a chunk of the very power that the reactor produces.) During fusion operations, the combined flow rate of all the cooling water will be as large as 12 cubic metres per second, or more than one third of the flow rate of the Canal de Provence. The actual demand on the Canal's water will be only a fraction of that value because ITER's power pulse will be just 400 s long with at most 20 such pulses daily, and ITER's cooling water is recirculated. But, the important point here is that, while ITER is producing nothing but neutrons and no power, its maximum coolant flow rate will still be nearly half that of a fully functioning coal-burning or nuclear plant that generates 1000 MWe of electric power. Operation of any large fusion facility such as ITER is only possible in a location such as the Cadarache region of France, where there is access to many high-power electric grids as well as a high-throughput cool water system. In past decades, the great abundance of freshwater flows and unlimited cold ocean water made it possible to implement large numbers of gigawatt-level thermoelectric power plants. In view of the decreasing availability of freshwater and even cold ocean water worldwide, the difficulty of supplying coolant water would by itself make the future wide deployment of fusion reactors impractical. It would be hard to meet the cooling water demands of actual nuclear fusion power plants, i.e. the 2760 power plants hypothesized in Chap. 19.

The cooling water also contains radionuclides, because impurities (e.g. tritium) diffuse from in-vessel components and the vacuum vessel and should be cleaned before returned to the Canal.

20.7 Earthquakes, Floods, Storms

The ITER facility is designed to resist an earthquake of amplitude 40 times higher and with energy 250 times higher than any earthquake for which there are historical or geological references in the area of Saint Paul-lez-Durance, France. Although some protection against earthquakes seems sensible, this precaution seems to be completely over the top and one wonders why this has been done (window-dressing?), as an earthquake would just bring the operation to a standstill and apart from the possible release of some tritium not much else can be expected. The ITER tokamak building will be made of specially reinforced concrete, and will rest upon bearing pads, or pillars, that are designed to withstand earthquakes (this technology is also used to protect other civil engineering structures such as electric power plants from the risk of earthquakes).

The risk of flooding, too, has been considered in ITER's design and Preliminary Safety Report. In the most extreme hypothetical situation—that of a cascade of dam failures north of the ITER site—more than 30 m remains between the maximum height of the water and the base of the nuclear buildings. Again a rather superfluous precaution, which has only added to the cost.

Following the Fukushima disaster in Japan in 2011, and the resulting tsunami and nuclear accident, the French government requested that the French Nuclear Safety Authority (ASN) carry out complementary safety assessments. The decision was made to assess not only nuclear power plants, as requested at the European level, but also research infrastructures such as ITER in order to examine the resistance of a facility in the face of a set of extreme situations leading to the sequential loss of lines of defence, such as very severe flooding, a severe earthquake beyond that postulated in the ITER safety case, or a combination of both.[10] It seems to be an exercise in futility to study the effect of extreme climatic conditions such as tornados, hailstorms, etc. on ITER in the south of France where such violent events never happen. It may be that lessons can be learned from this for similar devices at other places, but for ITER, whose lifetime will only be 25 years, there are more serious matters that deserve attention.

20.8 Aesthetics

This is a very minor point, hardly worth mentioning. A nuclear fusion power plant is neither less, nor more beautiful or attractive to look at than any other power plant. Some may prefer such a localised structure over large solar or windmill parks, which when constructed on land are indeed an eyesore. It becomes a different matter if in the decommissioning phase it were decided to entomb the plant on site. Entombment seems to be the cheapest option, but it leaves a carbuncle on the face of the Earth. The site will be unsuitable for other use and perhaps a hazard for the future. Mechanical disassembling and recycling is by far the preferable option but will be costly and time consuming.

20.9 Conclusion

From the above it can be concluded that power generation from fusion is not as environmentally clean and safe as claimed. Tritium and radioactive waste pose considerable problems in this respect, vastly bigger problems than any problem posed by solar and wind energy. Disruptions and quenching must at all cost be avoided. The only thing that is perfectly clear is that, when in operation, a fusion power plant will

---

[10]https://www.iter.org/faq#What_would_be_the_danger_of_an_earthquake_occurring_near_I TER_or_a_double_disaster_like_earthquake_and_flooding.

not produce any carbon dioxide and will therefore not contribute to global warming. Its construction will however involve a large carbon footprint and how this will work out for maintenance and downtime periods, during which huge amounts of energy, generated perhaps by fossil-fuel burning plants, will be consumed, is anybody's guess at present. Whether the benefit of being carbon free is bigger than the disadvantages stated above is difficult to say, but should be assessed with a clear head, not by making some flippant remarks about "unlimited, inexhaustible, clean energy", certainly not when such remarks are untrue.

In this respect it is worrying that fusion websites (from official organisations) are not clear about the environmental impact of fusion. At one time they compare fusion with coal stations when that is convenient for their purpose and it seems advantageous (and/or fashionable) to stress the point that fusion is carbon free. The fact that fusion power has clear environmental benefits compared to coal-fired power (no gaseous carbon dioxide emission, no major mining and transport operations) is actually completely irrelevant for the future of fusion. Coal-fired power stations are being phased out almost everywhere and will probably be all but gone before the first nuclear fusion power station is working, if that ever is going to be the case. At another time they use a comparison with fission plants since fusion power plants are inherently safe, with no possibility of a "meltdown" or "runaway reactions", no fission products and no plutonium production. Comparisons with solar and wind energy generation are seldom made, as fusion is then obviously at a disadvantage, certainly so long as steady-state power generation from fusion is still a faraway dream. Plainly wrong statements are often made. For instance, on the Fusion for Energy website[11] it was stated that "a fusion reactor is like a gas burner with all the fuel injected being 'burnt' in the fusion reaction." This statement is actually untrue as only a small part of the fuel is burned (see above) and worrying tritium contamination builds up in the reactor structure.

The fuel needed for fusion is normally called inexhaustible and readily available from seawater, for instance: "The fuel it requires is abundant everywhere on the planet reducing the risk of any geopolitical tension; it is extracted from sea water and the crust of the Earth."[12] That may be true for deuterium, but is not so for tritium and the lack of tritium may very well increase geopolitical tensions. The tritium produced on Earth in two or three decades from today will perhaps not even be sufficient for operating two or three DEMO plants (Ni et al. 2013). Moreover, fusion will probably also lead to a scarcity of lithium and other materials and to fierce competition with demands for other uses. You will not find any information about this on the websites of the fusion organisations.

Another statement from the Fusion for Energy website is: "Fusion machines are inherently safer (*sic*) posing no risk to populations in the vicinity, generating no long-lasting waste". Most people would consider 100 years (four generations) long-lasting; imagine that you would still have to look after the waste produced by your

---

[11] https://fusionforenergy.europa.eu/understandingfusion/merits.aspx, accessed April 2020; has now been changed and is much less explicit.

[12] https://fusionforenergy.europa.eu/merits-of-fusion-energy/ accessed 8 April 2020.

grandfather or great-grandfather, on top of the waste you yourself are producing. It is of course short compared to the millions of years some waste from nuclear fission power stations must be stored, but that is irrelevant. Having huge quantities of low-level radioactive waste in your backyard for 100 years is not something to look forward to. When first made, such statements of "no long-lasting waste" are embedded in comparisons with nuclear fission stations and then of course serve a purpose. Later, as in the above quotation, such statements start a life of their own and are routinely being made without any reference to nuclear fission, but then they are no longer true. It has just become a subtle way to mislead. The use of the word 'safer' without saying safer than what, also shows that the statement was earlier part of a longer one. Fusion power stations are in any case less safe than wind parks at sea or solar panels on a roof.

Wind and solar are undoubtedly the cleanest forms of energy, both as regards energy generation itself and the construction/manufacture of the necessary equipment. They also have no safety issues to speak of. Fusion energy will probably also beat fission in that respect, and it will win hands down in the safety compartment. As regards coal-fired power stations one wonders if fusion is cleaner than coal-burning power stations equipped with carbon capture installations, or than gas-burning power stations with carbon capture, after all if no carbon dioxide is released into the environment there is hardly anything against burning fossil fuels, with natural gas the prime candidate. The cost of carbon capture is rapidly decreasing. Even siphoning carbon dioxide from the atmosphere seems to be within reach with the cost of pulling a tonne of carbon dioxide from the atmosphere ranging between \$94 and \$232 (figures from 2017).[13] Burning natural gas for electricity generation produces 0.2 kg of CO_2 per kWh.[14] If the capture of a tonne of CO_2 would cost \$100, it would add \$0.02 to the price of 1 kWh of electricity generated by burning natural gas, adding less than \$100 to the annual electricity bill of a typical Dutch family. In any case, the difference in construction costs between a coal- or gas-burning plant and a nuclear fusion plant lets you capture quite a lot of carbon before running out of funds. It is also a technology that seems to progress faster than the cumbersome route of nuclear fusion, whereby construction times are so long that it cannot keep abreast of technological developments.

---

[13] Sucking carbon dioxide from air is cheaper than scientists thought, *Nature News* 7 June 2018; https://www.nature.com/articles/d41586-018-05357-w.

[14] https://www.volker-quaschning.de/datserv/CO2-spez/index_e.php.

Chapter 21
Applications and Spin-Offs

21.1 Introduction

It is common practice among embattled science practitioners to point at spin-offs, useful ancillary technology emerging from their efforts, as an extra justification for the huge amounts of taxpayers' money they are wasting. The argument is mainly brought out in favour of an expensive project that threatens to become a failure and when critical questions about its existence are being asked. It is certainly true that building huge and expensive things has in the past led to the development of some ancillary technology that turned out to be hugely important, but this can hardly be an argument for building such huge and expensive things just for the sake of the odd chance that some important ancillary technology will come out of it. It has been tried before with CERN until it was realised that it was actually weakening their science case. Spending money on science is something that should not and cannot be justified by its by-products. If the case for the science itself is not strong enough, just don't do it. Any (possible and per definition uncertain) spin-off can never be an argument. There certainly is other science for which a strong case can be made and that should then get the funds. As far as CERN is concerned it can for instance be asserted that the Internet is a spin-off of its endeavours. It is certainly beyond doubt that the first steps towards the Internet were made at CERN. But the fact that "CERN invented the web" can of course not be used as a justification to fund or increase funding for particle physics. It is the argument of a drowning man clutching at a straw.

But for CERN it was easy to leave that avenue behind. It has answered numerous science questions and can point at quite a number of scientific successes to justify its existence. You may not find the science it pursues that important or that exciting, but it does what it promises. More cannot be asked. In that case a spin-off is a pleasant aside, but no more than that. But what should you do if you have nothing to point at, just empty promises and failures? In that case you upload a number of videos to YouTube on "Fusion Spinoffs in Europe" as the ITER Organisation has done to highlight the (trivial) fact that products and methods developed for ITER have also applications elsewhere or you set up the Fusion Communications Group (a

L. J. Reinders, *The Fairy Tale of Nuclear Fusion*,
https://doi.org/10.1007/978-3-030-64344-7_21

collaborative venture with an all-star cast: the Office of Science of the US Department of Energy, ITER US, the ITER Organisation, PPPL, Oak Ridge National Laboratory, Savannah River National Laboratory, General Atomics and the MIT Plasma Science and Fusion Center) and bring out glossy brochures hailing all the beautiful spin-offs of fusion. If it does not help, it doesn't hurt either.[1]

And what a relief it is to hear, for instance, that nuclear fusion is actually catapulting the US Navy into the twenty-first century, as technological advances from nuclear fusion include the Electromagnetic Aircraft Launch System, an electromagnetic catapult that will replace steam catapults used on prior generations of aircraft carriers. Quite a roundabout route for the Navy to take, one would think, for arriving at such an innovation. Is it correct to assume that, since it has been agreed to share technological advances among ITER members, the aircraft carriers of the potential enemies of the US Navy will have access to the same system? Is ITER than also feeding the arms race?

Or you can read drivel like: "The benefits of over 57 years of fusion research are not simply deferred to the future when fusion power will be commercially available. Other technologies have developed out of fusion research that have important applications across American society. Because of the very high requirements of confining fusion, many of the technological developments we enjoy because of the fusion program are in laser or magnet technology. Sectors that have benefitted from technological spin-offs from fusion research include medicine, manufacturing, electricity transmission, environmental clean-up, and even national security. Some of the benefits we see every day, some are hidden, and still others have not yet achieved their promise. Even though fusion reactions have not yet powered so much as a light bulb, the impact of decades of fusion research is apparent across the American economy."[2] The implicit conclusion, it seems, is that even if fusion will never power so much as a light bulb, we nonetheless should continue to fund it, just for the spin-offs.

The Office of Science of the US Department of Energy has put all these spin-offs neatly together in a glossy brochure of its own.[3] It starts with the usual advertisement text: "The reaction that fuels the stars—fusion—would be a source of energy that is unlimited, safe, environmentally benign, available to all nations, and not dependent on climate or the whims of the weather." The final part of this statement is true, the rest is just fantasy. It then continues with crediting fusion with silicon wafers, space craft propulsion, annihilating tumours, making GPS more reliable, making lights brighter, understanding our universe, with safer and more efficient jet engines, cleansing water, saving accidents at nuclear facilities, as well as the launch system already mentioned. In short, without all this fusion research we would still be living in a rather primitive world.

---

[1] https://www.iter.org/doc/www/content/com/Lists/Stories/Attachments/624/spinoffs.pdf.

[2] https://www.americansecurityproject.org/wp-content/uploads/2011/12/Technology-from-Fusion-Research-Benefits-Today.pdf.

[3] FES Brochure, accessible at https://www.diif.de/en/deutsches-iter-industrie-forum-e-v-.890.0.0.0.0.html (accessed 22.04.20).

European researchers into fusion energy are now also hailing breakthroughs in magnetic resonance spectroscopy and advances in explosive metal construction that have helped the aeronautics industry. "The breakthroughs we have made in magnetic resonance spectroscopy have helped to save lives and create an industry worth almost US$20 billion," it is claimed. Well, well, that might cover almost the entire fusion bill. "The development of remote handling techniques has led to major advances in power decommissioning in other hostile or mega- and micro-scale processes where the human hand is insufficiently accurate" and "Advances in explosive metal construction have helped the aeronautics industry to create the next generation of aircraft cockpits."[4] Imagine to have missed out on a whole generation of aircraft cockpits; quite scary frankly.

When speaking to the military or warmongering politicians they might even be able to point at some development that could be useful for a hideous, but effective weapon against terrorism. Thinking out of the box is, after all, almost second nature to fusion scientists. Increase of funding would be assured.

When talking in this chapter about applications and spin-offs of nuclear fusion, we do not so much mean inventions and developments as quoted above, that just come along as by-products of the fusion effort, but processes and applications that actually make use of fusion itself, even in the event that its main application, making electricity in a commercially attractive way, will remain a pipe dream. In fact, it would concern applications that cost energy. The application that first comes to mind in this respect is to use the fusion reactor as a neutron source, either as a source as such or in a fusion-fission hybrid reactor.

21.2   Neutron Sources

A neutron source is any device that emits neutrons, irrespective of the mechanism used to produce them. They are very useful and find application in physics, engineering, medicine, nuclear weapons, petroleum exploration, biology, chemistry, and nuclear power. Neutron scattering for instance provides a powerful probe of the structure and dynamics of condensed matter, complementary to X-ray and other techniques, across a wide range of science from solid state physics to biology. In Chap. 16 when discussing the IFMIF we have already seen that this facility will essentially be a neutron source that generates neutrons with a fusion-relevant spectrum as an aid in the development of materials that withstand the onslaught of the neutrons unleashed in the nuclear fusion processes.

Neutron source variables include the energy of the neutrons emitted by the source, the rate of neutrons emitted by the source, the size of the source, the cost of owning and maintaining the source, and government regulations related to the source.[5] Fusion systems produce neutrons and hence can be used as neutron source. Small scale fusion

---

[4]https://www.energy-reporters.com/industry/nuclear-fusion-researchers-hail-spinoffs/.

[5]*Wikipedia,* https://en.wikipedia.org/wiki/Neutron_source.

systems exist for (plasma) research purposes at many universities and laboratories around the world, but none are as yet used as a neutron source. Apart from these there are a few dozen large neutron sources in the world.[6]

For most applications, a higher neutron flux (defined as the number of neutrons travelling through a small sphere of radius R in a time interval, divided by πR^2 (the cross section of the sphere) and by the time interval, expressed as the number of neutrons per cm^2 and per second) is preferable (since it reduces the time required to conduct the experiment or acquire whatever the source has been used for). Amateur fusion devices, like the fusor (see Chap. 14), generate only about 300,000 neutrons per second. Commercial fusor devices can generate on the order of 10^9 neutrons per second, which corresponds to a usable flux of less than 10^5 n/(cm^2 s). In Chap. 15 we have mentioned a few companies that make such neutron generators. Large neutron beamlines around the world achieve much greater fluxes. Reactor-based sources now produce 10^{15} n/(cm^2 s), and spallation sources generate fluxes greater than 10^{17} n/(cm^2 s). In general terms, spallation is a process in which fragments of material (*spall*) are ejected from a body due to impact or stress. In nuclear physics it is a nuclear reaction in which a nucleus ejects a number of neutrons. A beam of particles (protons) accelerated in a particle accelerator is shot into a target consisting of a heavy metal. The target is excited and upon deexcitation neutrons are expelled. The European Spallation Source (ESS), currently under construction in Lund (Sweden) and scheduled to be completed in 2025, will be a multi-disciplinary research facility based on what will be the world's most powerful pulsed neutron source. It is a so-called European Research Infrastructure Consortium (ERIC) of the European Union and would provide fluxes 10 to 20 times as high as the ISIS facility at the Rutherford Appleton Laboratory in the United Kingdom, which is at present the largest neutron source in Europe with a flux of 10^{16} n/(cm^2 s) and was for many years the world's most powerful spallation neutron source. It has been in operation since 1984 and is considered such a success story that its expected original operational life of 20 years has been extended with an extra 25 years to 2030. At ISIS the neutrons are created by accelerating 'bunches' of protons in a synchrotron (a type of cyclic particle accelerator) and colliding them with a heavy tungsten metal target. The impacts cause neutrons to spall off the tungsten atoms. Each proton produces 15–20 neutrons and around 2×10^{16} neutrons per second are produced. These neutrons are channelled through guides, or beamlines, to around 20 instruments, each individually optimised for the study of different types of interactions between the neutron beam and matter. (*Wikipedia.*) A big advantage of spallation sources is that they produce neutrons with a wide spectrum of energies, ranging from a few eV to several GeV. Another advantage is their ability to generate neutrons continuously or in pulses as short as a nanosecond.

From the comparison shown in Fig. 21.1 between reactors, spallation sources, and a neutron source based on inertial confinement fusion, it can be seen that fusion neutron sources from inertial confinement fusion (ICF) can achieve a much higher neutron flux than current methods. The figure includes the Intense Pulsed Neutron

---

[6]See https://www.ncnr.nist.gov/nsources.html.

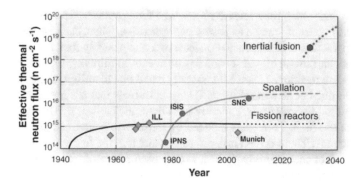

Fig. 21.1 Neutron source flux versus year (*from* Taylor et al. (2007))

Source (IPNS) at Argonne, the ISIS facility in the UK, the Spallation Neutron Source (SNS) at Oak Ridge National Laboratory, the neutron source at the Institut Laue-Langevin in Grenoble (France) and the new high-flux reactor FRM-II in Munich.

The figure shows that currently available neutron sources are reaching the limits of existing technologies, which consist of two distinct approaches: pulsed spallation sources, in which accelerated protons smash neutrons out of a heavy-metal target, and nuclear fission reactors. As can be seen, ICF has the potential to produce orders of magnitude more neutrons than spallation, with a flux that is expected to far exceed the flux of any facility currently in use, including the future ESS facility. Applications include neutron radiography (the process of producing a neutron image that is recorded on film; the highest resolution form of neutron imaging) which can be used to locate hydrogen atoms in structures, resolve atomic thermal motion and study collective excitations of nuclei more effectively than X-rays (Taylor et al. 2007). Without ignition such a facility will however remain a dream, as does any other neutron source based on nuclear fusion. However, when finally the energy generating option of nuclear fusion has foundered, some may be salvaged by application as a neutron source.

21.3 Fusion-Fission Hybrids

Fusion as a neutron source also finds application in fusion-fission hybrids that have already come along in this book in Chap. 12 with the LIFE project (Laser Inertial Fusion Energy), and in Chap. 17 where we spoke of the Russian DEMO plant.

The basic idea of this type of device is to use high-energy fast neutrons from a fusion reactor to trigger fission in fissile or non-fissile fuels like U-238 or Th-232. The collision of such a neutron with a thorium or uranium nucleus is likely to knock one or two additional neutrons off the nucleus. These neutrons have enough energy to cause other fission events. Alternatively, the fusion neutron can itself also cause a fission event, which may release as many as four neutrons. This is the key point in the hybrid concept, known as *fission multiplication*. Thus, each 14-MeV neutron generated by a fusion event provides at least two, and possibly four, neutrons that can be captured, and so can trigger several fission events, multiplying the energy released by each fusion reaction. This would not only make fusion designs more economical in power terms, but also make it possible to burn fuels that are not suitable for use in conventional fission plants, and even burn their nuclear waste. The energy would come from the fission reactions induced by the 14 MeV neutrons produced in the fusion reactions. There is then no need to demand that the fusion part of the hybrid on itself is a net energy producer, as it is only used as a neutron source. The fission part of the hybrid must generate the energy.

There are also much more neutrons to play with than in a fusion reactor, not just the single one produced by the D-T fusion reaction, but four times as many. Some of these can be used to replenish the tritium consumed in the initial fusion reaction by also including ^{6}Li in a blanket, and the others for capture by fissile material to produce energy.

A further important advantage is that in contrast to current fission reactors, hybrid reactors are potentially inherently safe because the fission part is driven by neutrons from the fusion events and is consequently not self-sustaining (subcritical). If the fusion process is shut off or the process is disrupted, the fission damps out and stops nearly instantly. In a hybrid configuration the fission and fusion reactions are decoupled, i.e. while the fusion neutron output drives the fission, the fission output has no effect whatsoever on the fusion reaction, completely eliminating any chance of a positive feedback loop. Thus, a criticality accident in a hybrid is physically impossible. But such accidents are also extremely unlikely to occur in a properly designed light water reactor (LWR) with negative temperature coefficients. (Freidberg and Kadak 2009).

The hybrid concept dates back to the 1950s, when apparently Andrei Sakharov was already thinking along these lines, although the claim that he actually proposed such a device is not strong (Sakharov 1990, p. 149). Hybrids were strongly advocated by Hans Bethe during the 1970s (Bethe 1979). Lawrence Lidsky, whom we have encountered as a critic of the tokamak magnetic confinement fusion option, was also a proponent of hybrids and neatly encapsulated the advantages and disadvantages of the two components of the hybrid as "fusion reactors are 'neutron rich' and 'power poor' while fission reactors are 'neutron poor' and 'power rich'" (Lidsky 1975).

Detailed studies of the economics of the systems carried out at the time however suggested that they could not compete with existing fission reactors. The idea was abandoned, at least in the West (we will come to speak about developments in the Soviet Union and Russia shortly), and lay dormant until the 2000s, when the continued delays in reaching breakeven for fusion led to a brief revival around 2009. Fusion's neutrons, it was thought, could almost completely burn up the radioactive waste produced in fission reactors, leaving a greatly reduced residue in both volume and radioactivity. So, why not combine the two into a fusion–fission hybrid reactor, and let each technology solve the problems of the other (Gerstner 2009)?

Fusion–fission designs essentially replace the lithium blanket that surrounds the vacuum vessel with a blanket of fission fuel, either natural uranium ore or even nuclear waste. The fusion core acts as a source of high-energy neutrons that can then be used for various applications in the fission blanket, depending on the design. The neutrons have more than enough energy to cause fission in U-238, as well as in many of the other elements in the fuel, including some of the transuranic waste elements. The applications are (1) the production of fissile fuel; (2) the production of energy; or (3) the management of nuclear waste, whereby this latter application is the most interesting.

When a neutron is captured by non-fissile material such as U-238 or Th-232, it ultimately produces fissile fuel in the form of Pu-239 or U-233. The fusion-fission hybrid would act here as a fast breeder reactor. Fast breeder reactors are reactors that breed fissile fuel with the help of fast neutrons. In the pure fission breeder reactors currently in use neutrons originating from fission convert non-fissile U-238 in a blanket surrounding the fission reactor into fissile Pu-239.

The second process of interest is nuclear fission whereby a neutron collides with the nucleus of some fissile material and splits this into several smaller pieces, generating energy that can be used to produce electricity.

But the most interesting application is the third one: the management of waste produced by nuclear fission reactors. When fuel is introduced into a pure fission light-water reactor (LWR) it consists of 4% U-235 (the fissile isotope) and 96% U-238. The fuel remains in the reactor for 3–4 years after which it must be replaced. The spent fuel contains approximately 92% U-238, 1% U-235, 1% Pu-239, 0.1% minor actinides (the 15 metallic chemical elements with atomic numbers from 89 to 103, actinium through lawrencium) and 5% fission by-products. A large fraction of this spent fuel has half-lives on the order of 30 years or less. Thus, after storing it for a time of 50–100 years a large fraction will have decayed into harmless elements, but unfortunately not all. The troublesome components consist of plutonium, some of the actinides and some fission by-products. This is a very small part of the total waste: a volume of about 4 L for a 1 GWe LWR reactor during one year of operation, but nonetheless a great nuisance and according to many essentially a showstopper for nuclear fission reactors. There are two approaches to further reduce this waste in a hybrid reactor. The first one is to reprocess it chemically and physically separate it into its separate components, which can then be bombarded by fusion neutrons and split into harmless elements, in addition to producing energy. The fission processes initiated by the neutron bombardments will however always produce some new actinides and although the volume of harmful actinides can be greatly reduced in this way, they can never be completely eliminated. Some waste will still have to be buried for a very long time, and there are several other reasons why this method is not the ideal one.

The second approach to deal with the long-lived actinides involves so-called deep burn. The fusion reactor in the core of the device is an independent neutron source. In ordinary fission reactors the fission processes supply the neutrons to keep the process going which requires a certain concentration of the correct fissile isotopes (typically U-235 or Pu-239) to be present in the reactor's fuel rods. In a hybrid configuration the neutrons come from the independent fusion core and the reaction can continue even when all of the U-235 has been burned off; the rate is controlled not by the neutrons from the fission events, but by the neutrons supplied by the fusion reactor. The fission fuel can be left in the reactor for a much longer period, up to twenty years, and the reactor can burn much longer. During burn the overall reactivity decreases and many of the undesirable actinides and fission products will be converted into harmless elements without any reprocessing being necessary. The waste that is left, comparable to the amount produced in the usual 3–4-year fuel lifetime, will contain much less long-lived isotopes (Freidberg 2009). The conversion of one chemical element or an isotope into another chemical element is called transmutation. It has the potential to make the nuclear waste from nuclear fission reactors more manageable by a considerable decrease in volume, but does not rule out that a deep geological repository (a radioactive waste repository excavated deep within a stable geologic environment) will still be needed (Stacey et al. 2002).

The snag in all these seemingly ideal scenarios is economics, as is often the case. A series of studies starting in the late 1970s provided a much clearer picture of the hybrid and indicated that there was no case for building a hybrid as the fusion component of such a system was technologically complex, scientifically risky and significantly more expensive than alternatives. All these three reasons seem still to be valid today. The fact is also that there are various other ways to deal with nuclear waste, e.g. non-breeding sodium-cooled fast-spectrum reactors (Generation IV fission reactors cooled by liquid metal, in this case sodium, which have the primary advantage of being weak neutron moderators, much weaker than water) or particle-accelerator-driven spallation hybrids. Both of these options are more developed than the fusion–fission hybrid. As must be clear from the previous chapters in this book, fusion, and thus the hybrid requires considerable additional research and development before it will be known if it could even work, and even if it were demonstrated to work, the end result would be a system essentially identical to ordinary breeder reactors, which have already been built for a long time. Moreover, from a purely economic point of view the best approach is not transmutation, but disposal of waste in a permanent repository, storage in an interim repository or burial in deep bore holes.

The investment of time and money required to commercialize the hybrid cycle can only be justified by a real or perceived advantage of the hybrid over the classical fast breeding reactor. Such an advantage would exist if fusion-fission hybrids were able to totally do away with long-lived isotopes, instead of just partially. That would still leave the problem of the radioactive waste of the fusion reactor, as discussed in Chap. 20, but might be sufficient progress to reconcile the public to nuclear energy from fusion *and* fission, working in harmony.

Hybrids have been designed with both inertial confinement fusion devices and tokamaks. The primary difference between an inertial confinement fusion neutron source and a magnetic confinement fusion source is that the ICF source is essentially a point-source of neutrons while tokamaks are more diffuse sources. In Chap. 12

the LIFE project at Lawrence Livermore National Laboratory has been discussed. It started life as a fusion-fission hybrid but was soon turned into a pure fusion project. When NIF failed to achieve ignition, the entire project was cancelled.

This leaves the Russian fission–fusion hybrid, called DEMO fusion neutron source, mentioned in Chap. 17, which is currently being designed at the Kurchatov Institute in Moscow. It is a conventional tokamak with a superconducting magnet system to harvest the fusion-produced neutrons to produce fissile materials and to destroy (break down) radioactive waste. Its primary aim is not to generate energy for the market, just 200 MW for its own use. It is supposed to be a successor to ITER, but construction is scheduled to start already in the early 2030s before ITER will start its D-T operations. It is the only hybrid fission–fusion system currently in an advanced drawing-board stage.

The interest in fusion-fission hybrids in Russia goes back to the time of the Soviet Union where this possibility was already mentioned in 1976 at the 25th Congress of the Communist Party of the Soviet Union. The T-10 tokamak (see Chaps. 4 and 5) was just in operation and it was casually mentioned that creating a reactor would not involve much more than increasing size: "Scientists also know how to increase the length of time the plasma energy is retained and, consequently, how to create a reactor. As theory predicts and experiments show, all it takes is to increase the device's dimensions." Nothing could be simpler, it seems, certainly for the Soviet Union which revelled in creating the biggest of this, that or the other, but nonetheless quite a few steps would be needed and in the meantime as the report of the 25th Congress announced: "[t]he following consideration should be taken into account in this connection: The energy from thermonuclear reactions is released in the form of a flow of high-energy neutrons. A question arises as to whether this quality could not be used more efficiently than by simply transforming the neutrons' energy into heat. For example, the reacting plasma might be surrounded by a layer of uranium in which the neutrons would produce nuclear fission, thereby increasing the amount of energy set free. Either natural uranium or even uranium poor in the more valuable U-235 isotope could be used. This would constitute a so-called hybrid, i.e., mixed, thermonuclear-atomic reactor in which the energy is provided by the uranium, and the thermonuclear portion serves only as a source of neutrons." (Velikhov and Kadomtsev 1976). Financial pressures experienced by fusion research in general prevented these plans from being carried out at the time, but under Evgeny Velikhov's leadership the notion of the hybrid was advanced to produce plutonium. It is of course shameful that scientists were involved in such odious plans, but military planners liked the possibility as it would provide yet another source of weapons fuel by producing large amounts of plutonium. The hybrid reactor proposed by Velikhov and his collaborators was a typical product of Soviet gigantomania. It was to produce 7000 MW of thermal energy with a 1000-ton uranium blanket, capable of producing 4200 kg of plutonium per year (Josephson 2000, p. 192). With the collapse of the Soviet Union these plans had to be shelved.

21.4   Actinide Burners

Various possibilities of incinerating or transmuting nuclear waste from fission reactors are at present being studied in various countries. Most of these studies are based on burning such waste using neutrons from fission reactors. There have also been theoretical studies involving the use of fusion reactors as so called "actinide burners" where a fusion reactor plasma, e.g. in a tokamak, could be "doped" with a small amount of the "minor" transuranic atoms, which would be transmuted to lighter elements upon successive bombardment by the neutrons produced in the fusion reactions. A study at MIT found that 2–3 fusion reactors with the general characteristics of ITER could transmute the entire annual minor actinide production from the present US fleet of light water reactors, removing a considerable part of the headache caused by long-lived radioactive waste from nuclear fission power plants. At the same time, the thermal power generated by fission of the minor actinides in each reactor will be sufficient to generate about 1000 MW of electricity.[7] For this of course a fusion reactor must first be made to work, but does not necessarily have to be a net energy producer. It could have an extremely useful life as an actinide burner.

21.5   Military Aspects

There are few, if any, outright military applications of fusion. That was one of the reasons that declassification of fusion research happened rather smoothly in 1958 (see Chap. 3). As related in Chap. 12, inertial confinement fusion has always been part of weapons research and consequently such research was excluded from the declassification. There is however not much reason to keep research on ICF secret, as it seems unlikely than ICF will ever result in any useful military hardware, but secrecy is a much-loved part of military culture and so it has partly remained classified research ever since.

The fusion-fission hybrid discussed above is however not without military application as it can be used to produce materials for building a nuclear bomb. Soviet fusion physicists, as we have seen above, advanced it as a means to produce plutonium for bombs, and Russia is possibly reviving it for precisely that purpose. Fortunately, the Soviet Union collapsed before this ill-begotten plan could be implemented, but the most disappointing aspect of the whole business is that the proposals came from scientists. On the other hand, if a country can build a fusion-fission hybrid it can also build a nuclear fission breeder reactor and obtain bomb-grade material in that way. The nuclear fusion component in the hybrid does not make it any easier.

It seems a little farfetched, but Lockheed Martin's compact fusion reactor, discussed in Chap. 14, might be capable of powering a Nimitz-class aircraft carrier, and could also be deployed, where needed, to locally provide cheap electricity at bases located in the US or on foreign territory, so it is claimed.

[7]MIT reports to the President 2000–2001, https://web.mit.edu/annualreports/pres01/13.07.html.

21.6 Involvement of Industry

In the early days of nuclear fusion until the time that the big tokamaks emerged, the involvement of industry in nuclear fusion was mostly as a manufacturer of products, equipment, or components according to the customer's exact specifications (so-called build-to-print products). Very few industrial corporations were directly involved in fusion research, the most auspicious exception being General Atomics. Its history has been described in Chap. 5, but also nowadays General Atomics is still very much involved in nuclear fusion and running the DIII-D tokamak. In this role it serves more as a 'laboratory' than as 'industry'. With its close ties to the Department of Energy to design, build, and conduct experiments while it also operates as a privately owned industry with a variety of commercial products, it has a unique position. The advantages for General Atomics were clear. For instance, the experience it gained during the Doublet tokamak experiments, paid for by the Department of Energy, and the subsequent development of superconducting magnet technologies for fusion, led to their current commercial manufacture of superconducting magnets for medical magnetic resonance imaging (MRI) systems. The technical demands of achieving 11 T fields in NbTi magnets with superfluid helium at $2°$ above absolute zero required General Atomics to develop design methods and technologies that were readily adapted to achieve the extreme precision needed for medical MRI magnet development and fabrication (Waganer et al. 2002). There are many examples of this symbiosis, which is of course rather obvious and true for many fields of science. Companies working for fusion experiments and developing special skills and technologies in the process will also use this expertise in developing other commercial projects. Nuclear fusion research is not at all special in this.

As the complexity of the fusion experiments increased (e.g. the big tokamaks, like TFTR) specific high technology industries were also drawn into these projects. Westinghouse's design expertise, for instance, played a role in the conceptual and preliminary design stages of TFTR, while in the later stages Ebasco, a designer and constructor of energy infrastructure, most notably nuclear fission power plants, provided architect and engineering support and Grumman, a leading producer of military and civilian aircraft (now Northrop Grumman after its merger with Northrop in 1994), provided engineering support for the power core. Various high technology companies, especially in aerospace supported the TFTR effort and joined the national fusion development team.

The role and involvement of industry further increased during the design and development of ITER, and especially during its construction. All ITER members have involved industry in their home countries in designing and constructing their part of ITER's components. It will be no surprise that many industries show an interest in the fat contracts that can be procured when working for an organisation like ITER. After all, during ITER's construction stage alone more than €20 billion will be spent, a large part in contracts with industry. In Germany, for instance, the German Industrial Forum for ITER (abbreviated dIIF (Deutsches ITER Industrie Forum)[8])

---

[8]https://www.diif.de/en/deutsches-iter-industrie-forum-e-v-.12.0.0.0.0.html.

was established to inform German companies on everything concerning the ITER project. The forum coordinates the companies' interests and their contributions to the project. Most countries in the European Union or associated with some of its programmes have established organisations that must help industry benefit from programmes like ITER (Bogusch et al. 2002). There is nothing wrong with that, of course, but it makes people and jobs dependent on such projects and creates a certain inertia against change. Everybody involved will continue to profit if the funds keep flowing.

21.7   Conclusion

From what has been discussed in this chapter it can be inferred that fusion reactors might still be useful, even if the production of commercial energy fails, although prospects are slim. The clean, safe, and efficient management of long-lived nuclear waste, such as minor actinides and some fission products, is one of the major issues in the nuclear field. As things stand at present, it does not seem to be possible to decarbonize electricity generation without a sizable contribution from nuclear fission power. If the waste problem could be solved by using neutrons generated by nuclear fusion for the transmutation of such waste, it might, in addition to inherently safe Generation IV reactors, erase the existing public-perception problem for nuclear fission power and convince the public that such power, in spite of the still existing dangers of freak accidents like the 2011 Fukushima disaster, is a safe enough option and in view of the impending disasters from climate change a risk well worth taking. It seems though that as regards waste the best that can be achieved is that high-level storage facilities can be reduced by a factor of 100, which is of course considerable progress, but not enough to convince the die-hard opponents, I am afraid.

Proponents of fusion-fission hybrids claim that such devices can make fission safer and are at the same time a faster route to operational fusion. It indeed seems that the obstinate fascination with fusion solely as a power generation option is a dead-end street and that such applications might present fusion with an escape route from an almost certain death. Sceptics argue though that such hybrid reactors would be extremely expensive and difficult to design, and would detract from the main objective of developing pure fusion, to which the reply can be that pure fusion is likely to be pure fantasy (Chen 2011, p. 168).

Summary and Final Conclusion

In the preceding chapters we have journeyed from the early days of nuclear fusion just after World War II all the way to the present day. In two introductory chapters (Chaps. 1 and 2) the basic science behind nuclear fusion has been introduced, including its connection with stellar processes and quantum mechanics, as well as the basic principles of plasma physics with special emphasis on plasmas in nuclear fusion devices and the conditions for fusion, such as the Lawson criterion.

Parts II and IV (Chaps. 3–17), interrupted by an intermezzo (Part III, Chap. 9) after the shutdown of JET to take stock of the developments up to that point and the prospects for the future, are devoted to a review of seventy years of efforts from the earliest attempts with pinch effects at Oxford and Imperial College in Britain in the 1940s until the latest devices like the International Thermonuclear Experimental Reactor (ITER) and plans for a follow-up. Almost from the very beginning it was realised that a configuration of magnetic fields offered the best chance of confining a plasma, with torus-like devices showing the most promise. Research was at first kept secret and only three countries, United States, Britain and the Soviet Union, had nuclear fusion research programmes. Nonetheless, already in 1955 predictions were made that commercial power generation from nuclear fusion would be realised in two decades. Such rash and unwarranted predictions have remained a constant factor, perhaps the only constant one, in the fusion saga ever since.

Multiple breakthroughs were announced over the years, but so far none has led to any really significant progress. The early history of nuclear fusion until about 1970 was shaped by two pivotal moments. The first one was the declassification set in motion by the second Geneva conference in 1958, the result of a lack of both success and military applications. It informed the world about the secret work carried out by the US, Great Britain and the Soviet Union since the end of World War II. These early developments have been described in some detail in Chap. 3.

The second pivotal moment was the apparent triumph of the Soviet designed tokamak in 1968 (Chap. 4), which suddenly after a time of frustrating deadlock seemed to provide a route forward and caused a veritable global stampede into tokamak research (Chaps. 5–7). So-called Bohm diffusion which had plagued all designs of nuclear fusion devices until this time was finally overcome and has now largely been forgotten. A large number of other countries (notably Japan and various European countries) got convinced that they also wanted to get involved out of fear of missing out on perhaps the greatest bonanza of all time: cheap and abundant energy. They all scrambled to develop their own programmes without any specific guidance or plans. Just do what the rest had already been doing for some time. Tokamak research in other Asian countries on any significant scale started a few decades later.

Attempts with other designs like stellarators, mirrors and various pinches were summarily (and probably foolishly) abandoned and since roughly 1970 all were steadfastly trotting in the same direction, almost like lemmings heading towards a cliff, ignoring failures on the way or declaring them triumphs, and just hoping for the best.

A further momentous development that took place around the same time due to the overly optimistic attitude after the 1968 tokamak results was the switch from a research-oriented approach to the goal-oriented approach of commercial power generation, i.e. from a scientific research project fusion was suddenly turned into an engineering project. It is clear that this switch came too early. Fusion was not ready for this. The behaviour of a plasma was (and still is) insufficiently understood to embark on such ventures. It was nevertheless done, but soon ran into new problems. This time with the heating of the plasma: confinement got worse when the plasma temperature was increased by external heating. Another breakthrough was needed, which came in 1982 when the so-called high-confinement mode was (accidentally) discovered on the ASDEX tokamak in Germany. This gave a renewed perspective to the big tokamaks, the European JET and the Japanese JT-60, which came online in the 1990s. They had already been planned and designed before the H-mode discovery was made on ASDEX but would have been doomed without it. The American TFTR, which could not access H-mode, found its own supershot regime which apparently was particular to this specific machine as it has never been reproduced on any other device. Although it was realised very early on in the game that plasmas of deuterium and tritium offered the best prospects for fusion, TFTR and JET, which operated during the 1990s, have so far been the only tokamaks that carried out a rather limited number of experiments with such plasmas. The main reason for this is that tritium is very unpleasant to work with. They also are the only tokamaks that managed to generate some fusion power, albeit much less power than needed to run the system (a paltry 2% in the case of JET and even less for TFTR). The story of the big tokamaks has been described in Chap. 8.

Then politics got involved and the design of a follow-up device became mixed up with détente politics between East and West and Gorbachev's vain attempts to save the Soviet Union. It is very questionable that from the results of TFTR and JET, which both were failures as they did not achieve scientific breakeven ($Q = 1$), a case for ITER can be made on grounds other than political ones. After much toing and froing, it came about in the form of an impossibly large international collaboration, spanning more than half the Earth, a voracious monster that gobbles up most of the world's public funds for nuclear fusion. Its story, which is still unfolding, has been related in Chap. 10. After a long gestation period from the early 1980s, ITER has been under construction in the south of France since 2013 with first plasma expected at the very end of 2025. The project has been dogged by huge cost overruns and delays. A few years before 2040 a deuterium-tritium plasma must then start to burn and prove that controlled nuclear fusion is in principle possible.

All this activity only relates to magnetic confinement fusion whereby magnetic fields are used to try to confine the plasma. An alternative and completely different approach is inertial confinement fusion (ICF), reviewed in Chap. 12, whereby laser or particle beams are used to unleash fusion reactions by compressing and imploding a small (spherical) target filled with deuterium-tritium fuel. Success, in spite of considerable investment, has so far been limited. ICF has always been part of weapons research, and with a few exceptions has been and still is mainly an American affair. Since the National Ignition Facility failed to reach ignition (i.e. produce a self-sustaining nuclear fusion reaction) by the scheduled 2012 date and is unlikely to do so in the near future, it is unlikely that ICF will soon be an energy generating option, at any rate this will not happen in this century.

As regards magnetic confinement fusion, due to the exceedingly long time of trial and error with tokamaks without any solid result to show for, doubt is now also slowly creeping in that standard tokamaks, like ITER and most of the other tokamaks built since the 1970s, are the right solution to the problem of generating energy from nuclear fusion. Some are now putting their money on spherical tokamaks (Chap. 11), are reviving older approaches (Chap. 13), of which the stellarator seems to stand the best chance, or relying on non-mainstream approaches (Chap. 14) that were already discarded in the past. The recent flow of venture capital has brought a return of the almost frivolous atmosphere of the 1950s when the most outlandish proposals managed to get funded. Chapter 15 reviews this private investment.

Chapter 16 of the book, before embarking in Chap. 17 on a discussion of the rather premature plans that are being made all over the world for DEMOs and fusion reactors, contains a discussion of the challenging engineering and materials issues fusion is faced with and have to be solved before any of these plans can become reality. It involves issues that can make or break the future for fusion. In spite of all the hullabaloo and the brave optimistic face that is put up by the fusion proponents, nothing is certain yet, far from it. New materials have to be developed that can withstand the onslaught of the 14 MeV neutrons produced in the fusion reactions, but especially the self-sufficiency of tritium is an issue that may well be insoluble. Tritium is very rare on Earth and the breeding of sufficient tritium in a blanket surrounding the reaction chamber is a must for fusion with deuterium-tritium plasmas to succeed.

In the final part, Part V, of the book we discuss a variety of issues that get little attention as they are normally drowned out by the noise made by fusion enthusiasts and their supporters in the main press, which appears to be very reluctant to assess the fusion claims on their merits. We start in Chap. 18 with an overview of the criticism of the fusion enterprise that is becoming ever louder, followed in Chap. 19 by the shaky economics and doubtful sustainability of fusion, and in Chap. 20 by the equally not so rosy safety and environmental picture. Fusion is claimed to be safe and environmentally friendly, a claim that is built on shaky grounds and blatantly untrue as regards the huge volumes of radioactive waste produced by fusion reactors. If fusion ever starts to generate power in quantities that meet a significant part of the electricity needs of the world, thousands and thousands of tonnes of radioactive waste will have to be stored for up to 100 years. Since the activation of parts of the reactor by the highly energetic neutrons is an ongoing process, such waste will be produced year after year, not just at the end of the lifetime of the reactor. Fusion proponents tend to hide this issue behind the irrelevant comparison with the radioactive waste of nuclear fission power plants. And indeed, the waste is admittedly less dangerous than some of the waste from nuclear fission power stations but is nevertheless very real and a problem the general public is probably not at all aware of. The comparison with the waste from nuclear fission stations is just an attempt to be little this problem: "if you wish to make grey look white, put it against black".[1] None of the other ways of generating energy has any nuclear waste problem.

This book's central thesis, if there is one, is that there is no chance whatsoever that in this century nuclear fusion will make a contribution of any significance to the carbon-free energy mix, implying that it will not play a role in the urgent decarbonization of energy production, in spite of the claims to the contrary made by the fusion community. No commercially viable nuclear fusion power station will be working in this century, and probably never will, quite a climbdown from its original boast of "too cheap to meter", and in spite of all the "breakthroughs" in the time between. With every breakthrough the generation of energy by nuclear fusion, cheap or expensive, was delayed by a further year or two. In addition, the expected price of the energy went up, quite contrary to the ordinary behaviour of the price of a commodity when a breakthrough is achieved. It is clear that the emperor has no clothes and never had any. This conclusion stands, even if everything goes according to plan, which it never does (just think about the rather simple mishap of a magnet shorting out with the upgrade of the NSTX in 2016, a problem that now, four years later, has still not been sorted).

The main conclusion of the book can be summarised as follows:

"If ITER fulfils all its promises, meaning that it:

shows that energy production by controlled nuclear fusion is *in principle* possible (i.e. that it succeeds in achieving $Q = 10$, without yet producing any net energy as more energy will be put in than comes out);

shows that tritium production in a lithium or other blanket is possible;

shows that this tritium can be collected in sufficient quantities;

shows that disruptions and instabilities can be kept at bay;

shows that the structure of the facility can withstand the onslaught of the neutrons (not such a problem at ITER, but a very real one for any follow-up device);

---

[1]Garton Ash (1997, p. 239), in a wholly different context.

then a still bigger DEMO device must first be built in which all this experience will have to be incorporated, resulting in some net energy production, plus a surplus amount of tritium (self-sufficiency of tritium is a must and nobody has so far shown in any convincing way that this is possible). The DEMO must also show that the materials used, some of them still to be developed, will not too easily be activated.

Such a DEMO can be built at the earliest after 2040, will need a further twenty odd years to show its viability and resolve the problems and difficulties that will undoubtedly arise, before the construction of the first pilot power plant can be contemplated, constructed and set to work. Results from DEMO cannot be expected before the 2060s, after which this pilot plant will have to show that commercial energy production from fusion is indeed possible. This will take us at least into the 2080s, if not a considerable time later still, after which the construction of real commercial power stations can possibly start. As discussed in this book, hundreds, if not thousands, of such power stations are needed for nuclear fusion to make a sizable contribution to energy generation. The scenario in Chap. 19 requires 2760 fusion power plants to provide 30% of baseload electricity by the end of this century; so, a paltry 3% would already require 276 power plants! There is simply not enough time to build a sizable number of power plants in the at most two decades left for this.

The experience laid out in this book shows beyond any reasonable doubt that all the remaining eighty years of this century will be far too short for all this to happen.

Moreover, when at the end of the century we take stock of all these 'success' stories, it will turn out that a nuclear fusion power station will be hideously complex, hugely expensive and extremely unreliable, and can never compete with any of the other carbon-free or low-carbon options that will then be at a mature stage. None of the forms in which nuclear fusion is presently presented to the world will contribute to a solution of the climate change problems mankind is faced with. Also then, in the twenty-second century and beyond, there will be little prospect for nuclear fusion as an energy generating option. Nothing, apart from a miracle, can change this."

Although there is no possibility that energy from nuclear fusion will make a tangible contribution to electricity generation in this century, the proponents of nuclear fusion nonetheless steadfastly try to fool us in believing that this could be the case, but they are actually fooling themselves. As Richard Feynman said "the first principle is that you must not fool yourself–and you are the easiest person to fool." The history of fusion shows that the human capacity for self-deception knows no bounds.

The question is whether they have crossed the red line from foolishness to fraud. Is fusion now at a stage where they are trying to deliberately fool the entire world? Does nuclear fusion deserve a place in the gallery of infinite-energy fraudsters and pranksters portrayed so eloquently and humorously, but also with repressed anger by Robert Park in his book *Voodoo Science* (Park (2000))? The ones Park describes are rather innocent, misguided individuals who were at first genuinely convinced they had hit a hitherto unknown jackpot. ITER and everything around it is much more serious and much more costly. If you are watching the videos of Europe's Fusion for

Energy Organisation on YouTube[2] with the solemn voice of the speaker spelling out the huge numbers involved in the ITER exercise, the thousands of people working on its 24,000 tons, three times the Eifel tower, the many references to the power of the Sun, while giving you a close look at the advanced scientific process of … pouring concrete, and a tokamak building which like a true Tower of Babel is partly shrouded in swirling clouds, against the backdrop of slowly swelling almost liturgical music, you get the impression of attending a religious service, where man is in awe of and worshipping his own creation, like the Jewish people on their flight from Egypt worshipping the golden calf in the desert in Moses's absence on Mount Sinai. Only now no Moses will be coming down to command them to mend their ways. The video is clearly meant to impress the public and has hardly any further content, just empty rhetoric, enumerating numbers to show off the massiveness of the project, the larger the better, gigantomania at its peak; numbers the ITER organisation foolishly seems to be excessively proud of. The connection with science is lost. Also here, it is apt to quote Richard Feynman who said that "for a successful technology, reality must take precedence over public relations, for Nature cannot be fooled".

With this I do not mean to say that the research and efforts into developing nuclear fusion are a complete waste of money, time and energy. It is often brilliant research and much of it should be continued, but no longer directed at building commercial power stations for generating electricity. In the early 1970s the original research programs oriented at plasma physics and at understanding its intricacies, were hijacked by pragmatists with too little understanding of science and of how science works and were forcefully bent into a vain march towards a perceived cornucopia of unlimited, cheap energy, heaping hype upon hype, declaring breakthrough after breakthrough without encountering much criticism or in-depth questioning from the media or from fellow scientists (and those who vented criticism were beaten into silence). By now it has become a juggernaut that thunders on and will only be stopped with a hard landing of ITER around 2035–2040. It also has captured a (too) large part of the science budgets in many countries, especially in the European Union with its huge contribution to ITER, leaving less and less for others. Many people and industries are partly dependent on the fusion fleshpots, will resist any change and will continue to argue that commercial energy production by nuclear fusion is just around the corner. Well, it isn't and will never be.

---

[2] A beautiful example is https://www.youtube.com/watch?v=zqEkkN0f59E, accessed August 2020.

Glossary

Ablation the removal of material from a solid surface by evaporation, erosion or other means, usually by very high-power heating. Occurs in inertial confinement fusion when the outer layer of the fusion fuel pellet is heated by the laser(s).

Ablator Outermost layer of the target capsule used in inertial confinement fusion that is rapidly heated and vaporized, compressing the rest of the target.

Adiabatic compression heating of the plasma by compression without heat exchange in a magnetic field that rises on a timescale that is short compared with the energy-loss time, but not so short as to induce shock waves.

Alpha particle nucleus of a helium atom; emitted in the radioactive decay of heavy elements (alpha radiation), and the stable non-radioactive end product of many fusion reactions.

ARIES a series of design studies for fusion power plants by a consortium of US university and national laboratories and private enterprise (Boeing and General Atomics).

Aspect ratio ratio of the major and minor axis of the torus of a magnetic confinement device like a tokamak or stellarator.

Atomic mass the mass of an atom, indicated as A, relative to the standard of the atom of carbon, defined as 12 units.

Atomic number the atomic number, indicated as Z, of an element is equal to the number of protons found in the nucleus of the atom of such element. It is identical to the charge number of the element. It uniquely identifies an element.

Auxiliary heating the application of neutral particle beams and/or high-frequency microwave radiation to the plasma from external sources, in order to provide the input heating power necessary to reach the temperatures required for fusion. Additional heating bridges the gap between resistive (or ohmic) heating due to the plasma toroidal current (which gets weaker with increased temperature) and alpha-particle heating due to the slowing down of the helium reaction products in the plasma (which gets stronger with higher temperature).

Axisymmetric(al) refers to an object having cylindrical symmetry, or axisymmetry (i.e. rotational symmetry with respect to a central axis).

Banana orbit in tokamaks and similar toroidal arrangements the magnetic field is stronger on the inside curve than the outside simply due to the magnets being closer together in that area. To even out these forces, the field as a whole is twisted into a helix, so that the particles alternately move from the inside to the outside of the reactor. As the particle transits from the outside to the inside, it sees an increasing magnetic force. If the particle energy is low, this increasing field may cause the particle to reverse direction, as in a magnetic mirror. The particle now travels in the reverse direction through the reactor, to the outside and then towards the inside where the same process occurs. This leads to a population of particles bouncing back and forth between two points, tracing out a path that looks like a banana from above, the so-called banana orbits.

Beta the *beta* value of a fusion plasma is defined as the ratio of the thermal plasma pressure to the pressure of the externally applied magnetic field. Since for fusion the plasma pressure has to be kept as high as possible and for cost reasons the external magnetic field should be as small as possible, a high *beta* value is desirable.

Blanket the blanket covers the interior surfaces of the vacuum vessel, providing shielding to the vessel and the superconducting magnets from the heat and neutron flux of the fusion reactions. The neutrons are slowed down in the blanket, where their kinetic energy is transformed into heat energy and collected by the coolants. In a fusion power plant, this energy will be used for electric power production. In ITER, some of the 440 individual blanket modules will be used to test materials for tritium-breeding concepts.

Bolometer a device for measuring the power of incident electromagnetic radiation via the heating of a material with a temperature-dependent electrical resistance.

Bootstrap current a self-generated current that spontaneously arises due to collisions between trapped particles and passing particles within the plasma confined in a toroidal fusion device, if the gas pressure of the plasma varies across the radius of the cylinder. It is commonly found in tokamaks. The tokamak uses a combination of external magnets and a current driven in the plasma to create a stable confinement system. One goal of advanced tokamak designs is to maximize the bootstrap current, and thereby reduce or eliminate the need for an external current driver. This could dramatically reduce the cost and complexity of the device.

Breakeven the situation in which the power released by nuclear fusion reactions is equal to the external power used in heating the plasma. See also **scientific breakeven**.

Bremsstrahlung electromagnetic radiation emitted by a plasma due to the acceleration of charged particles or their deceleration when deflected by other charged particles.

Capacitor an electrical device that is designed to store an electric charge and release it instantaneously. It is built up of two electrical conductors with a relatively large surface area that are close to each other and separated by a non-conducting material or vacuum, the dielectric medium. When two conductors experience a potential difference, for example, when a capacitor is attached across a battery, an electric field develops across the dielectric, causing a net positive charge to collect on one plate and net negative charge to collect on the other plate. The capacitor was originally known as a condenser or condensator, a name still widely used in many languages but rarely in English.

Capsule small pellet of fusion fuel, usually frozen Deuterium-tritium enclosed in a plastic shell, for inertial confinement fusion.

Central solenoid part of the magnet system in a tokamak; the central solenoid acts like a large transformer, producing and sustaining the plasma current which heats and shapes the plasma.

Charge reservoirs in superconductors, charge reservoirs are the layers that may control the oxidation state of adjacent superconducting planes (even though they themselves are not superconducting). In the layered cuprates, these consist of copper-oxide chains.

Compact torus a class of toroidal plasma configurations that are self-stable, and whose configuration does not require magnet coils running through the centre of the toroid. The lack of complex magnets and a simple geometry may allow the construction of dramatically simpler and less expensive fusion reactors. The two best studied compact torus systems are the spheromak and field-reversed configuration (FRC).

Conductivity the ability of a substance to conduct heat or electricity. Plasmas are exceptionally good conductors of electricity.

Confinement time the characteristic time of energy loss from a fusion plasma; for a plasma in thermal equilibrium defined as the total energy content divided by the total energy loss.

Convection the transfer of heat by motion of the hotter parts of a fluid to the colder parts.

Correction coils coils whose purpose is to compensate small errors in the confining magnetic field arising from fabricating misalignments.

Critical current (solid-state physics) the current in a superconductive material above which the material is normal and below which the material is superconducting, at a specified temperature and in the absence of external magnetic fields.

Critical density the density above which an electromagnetic wave (e.g. from a laser) cannot propagate through a plasma because the wave is reflected or absorbed.

Cryostat a vacuum vessel built around a superconducting tokamak, capable of being evacuated at room temperature, which provides thermal insulation to maintain the magnets at low temperature.

Current drive a means of producing the toroidal plasma current.

Current limit the maximum current allowed in a tokamak, determined mainly by the magnetic field and the plasma dimensions.

Cyclotron frequency the rotational frequency of a charged particle (i.e. the number of cycles a particle completes per second around its circular circuit) moving perpendicular to the direction of a uniform magnetic field. The cyclotron frequency is independent of the radius and velocity and therefore independent of the particle's kinetic energy; all particles with the same charge-to-mass ratio rotate around magnetic field lines with the same frequency.

Cyclotron motion the circular motion of a charged particle in a uniform magnetic field, as described in a cyclotron, a particle accelerator that accelerates charged particles outwards from the centre along a spiral path. The particle will emit so-called cyclotron radiation when describing such motion.

Diode-pumped lasers lasers which contain laser diodes illuminating a solid gain medium (such as a crystal or glass).

Dipole magnet magnet with two poles (north and south poles) that form a closed field loop. The simplest example of a dipole magnet is a bar magnet.

Direct drive inertial confinement fusion technique whereby the driver energy strikes the fuel capsule directly.

Divertor a device within a tokamak that allows the online removal of waste material from the plasma while the reactor is still operating. It deflects magnetic field lines at the periphery of the plasma chamber so that particle and energy fluxes can be controlled. It is usually located at the bottom of the torus. The tiles in the divertor have to withstand the highest heat loads. For JET the tiles are made of tungsten, also called wolfram, which is very robust and has the highest melting point ($3422\,^0C$) of all elements discovered.

Driver mechanism which delivers energy to the fuel capsule in inertial confinement fusion. Typical techniques use lasers, heavy-ion beams, and Z-pinches.

Edge Localised Mode (ELM) a relaxation instability of the steep edge density gradient in the H-mode of a tokamak. Regular, energetic bursts of energy and particles that escape from the magnetic field surrounding the plasma and cause loss of energy. It is a periodic distortion of the plasma boundary which rotates with the velocity of several kilometres per second and exists for about half a millisecond. The mitigation of this phenomenon is an important preoccupation of tokamak physicists.

Effective temperature of a body such as a star or planet is the temperature of a black body that would emit the same total amount of electromagnetic radiation.

Electron particle with one unit of negative electric charge, its mass is 1/1836 of the mass of a **proton**. It is not affected by the strong nuclear force.

Electron Bernstein waves a type of high-frequency electron cyclotron waves, which are electromagnetic waves with a frequency in the range of the electron **cyclotron frequency**. EBW are radio-frequency waves at one to three times the frequency with which electrons gyrate around the magnetic field in a magnetized 'hot' plasma and can be used for heating the plasma.

Electron cyclotron resonance heating (ECRH) an external mode of heating the plasma through resonant absorption of energy by introducing electromagnetic waves into the plasma at the cyclotron frequency of electrons. The frequencies are in the gigahertz range (170 GHz in ITER), for which huge microwave generators are needed. They do not need an antenna inside the vacuum chamber but must be launched from the cramped space also occupied by the central solenoid and the inside legs of the toroidal magnets.

Engineering breakeven in (inertial confinement) fusion the situation in which the driver energy output equals the fusion energy production, so the latter equals the energy the driver requires. Engineering breakeven is therefore harder to meet than scientific breakeven.

EUROfusion the consortium of national fusion research institutes in the European Union, Switzerland and Ukraine, established in 2014 to succeed the European Fusion Development Agreement (EFDA) as the umbrella organisation of Europe's fusion research laboratories.

eV or electron volt is the amount of energy gained by a single electron moving across an electric potential difference of one volt. It is equal to 1.6×10^{-19} J and equivalent to $12,000°$.

Extrapolated breakeven (scientific) breakeven projected for the actual reactor fuel (e.g. deuterium and tritium) from experimental results using an alternative fuel (e.g. deuterium only) by scaling the reaction rates for the two fuels; slightly higher than **(scientific) breakeven**.

Fast ignition inertial confinement fusion technique whereby the driver gradually compresses the fuel capsule, followed by a high-intensity, ultrashort-pulse laser that strikes the fuel to trigger ignition.

Field-reversed configuration a magnetic confinement concept with no toroidal field, in which the plasma is essentially cylindrical in shape.

First wall first surface of the fusion reactor chamber encountered by radiation and/or debris emitted from the fusion reactions and in inertial confinement fusion from the target implosion. These walls may vary in composition and execution such as dry, wetted, or liquid jets.

Fuel capsule see capsule.

Fusion energy gain factor expressed by the symbol Q, is the ratio of fusion power produced in a nuclear fusion reactor to the power required to maintain the plasma in steady state. The condition of $Q = 1$, when the power released by the fusion reactions is equal to the required heating power, is referred to as **breakeven**, or in some sources, **scientific breakeven**.

Fusion triple product the "triple product" of density, energy confinement time and plasma temperature is used by researchers to measure the performance of a fusion plasma. The triple product has seen an increase of a factor of 10,000 in the last thirty years of fusion experimentation; another factor of six is needed to arrive at the level of performance required for a power plant.

Gas puffing fuelling the outer regions of the plasma by the controlled injection of additional gas, puffs of fuel or impurity gas from valves during the discharges into the **scrape-off layer** (SOL). Fuel pellets are used to fuel deeper into the plasma.

Heavy-ion fusion inertial confinement fusion technique whereby ions of heavy elements are accelerated and directed onto a target.

High-harmonic fast-waves radio-frequency waves at many times the frequency with which ions gyrate around the magnetic field, which can heat electrons and, in theory, drive a plasma current.

H-mode or high-confinement regime a regime of improved confinement in toka-maks characterised by steep density and temperature gradients at the plasma edge. **L-mode** denotes the normal confinement in tokamaks.

Hohlraum hollow container (radiation case) in which an ICF target may be placed, the walls of which are used to reradiate incident energy to drive the fuel capsule's implosion in ICF.

IFMIF-DONES and A-FNS the International Fusion Materials Irradiation Facility (IFMIF) is a proposed device that shall test and validate the structural integrity of fusion power plant materials under appropriate neutron spectrum and fusion irra-diation damage conditions. The detailed design and prototyping are being under-taken by Europe and Japan as a Broader Approach project. DONES (DEMO oriented Neutron Source) and A-FNS (Advanced Fusion Neutron Source) are rather similar, reduced-scope fusion materials test facilities with the potential to be upgraded to "full" IFMIF.

Ignition in magnetic confinement, the condition where the α-particle heating (20% of the fusion energy output) balances the energy losses, and the nuclear fusion reaction becomes self-sustaining; at ignition, fusion self-heating is sufficient to compensate for all energy losses; external sources of heating power are no longer necessary to sustain the reaction. In inertial confinement, it is the point at which the core of a compressed fuel capsule starts to burn. See **scientific breakeven**.

Indirect drive inertial confinement fusion technique whereby the driver energy strikes the fuel capsule indirectly—for example, by the X-rays produced by heating the high-Z enclosure (hohlraum) that surrounds the fuel capsule.

Inertia the resistance of any physical object to any change in its motion. This includes changes to the object's speed, or direction of motion. An aspect of this property is the tendency of objects to keep moving in a straight line at a constant speed, when no forces act upon them.

Inertial confinement fusion concept in which a driver delivers energy to the outer surface of a fuel capsule (typically containing a mixture of deuterium and tritium), heating and compressing it. The heating and compression then initiate a fusion chain reaction.

Ion Bernstein Waves type of high-frequency ion cyclotron waves, with somewhat similar properties to **electron Bernstein waves**. However, while the ion contri-bution to the dispersion relation can be neglected for high-frequency waves, the electron contribution cannot be neglected for low frequencies, so there is not a complete symmetry between the two types of Bernstein waves.

Ion cyclotron resonance heating (ICRH) an external mode of heating the plasma through resonant absorption of energy by introducing electromagnetic waves into the plasma at the cyclotron frequency of the ions. It is in the MHz range (around 50 MHz in ITER) and depends on the magnetic field strength. An antenna has to be placed inside the vacuum chamber for this, so close to the plasma that it will be bombarded by ions and antenna material will sputter into and contaminate the plasma.

Isotopes variants of a particular element that have the same chemical properties (and the same atomic number), but differ in the number of neutrons in the nucleus;

for hydrogen the most common isotope has a nucleus of only one proton and no neutrons, deuterium has an extra neutron in its nucleus, and tritium has two neutrons.

ITER Council the supervising body of the ITER Organization, responsible for the promotion and overall direction of the ITER Organization. It appoints the Director-General and senior staff, adopts and amends the Project Resources Management and Human Resources Regulations, and approves the annual budget of the ITER Organization. It also decides the total budget for the ITER Project and the participation of additional states or organizations in the project. It meets at least twice a year. Representatives of the seven Members sit on the ITER Council.

Kinetic ballooning modes (KBM) a type of internal pressure-driven driven plasma instability. The name refers to the shape and action of the instability, which acts like the elongations formed in a long balloon when it is squeezed.

Krypton fluoride (KrF) laser gas laser that operates in the ultraviolet at 248 nm.

Laser acronym for light amplification by stimulated emission of radiation, a light source that can produce a small coherent intense bundle of light. If the electrons in special atoms in glasses, crystals, or gases (the 'gain medium') are energized into excited (metastable) atomic states (pumped) by an external energy source e.g. a flashlamp, they will, when falling back to the lower energy level, emit photons. The photons will all be at the same wavelength and will also be "coherent," meaning the light wave's crests and troughs are all in lockstep. In contrast, ordinary visible light comes in multiple wavelengths and is not coherent. The first laser was built in 1960.

Lawson criterion the minimum value, at the ignition temperature, of the product of energy confinement time and density that is required, such that the power produced by the fusion processes in the plasma exceeds the energy losses of the plasma (hence to achieve breakeven). For fusion plasma with a temperature of 10 keV (about 120 million degrees) the Lawson criterion demands that density times confinement time is larger than 10^{14} s/cm^3. Equivalent to the **(fusion) triple product**.

Limiter plasma configuration in which the plasma is bounded by a material limiter inside the vessel. Here magnetic field lines from the hot plasma impinge directly on the limiter. The limiter configuration has become less important because the divertor configuration has proved to be more favourable for good plasma confinement .

Liquid wall fusion reactor chamber's first wall that features thick jets of liquid coolant. This design may also shield the solid chamber walls from neutron damage.

L-mode or low-confinement mode ordinary confinement regime in tokamaks.

Lorentz force the combination of electric and magnetic force on a point charge due to electromagnetic fields: $F = q(\mathrm{E} + v \times B)$, with q the charge of the particle and v its velocity. E and B are the electric, respectively the magnetic field.

Loss cone the set of angles at which a particle will no longer be trapped in a magnetic field while particles with pitch angles outside the loss cone will be reflected and continue to be trapped (magnetic mirror device).

Lower Hybrid Current Drive a technique to drive a non-inductive current in a toroidal plasma using microwave radiation at the lower hybrid resonance frequency, which lies between the cyclotron frequencies of the ions and electrons (about 5 GHz in ITER). Klystrons are used to generate frequencies in this range.

Magnetic flux tube a region of space bounded by a flux surface, i.e. a surface such that the magnetic field is everywhere perpendicular to the normal to the surface, hence the magnetic field does not cross the surface anywhere.

Magnetic island a localised region of the magnetic field where the field lines form closed loops unconnected with the rest of the field lines.

Magnetic mirror machine open-ended system where the plasma is confined in a solenoidal magnetic field, the strength of which increases at the ends: the "magnetic mirrors", which reflect the charged particles.

Magnetic perturbation an externally-induced small perturbation of the equilibrium magnetic field, such that it is resonant with the field at a given magnetic flux surface, usually located in the plasma edge region, with the goal to stabilize a specific, mainly ELM, instability.

Magnetic reconnection a fundamental process that converts the magnetic energy of reconnecting fields to kinetic and thermal energy of plasma through the breaking and topological rearrangement of magnetic field lines. It is most prominent in the form of solar flares on the surface of the sun, and also powers the northern lights, gamma-ray bursts, and other violent natural phenomena. It can also occur in magnetically confined plasmas where it comes into play with **magnetic islands**.

Magnetized target fusion inertial confinement fusion technique whereby a magnetic field is created surrounding the target; the field is then imploded around the target, initiating fusion reactions.

Merging compression a technique for starting up the plasma in a tokamak without use of a central solenoid. Two symmetric magnet rings are constructed inside the torus chamber. Running high currents through these magnetic coils creates two rings of plasma around them, and as the coil current is reduced to zero the plasma rings attract and combine. When the rings combine, their magnetic fields reconfigure in a process called **magnetic reconnection**. Stretched field lines break and release huge amounts of energy. This energy heats the plasma.

Micro-tearing mode (MTM) instability that produces magnetic reconnection (into a magnetic island) in a spherical tokamak with length scales of a few ion Larmor radii perpendicular to the magnetic field lines.

Neoclassical tearing mode (NTM) a metastable mode. In certain plasma configurations, a sufficiently large deformation of the bootstrap current produced by a "seed island" can contribute to the growth of the island. The NTM is an important performance-limiting factor in many tokamak experiments, leading to degraded confinement or disruption. The stability of NTMs is a key issue for magnetic configurations with a strong bootstrap current.

Neodymium a rare-earth metal with atomic number 60 belonging to the lanthanides. Neodymium-doped crystals can generate high-powered infrared laser beams.

Neodymium glass laser neodymium is a rare-earth metal with atomic number 60 belonging to the lanthanides. Certain transparent materials (like glass) with a small concentration of neodymium ions can be used in lasers as gain media (see **laser**) for infrared wavelengths (1054–1064 nm). Neodymium-doped crystals generate high-powered infrared laser beams. Neodymium glass lasers are used in extremely high power (terawatt scale), high energy (megajoules) multiple beam systems for ICF.

Neutral-beam heating a method to heat a plasma beyond ohmic heating inside a fusion device by injecting a beam of highly energetic neutral particles into the core of the plasma, where they become ionized.

Neutron activation the process in which neutrons induce radioactivity in materials. It occurs when atomic nuclei capture free neutrons. All naturally occurring materials, including air, water, and soil, can be induced (activated) by neutron capture to become radioactive to some degree.

Neutron wall loading energy flux carried by fusion neutrons into the first wall of the vessel surrounding the plasma.

Neutron constituent particle of atomic nuclei with no electrical charge and mass similar to that of a proton.

Nuclear fission the process in which a neutron strikes a nucleus, e.g. of uranium, and splits it into two or more lighter nuclei whose combined mass is less than the mass of the initial nucleus. In the fission process the missing mass is converted into energy (in the form of high-speed neutrons), which in a nuclear reactor is used to generate electricity.

Nuclear fusion the process in which the nuclei of light elements combine, or fuse, to form heavier nuclei, releasing energy.

Ohmic heating heating of a plasma by means of an electric current, making use of the ohmic resistance of the plasma. This resistance is described by the **Spitzer conductivity**, which decreases as the temperature increases, thus reducing the efficiency of ohmic heating at high temperatures.

Overnight capital cost a term used in the power generation industry to provide a simplistic comparison of the costs of building power plants. It is usually computed by dividing the overnight cost of building the plant by the maximum instantaneous power the plant can deliver, whereby overnight cost is the cost of a construction project if no interest was incurred during construction, as if the project was completed "overnight." Overnight capital cost does not take into account the life span of a plant or its key components; the capacity factor, i.e. the ratio between the effective mean power (actually delivered through the year) and the maximum power (maybe reached only a few hours per year); and the financing costs, hence it is not an actual estimate of construction costs.

Pedestal region a narrow transport barrier formed in the outer few percent of a confined plasma in the transition from **L-mode** to **H-mode**. Predicted fusion power scales with the square of the pressure at the top of the pedestal.

Photovoltaics the conversion of light into electricity using semiconducting materials that exhibit the photovoltaic effect, i.e. the generation of voltage and electric current by the absorption of light which excites an electron or other charge carrier into an excited state. This technique is applied in solar cells in solar panels, which are therefore also called photovoltaic of PV modules. An alternative technology is concentrated solar power (CSP) whereby solar power is generated by using mirrors or lenses to concentrate a large area of sunlight onto a receiver.

Pinch effect the compression of a plasma due to the inward radial force of the azimuthal magnetic field associated with a longitudinal current in a plasma.

Pitch angle angle between the velocity vector of a particle and the local magnetic field.

Plasmoid a coherent structure of plasma and magnetic fields (a plasma-magnetic entity). They have been proposed to explain natural phenomena such as ball lightning, magnetic bubbles in the magnetosphere, and objects in comet tails, in the solar wind, in the solar atmosphere, and in the heliospheric current sheet. Plasmoids produced in the laboratory include field-reversed configurations, spheromaks and dense plasma focuses .

Poloidal field magnetic field generated by the toroidal current (current flowing around in the torus) flowing in the plasma.

Proton constituent particle of atomic nuclei with one unit of positive electrical charge and mass similar to that of the neutron.

Pulsed-power fusion inertial confinement fusion technique that uses a large electrical current to magnetically implode a target.

Pump-out the phenomenon that particles drift to the edge of the confinement vessel faster than predicted by theory and by doing so take heat out of the plasma. It was a problem that could not be solved in the Model-C stellarator at Princeton and one of the reasons of its early demise.

Pyrotron in a pyrotron device a relatively low-energy plasma is injected into the straight discharge chamber at a time when the confining field is low. The field is then increased in magnitude at a reasonably slow rate (i.e., milliseconds, instead of microseconds), thereby trapping the particles between the magnetic mirrors and heating them by adiabatic compression. At the end of the heating operation the plasma has become sufficiently hot to generate X-rays, strong microwave noise signals and other high temperature phenomena. Under the proper conditions it can be shown that heated components of the plasma remain stably confined for many milliseconds, a time which increases with the degree of heating.

Q-value the ratio of the power released by the fusion reactions to the net heating power put into the plasma in a fusion device; $Q = 1$ is referred to as breakeven.

Radio-frequency heating a method used for plasma heating in a nuclear fusion device, whereby waves tuned to various plasma resonance frequencies of the ions and electrons are sent into the plasma. These are absorbed and energy can be transferred to the charged particles in the plasma, which then collide with other plasma particles and increase the temperature of the plasma as a whole.

Reversed field pinch (RFP) a closed magnetic confinement concept having toroidal and poloidal magnetic fields that are approximately equal in strength, and in which the direction of the toroidal field at the outside of the plasma is opposite from the direction at the plasma centre.

Safety factor a measure of the helical twist of magnetic field lines in a torus (tokamak), denoted by the symbol q, the ratio of the number of times a toroidal field line goes around the long way in the torus to the number of times a poloidal field line goes around the short way. Suppression of the kink instability requires that $q > 1$ (the Kruskal-Shafranov limit). In practice q must be greater than about 3 at the plasma edge for stability against disruptions.

Scientific breakeven in inertial confinement fusion the situation in which the energy produced by the fusion products equals the energy delivered by the laser to the target. In magnetic confinement fusion it means that the fusion power produced in a plasma matches the external heating power applied to the plasma to sustain it. To achieve this is the main goal of present-day fusion research.

Scientific feasibility producing and containing a high temperature plasma that exceeds the Lawson criterion.

Scrape-off layer (SOL) the outer edge of the plasma, characterised by open magnetic field lines that guide particles and energy from the core plasma into the divertor. In divertor plasmas, the SOL absorbs most of the plasma exhaust (particles and heat) and transports it along the field lines to the divertor plates.

Shock ignition inertial confinement fusion technique that uses hydrodynamic shocks to ignite the compressed hot spot.

Spallation a process in which fragments of material (*spall* are flakes of a material that are broken off a larger solid body) are ejected from a body due to impact or stress. In this connection a spallation neutron source is an accelerator-based neutron source facility whereby high energy protons strike a target of some liquid metal (e.g. mercury), where spallation occurs in the form of neutrons. The spalled neutrons are then slowed down and can be used in a wide variety of experiments.

Spheromak a compact toroidal magnetic confinement concept in which a large fraction of the confining magnetic fields is generated by currents within the plasma.

Spitzer conductivity the electrical resistance in a plasma is approximately given by the so-called Spitzer conductivity, first formulated by Lyman Spitzer, the inventor of the **stellarator**, according to which the resistance decreases in proportion to the electron temperature as $T_e^{-3/2}$. To calculate the resistance of a plasma is not easy since the collisions are not billiard-ball collisions. Energy is transferred though many glancing collisions as the particles pass each other at a distance, acting on each other through their magnetic fields.

Stellarator a class of toroidal confinement systems where rotational transform of the basic toroidal field is produced by twisting the magnetic axis, by adding an external helical field or by distorting the toroidal field coils. It includes the early figure-8 and helical windingstellarators developed in the 1950s by Lyman Spitzer at Princeton as well as torsatrons, heliacs and modular stellarators.

Stockpile Stewardship and Management Program a program of the US Department of Energy to ensure that the nuclear capabilities of the United States are not eroded as nuclear weapons age. It carries out reliability testing of nuclear weapons for which advanced science facilities, like the National Ignition Facility, have been built.

Super-X divertor a divertor design in which the power that strikes material surfaces per unit area is greatly reduced. It requires a set of divertor coils that extends and controls a long plume of exhaust plasma. The length of the plume allows high radiative cooling before the plasma reaches the target. Also, the radius of the target is higher than in other designs, which increases the target area.

Tandem mirror an arrangement of several magnetic mirrors in tandem. The outer mirror cells help confine the plasma within the central mirror.

Target fuel capsule, together with a holhraum or other energy-focusing device (if one is used), that is struck by the driver's incident energy in order to initiate fusion reactions.

TBM programme the Test Blanket Module (TBM) Programme is a specific programme for the development of blanket modules for application in fusion power plants. ITER will test a number of concepts through the implementation of the Test Blanket Module Programme under the ITER agreement.

Theta pinch or θ-pinch a confinement system, based on the pinch effect and generally open-ended, where a rapidly-pulsed axial magnetic field generates a plasma current in the azimuthal (i.e. the θ) direction.

Tokamak the most investigated and furthest advanced configuration for the magnetic cage of a fusion plasma. In a tokamak the plasma is confined by two superposed magnetic fields: a toroidal field produced by external coils and the field of a ring current flowing in the plasma. The field lines in the combined field are then helical. In addition, the tokamak requires a third, vertical field that fixes the position of the current in the plasma vessel and shapes the plasma edge.

The plasma current is normally induced by a transformer coil in the plasma. A tokamak therefore does not operate continuously, but in pulsed mode like the transformer. Pulse times of a few hours are anticipated in a future power plant. For technical reasons, however, a power plant has to operate in continuous mode and so methods of producing a continuous current are being investigated. In contrast, in the **stellarator** the entire field is externally produced.

Toroidal field the main magnetic field usually created by large magnetic coils wrapped around the vessel that contains the plasma.

Z-pinch a confinement system, based on the pinch effect where a toroidal magnetic field generates a plasma current in the axial (i.e. the z) direction.

Literature

Books

Badash L (1995) Scientists and the development of nuclear weapons. Humanity Books, Amherst

Basu D (ed) Dictionary of material science and high energy physics. CRC Press, Boca Raton

Bell MG (2016) The tokamak fusion test reactor. In: Neilson G (ed) Magnetic fusion energy: from experiment to power plant. Elsevier, Amsterdam

Bellan PM (2000) Spheromaks: a practical application of magnetohydrodynamic dynamos and plasma self-organization. Imperial College Press

Bishop AS (1958) Project Sherwood. Addison-Wesley

Bobin JL (2014) Controlled thermonuclear fusion. World Scientific

Boenke S (1991) Entstehung und Entwicklung des Max-Planck-Instituts für Plasmaphysik 1955–1971. Campus Verlag

Braams CM, Stott PE (2002) Nuclear fusion: half a century of magnetic confinement fusion research. Taylor & Francis, New York

Bromberg JL (1982) Fusion—science, politics, and the invention of a new energy source. MIT Press, Cambridge

Brunelli B, Leotta GG (eds) (1982) Unconventional approaches to fusion. Plenum Press, New York

Brunelli B, Knoepfel H (eds) (1990) Safety, environmental impact, and economic prospects of nuclear fusion. Plenum Press, New York

Cairns RA (1993) Radio-frequency plasma heating. In: Dendy R (ed) Plasma physics: an introductory course. Cambridge University Press, Cambridge

Cassidy DC (1992) Uncertainty: the life and science of Werner Heisenberg, New York

Chen FF (2011) An indispensable truth, Springer, Berlin

Chen FF (2016) Introduction to plasma physics and controlled fusion. Springer, Berlin

Claessens M (2019) ITER: the giant fusion reactor. Springer, Berlin

Clery D (2013) A piece of the sun. Overlook Duckworth, New York

Close F (1990) Too hot to handle. Penguin Books

Curli B (2017) Italy, Euratom and early research on controlled thermonuclear fusion (1957–1962). In: Bini E, Londero I (eds) Nuclear Italy. An international history of Italian nuclear policies during the Cold War. EUT Edizioni Università di Trieste

Davidson PA (2010) An introduction to magnetohydrodynamics. Cambridge University Press, Cambridge

Dean SO (2013) Search for the ultimate energy source: a history of the U.S. Fusion energy Program. Springer, Berlin

Dendy R (ed) (1993) Plasma physics: an introductory course. Cambridge University Press

Dinan R (2017) The fusion age: modern nuclear fusion reactors. Applied Fusion Systems, Abingdon

Dolan T (2000) Fusion research. Pergamon Press

Dolan T (ed) (2013) Magnetic fusion technology. Springer, Berlin

Dolan TJ (2013) Materials issues. In: Dolan TJ (ed) Magnetic fusion technology. Springer, Berlin

Eckert M (1989) Vom 'Matterhorn' zum 'Wendelstein': Internationale Anstöße zur nationale Groß-
forschung in der Kernfusion. In: Eckert M, Osietzki M (eds) Wissenschaft für Macht und Markt.
Beck, München

Elliott JA (1993) Plasma kinetic theory. In: Dendy R (ed) Plasma physics: an introductory course.
Cambridge University Press, Cambridge

Fowler TK (1997) The fusion quest. The Johns Hopkins University Press

Friedberg J (2007) Plasma physics and fusion energy. Cambridge University Press, Cambridge

Gamow G (1970) My World Line. New York.

Garton Ash T (1997) The file. Vintage Books

Green L (2016) 15 million degrees: a journey to the centre of the sun. Penguin Books

Haas FA (1993) Turbulence in fluids and fusion plasmas. In: Dendy R (ed) Plasma physics: an
introductory course. Cambridge University Press, Cambridge

Heppenheimer TA (1984) The man-made sun: the quest for fusion power. Omni Press, Boston

Herman R (1990) Fusion—the search for endless energy. Cambridge University Press, Cambridge

Hirsch RL, Bedzek RH, Wendling RM (2010) The impending world energy mess. Apogee Prime,
Ontario

Hirsch RL (2013). In: Dean SO, Search for the ultimate energy source. Springer, Berlin

Jaffe RL, Taylor W (2018) The physics of energy. Cambridge University Press, Cambridge

Josephson PR (2000) The red atom, Russia's nuclear power program from Stalin to Today. Freeman
and Company, New York

Jursa AS (ed) (1985) Handbook of geophysics and space environment. United States Air Force

Kawabe T (1983) Introduction to the tandem mirror. In: Post RF et al (eds) Mirror-based and
field-reversed approaches to magnetic fusion. International School of Plasma Physics, Varenna

Knight A (1993) Beria—Stalin's first Lieutenant. Princeton University Press

Kragh H (1999) Quantum generations. Princeton

Küppers G (1979) Fusionsforschung – Zur Zielorientierung im Bereich der Grundlagen Forschung.
In: van den Daele W, Krohn W, Weingart P (eds) Geplante Forschung. Suhrkamp, Frankfurt

Leontovich MA (ed) Plasma physics and the problem of controlled thermonuclear reactions.
Pergamon Press (four volumes)

Mannone F (ed) Safety in tritium handling technology. Kluwer Academic Publishers

McCracken G, Stott P (2013) Fusion: the energy of the universe. Academic, New York

Mirnov SV (2007) Energiya iz vody. MIFI, Moscow

Miyamoto K (2011) Fundamentals of plasma physics and controlled fusion. National Institute for
Fusion Science, Japan

Murray RL, Holbert KE (2020) Nuclear Energy—an introduction to the concepts, systems and
applications of nuclear processes. Elsevier, Amsterdam

National Academy of Sciences (2013) An assessment of the prospects for inertial fusion energy.
The National Academies Press

National Academy of Sciences (2018) Final report of the Committee on a Strategic Plan for U.S.
Burning Plasma Research. The National Academies Press

Parisi J, Ball J (2019) The future of fusion energy. World Scientific

Park RL (2000) Voodoo science: the road from foolishness to fraud. Oxford University Press, Oxford

Pease RS (1993) Survey of fusion plasma physics. In: Dendy R (ed) Plasma physics: an introductory
course. Cambridge University Press, Cambridge

Perlado JM, Sanz J (1992) Irradiation effects and activation in structural material. In: Velarde GJ,
Ronen Y, Martinez-Val JM (eds) Nuclear fusion by inertial confinement: a comprehensive treatise.
CRC Press, Boca Raton

Pfalzner S (2006) An introduction to inertial confinement fusion. Taylor & Francis, New York

Pinker S (2018) Enlightenment now. Penguin Books

Phillips JA (1983) Magnetic fusion. Los Alamos Science

Porfiri MT, Pinna T, Di Pace L (2013) Tritium in tokamak devices: safety issues. In: Tosti S, Ghirelli N (eds) Tritium in fusion. Nova Science Publishers

Read BC (2014) The history and science of the manhattan project. Springer, Berlin

Rebut P-H, JET Design Team (1976) The JET project: design proposal for the JOINT European Torus. Commission of the European Communities, EUR-JET-R5

Rossiter M (2014) The Spy who changed the World. Headline Publishing Group, London

Sagdeev RZ (1994) The making of a Soviet scientist. Wiley, New York

Sakharov A (1990) Memoirs. Vintage Books, New York

Schatzkin P (2002) The boy who invented television. TeamCom Books

Segrè G, Hoerlin B (2016) The pope of physics: Enrico Fermi and the birth of the atomic age. New York

Seife C (2008) Sun in a bottle: the strange history of fusion and the science of wishful thinking. Penguin Books

Shaw EN (1990) Europe's experiment in fusion—The JET Joint Undertaking. North-Holland, Amsterdam

Sivaram V (2018) Taming the Sun—innovations to harness solar energy and power the planet. MIT Press, Cambridge

Speake J, LaFlaur M (2002) The Oxford essential dictionary of foreign terms in English. Oxford University Press, Oxford

Stacey WM (2009) Fusion—an introduction to the physics and technology of magnetic confinement fusion. Wiley, London

Stacey WM (2010) The quest for a fusion reactor: an insider's account of the INTOR workshop. Oxford University Press, Oxford

Stringer TE (1993) Transport in magnetically confined plasmas. In: Dendy R (ed) Plasma physics: an introductory course. Cambridge University Press, Cambridge

Tamm IE (1929) Osnovy Teorii Elektrichestva. Leningrad

Tosti S, Ghirelli N (eds) Tritium in fusion. Nova Science Publishers

Tucker C (2019) How to drive a nuclear reactor. Springer, Berlin

van den Daele W, Krohn W, Weingart P (1979) Geplante Forschung. Suhrkamp, Frankfurt

Voronov GS (1988) Storming the fortress of fusion. MIR, Moscow

Weisskopf V (1991) The joy of insight. Basic Books, New York

Wesson J (1999) The science of JET. JET–R(99)13

White RB (1989) Theory of tokamak plasmas. North-Holland, Amsterdam

Wilhelmsson H (2000) Fusion—a voyage through the plasma universe. IOP Publishing, Bristol

Windridge M (2012) Star chambers: the race for fusion power. White Label Books

Zohuri B (2017a) Inertial confinement fusion driven thermonuclear energy. Springer, Berlin

Zohuri B (2017b) Magnetic confinement fusion driven thermonuclear energy. Springer, Berlin

Newspaper Articles

Becker M (2019) EU-Kommission erklärt Fusionsreaktor zum Klimaschutzprojekt, Spiegel Wissenschaft, 01 Mar 2019. https://www.spiegel.de/wissenschaft/technik/eu-kommission-erk laert-fusionsreaktor-mit-rechentrick-zum-klimaschutz-projekt-a-1255885.html

Charpak G, Treiner J, Balibar S (2010) Nucléaire: arrêtons Iter, ce réacteur hors de prix et inutilisable (Nuclear: Let's stop ITER, this overpriced and useless reactor), Libération, 10 Aug 2010. https://www.liberation.fr/sciences/2010/08/10/nucleaire-arretons-iter-ce-reacteur-hors-de-prix-et-inutilisable_671121

De Clercq G (2018) ITER nuclear fusion project avoids delays as U.S. doubles budget, Reuters, 26 Mar 2018

Mihai A (2016) New record gets us closer to fusion energy. ZME Science, 17 Oct 2016

Morris C (2016) Germany—a nuclear fusion leader, Feb 2016. https://energytransition.org/2016/02/germany-a-nuclear-fusion-leader/

New York Times, 3 Sept 1958, p 3

New York Times, 28 Mar 2017

On the Way: Genuine Fusion, Time Magazine 75 (14) (1960), p 63

Plumer B, Popovich N (2019) These countries have prices on carbon. Are they working? New York Times, 2 April 2019

The Guardian Weekly, Fusion power to be 'on grid in 15 years', 16 Mar 2018, p 8

Scientific Articles

Abdou M (2019) Lessons learned from 40 years of fusion science and technology research. In: 14th international symposium on fusion nuclear technology, Budapest, Hungary, Sept 2019

Adamenko S et al (2015) Exploring new frontiers in the pulsed power laboratory: recent progress. Results Phys 5:62–68

Akers RJ et al (2002) Neutral beam heating in the START spherical tokamak. Nucl Fusion 42:122–135

Amendt P et al (2019) Ultra-high (>30%) coupling efficiency designs for demonstrating central hot-spot ignition on the National Ignition Facility using a Frustraum. Phys Plasmas 26:082707

Amrollahi R et al (2019) Alborz tokamak system engineering and design. Fusion Eng Design 141:91–100

Ananyev SS, Spitsyn AV, Kuteev BV (2015) Fuel cycle for a fusion neutron source. Phys Atomic Nuclei 78:1138–1147

Anklam TM et al (2011) LIFE: the case for early commercialization of fusion energy. Fusion Sci Technol 60:66–71

Argenti D et al (1981) The THOR tokamak experiment. Il Nuovo Cimento 63:471–486

Artsimovich LA (1958) Research on controlled thermonuclear reactions in the USSR. In: Proceedings of the second international conference on peaceful uses of atomic energy, Geneva, vol 31, pp 6–20

Artsimovich LA, Shafranov VD (1972) Tokamak with non-round section of the plasma loop. JETP Lett 15:51–54

Atkinson R, Houtermans F (1929) Zur Frage der Aufbaumöglichkeit der Elementen in Sternen. Zeitschrift für Physik 54:656–665

Aymar R (1999) ITER achievements by July 1998 and future prospects. Fusion Eng Design 46:115–127

Azizov EA (2012) Tokamaks: from A D Sakharov to the present (the 60-year history of tokamaks). Physics-Uspekhi 55:190–203

Bahcall JN, Salpeter E (2005) Stellar energy generation and solar neutrinos. Phys Today 58:44

Barbarino M (2020) A brief history of nuclear fusion. Springer Nature. https://www.nature.com/articles/s41567-020-0940-7

Basko MM (2005) Inertial confinement fusion: steady progress towards ignition and high gain. Nucl Fusion 45:S38–S47

Batistoni P (2016) Neutrons: little particles with powerful impact. Fusion Europe 1:4

Baylor LR et al (2019) Shattered pellet injection technology design and characterization for disruption mitigation experiments. Nucl Fusion 59:066008

Beidler CD et al (2001) The Helias reactor. Nucl Fusion 41:1759–1766

Belloni F et al (2018) On the enhancement of p-^{11}B fusion reaction rate in laser-driven plasma by $\alpha \to$ p collisional energy transfer. Phys Plasmas 25:020701

Berkowitz J et al (1958) Cusped geometries. In: Proceedings of the second international conference on the peaceful uses of atomic energy, Geneva, vol 31, p 171. J Nucl Energy 7:292–293

Berkowitz J, Grad M, Rubin H (1958) Magnetohydrodynamic stability. In: Proceedings of the second international conference on the peaceful uses of atomic energy, Geneva, vol 31, p 177

Berzak Hopkins L et al (2019) Toward a burning plasma state using diamond ablator inertially confined fusion (ICF) implosions on the National Ignition Facility (NIF). Plasma Phys Controlled Fusion 61:014023

Bethe HA (1979) The fusion hybrid. Phys Today 32:44–51

Bickerton RJ (1999) History of the approach to ignition. Philos Trans R Soc Lond A 357:397–413

Biermann L (1958) Recent work on controlled thermonuclear fusion in Germany (Federal Republic). In: Proceedings of the second international conference on the peaceful uses of atomic energy, Geneva, vol 31, pp 21–26

Bigot B (2019) ITER construction and manufacturing progress toward first plasma. Fusion Eng Design 146:124–129

Bodner SE (1995) Time-dependent asymmetries in laser-fusion hohlraums. Comments Plasma Phys Controlled Fusion 16:351–374

Bogusch E et al (2002) Benefits to European industry from involvement in fusion. Fusion Eng Design 63–64:679–687

Bombarda F et al (2004) Ignitor: physics and progress towards ignition. Braz J Phys 34:1786–1791

Bondarenko BD (2001) Role played by O.A. Lavrent'ev in the formulation of the problem and the initiation of research into controlled nuclear fusion in the USSR. Physics-Uspekhi 44:844–851

Bongard MW et al (2019) Advancing local helicity injection for non-solenoidal tokamak start-up. Nucl Fusion 59:076003

Bowman C et al (2018) Pedestal evolution physics in low triangularity JET tokamak discharges with ITER-like wall. Nucl Fusion 58:016021

Braams CM (1987) Fusion research in the smaller associations. Plasma Phys Controlled Fusion 29:1457–1464

Bradshaw AM et al (2011) Is nuclear fusion a sustainable energy form? Fusion Eng Design 86:2770–2773

Bradshaw AM, Hamacher T (2013) Nuclear fusion and the helium supply problem. Fusion Eng Design 88:2694–2697

Bromberg L (1982) TFTR: the anatomy of a programme decision. Social Stud Sci 12:559–583

Brown IG et al (1978) Tormac reactor considerations. Nucl Fusion 18:761–768

Bruhns H et al (1987) Study of the low aspect ratio limit tokamak in the Heidelberg Spheromak Experiment. Nucl Fusion 27:2178–2182

Bruzzone P (2018) High temperature superconductors for fusion magnets. Nucl Fusion 58:103001

Bussard RW (1991) Some physics considerations of magnetic inertial electrostatic confinement: a new concept for spherical converging flow fusion. Fusion Technol 19:273

Bustreo C et al (2019) How fusion power can contribute to a fully decarbonized European power mix after 2050. Fusion Eng Design 146:2189–2193

Buxton PF et al (2019) Overview of ST40 results and planned upgrades. In: 46th EPS conference on plasma physics

Cabal H et al (2012) Analysing the role of fusion power in the future global energy system. EPJ Web Conf 33:01006

Campbell DJ et al (2019) Innovations in technology and science R&D for ITER. J Fusion Energy 38:11–71

Cartlidge E (2017) Europe pauses funding for €500 million fusion research reactor. Nature, 10 July 2017

Causey RA et al (2012) Tritium barriers and tritium diffusion in fusion reactors. Comprehensive Nucl Mater 4:511–549

Chen D et al (2015) Preliminary cost assessment and compare of China fusion engineering test reactor. J Fusion Energy 34:127–132

Chen F (1979) Alternate concepts in magnetic fusion. Phys Today 32:36–42

Clery D (2015) ITER fusion project to take at least 6 years longer than planned. Science, 19 Nov 2015

Clery D (2020) 'Got my fingers crossed.' As ITER fusion project marks milestone, chief ponders pandemic impact. Science, 27 May 2020. https://www.sciencemag.org/news/2020/05/got-my-fingers-crossed-iter-fusion-project-marks-milestone-chief-ponders-pandemic

Coleman ER, Cohen SA, Mahoney MSS (2011) Greek Fire: Nicholas Christofilos and the Astron Project in America's early fusion program. J Fusion Energy 30:238–256

Costley AE (2016) On the fusion triple product and fusion power gain of tokamak pilot plants and reactors. Nucl Fusion 56:066003

Costley AE (2019) Towards a compact spherical tokamak fusion pilot plant. Philos Trans R Soc A 377:20170439

Cousins SW, Ware AA (1951) Pinch effect oscillations in a high current toroidal ring discharge. Proc Phys Soc B 64:159–166

Cox M, The MAST Team (1999) The mega amp spherical tokamak. Fusion Eng Design 46:397–404

Craxton RS et al (2015) Direct-drive inertial confinement fusion: a review. Phys Plasmas 22:11051

Crisanti F et al (2017) The divertor tokamak test facility proposal: physical requirements and reference design. Nucl Mater Energy 12:1330–1335

Danani C et al (2019) One-dimensional nuclear design analyses of the SST-2. Pramana J Phys 92:15

Davies JR et al (2017) Laser-driven magnetized liner inertial fusion. Physics of Plasmas 24:062701

Dean SO (1998) Lessons drawn from ITER and other fusion international collaborations. J Fusion Energy 17(1998):155–175

Dean SO (1998) Fusion power by magnetic confinement program plan. J Fusion Energy 17(1998):263–286

Derzon MS et al (2000) An inertial-fusion Z-pinch power plant concept, SAND 2000-3132J

de Villiers JAM, Hayzen AJ, O'Mahony JR, Roberts DE, Sherwell D (1979) Tokoloshe—the South African tokamak. South African J Sci 75:155–157

de Vries PC et al (2011) Survey of disruption causes at JET. Nucl Fusion 51:053018

Dimov GI, Zakajdakov VV, Kishinevsky ME (1976) Open trap with ambipolar mirrors. In: Sixth international conference on plasma physics and controlled nuclear fusion research, Berchtesgaden

Dnestrovskij YuN (2001) Physical results of the T-10 tokamak. Plasma Phys Rep 27:825–842

Donné AJH et al (2018) European research roadmap to the realisation of fusion energy (EUROfusion, 2018). https://www.euro-fusion.org/fileadmin/user_upload/EUROfusion/Documents/2018_Research_roadmap_long_version_01.pdf

Drozdov AP et al (2019) Superconductivity at 250 K in lanthanum hydride under high pressures. Nature 569:528–531

Duckworth R et al (2017) Progress in magnet design activities for the material plasma exposure experiment. Fusion Eng Design 124:211–214

Eckhartt D et al (1965) Comparison of alkali plasma loss rates in a stellarator and in a toroidal device with minimum mean-B properties. In: Proceedings of the 2nd IAEA conference on plasma physics and controlled fusion research, Culham, UK, vol II, pp 719–731

Eckhartt D (1995) Nuclear fuels for low-beta fusion reactors: lithium resources revisited. J Fusion Energy 14:329–341

Eddington A (1920) The internal constitution of stars. Science 52:233–240

El-Guebaly LA (2009) 40 years of power plant studies: brief historical overview and future trends. In: Proceedings of the 3d IAEA technical meeting on first generation of fusion power plants: design and technology

El-Guebaly LA (2010) Fifty years of magnetic fusion research (1958–2008): brief historical overview and discussion of future trends. Energies 3:1067–1086

Eliezer S et al (2016) Avalanche proton-boron fusion based on elastic nuclear collisions. Phys Plasmas 23:050704

Elmore WC, Tuck JL, Watson KM (1959) On the inertial-electrostatic confinement of a plasma. Phys Fluids 2:239–246

Entler S et al (2018) Approximation of the economy of fusion energy. Energy 152:489–497

Fasel D, Tran MQ (2005) Availability of lithium in the context of future D-T fusion reactors. Fusion Eng Design 75–79:1163–1168

Feder T (2004) ITER impasse illustrates challenge of site selection. Phys Today 57:28–29

Feder T (2005) Barring ITER site consensus, Europe will forge ahead. Phys Today 58:30

Federici G et al (2019) Overview of EU DEMO design and R&D activities. Nucl Fusion 59:066013

Fermi E (1949) On the origin of the cosmic radiation. Phys Rev 75:1169–1174

Fietz WH et al (2013) Prospects of high temperature superconductors for fusion magnets and power applications. Fusion Eng Design 88:440–445

Fraunhofer ISE (2015) Current and future cost of photovoltaics. Long-term scenarios for market development, system prices and LCOE of utility-scale pv systems, study on behalf of Agora Energiewende

Freidberg JP, Kadak AC (2009) Fusion–fission hybrids revisited. Nat Phys 5:370–372

Fujita T et al (1998) High performance reversed shear plasmas with a large radius transport barrier in JT-60U. Nucl Fusion 38:207–221

Freidberg J (2009) Fusion-fission hybrids—what can they do. In: Report of the Conference on Hybrid Fusion Systems. Washington

Fujita T et al (1999) High performance experiments in JT-60U reversed shear discharges. Nucl Fusion 39:1627–1636

Galambos JD et al (1995) Commercial tokamak reactor potential with advanced tokamak operation. Nucl Fusion 35:551–573

Gamow G (1928) Zur Quantentheorie des Atomkernes. Zeitschrift für Physik 51:204–212

Gamow G (1938) Nuclear energy sources and stellar evolution. Phys Rev 53:595–604

Garnier D et al (1999) Overview of the levitated dipole experiment. In: Report at the 41st annual meeting of the Division of Plasma Physics of the American Physical Society, Seattle, 15 Nov 1999

Gascon JC et al (2012) Design and key features for the ITER electrical power distribution. Fusion Sci Technol 61:47–51

Gerstner E (2009) Nuclear energy: the hybrid returns. Nature 460(7251):25–28

Giegerich T et al (2019) Development of a viable route for lithium-6 supply of DEMO and future fusion power plants. Fusion Eng Design 149:111339

Gilbert M et al (2018) Waste assessment of European DEMO fusion reactor designs. Fusion Eng Design 136:42–48

Gilleland JR et al (1989) ITER: concept definition. Nucl Fusion 29:1191–1231

Glanz J, Lawler A (1998) Planning a future without ITER. Science 279:20–21

Glanz J (1998) Requiem for a heavyweight at meeting on fusion reactors. Science 280:818–819

Glenzer SH et al (2012) Cryogenic thermonuclear fuel implosions on the National Ignition Facility. Phys Plasmas 19:056318

Goncharov GA (2001) The 50th anniversary of the beginning of research in the USSR on the potential creation of a nuclear fusion reactor. Phys-Uspekhi 44:851–858

Gota H et al (2019) Formation of hot, stable, long-lived field-reversed configuration plasmas on the C-2W device. Nucl Fusion 59:112009

Gourdon C et al (1968) Configurations du type stellarator avec puits moyen et cisaillement des lignes magnetiques, In: Proceedings of the IAEA conference on plasma physics and controlled nuclear fusion research, Novosibirsk, vol 1, pp 847–861

Granetz RS et al (1996) Disruptions and Halo currents in Alcator C-Mod. Nuclear Fusion 36:545

Grant P (1987) Do it yourself superconductors. New Scientist 115(1571):36–39

Gryaznevich M et al (1998) Achievement of record β in the START spherical tokamak. Phys Rev Lett 80:3972–3975

Gryaznevich MP, Sykes A (2017) Merging-compression formation of high temperature tokamak plasma. Nucl Fusion 57:072003

Gryaznevich M (2019) Presentation at the International Spherical Torus Workshop, Frascati, Italy, Oct 2019

Haan SW et al (2011) Point design targets, specifications, and requirements for the 2010 ignition campaign on the National Ignition Facility. Phys Plasmas 18:051001

Haines MG (1977) Plasma containment in cusp-shaped magnetic fields. Nucl Fusion 17:811–858

Haines MG (1996) Fifty years of controlled fusion research. Plasma Phys Controlled Fusion 38:643–656

Han W, Ward D (2009) Revised assessments of the economics of fusion power. Fusion Eng Design 84:895–898

Hansen JR (1992) Secretly going nuclear. http://www.inventionandtech.com/content/secretly-going-nuclear-1

Hawryluk RJ, Batha S et al (1998) Fusion plasma experiments on TFTR: A 20 year retrospective. Phys Plasmas 5:1577–1589

Hawryluk RJ (1998) Results from deuterium-tritium tokamak confinement experiments. Rev Mod Phys 70:537–587

Hedditch J, Bowden-Reid R, Khachan J (2015) Fusion energy in an inertial electrostatic confinement device using a magnetically shielded grid. Phys Plasmas 22:102705

Helsley CE, Burke RJ (2014) Economic viability of large-scale fusion systems. Nucl Instrum Methods Phys Res A 733:51–56

Henning CD, Logan BG (1987) Overview of Tiber- II—evolution toward an engineering test reactor. J Fusion Energy 1:241–256

Hirsch RL (1986) Whither fusion research. J Fusion Energy 5:101–105

Hirsch RL (2015) Fusion research: time to set a new path. Issues Sci Technol 31(4). https://issues.org/fusion-research-time-to-set-a-new-path/

Hirsch RL (2017) Necessary and sufficient conditions for practical fusion power. Phys Today 70:11–13

Hofmann I (2018) Review of accelerator driven heavy ion nuclear fusion. Matter Radiation Extremes 3:1–11

Honda T et al (2000) Analyses of loss of vacuum accident (LOVA) in ITER. Fusion Eng Design 47:361–375

Hora H et al (2017) Road map to clean energy using laser beam ignition of boron-hydrogen fusion. Laser Particle Beams 35:730–740

Horacek J et al (2008) Multi-machine scaling of the main SOL parallel heat flux width in tokamak limiter plasmas. Plasma Phys Controlled Fusion 58:074005

Horton W et al (2014) Tandem mirror experiment for basic fusion science. World J Nucl Sci Technol 4:53–58

Hsu SC, Francis Thio YC (2018) Physics criteria for a subscale plasma liner experiment. J Fusion Energy 37:103–110

Huang Q, FDS Team (2014) Development status of CLAM steel for fusion application. J Nucl Mater 455:649–654

Huld T et al (2014) Cost maps for unsubsidised photovoltaic electricity. In: Scientific and policy report by the Joint Research Centre of the European Commission

Hurricane O et al (2014) Fuel gain exceeding unity in an inertially confined fusion implosion. Nature 506(7488):343–349

Imazawa R et al (2012) First plasma experiment on spherical tokamak device UTST. Electr Eng Japan 179:20–26

International Fusion Research Council (IFRC) (2005) Status report on fusion research. Nucl Fusion 45:A1

IRENA (2019) Renewable power generation costs in 2018. International Renewable Energy Agency, Abu Dhabi

Ivanov AA, Prikhodko VV (2013) Gas-dynamic trap: an overview of the concept and experimental results. Plasma Phys Controlled Fusion 55:063001

Ivanov AA, Prikhodko VV (2017) Gas dynamic trap: experimental results and future prospects. Phys Uspekhi 60:509–533

Jackson GL et al (1991) Regime of very high confinement in the boronized DIII-D tokamak. Phys Rev Lett 67:3098–3101

Jacquinot J (2005) Steady-state operation of tokamaks: key experiments, integrated modelling and technology developments on Tore Supra. Nucl Fusion 45:S118–S131

Jacquemot S (2017) Inertial confinement fusion for energy: overview of the ongoing experimental, theoretical and numerical studies. Nucl Fusion 57:102024

Jarboe TR et al (2012) Imposed-dynamo current drive. Nucl Fusion 52:083017

Jassby DL (1978) A small-aspect-ratio torus for demonstrating thermonuclear ignition. IAEA-TC-145/9

Jassby DL (1978) Small-aspect-ratio, small-major-radius tokamak reactors. Comments Plasma Phys Controlled Fusion 3:151–158

Jassby DL (2017) Fusion reactors: not what they're cracked up to be. Bull Atomic Sci, 19 April 2017. https://thebulletin.org/2017/04/fusion-reactors-not-what-theyre-cracked-up-to-be/

Jassby DL (2018) ITER is a showcase … for the drawbacks of fusion energy. Bull Atomic Sci. https://thebulletin.org/2018/02/iter-is-a-showcase-for-the-drawbacks-of-fusion-energy/

Jones IR et al (1998) Operation of the rotamak as a spherical tokamak: the flinders Rotamak-ST. Phys Rev Lett 81:20172–20175

Kammen DM (2010) 2020 visions. Nature 463, 7 Jan 2010

Keilhacker M et al (2001) The scientific success of JET. Nucl Fusion 41:1925–1966

Kessel CE et al (2018) Overview of the fusion nuclear science facility, a credible break-in step on the path to fusion energy. Fusion Eng Design 135:236–270

Key MH (2000) Status of and prospects for the fast ignition inertial fusion concept. Phys Plasmas 14:5

Khvesyuk VI et al (2001) D-3He field reversed configuration fusion power plant. Fusion Technol 39:410–413

Khvostenko PP et al (2019) Tokamak T-15MD—two years before the physical start-up. Fusion Eng Design 146:1108–1112

Kikuchi M (2018) The large tokamak JT-60: a history of the fight to achieve the Japanese fusion research mission. Eur Phys J H 43:551–577

Kim K et al (2015) Design concept of K-DEMO for near-term implementation. Nucl Fusion 55:053027

Kirk A (2016) Nuclear fusion: bringing a star down to Earth. Contemp Phys 57(1):1–18

Kirk A et al (2017) Overview of recent physics results from MAST. Nucl Fusion 57:102007

Knaster J et al (2015) The accomplishment of the engineering design activities of IFMIF/EVEDA: the European-Japanese project towards a Li(d, xn) fusion relevant neutron source. Nucl Fusion 55:086003

Knaster J et al (2016) Materials research for fusion. Nature 12:424–434

Knaster J et al (2017) Overview of the IFMIF/EVEDA project. Nucl Fusion 57:102016

Kolbasov BN et al (2016) On use of beryllium in fusion reactors: resources, impurities and necessity of detritiation after irradiation. Fusion Eng Design 109–111:480–484

Kovari M et al (2018) Tritium resources available for fusion reactors. Nucl Fusion 58:026010

Koyanagi T et al (2018) Recent progress in the development of SiC composites for nuclear fusion applications. J Nucl Mater 511:544–555

Kramer D (2015) New MIT design revives interest in high-field approach to fusion. Phys Today 68:23–24

Kramer D (2016) NIF may never ignite, DOE admits. Phys Today. https://physicstoday.scitation.org/do/10.1063/PT.5.1076/full/

Kramer D (2018) Laser program at University of Rochester targeted for shutdown. Phys Today, 16 Feb 2018. https://physicstoday.scitation.org/do/10.1063/PT.6.2.20180216b/full/

Kramer D (2018) NIF achieves new fusion output milestone. Phys Today, June 2018. https://physicstoday.scitation.org/do/10.1063/PT.6.2.20180615b/full/

Kruskal M, Schwarzschild M (1954) Some instabilities of a completely ionized plasma. Proc R Soc A 223:348–360

Kurchatov IV (1957) On the possibility of producing thermonuclear reaction in a gas discharge. J Nucl Energy II 4:193–202

Kuteev BV et al (2015) Development of DEMO-FNS tokamak for fusion and hybrid technologies. Nucl Fusion 55:073035

Kuteev BV et al (2017) Status of DEMO-FNS development. Nucl Fusion 57:076039

Laberge M (2019) Magnetized target fusion with a spherical tokamak. J Fusion Energy 38:199–203

Lampasi A et al (2017) The European DEMO fusion reactor: Design status and challenges from balance of plant point of view. In: IEEE international conference on environment and electrical engineering

Lapp RE (1956) Limitless power of the seas drive future machines. Life Mag 176–190

Lawson JD (1957) Some criteria for a power producing thermonuclear reactor. Proc Phys Soc 70:6–10

LePape S et al (2018) Fusion energy output greater than the kinetic energy of an imploding shell at the National Ignition Facility. Phys Rev Lett 120:245003

Lerner E et al (2017) Confined ion energy >200 keV and increased fusion yield in a DPF with monolithic tungsten electrodes and pre-ionization. Phys Plasmas 24:102708

Levine MA (1978) Tormac concept. LBL-7975

Li J, Wan Y (2019) Present state of chinese magnetic fusion development and future plans. J Fus Energy 38:113–124

Libeyre P et al (2019) From manufacture to assembly of the ITER central solenoid. Fusion Eng Design 146:437–440

Lidsky LM (1975) Fission-fusion systems: hybrid, symbiotic and augean. Nucl Fusion 15:151–173

Lidsky LM (1982) End product economics and fusion research program priorities. J Fusion Energy 2:269–292

Lidsky LM (1983) The trouble with fusion. MIT Technol Rev 1–12

Lindl J et al (2014) Review of the National Ignition Campaign 2009–2012. Phys Plasmas 21:020501

Litaudon X et al (2017) Overview of the JET results in support to ITER. Nucl Fusion 57:102001

Llewellyn Smith C, Cowley S (2010) The path to fusion power. Philos Trans R Soc A 368:1091–1108

Lloyd B (2019) Presentation at the International Spherical Torus Workshop, Frascati, Italy, Oct 2019

Logan BG (1994) Inertial fusion commercial power plants. J Fusion Energy 13:171–172

Lopes Cardozo NJ et al (2016) Fusion: expensive and taking forever? J Fusion Energy 35:94–101

Lopes Cardozo NJ (2019) Economic aspects of the deployment of fusion energy: the valley of death and the innovation cycle, Philos Trans R Soc A 377:20170444

Lubkin GB (1977) Sandia and Kurchatov groups claim electron-beam fusion. Phys Today 30:17–19

McCray WP (2010) 'Globalization with hardware': ITER's fusion of technology, policy, and politics. History Technol 26:283–312

McGuire KM et al (1995) Review of deuterium-tritium results from the tokamak fusion test reactor. Phys Plasmas 2:2176–2188

McGuire TJ et al (2016) Lockheed Martin compact fusion reactor concept, confinement model and T4B experiment

McNamara S et al (2019) Ohmic operations of the ST40 high field spherical tokamak. In: 46th EPS conference on plasma physics

Marbach G, Cook I, Maisonnier D (2002) The EU power plant conceptual study. Fusion Eng Design 63–64:1–9

Marshall E (1984) Congress turns cold on fusion. Science 224:1322–1323

Meade DM (1988) Results and plans for the tokamak fusion test reactor. J Fusion Energy 7:107–114

Meade DM (1998) Q, break-even and the $n\tau_E$ diagram for transient fusion plasmas. Office of Scientific & Technical Information Technical Report, April 1998

Melnikov AV et al (2015) Physical program and diagnostics of the T-15 upgrade tokamak. Fusion Eng Design 96–97:306–310

Menard JE et al (2016) Fusion nuclear science facilities and pilot plants based on the spherical tokamak. Nucl Fusion 56:106023

Menard JE et al (2017) Overview of NSTX Upgrade initial results and modelling highlights. Nucl Fusion 57:102006

Menard JE et al (2019) Presentation at the International Spherical Torus Workshop, Frascati, Italy, Oct 2019

Meridian International Research (2008) The trouble with lithium 2—Under the Microscope (2008).

Miller R et al (1979) Parametric study of the Tormac fusion reactor concept. Los Alamos Report

Mitchell TB, Schauer MM, Barnes DC (1997) Observation of spherical focus in an electron Penning trap. Phys Rev Lett 78:58–61

Mohr S et al (2010) Lithium resources and production: a critical global assessment. Cluster Research Report 1.4, Oct 2010

Mohr S, Ward J (2014) Helium production and possible projection. Minerals 4:130–144

Moiseenko VE et al (2016) Progress in stellarator research at IPP-Kharkov. Nukleonika 61:91–97

Molvik AW et al (2010) A gas dynamic trap neutron source for fusion material and subcomponent testing. Fusion Sci Technol 57:369–394

Moon PB (1977) Biogr Memoirs Fellows R Soc 33:529–556

Morland H (2005), Born secret. Cardozo Law Rev 26:1401–1408

Morris AW et al (2005) Spherical tokamaks: present status and role in the development of fusion power. Fusion Eng Design 74:67–75

Moses EI (2009) Ignition on the national ignition facility: a path towards inertial fusion energy. Nucl Fusion 49:104022

Nagao S et al (1974) A tokamak of small aspect ratio with multipole field (Asperator T-3). In: Proceedings of plasma physics and controlled nuclear fusion research, Tokyo, vol II, pp 123–128

Najmabadi F, The ARIES Team (2003) Spherical torus concept as power plants: the ARIES-ST study. Fusion Eng Design 65:143–164

Najmabadi F et al (2008) The ARIES-CS compact stellarator fusion power plant. Fusion Sci Technol 54:655–672

Nakai S, Mima K (2004) Laser driven inertial fusion energy: present and prospective. Reports Progress Phys 67:321–349

Narushima Y et al (1999) Highly elongated low aspect ratio tokamak produced by negative-biased theta-pinch. In: Proceedings of the 26th EPS conference on plasma physics and controlled fusion, Maastricht, 14–18 June 1999, p 1.082

Nehl CL et al (2019) Retrospective of the ARPA-E ALPHA Fusion Program. J Fusion Energy 38:506–521

Nevins W (1998) A review of confinement requirements for advanced fuels. J Fusion Energy 17:25–32

Nevins WM, Swain R (2000) The thermonuclear fusion rate coefficient for p-^{11}B reactions. Nuclear Fusion 40:865–872

Ni Muyi et al (2013) Tritium supply assessment for ITER and DEMOnstration power plant. Fusion Eng Design 88:2422–2426

Nuckolls J et al (1972) Laser compression of matter to super-high densities: thermonuclear (CTR) applications. Nature 239:139–142

Nuckolls JH (1998) Early steps toward inertial fusion energy (IFE) (1952 to 1962). UCRL-ID-131075, June 1998

Okano K (2019) Review of strategy toward DEMO in Japan and required innovations. J Fusion Energy 38:138–146

Olson C, The Z-IFE Team (2005) Z-pinch inertial fusion energy. In: Fusion power associates annual meeting and symposium, Washington DC, 11–12 Oct 2005

Ongena J, Ogawa Y (2016) Nuclear fusion: status report and future prospects. Energy Policy 96:770–778

Ongena J et al (2016) Magnetic-confinement fusion. Nature 12:398–410

Ono M, Kaita R (2015) Recent progress on spherical torus research. Phys Plasmas 22:040501

Ono M et al (2004) Next-step spherical torus experiment and spherical torus strategy in the course of development of fusion energy. Nucl Fusion 44:452–463

Ono M et al (2015) Progress toward commissioning and plasma operation in NSTX-U. Nucl Fusion 55:073007

Ono Y et al (2015) High power heating of magnetic reconnection in merging tokamak experiments. Phys Plasmas 22:055708

Osborne TH et al (1995) Confinement and stability of VH mode discharges in the DIII-D tokamak. Nucl Fusion 35:23–37

Palumbo D (1987) The work of the European Commission in promoting fusion research in Europe. Plasma Phys Controlled Fusion 29:1465–1473

Palumbo D (1988) The growth of European fusion research. Plasma Phys Controlled Fusion 30:2069–2072

Parail VV, Pogutse OP (1978) The kinetic theory of runaway electron beam instability in a tokamak. Nuclear Fusion 18:303

Parail VV, Pogutse OP (1976) Instability of the runaway-electron beam in a Tokamak. Soviet J Plasma Phys 2:228

Park J et al (2005) Periodically oscillating plasma sphere. Phys Plasmas 12:056315

Park J et al (2015) High energy electron confinement in a magnetic cusp configuration. Phys Rev X 5:021024

Parry I (2019) Putting a price on pollution. Finance Develop 56(4):16–19

Pashley DW (1987) Biogr Memoirs Fellows R Soc 33:47–64

Peng YKM, Dory RA (1978) Very small aspect ratio tokamaks. Oak Ridge National Laboratory, Report ORNL/TM-6535

Peng Y-KM, Strickler DJ (1986) Features of spherical torus plasmas. Nucl Fusion 26:769–777

Peng Y-KM (2000) The physics of spherical torus plasmas. Phys Plasmas 7:1681–1692

Peng Y-KM et al (2005) A component test facility based on the spherical tokamak. Plasma Phys Controlled Fusion 47:B263–B283

Peng YKM, Yuan BS, Liu MS, The EXL-50 Team (2019) 20th international ST workshop—ENEA Frascati Research Center, Oct 2019

Perrault D (2019) Nuclear safety aspects on the road towards fusion energy. Fusion Eng Design 146:130–134

Petrov Y et al (2010) Experiments on rotamak plasma equilibrium and shape control. Phys Plasmas 17:012506

Pitts RA et al (2013) A full tungsten divertor for ITER: physics issues and design status. J Nucl Mater 438:S48–S56

Pollock JA, Barraclough S (1905) Proc R Soc New South Wales 39:131

Post RF (1956) Controlled fusion research—an application of the physics of high temperature plasmas. Rev Mod Phys 28:338–362

Post RF (1959) High-temperature plasma research and controlled fusion. Ann Rev Nucl Sci 9:367–436

Post RF (1976) Physics of mirror fusion systems. In: Proceedings of the 2nd ANS topical meeting on the technology of controlled nuclear fusion, Richland, Washington

Post RF (1978) The magnetic mirror approach to fusion. Nucl Fusion 27:1579–1739

Raj B et al (2010) Progress in the development of reduced activation ferritic-martensitic steels and fabrication technologies in India. Fusion Eng Design 85:1460–1468

Raman R (2019) Presentation at the international spherical torus workshop. Frascati, Italy

Rayleigh L (1883) Investigation of the character of the equilibrium of an incompressible heavy fluid of variable density. Proc R Math Soc 14:170

Rebut P-H et al (1985) The joint European Torus: installation, first results and prospects. Nucl Fusion 25:1011–1022

Rebut P-H, Lallia PP (1989) JET results and the prospects for fusion. Fusion Eng Design 11:1–21

Rebut P-H (2018) The joint European Torus (JET). Eur Phys J H 43:459–497

Redfearn J (1999) Europe, Japan finalizing reduced ITER design. Science 286:1829–1831

Reece Roth J (1976) Alternative approaches to fusion. Nasa Technical Memorandum, NASA TM X-73429

Rider TH (1997) Fundamental limitations on plasma fusion systems not in thermodynamic equilibrium. Phys Plasmas 4:1039–1046

Rinderknecht HG et al (2018) Kinetic physics in ICF: present understanding and future directions. Plasma Phys Controlled Fusion 60:064001

Roberts M (1974) The birth of Ormak.... Oak Ridge Nat Lab Rev 7:12–17

Romanelli F et al (2006) Assessment of open magnetic fusion for space propulsion. In: AIAA 57th International Astronautical Congress.

Rosanvallon S et al (2016) Waste management plans for ITER. Fusion Eng Design 109–111:1442–1446

Rosenbluth MN, Garwin R, Rosenbluth A (1954) Infinite Conductivity Theory of the Pinch. Los Alamos Scientific Laboratory Report LA-1850

Rosenbluth MN, Hinton FL (1994) Generic issues for direct conversion of fusion energy from alternative fuels. Plasma Phys Controlled Fusion 36:1255–1268

Rowcliffe AF et al (2018) Materials-engineering challenges for the fusion core and lifetime components of the fusion nuclear science facility. Nucl Mater Energy 16:82–87

Sagdeev R (2018) His story of plasma physics in Russia, 1956–1988. Eur Phys J H 43:355–396

Sakharov AD (2001) Referee report on O.A. Lavrentiev's paper. Physics-Uspekhi 44:865

Santini F (2006) Non-thermal fusion in a beam plasma system. Nucl Fusion 46:225–231

Schiermeier Q (2000) German Greens go cold on nuclear fusion. Nature 405:107

Schlüter A (1964) Controlled nuclear fusion research in Western Europe. In: Proceedings of the third international conference on the peaceful uses of atomic energy, Geneva, vol 15, pp 35–40

Schmitt AJ et al (2010) Shock ignition target design for inertial fusion energy. Phys Plasmas 17:042701

Schmitt JC et al (2013) Results and future plans of the lithium tokamak eXperiment (LTX). J Nucl Mater 438:S1096–S1099

Sekiguchi T (1983) History and organizations. J Fusion Energy 3:313–321

R.F. Service (2020) Seawater could provide nearly unlimited amounts of critical battery material. Science, 13 July 2020. https://www.sciencemag.org/news/2020/07/seawater-could-provide-nearly-unlimited-amounts-critical-battery-material

Shafranov VD (2001) The initial period in the history of nuclear fusion research at the Kurchatov Institute. Physics-Uspekhi 44:835–843

Shimomura Y (1994) Overview of international thermonuclear experimental reactor (ITER) engineering design activities. Phys Plasmas 1:1612–1618

Shmatov ML (2016) Comment on "Avalanche proton-boron fusion based on elastic nuclear collisions". Phys Plasmas 23:094703

Shpanskiy YuS, The DEMO-FNS Project Team (2019) Progress in the design of the DEMO-FNS hybrid facility. Nucl Fusion 59:076014

Shumlak U, The FuZE Team (2017) A compact fusion device based on the sheared flow stabilized Z-Pinch. https://arpa-e.energy.gov/sites/default/files/03_SHUMLAK.pdf

Simonen TC (2016) Three game changing discoveries: a simpler fusion concept? J Fusion Energy 35:63–68

Slough J, Votroubek G, Pihl C (2011) Creation of a high-temperature plasma through merging and compression of supersonic field reversed configuration plasmoids. Nucl Fusion 51:053008

Slutz SA, Vesey RA (2012) High-gain magnetized inertial fusion. Phys Rev Lett 108:025003

Smirnov VP (2010) Tokamak foundation in the USSR/Russia 1950-1990. Nucl Fusion 50:1–8

Sorbom BN et al (2015) ARC: A compact, high-field, fusion nuclear science facility and demonstration power plant with demountable magnets. Fusion Eng Design 100:378–405

Spano AH (1975) Large tokamak experiments. Nucl Fusion 15:909–931

Spitzer L, Grove DJ, Johnson WE, Tonks L, Westendorp WF (1954) Problems of the stellarator as a useful power source. USAEC NYO-6047, Aug 1954

Spitzer L (1958) The stellarator concept. Phys Fluids 1:253–264

Spitzer L (1965) Controlled nuclear fusion research 1965: review of experiment. In: Proceedings of the 2nd IAEA conference on plasma physics and controlled fusion research, Culham, UK, vol I, p 9, 6–10 Sep 1965

Srinivasan R, The Indian DEMO Team (2016) Design and analysis of SST-2 fusion reactor. Fusion Eng Design 112:240–243

Stacey WM (1978) The INTOR workshop: a unique international collaboration in fusion. Progress Nucl Energy 22:119–172

Stacey WM et al (2002) A fusion transmutation of waste reactor. Fusion Eng Design 63–64:81–86

Stambaugh RD et al (1998) The spherical tokamak path to fusion power. Fusion Technol 33:1–21

Steiner D et al (1981) The engineering test facility. Nucl Eng Design 63:189–198; J Fusion Energy 1:5–13

Stix TH (1998) Highlights in early stellarator research at Princeton. J Plasma Fusion Res 1:3–8

Stott PE (2005) The feasibility of using D–3He and D-D fusion fuels. Plasma Phys Controlled Fusion 47:1305–1338

Strait EJ et al (2019) Progress in disruption prevention for ITER. Nucl Phys 59:112012

Strelkov VS (1985) Twenty-five years of tokamak research at the I.V. Kurchatov Institute. Nucl Fusion 25:1189–1194

Strelkov VS (2001) History of the T-10 tokamak: creation and development. Plasma Phys Rep 27:819–824

Sutherland D et al (2014) The dynomak: an advanced spheromak reactor concept with imposed-dynamo current drive and next-generation nuclear power technologies. Fusion Eng Design 89:412–425

Sutherland D (2014) The Dynomak reactor system. In: Talk at the EPR conference, Madison, Wisconsin, Aug 2014

Sykes A (2008) The development of the spherical tokamak. ICPP, Fukuoka

Sykes A et al (1979) Proceedings of the 7th international conference on controlled fusion and plasma physics, Innsbruck, vol 1, p 593

Sykes A et al (1992) First results from the START experiment. Nucl Fusion 32:694–699

Sykes A et al (1999) The spherical tokamak programme at Culham. Nucl Fusion 39:1271–1281

Sykes A et al (2001) First results from MAST. Nucl Fusion 41:1423–1433

Sykes A et al (2018) Compact fusion energy based on the spherical tokamak. Nucl Fusion 58:016039

Tabak M et al (2014) Alternative ignition schemes in inertial confinement fusion. Nucl Fusion 54:054001

Tahill W (2007) The trouble with lithium: implications of future PHEV production for lithium demand, Meridian International Research

Tanaka S (2006) Overview of research and development activities on fusion nuclear technologies in Japan. Fusion Eng Design 81:13–24

Taylor GI (1950) The instability of liquid surfaces when accelerated in a direction perpendicular to their plane. Proc R Soc Lond A 201:192

Taylor A et al (2007) A route to the brightest possible neutron source? Science 315(5815):1092–1095

Teller E (1958) Peaceful Uses of Fusion. In: Proceedings of the second international conference on peaceful uses of atomic energy, Geneva, vol 31, pp 27–33

Thio YCF (2008) Status of the U.S. program in magneto-inertial fusion. J Phys: Conf Ser 112:042084

Thonemann PC (1958) Controlled thermonuclear research in the United Kingdom. In: Proceedings of the second international conference on peaceful uses of atomic energy, Geneva, vol 31, pp 34–38.

Tikhonchuk VT (2019) Physics of laser plasma interaction and particle transport in the context of inertial confinement fusion. Nucl Fusion 59:032001

Tobita K et al (2019) Japan's efforts to develop the concept of JA DEMO during the past decade. Fusion Sci Technol 75:372–383

Tolok VT (2001) A history of stellarators in the Ukraine. J Fusion Energy 20:117–129

Tomabechi K et al (1991) The ITER conceptual design. Nucl Fusion 31:1135–1224

Toschi R (1986) The next European Torus. Nucl Eng Design/Fusion 3:325–330

Toschi R (1989) NET and the European Fusion Technology Programme. Fusion Eng Design 11:47–62

Tuck J. Curriculum vitae and autobiography. http://bayesrules.net/JamesTuckVitaeAndBiography.pdf

van Dijk J, Kroesen GMW, Bogaerts A (2009) Plasma modelling and numerical simulation. J Phys D: Appl Phys 42:1–14

Velikhov EP, Kartishev KB (1989) The USSR fusion program. Fusion Eng Design 8:23–26

Velikhov EP, Kadomtsev BB (1976) A step toward thermonuclear energy. Curr Digest Soviet Press 28(10):26–27

Vidal O, Goffe B, Arndt N (2013) Metals for a low-carbon society. Nat Geosci 6:894–896

Vollset SE et al (2020) Fertility, mortality, migration, and population scenarios for 195 countries and territories from 2017 to 2100: a forecasting analysis for the Global Burden of Disease Study. Lancet, July 2020

Voss GM et al (2000) A conceptual design of a spherical tokamak power plant. Fusion Eng Design 51–52:309–318

Voss GM et al (2002) Development of the spherical tokamak power plant. Fusion Eng Design 63–64:65–71

Waganer LM, Davis JW, Schultz KR (2002) Benefits to US industry from involvement in fusion. Fusion Eng Design 63–64:673–678

Wagner F et al (1982) A regime of improved confinement and high beta in neutral beam heated divertor discharges of ASDEX. Phys Rev Lett 49:1408

Wagner F (2013) Physics of magnetic confinement fusion. EPJ Web Conf 54:01007

Wan Y et al (2017) Overview of the present progress and activities on the CFETR. Nucl Fusion 57:102009

Wang B (2019) Update on ZAP Energy compact Z-Pinch nuclear fusion (September 2019). https://www.nextbigfuture.com/2019/09/update-on-zap-energy-compact-z-pinch-nuclear-fusion.html

Ward DJ et al (2005) The economic viability of fusion power. Fusion Eng Design 75–79:1221–1227

Weisel GJ (2001) Containing plasma physics: a disciplinary history, 1950–1980. Dissertation, University of Florida

Williams MD (1997) D-T operation on TFTR. In: Proceedings of the 19th symposium on fusion technology, vol 1. Elsevier Science, Amsterdam, pp 135–142

Wilson HR et al (2004) Integrated plasma physics modelling for the Culham steady state spherical tokamak fusion power plant. Nucl Fusion 44:917–929

Wilson DC et al (2015) The first cryogenic DT layered, beryllium capsule implosion at the National Ignition Facility. In: NEDPC proceedings

Windsor C (2019) Can the development of fusion energy be accelerated? Philos Trans R Soc A 377:20170446

Xie H et al (2018) Evaluation of tritium burnup fraction for CFETR scenarios with core-edge coupling simulations. Nucl Fusion 60:046022

Xu Y (2016) A general comparison between tokamak and stellarator plasmas. Matter Radiation Extremes 1:192–200

Yonas G et al (1974) Electron beam focusing and application to pulsed fusion. Nucl Fusion 14:731

Yoshikawa M (1985) The JT-60 project and its present status. Nucl Fusion 25:1081–1085

Zakharov LE, Shafranov VD (2019) Necessary conditions for fusion energy. Plasma Phys Rep 45:1087–1092

Zhang Y et al (2019) Sustained neutron production from a Sheared-Flow stabilized Z Pinch. Phys Rev Lett 122:135001

Zhou RJ et al (2015) Microwave experiments on Prairie View Rotamak. Phys Plasmas 22:054501

Zucchetti M et al (2018) Fusion power plants, fission and conventional power plants. Radioactivity, radiotoxicity, radioactive waste. Fusion Eng Design 136:1529–1533

Index

Printed in the United States
by Baker & Taylor Publisher Services